Historical Perspectives on East Asian Science, Technology and Medicine

T0192267

Historical Perspectives on East Asian Science, Technology and Medicine

Edited by

ALAN K.L. CHAN
Department of Philosophy
National University of Singapore

GREGORY K. CLANCEY
Department of History
National University of Singapore

HUI-CHIEH LOY
Department of Philosophy
National University of Singapore

SINGAPORE UNIVERSITY PRESS
NATIONAL UNIVERSITY OF SINGAPORE

World Scientific
Singapore • New Jersey • London • Hong Kong

© Singapore University Press
Yusof Ishak House, NUS
31 Lower Kent Ridge Road
Singapore 119078

and

World Scientific Publishing Co. Pte. Ltd
PO Box 128, Farrer Road, Singapore 912805
USA office: Suite 1B, 1060 Main Street, River Edge, NJ 07661
UK office : 57 Shelton Street, Covent Garden, London WC2H 9HE

ISBN 9971-69-259-7 (Paper)

Typeset by: International Typesetters Pte. Ltd.
Printed by: Vetak Services

CONTENTS

PREFACE, *Alan K.L. Chan*   xi
FOREWORD, *Catherine Jami*   xiii
FOREWORD, *Ho Peng Yoke*   xvii

I.   PHILOSOPHY / CULTURAL STUDY OF SCIENCE &
**   TECHNOLOGY**

Nakayama Shigeru   The Digital Revolution and East
                       Asian Science   3

Kim Yung Sik   "Fossils", "Organic World-View",
                       "The Earth's Motions", etc.:
                       Problems of Judging East Asian
                       Scientific Achievements from
                       Western Perspectives   14

Gregory Clancey   The Science of Eurasia:
                       Meiji Seismology as Cultural
                       Critique   27

Sugiyama Shigeo   "Good Luck" in the History of
                       Science in Japan   43

Gao Xuan   Knowledge of the Dual Character
                       of Technology: Viewpoints on
                       Technology in Ancient China   51

Lee Sung Kyu   Traditional East Asian Views of
                       Nature Revisited   60

II. SCIENCE & TECHNOLOGY POLICY

Xi Zezong   On the Mistakes of Emperor
                       Kangxi's Scientific Policy   69

Li Di   The Late Qing Government's
                       Policy Towards Science and
                       Technology: The Case of
                       Jiangnan Manufactory   79

Zhang Li   The Institutionalization of Higher
                          Science Education in China:
                          A Case Study of Higher
                          Chemistry Training before 1937   87

Duan Yibing &   A Shift in Academia Sinica's
Fan Hongye   Mission in 1935   99

Chen Shiwei   Enhancing China's International
                          Prestige: The 1948 Sino-American
                          Scientific Expedition to
                          Mount Amne Machin   103

Kuo Wen-Hua   When State and Policies
                          Reproduce Each Other:
                          Making Taiwan a Population
                          Control Policy; Making a
                          Population Control Policy for
                          Taiwan   121

Liu Yidong   The Internet in China:
                          Policy and Problems   138

Liao Boqin,   The Impact of Local Government
Wang Liming,   Policies on the Development of
John Davis &   Traditional Technology:
Jia Bosheng   A Case Study of the Daning Salt
                          Factory   148

III. MEDICINE AND THE LIFE SCIENCES

Mercedes G. Planta   Traditional Medicine and
                          Pharmacopoeia in the Philippines,
                          16th and 17th Centuries   157

Luo Guihuan   Biology in Song China   171

Paul U. Unschuld   Diseases in the *Huang Di Neijing
                          Suwen*: Facts and Constructs   182

Ye Xiaoqing   Regulating the Medical
                          Profession in China:
                          Health Policies of the Nationalist
                          Government   198

Ma Bo-Ying & The Transmission of Traditional
Alicia Grant Chinese Medicine (TCM) to England
(Outline) 214

Hokari Hiroyuki The Presentation of Traditional
Chinese Medicine (TCM)
Knowledge in Hong Kong 222

Erling Høg Traditional Chinese Medicine (TCM)
in Denmark: Conditions in a
New Host Culture 236

Kim Taylor 'Improving' Chinese Medicine:
The Role of Traditional Medicine
in Newly Communist China,
1949–53 251

Lan Tuyet Chu An Introduction to the History of
Traditional Medicine and
Pharmaceutics in Vietnam 264

IV. WESTERN SCIENTIFIC INFLUENCE

Tsukahara Togo Westernization from Different
Angles: Review of the
Historiography of Science from the
Viewpoint of Colonial Science 279

Luis Saraiva & The College of São Paulo in
Henrique Leitão Macao: A Background
(16th and 17th Centuries) 285

Isaia Iannaccone A Jesuit Scientist and Ancient
Chinese Heavenly Songs:
Antoine Gaubil's Research and
Translation of *Bu Tian Ge* 299

José Antonio Cervara Two Spanish Cosmographers in
the Philippines:
Andrés de Urdaneta and Martín
de Rada 317

V. MATHEMATICS

Ji Zhigang   The Development of Interpolation
                     Methods in Ancient China:
                     From Liu Zhuo to Hua Hengfang   327

Qu Anjing   Why Interpolation?   336

Chen Cheng-Yih &   An Investigation of the *Yíng-*
Li Zhao-Hua   *Bùzú* Method   345

Guo Shirong   The Influence of Yang Hui's
                     Works on the Mathematical
                     Mainstream in the Ming Dynasty   358

Chen Meidong   On the Basic Rules for
                     Reconstruction of the Calendar
                     Used in the State of Lu During
                     the Spring-Autumn Period   368

Wang Rongbin   A Reversion Study on the Solar
                     Shadow Algorithm (SSA) of the
                     Qintian Calendar   376

VI. ASTRONOMY

Helmer Aslaksen   Fake Leap Months in the Chinese
                     Calendar: From the Jesuits
                     to 2033   387

Ôhashi Yukio   Originality and Dependence of
                     Traditional Astronomies in the
                     East   394

Hashimoto Keizō   A Cartesian in the Kangxi Court
                     (as Observed in the *Lifa Wenda*)   406

Niu Weixing &   On the Naksatras in the Chinese
Jiang Xiaoyuan   Translated Sutras   415

Hu Tiezhu   The Accuracy of *Da Yan Li*'s
                     Calculation of the Solar Eclipse   431

Ahn Young Sook,   A Date Conversion Table for
Yang Hong Jin,   the Lunisolar and Julian Calendars
Sim Kyung Jin,   During the Koryo Dynasty
Han Bo Sik &   (A.D. 918–1392) in Korea
Song Doo Jong   438

Fung Kam-Wing   Mapping the Universe:
Two Planispheric Astrolabes in
the Early Qing Court   448

VII. TECHNOLOGY / TECHNIQUE

Zhou Zhongfu,   A Study of the Lacquerish Patina
Sun Shuyun,   (Qi Gu) on Ancient Bronze
Han Rubin & T. Ko   Mirrors   465

Sergey Lapteff   Chinese Influence on the
Technological Development of
Yayoi Period Japan: Problems of
Metal Casting and the Production
of Bronze Mirrors   470

Guan Zengjian &   On Liu Xin's Metrological Theory   477
Fu Guimei

Liu Keming,   The Preliminary Study of
Zhou Zhaoying &   Mechanical Design Methodology
Yang Shuzi   in Ancient China   489

Jin Qiupeng   Summary of Chinese Ship
Navigation in Ancient Southeast
Asia   496

Guo Qinghua   A Traditional Architectural
Heating System in China:
The Manchurian *kang* and *huodi*   505

Mau Chuan-Hui   The Influence of Chinese
Techniques on the French Silk
Industry as Shown by French
Patents from 1791 to 1860   515

H.T. Huang   Takamine Jokichi and the
Transmission of Ancient Chinese
Enzyme Technology to the West   525

Ishida Fumihiko &   The Modernization of the
Ishii Taro   Petroleum Industry in Japan   533

VIII. SCIENTIFIC LITERATURE

Nguyen Dien Xuan   Ancient Vietnamese Manuscripts
                             and Printed Books Related to
                             Science, Medicine and
                             Technology (Inventory,
                             Classification and Preliminary
                             Assessment)   547

Evgeny Torchinov   External and Internal in
                             Ge Hong's Alchemy   555

Colette Diény   On Some Trends in Contemporary
                             Critiques of Shen Gua and His
                             Works   560

Wang Xiaoqin   A History of *Dan-Yan Qiu-Yi Shu*
                             in the West   570

PREFACE

The papers presented in this volume arose from the 9th International Conference on the History of Science in East Asia (ICHSEA), which was held in Singapore in August 1999. Held once every three years, under the auspices of the International Society for the History of East Asian Science, Technology and Medicine (ISHEASTM), the ICHSEA provides a major forum for scholars of East Asian science to share their latest findings, to reinforce existing ties, and to explore new areas of research collaboration. The 9th ICHSEA was hosted by the Faculty of Arts and Social Sciences, National University of Singapore (NUS).

The Local Organizing Committee was assisted by a distinguished panel of international advisors, chaired by Dr Lam Lay Yong, formerly of the Mathematics Department, NUS. Professor Ho Peng Yoke, Director of the Needham Research Institute, and Professor Catherine Jami, then President of the ISHEASTM, offered invaluable guidance and support. Within NUS, the Office of the Dean, Faculty of Arts and Social Sciences, the Department of Chinese Studies, the Department of History, the Centre for Advanced Studies, and the East Asian Institute, generously gave their support to bring this multidisciplinary conference to Singapore. Outside the University, the Kwan Im Thong Hood Cho Temple, the Lee Foundation, and the Tan Foundation provided much appreciated financial assistance. A grant from the Chiang Ching Kuo Foundation for Intellectual Scholarly Exchange enabled several scholars from Taiwan to participate in the conference. On behalf of the Local Organizing Committee, I would like to place on record our sincere thanks to all our sponsors and advisors. We would also like to thank Professor Chong Chi Tat, Deputy Vice-Chancellor, NUS, for giving the opening address at the 9th ICHSEA.

Some 150 scholars from 21 countries attended the conference. Notably, for the first time in the history of the ICHSEA, there were participants from Vietnam, India, and the Philippines. From the start, we were keen to expand on the remit of the conference to include Southeast Asia and to accommodate discussions of contemporary concerns. The multidisciplinary nature of the discipline seems ideally suited to crossing boundaries, both historical and geographic. We were thus pleased that the 9th ICHSEA featured several papers on aspects of the history of science in Vietnam and the Philippines, three of which have been revised for publication in this volume. These papers not only address topics of intrinsic interest but also afforded an important comparative perspective. There were also two special symposia organized by NUS colleagues on science and technology policy in East and Southeast Asia.

Over a period of four and a half days, some 120 papers on a rich array of topics were presented, including astronomy, Chinese traditional medicine, history of mathematics, and Western science in East Asia. The discussions on traditional Chinese medicine attracted considerable public attention. Participants benefited also from four plenary addresses, given by Professor Xi Zezong, Institute for the History of Natural Sciences, Chinese Academy of Science, PRC; Professor Nathan Sivin, University of Pennsylvania; Professor Nakayama Shigeru, Kanagawa University; and Professor Ho Peng Yoke, Director, Needham Research Institute. Professors Xi and Nakayama have kindly agreed to make their papers available for publication in this volume. Professors Ho Peng Yoke and Catherine Jami have each graciously contributed a Foreword.

The task of publishing a volume of selected papers from the conference proved daunting. The sheer number of papers was one concern. The diverse topics they addressed contribute also to the difficulty. From "Japanese Science as Cultural Critique" and the "Mistakes of Emperor Kangxi's Scientific Policy" to "Traditional Medicine and Pharmacopoeia in the Philippines" and the "Basic Rules for Reconstruction of the Calendar", the range was truly remarkable and rendered selection and classification a challenging process. Some of the papers were highly theoretical, while others were technical or pursued subjects of historical interest. In the end, we could only include some 53 papers grouped under eight general headings: Philosophy/Cultural Study of Science and Technology; Science and Technology Policy; Medicine and the Life Sciences; Western Scientific Influence; Mathematics; Astronomy; Technology/Technique; and Scientific Literature.

The papers included have been extensively revised or edited. They give a sense of the range of presentations at the 9th ICHSEA, but more importantly they make significant contributions to our understanding of the History of Science in East and Southeast Asia. In matters of transliteration, translation, spelling, and citation, the editorial team has decided to respect the authors' choices. In many instances, these represent an academic judgment, which have not been made to conform to one standard for convenience.

It remains my pleasant duty to thank the authors for their patience in answering our editorial queries, and many others who have helped in different ways in bringing this volume to print. Ms. Patricia Tay, Managing Editor, Singapore University Press, Ms. Kim Tan, Director (Publishing), World Scientific Publishing Co. Pte. Ltd., and Dr Tan Ern Ser, the Centre for Advanced Studies, have been supportive throughout the project. Professor Kim Yung Sik, current President of the ISHEASTM and chief organizer of the 8th ICHSEA, gave valuable advice. We are pleased that we were able to contribute to the work of the ICHSEA and wish the 10th ICHSEA to be held in China in 2002 every success.

Alan K.L. Chan
Singapore

FOREWORD

This volume is the final outcome of the 9th International Conference on the History of Science in East Asia (23–27 August 1999), which was organized by the Faculty of Arts and Social Sciences and the East Asian Institute of the National University of Singapore, on behalf of the International Society for the History of East Asian Science, Technology and Medicine (ISHEASTM). The Conference was the third plenary meeting of ISHEASTM, and thanks to the organizers' and editors' effort, it is also the third time that a conference volume has been produced.[1] The tradition of discussion and circulation of research on East Asian history of science is thus consolidated. For this achievement, I should like to express my gratitude to our colleagues in Singapore.

Fifty-three of the 115 papers read at the Conference are presented here in the form of concise research reports: they aim at giving the reader an idea of the state of the field. The variety of the topics discussed reflects the diversity of research, while the structure of the volume highlights the different, albeit complementary, approaches taken by the contributors.

China remains the main focus of research (forty papers). Studies on Japan and Korea had already taken a significant place in the two previous meetings. At the Singapore Conference, however, research on Southeast Asian countries was for the first time presented by scholars from these countries: Vietnam, the Philippines, and Singapore. Some of these contributions are presented here, and provide a more complete picture of science, technology, and medicine as they developed in that part of the world where Chinese characters have been a major vehicle for the circulation of knowledge.

Similarly, while the pre-modern period still occupies a majority of researchers, investigations on the 20th century, which have their specific methods and questions, are growing steadily. In particular, most of the papers on science policy are devoted to it. That our field should bring ancient and modern together nourishes reflection on science in today's world. For one thing, all of us are concerned, as scholars, but also as citizens of this world, with the conflict, apparent in academia as well as in newspapers, between two assumptions: on the one hand, the assertion that science as it is is universal by essence, and, on the other hand, claims that each culture, or even each religion, produces its own version of science, allegedly incompatible with the versions produced elsewhere. History of science can and should contribute to defusing this absurd and dangerous conflict.

Specialized disciplinary history, which reconstructs and assesses ancient methods (all deal in pre-1800 sources) using today's knowledge and language, is still dominant in the study of mathematics and astronomy, the two fields of "exact science" most developed in imperial China. Their technical complexity still needs to be better grasped. The scholars working along these lines, while relying, in most cases, on their own scientific training, are also continuing two legacies: that of evidential scholarship (*kaozhengxue*考證學) and that of Joseph Needham. In the study of medicine and technology, this "internalist" approach more and more tends to coexist with the methods of social history. The same can be said about papers that discuss broader issues concerning scientific conceptions, or the role and status of the sciences.

The issue of transmission is raised in a significant number of papers. Whether across time or across space, transmission has come to be recognized as an essential process in the history of science, as old clichés such as "the Greek miracle", or "the great geniuses", gradually give way to more subtle — and more accurate — analyses of the formation, circulation, and transformation of ideas, texts, instruments, and practices. It is perhaps an asset to the research on East Asian science that it has long integrated the study of transmission — the work of Yabuuti Kiyosi is an outstanding example.

Finally, I would like to point out an interesting feature of this volume, compared to its two predecessors: a relatively high number of its papers (twenty-two) are authored by scholars working in the People's Republic of China. East Asian history of science has long faced difficulties as to the diffusion of its research in the wider community of historians of science. These difficulties have always been a concern of ISHEASTM. As most of the research in our field is done in the PRC, it seems quite natural that the main language of publication should be Chinese. On the other hand, there is an urgent need to make our "Occidentalist" colleagues — that is, the vast majority of historians of science — aware that a meaningful history of science can only be written if all parts of the world are taken into account, and that East Asia does produce its indispensable contribution towards that goal. It is indeed a great merit of this collection that it makes work by some of the most eminent representatives of our field in the PRC available in English. This is all the more timely as the next International Congress of the History of Science is to take place in Beijing, in 2005. Just as it no longer makes sense to wonder "whether East Asia had science", it is no longer possible to ignore the history of science produced there nowadays. The present volume takes us one step further away from this ignorance.

Catherine Jami
Paris, August 2001

Note

[1] The two previous meetings resulted in Hashimoto, Keizō; Jami, Catherine and Skar, Lowell, eds., *East Asian Science: Tradition and Beyond*. Osaka: Kansai University Press, 1995; and Kim, Yung Sik and Bray, Francesca, eds., *Current Perspectives in the History of East Asian Science*. Seoul: National University Press, 1999.

FOREWORD

This book forms the Proceedings of the 9th International Conference of the History of Science in East Asia held by the National University of Singapore during 23–27 August 1999. The full story of the connection of Singapore with the field of the history of East Asian science, technology and medicine is known only to very few, perhaps not even to the organizers and the participants of the conference before its commencement. As an eyewitness, I owe it to the host country to add a brief historical perspective on Singapore's association with the history of East Asian science, one that has extended over a period of half-a-century.

In the years 1948 and 1949, Dr D.J. da Solla Price, known in those days as Dr. Derek Price, was a lecturer in the mathematics department of Raffles College, Singapore. He left in 1950 to do a second Ph.D. degree on history of science and collaborated with Dr. Joseph Needham and Dr. Wang Ling in the *Heavenly Clockwork* before he left for North America, where he first joined the Smithsonian Institute and later became Professor of History of Science and Medicine at Yale University. Mr. Kenneth Robinson, Needham's collaborator in the acoustic section as well as the main part of the concluding chapter in the forthcoming last volume of *Science and Civilisation in China*, headed the Chinese section in the Singapore Teachers Training College in 1953, when I first corresponded with Needham.

Although I later became a close friend of Needham's, I have to concede that the first person from Singapore who became a personal friend of the Needhams was Ivy Soh, who went to England to study English teaching after obtaining her first degree in Singapore. She became a houseguest of the Needhams whenever she visited Cambridge, and conversely, when the Needhams visited Singapore they would rely on her to arrange for hotel accommodations. The Needhams also had indirect links with Singapore. They were good friends of Professor Meredith Jackson and Mrs. Lianli Jackson, who were attracted to the better side of East Asian values. Professor Jackson was the teacher of Mr. Lee Kuan Yew, who later became the founding father of Singapore. Lianli Jackson told me in the presence of Needham that, in spite of his high social status, the Singapore Prime Minister used to call on his teacher whenever he visited Cambridge.

My professor of mathematics at the University of Malaya (Singapore), Sir Alexander Oppenheim, suggested that I do research in the history of Chinese science after he listened to a talk I gave on the subject at the Malayan Mathematical Society. He became the supervisor for my doctoral dissertation, and arranged for me to get in touch with Needham in 1953. Needham suggested that I should make a full and annotated translation of the Astronomical Chapters in the *Jinshu* (Official History of the Jin Dynasty), which would be very useful to him in writing the astronomical section of *Science and Civilisation in China,* and that for his part he would help me to go over my draft to render it appropriate for a thesis. In 1958 and 1959 Singapore granted me sabbatical leave on full salary to enable me to work in Cambridge in collaboration with Needham on his alchemical sections.

Towards the end of 1959 I returned from Cambridge to Singapore. The next year I was appointed Reader in History of Science within the Physics Department, which meant that I had to teach physics as well as to start a course on the history of science. In 1960 I had three visitors whose names are widely known among historians of East Asian science, namely Nakayama Shigeru, Wang Ling and H.T.Huang. The last-mentioned has a close family relationship with Singapore. He was Needham's personal secretary in Chongqing during World War II, later served as a Deputy Director in the Needham Research Institute and wrote the food and nutrition chapter of *Science and Civilisation in China.*

Dr. Rayson Huang, my senior colleague at the University of Malaya in Singapore, was the person who introduced me to Needham. Rayson Huang knew Needham when they were in Siquan province during World War II. He was Vice-Chancellor of the Nanyang University in Singapore from 1968 to 1972 and Vice-Chancellor of the University of Hong Kong from 1973 to 1986. Needham gave a public lecture at the Nanyang University in 1971 and received an Honorary Degree at the University of Hong Kong in 1974. He is a Life Member of the Croucher Foundation of Hong Kong that made a sizable donation to the East Asian History of Science Trust in the United Kingdom as endowment for the Needham Research Institute.

In my youth I received much encouragement and inspiration from a friend, Dr. Wu Lien-teh, who was some forty-five years my senior. Dr Wu was the world-renowned plague fighter in the early years of the twentieth century, who received his training from Emmanuel College, Cambridge and had studied under the same teacher (Sir Frederick Gowland Hopkins) as Needham, though over 20 years apart. He had an interest in the history of Chinese science as a hobby, having written the *History of Chinese Medicine* with K.C.Wong in 1932. To encourage his younger fellow-student of "Hopi" in his work, Dr Wu approached Dato Dr Lee Kong Chian for financial assistance for the *Science and Civilisation in China* project. Accordingly Dr. Lee Kong Chian sent a cheque to Needham, who acknowledged it in his Notes of *Science and Civilisation in China*, volume 4 part 1 as a 'splendid contribution'.

In 1962 the University of Malaya in Singapore became the University of Singapore, with Dr Lee Kong Chian as its first Chancellor. He was very kind to me and showed great personal interest in my research. He had invited me to lunch several times in his business dining room, known as the Garden Club. On one of these occasions I brought along with me Nathan Sivin, who was then a Harvard University graduate student doing research on the alchemical work of Sun Simiao for his doctoral degree, and was spending a few months with me in Singapore. Nathan Sivin also assisted me in some tutorial work for students doing the history and philosophy of science course. It was during the same luncheon meeting that Dr Lee Kong Chian introduced us to Tan Sri Tan Chin Tuan, who later became one of the greatest benefactors of the Needham Research Institute. The story of his magnificent gift, however, has to wait until I introduced another figure well known and well respected among modern historians of Chinese mathematics.

To many historians of East Asian science today the word 'Singapore' brings to mind only the name of Lam Lay Yong, who has distinguished herself in the history of Chinese mathematics and who served the National University of Singapore as Professor of Mathematics at the time of her retirement. Just before the arrival of Nathan Sivin in Singapore, Lam Lay Yong was working with me writing her doctoral thesis on the 13th-century Chinese mathematician Yang Hui. She wished to refer to a handwritten copy by Seki Kōwa of a book by Yang Hui that contains magic squares. The manuscript concerned was then in the possession of Li Yan, a distinguished historian of Chinese mathematics in Beijing. Lay Yong was thinking of asking her grandfather, the eminent Mr. Tan Kah Kee, to approach Li Yan for a photocopy, but hesitated to bother an old man in relative ill-health. I wrote on her behalf to Needham instead. There was a prompt response from Li Yan, who sent one set of microfilm to Cambridge and another copy to Singapore. This kind act together with her respect for Needham resulted in a large benefaction for the Needham Research Institute.

One of Tan Sri Tan Chin Tuan's daughters was Lay Yong's classmate. Lay Yong spoke to Tan Sri Tan Chin Tuan through that daughter of his and obtained his agreement to make a very substantial donation for the Needham Research Institute Library Building. Having paved the way she made arrangements for Needham and Lu Gwei-Djen to see Tan Sri Tan Chin Tuan in the year 1984 when they were on their way to Hong Kong, Beijing and Taipei. This is by far the largest single contribution to date given directly to the Needham Research Institute. Now Lay Yong and Dr. Lee Kong Chian's children are cousins. It was also through the good office of Lay Yong that the Lee Foundation offered to make an annual contribution to the Needham

Research Institute. The Lee Foundation has also supported the travelling expenses of Lay Yong and Professor Ang Tian Se to attend international conferences on the history of Chinese science on many occasions. Dr S.T. Lee of the Lee Foundation is an Honorary Fellow of the Needham Research Institute as well as an Honorary Fellow of the British Academy.

A participant at the conference, in the person of the present Librarian of the Needham Research Institute, John Moffett, also has a Singapore connection. He lived with his parents in Singapore for two years from 1969 when he was only seven years old. His father was serving in the British Navy in Changi. Then there are Mr. George Hicks and Mrs. Julia Hicks, who migrated from Hong Kong to Singapore in the early nineties. The Hicks were the greatest personal benefactors of the *Science and Civilisation and China* project and the Needham Research Institute from Hong Kong when they were there. Hicks was also responsible for persuading Mr. K.P. Tin to donate money for the K.P. Tin Hall of the Needham Research Institute. Last but not least comes Professor Wang Gungwu, whose Institute was one of the co-sponsors of the 9th International Conference of the History of Science in East Asia. He is the Director of the East Asian Institute in the National University of Singapore. When he was Vice-Chancellor of the University of Hong Kong he served as a director of the East Asian History of Science Foundation, Hong Kong. More recently he became a Life Member of the Advisory Committee for the Chinese Civilization Education Trust, Hong Kong that looks after the interest of the Needham Research Institute in Cambridge and the Institute for the History of Natural Science in Beijing. The Chinese Civilization Education Trust, under the leadership of Dr. Q.W. Lee of Hong Kong, has recently enriched the Needham Research Institute with the largest endowment to date.

It was with this tradition behind them that the organizers conducted the 9th International Conference of the History of Science in East Asia with great efficiency, and to the admiration of the participants. As one can see from this book, participants of the Conference included both the current international leading scholars as well as up-and-coming researchers in this field of study. In fact, the organizers had encouraged the participation of young scholars by offering them financial support.

This book does not include all the papers presented at the Conference nor does it give a list of the names of all the participants. However, in it one can find a good representative of both. It includes contributors by most senior scholars like H.T. Huang, Xi Zezong, Li Di and Nakayama Shigeru, in the order of their age, among those who are authorities in their respective fields. Another senior scholar, Paul U. Unschuld, hails as the authority in his field in Europe. Deserving special mention is Hashimoto Keizō, whose modesty hides the fact that he was Needham's one and only official Ph.D. research student in the history of science at Gonville and Caius College, Cambridge as well as one of Professor Yabuuti Kiyosi's disciples. Yabuuti, the counterpart of Needham in Japan in the history of Chinese science, spent much time with his last three disciples in his retirement years. Yano Michio, a participant in the conference but whose paper is not included in this book, was among this small group of Yabuuti's favourite disciples. Catherine Jami has translated Yabuuti's book on Chinese mathematics into French. Kim Yung Sik is another senior scholar, although not quite yet in term of age, but judging from his academic status and achievements. He was the organizer of the 8th International Conference of the History of Science in East Asia held in Seoul in 1996 and is the current President of the International Society for the History of East Asian Science, Technology, and Medicine. Examples of emerging scholars to take the batons from their forerunners are Fung Kam-Wing, Guo Shirong, Niu Weixing, Qu Anjing and Togo Tsukahara. Among the junior scholars is the name of Kim Taylor, who recently obtained her Ph.D. degree from the University of Cambridge. The names mentioned above are not exhaustive and are limited to those of whom I have personal knowledge.

One great contribution made by the conference was to bring scholars in Southeast Asia to the notice of the world community of scholars in the field. This book contains, for example, a contribution by Lan Tuyet Chu on Vietnamese medicine, another by Dien Xuan Nguyen on Vietnamese documents related to science, and Jose Antonio Cervera's account of cosmographers in the Philippines.

The book also shows a broad coverage, from the range of topics stretching from astronomy and mathematics to science policy and philosophy, from the ancient to the contemporary, and to cultures spanning East and West. Being asked to write this Foreword is to me a great privilege.

Ho Peng Yoke
Director, Needham Research Institute, Cambridge
June 2001

PART I

Philosophy/Cultural Study of Science & Technology

The Digital Revolution and East Asian Science

NAKAYAMA SHIGERU 中山茂

Kanagawa University, Japan

Figure or Number?

I recall that at the first conference of our Society in 1982 in Leuven, a young Chinese scholar from the Research Institute of Mathematics boldly argued that, looking back from the present-day age of the computer, the Chinese algebraic tradition, rather than the Western geometrical tradition, may have defined the main stream of the history of mathematics. This may or may not have been just a playful remark, but today it is worth considering a little more seriously.

The two major subjects of mathematics, geometrical figures and algebraic numbers, were unrelated until Descartes synthesized the two. For the beginner, geometry may be more attractive than algebra, as the former appeals directly to visual intuition, while the latter suggests tedious, mechanical, unimaginative computation.

Euclidean geometry, which occupied a privileged position in the Western educational world, is, from a non-European viewpoint, a very peculiar tradition. It lacks the practical merits of numerical calculation, while maintaining, to a painful extreme, logical consistency and unconcern for numerical values.

In what follows I wish to compare the Euclidean tradition in the West with the Chinese tradition of calendrical astronomy, which also enjoyed the highest prestige among the various mathematical techniques employed in China. Until medieval times, calendrical astronomy provided the most challenging mathematical problems. There was no independent mathematical discipline. It remained a handmaid of astronomy, just as in the West, mathematics was a handmaiden of science (in Sir Francis Bacon's phrase), culminating in the Laplace-Lagrange tradition.

In the 17th century, when Hsu Kuang-ch'i and Matteo Ricci translated part of the Euclidean Elements into Chinese, they used *chi-ho* for "geometry". *Chi-ho* became the standard translation of the word, but in those days it simply meant "how much". This reflected a typical Chinese tendency to see Western

3

geometry in numerical terms, far from the way occidentals viewed it. East Asians found the Western bias toward it incomprehensible. Perhaps, for instance, Japanese *wasan* mathematicians did not consider a demand for rigorous proof compatible with their ideal of playful elegance.

Scholars before the Han period speculated on the shape of the cosmos, but such discussions ended after the Sui dynasty. In the astronomical treatises of the dynastic histories, the debate on cosmology culminated in the History of the Chin (completed 644).

Generally, the East Asian elites gave a low status to the figure, which they thought of as merely superficial, while what is essential is within, and is not spatial. An 18th-century Japanese astronomer, Nishikawa Masayoshi, classified astronomy into two categories, *meiri* and *keiki*. *Meiri* astronomy, including astrological prognostications, was metaphysical; *keiki* (figure and instrument) astronomy corresponded to mathematical astronomy. While in Aristotelian thought formal cause is more important than material cause, in the East, figure or shape is not substantial but ostensible. In Greek sculpture the lifelike figure was greatly admired, in Eastern painting even portraits did not seek detailed likenesses.

We often find in the private notebooks of astronomers, or in popular treatises on astronomy, abundant illustrations to explain the principles of eclipses. In formal writings on calendrical science, connected with official practice, treatises contain nothing but verbal forms: technical terms for various astronomical parameters, their values, and often tables of solar, lunar and planetary motions. Although there was no place in an official treatise to discuss the shape of the universe, Japanese eclipse predictions, like those of the Chinese Season-Granting System (*Shou-shi li*), presupposed such a notion. This was important, for forecasting solar eclipses was the most important subject-matter and the ultimate test of a calendrical system. The lack of illustrations in official Chinese treatises seems to have been a matter of academic style before they were influenced by the Sino-Jesuits in the 17th century. Astronomical officials must have considered illustration a mark of amateurism. For professionals, numerals were the fruit of their labors, which only other professionals could assess.

Let us, just for fun, apply Wittfogel's thesis, despite its well-known flaws, to the "figure or number" problem. Under his "oriental" system of centrally controlled irrigation, algebra flourished because of its use in managing and accounting for the state enterprises needed to maintain efficient irrigation. In the small-scale town politics and economics of the ancient Greek city-states, geometry became a tool of public debate and thus attracted attention of the citizenry.

Chinese Numericism

In the East, as we have seen, there was not much development of geometry, at least in the Euclidean sense. On the other hand, numericism — a concern with numerical methods — is apparent throughout the history of Chinese astronomy. We are tempted to call it the Chinese numerical tradition, in contrast to the geometrical one of the West. In Chinese treatises of computational astronomy, neither illustrations nor algebraic symbols appeared. They merely set out the method for calculating the ephemeris, without discussing the underlying algorithms. This was also true of the *Almagest* and similar writings, but the Chinese algorithms were algebraic rather than geometric.

As early as the late Han dynasty, Chinese rejected their earlier simple cyclic view of cosmic processes and gave priority to observational data. They used numerical methods to calculate the equation of time, relying on complex interpolation in their data, until in the Fu-tian calendrical system of the 780s they adopted an algebraic method (a second-degree parabolic equation, i.e., the method of finite differences for each quadrant) to express the solar equation of center. The Uigurs used this same method, considering it a unique Chinese method.

Successive Chinese systems adopted the Fu-tian method, culminating in Guo Shou-jing's *Shou-shi* system of 1279. It firmly established the use of third-degree algebraic equations to represent the equation of center. Kuo replaced fractions by what amounts to a decimal notation. Behind the dry sequence of instructions we can imagine clerks in the Astronomical Bureau busy with decimal computation, first by arranging rods on the counting-board and, from about the 15th century on, by using the abacus. These were manual computers, digital like the computers of today.

Guo also did away with the conventional calendrical epoch based on a grand conjunction of the sun, moon, and planets placed far in the past. Instead he used an astronomically arbitrary recent point of origin. Determining a grand conjunction requires solving a sophisticated set of indeterminate equations. This was one of the most difficult problems in the history of Chinese mathematics. He also introduced a minute secular change in the length of the tropical year, which would have been unthinkable in the 13th-century West. All of these changes were the result of a conscious effort at numericism, avoiding all geometric and theoretical assumptions. It is quite remarkable to see him solving problems such as transformations from eliptic to equatorial coordinates using purely numerical and algebraic pseudo spherical-trigonometric methods, though working out such solutions involves a clear understanding of spatial relationships. Guo Shou-jing was truly a numerical fundamentalist.

Prior to Pierre Simon de Laplace (1749–1827), no one was prepared to appreciate and use ancient Chinese data for the variation of astronomical parameters over time, because European cosmologies assumed that they did not change. Western astronomers in 1700 still adhered in most respects to the old Greek notion that the heavens are eternal and free from change.

Because the Chinese had no such belief, they were free to explore any possible kind of secular change. Any change in the sky, at least any kind that astronomy accepted, could be expressed in numbers. They had no reason to sacrifice numerical precision for the sake of geometrical or theoretical coherence. This openness of approach was an aspect of Chinese numericism.

Western Appreciation of Chinese Numerical Tradition

Laplace appreciated the quality of Guo Shou-jing's observations. In 1811 he needed a dated ancient value with which to test his theory of the secular variation in the obliquity (the angle between the ecliptic and the equator). In addition to the observations of the dates of sygyzies in the ancient Greek *Almagest*, and those of the Muslims Ibn Yunis and Ulugh Beg, he had the Jesuit missionary Antoine Gaubil's reports on the history of Chinese astronomy, published ca. 1730. These told him about Kuo's measurements of noon solar shadow lengths at the summer and winter solstices around 1279. He compared them with his other sources and showed that the Chinese ones were by far the best.[1] Laplace simply presented Kuo Shou-ching's data alongside other historical observations to prove the validity of his theory. I have calculated the times of true sygyzies and compared them with these observations. I discovered, with far greater certainty than Laplace's method gives, that Kuo's solstice observations were astonishingly precise. They fall within several minutes of the values calculated by Simon Newcomb in the late 19th century, using a method still standard in 1950. The Greek values were accurate only to one or two days, and Islamic observations to several hours.[2] My findings would have pleased Newcomb, because he never trusted old Western astronomical observations. But by using only data since 1750 in his determination of tropical year length, he overlooked Kuo's results.

Half a century after Laplace, J. B. Biot, the many-sided astronomer and chemist, reached a more satisfactory understanding of Chinese astronomy with the aid of his son E. Biot, a young Sinologist. He commented in 1862 that without mastering spherical astronomy, the Chinese could not develop further toward numerical precision.[3] It is impossible to precisely transform ecliptic to lunar coordinates without a certain degree of geometrical expression. That is why Chinese numericistic astronomy, after culminating in the *Shou-shi* calendar, did not significantly innovate.

Numericism Revisited

In the tradition of pure mathematics, geometry is still considered 'elegant', much more so than algebraic solutions. Talented mathematicians appreciated the appeal to intuition more than tedious computation. In celestial mechanics and quantum mechanics, 'elegant' solutions meant analytical ones, more down to earth and tied to data in nature. Celestial mechanics in the style of Laplace, Lagrange, Todhunter, Whittaker, and the Cambridge Mathematical Tripos examination typified this academic notion of elegance.

1950, A Turning Point to Digital Revolution

Jumping to my own research experience, 1950 was significant as the year in which scientists first used a computer. In that year, I was doing dissertation research that required much hand calculation to achieve an analytical solution in quantum mechanics. I happened to read a recently arrived American journal, and found a short note informing readers that the same calculation as mine was being carried out with manually sorted punch cards four hundred times faster than could be done by hand. This was a predecessor of electronic computers, about which we in Japan then knew nothing. Those who had begun doing this work would have dismissed my analytical approach as manual labor.

I was quite shocked by this news from overseas. I asked my old professor whether we should continue with our approach. He insisted on continuing, as many others in celestial and classical mechanics were doing. I instinctively realized that once rapid numerical solutions were feasible, no one would pay any further attention to elegant solutions that had no practical significance. On the other hand, in Japan shortly after the war, we were too poor to afford the new American style of computational astronomy. At that point, I quit astronomy and turned to the history of science.

It was also in 1950 when D. Hartree of St. John's College, Cambridge, used the EDSAC computer to calculate Eigenfunctions of quantum mechanics.[4] St. John's College was famous as the old stronghold of Cambridge analytical dynamics, where the Todhunter tradition once dominated. Within the same college there had been a scientific revolution. From then on, the use of the computer for all sorts of "numerical" mathematics, such as fractals, fuzzy mathematics, chaos and non-linear equations, led to the truly digital scientific revolution.

Numericism was thus reborn 700 years after Kuo Shou-ching. It may be nonsensical to connect these two activities, ignoring their distance in time and space and the difference between decimal and digital. Still, since Descartes' invention of analytical geometry in the 17th century, it has been obvious that numbers are more fundamental than figures. Whether future generations will look at the history of mathematics in the same way is not at all clear. Does today's computer simulation belong to the algebraic or geometrical tradition? That question has no objective answer. All we can say is that some prefer, when working in this field, to begin with figures rather than digital formulas. Perhaps geometry is aesthetically more attractive but I personally believe that numbers are more basic than geometric figures, in view of all-pervasive digital revolution now ongoing.

Computer Use for our Discipline

Shou-shi Li Digitalized

Digitization of the history of East Asian science perhaps began in 1972, when Nathan Sivin and I began an English translation of the Season-Granting Astronomical Treatise of the Yuan dynasty (*Shou-shi li*). We translated the text directly into the MIT mainframe computer, to which, as a Professor, Sivin had access. At the time, of course, there were no personal computers — not even computer monitors — and no word processors, only simple text editors. Alphabetizing a list meant writing a fairly elaborate program. But the software did offer search and replace; we found it handy to change translations of technical terms throughout the whole text more or less instantaneously.

Thus we came to rely on the computer, although we had no idea that it would develop into a really versatile tool. Since I did not have access to one at the University of Tokyo, in fact, I did not give this question much thought. We completed the translation, but there was one column especially in the planetary tables in the treatise whose physical meaning we could not fully understand. We decided to delay publication until we had solved these enigmas, and intend to return to them in due course.

Astronomical Computation

Because computers were, after all, invented for computation, we should mention their quantitative uses. In my own work on the Season-Granting System in the late 1960s and early 1970s, I did all the computation by hand. After rather an exhaustive study of the solar motion, the next step was obviously the lunar motion. At that point, lacking electronic aids, I encountered insurmountable computational difficulties. The orbit of the moon, influenced heavily by the sun and the earth, is far more complicated than that of the sun. It must be treated as a three-body problem. At that point lack of access to computers prompted my second exodus from astronomy, this time from its computational history.

Among my own generation, Professor Kiichiro Furukawa at Tokyo Astronomical Observatory was the first in Japan to apply computational technology to the history of astronomy. Dr. Raymond Mercier told me that early in the 1970s, he had to go to the astronomical observatory to use a computer for work that he can now do at home. From the 1980s on, a group of young scholars led by Professor Yano Michio emerged in Japan to apply computer simulation and virtual experiments to historical problems of astronomy — just a quarter of a century after my exodus.

In the history of astronomy, it is now possible for any historian with a foundation in the exact sciences to analyze the techniques used at a given time for eclipse prediction, the main goal of traditional calendrical science. Specialists are clarifying, one by one, the complex parameters employed in eclipse prediction, such as the "east-west difference" and "south-north difference". Even the cheapest personal computer can now serve as a planetarium. Someone who knows no mathematics can now watch a solar eclipse happening, with fair accuracy, just as it could have been seen thousands of years ago from any point in East Asia.

The use of computers is not limited to the exact sciences. Statistics and graphing programs that have been available for some time also make it easy to apply numerical analysis and simulation to problems in medicine and technology.

Toward CJK Unification

Any scholar of East Asian science, even one who speaks only Vietnamese, Korean, or Japanese, should be able to comprehend early Chinese scientific texts, since all the traditions shared classical Chinese. Let us consider some capabilities useful to all East Asian scholars.

We can now enumerate, document, and cite classical evidence electronically, with a fraction of the effort it once took. This greatly extends the new possibilities of research on China that opened up in 1931, when concordances to classical texts were first published in modern form. It is now possible in a couple of minutes to gather and compare every occurrence of a given word in an enormous body of writings. My purpose now is to enumerate various possibilities of scholarship using computers, as well as to report on the state of affairs, as an invitation to a general discussion of future approaches.

Dynastic Histories Digitalized

As I stated in an article in *Osiris*,[5] I have worked on both the ancient and modern periods, and found that it is nearly impossible to bridge both with a coherent methodology. I should say that it is easier to work on the modern and contemporary period by making use of computer databases and the internet rather than the classical period, where many characters fall outside standard language usage. I must confess that one of the reasons why I spend more time on the study of modern and contemporary science involves the availability of such research tools.

Today, however, I find that ancient and modern are coming together in terms of digital research material. I was impressed when searching for references to Ts'ao Shih-wei, an obscure figure, perhaps Uighur, who computed the solar equation of center in his "Fu-tien li", I came across the Taiwanese digital text of the Chinese official histories. Within a few seconds, I had a list of all of their references to him. Most of the orthodox classics have been digitalized in Japanese code, and a variety of other texts are available in mainland China. Eventually electronic texts may be quite adequate for scholarship on traditional Chinese science. It may be appropriate for our Society to consider establishing a unified format and editing all the available digital texts into it.

Manuscript Studies and Collation

It seems that in China there are not many textual studies conducted, as scholars are all fed up with the very many existing printed works. On the other hand, Japanese scholars still spend lots of energy in digging up manuscripts and collating them. Joseph Needham wrote a review in the journal *Science* of my first book in English, *A History of Japanese Astronomy*. He was impressed by my frequent use of manuscripts. For instance, I collated more than ten manuscripts to establish the most authentic version of *Kyurikiron* by Shizuki Tadao (1760–1806), the first proponent of Newtonianism in East Asia. I am told that there are many manuscripts which escaped my eyes at the time.

Because of the larger literary audience, it is said that everything seems to be printed in China, excepting the modern period. As a matter of fact, however, Japanese historians working on the pre-modern period begin by searching for manuscripts in bookshops as well as libraries. In Japan and Korea, printed works mostly address popular audiences, and rare or specialized works are only available in manuscripts or hand-written copies, as their audiences are very limited. For instance, I think that one of the best commentaries on the *Shou-shi li* was that prepared by Takebe Katahiro (1664–1739). The next step is to begin with this text, post it on the internet, and invite scholars in China, Japan, and Korea to adduce other texts and establish a definitive edition.

Because in Japan, as well as in Korea, most academic writing used classical Chinese until the 19th century, many are qualified to participate in such a cooperative endeavor. How many young computer-literate scholars can accurately read classical Chinese is an open question.

Unicode and Tron

Existing proprietary language codes obviously are not designed to accommodate lovers of antiquity. Anyone who works on classical East Asia must use, or make for himself, special fonts for rare characters. For instance, the Taiwan government's CNS code already includes 48,000 characters and is still growing. Even so, in connection with the Zen Buddhist text project at Hanazono University in Kyoto, it was necessary to prepare a font of 600 more characters that often occur in Zen sources but are missing from both the CNS code and the Japanese JIS code. That is certainly not the end of it. The Japanese Tron code, developed in the 1980s, now supports 130,000 characters. This was a scholar's dream, not a commercial tool. At one point in time, both myself and the Japanese government supported it as an ideal system for scholarship.

Tron was disastrously involved in political battles over international trade regulations when the USA and Japan revised their science and technology agreement in 1988. The American negotiators treated Tron as if it contributed to trade barriers. The Japanese government eventually abandoned it, but some Japanese librarians still prefer it.

In a capitalistic society, commercial feasibility rules. The Unicode system, which includes every important language in the world in a character set of (potentially) 65,000 codes, has been officially adopted years ago by international bodies, and word processors that support it are now becoming widely available. By coding each symbol in 16 bits (or 2 bytes) instead of the established 8 bits (or 1 byte), Unicode has space in theory for over 65,000 characters, but only about 40,000 have been assigned. Of these, roughly half are for Chinese (old-style and simplified), Japanese, and Korean, and the other half include all the world's other important written languages.

This standard was planned for present-day business and information communication, not for scholarship. Chinese alone, if we wanted to include all the characters that occur in writings before the 20th century, would require space for considerably more than 40,000, and there would be no room for any other language. In order to make a character set large enough to include them all, each character would have to be recorded in 24 bits (or 3 bytes). That would mean that every text file and message in every language would be three times as long as today, or $1\frac{1}{2}$ times as long as in Unicode.

As the internet became a useful medium of communication, Sivin started several subscription lists, such as the one for East Asian science, EASCI, that began in the early 1990s. Japanese historians of science now have lists in their own language. Perhaps someone will inform us about other such lists in East Asia. It would be advisable to begin using Unicode, the single standard character code, which enables a worldwide network open to any language.

East Asians are accustomed to dealing with a multibyte system, in contrast to Western monobyte reductionist culture. It may be that in the future our multibyte culture will prove advantageous for dealing with complex systems.

Despite the small size of the East Asian history of science community, we are unique in including both scientific and humanistic constituencies. The former communicate in English and are strong in using computers, whereas the latter tend to use classical Chinese characters. Such choices are ultimately made on political, not academic, grounds. Still, historians of science with expertise in classical scholarship as well as modern technology are in a strategic position to provide informed opinions. I hope that our Society will organize a committee and start seriously working towards getting those opinions placed where they are needed.

The Digital Revolution: Final Words

It has generally been agreed by historians of science that experiment played a crucial role in creating modern science during the Scientific Revolution of the 17th century. Certain computer applications for simulation and virtual reality play a similar role today. The Digital Revolution seems to be contributing to the creation of postmodern science, as the Scientific

Revolution created modern science. While modernization directly affected, in the main, science and technology, the Digital Revolution is influencing the social sciences and humanities as well.

It may be useful to think of computer applications as 'kyoken', which could be translated as 'imaginary' or 'virtual', in contrast to 'jikken' ('experiment', a modern Japanese translation). This acknowledges that it is contributing to a scientific revolution arguably of the same magnitude as that of the 17th century. 'Kyoken's' implication of the all-pervasiveness of Yin-Yang, and its jitsu-kyo relativism, is difficult to express in terms of any Western dichotomy. I hope the jikken-kyoken dichotomy will make sense to the members of our Society.

Notes

[1] Additions in *La Connaissance des Temps*, 1811, pp. 429–37; also *Exposition du Systeme du Monde*, p. 398. Joseph Needham made a nice plot of ancient and medieval measurements in *Science and Civilisation in China*, vol. 3, fig. 113, p. 289 (Cambridge, 1959).

[2] "Premodern determination of tropical year length", *Japanese Studies in the History of Science*, no. 2, 1963, or figure 15 of *A History of Japanese Astronomy: Chinese Background and Western Impact* (Harvard, 1969), p. 132.

[3] *Etudes sur l'astronomie indienne et l'astronomie chinoise*, p. 297.

[4] P. A. Medwick, "Douglas Hartree and Early Computations in Quantum Mechanics", *Annals of the History of Computing*, vol. 10, number 2, 1988, pp. 105ff.

[5] "Needs and Opportunities of the History of East Asian Science", vol. 10, 1995, pp. 80–94.

"Fossils", "Organic World-View", "The Earth's Motions", etc.: Problems of Judging East Asian Scientific Achievements from Western Perspectives

KIM YUNG SIK 金永植
Seoul National University

From the very early days of the modern study of the history of East Asian science, there have been frequent attempts to find East Asian "forerunners" of Western scientific achievements. Most of these attempts have adopted Western science as the standard for evaluating East Asian scientific achievements, and studied those ideas of traditional East Asia that showed similarity to those of Western (usually modern) science, often taking these as signs of the priority or superiority of East Asian science. This paper discusses the problems with such approaches, and illustrates them with examples — ideas and characteristics of East Asian science — that have frequently been discussed as showing such priority or superiority: the idea of "fossils"; the organic character of the East Asian world-view; the idea of the earth's motions; the idea of the round earth. In doing this I am perhaps beating a dead horse, for no serious historian of science still holds the attitude that I criticize, especially concerning the examples I discuss below. But I have chosen these examples intentionally because they allow me to illustrate the problems more clearly, and to remind at least myself, if not anyone else, of the historiographical dilemma.

The Idea of Fossils

In 1941, in a query published in *Isis*, George Sarton cited Joinville's record (1309) of an observation of fossilized fish made by St. Louis in 1253, and wondered whether this could be the "earliest reference to fossil fishes".[1] This led many scholars to find "forerunners" in the discovery and the recording of fossils. Some found them in China. L. Carrington Goodrich in 1942 drew attention to Chu Hsi's 朱熹 (1130–1200) comment on the existence of shells on high mountains, and considered this as showing the great Sung philosopher's understanding of the nature of fossils.[2] Chu Hsi wrote:

> I have seen on high mountains shells of conches and oysters, some of them in the rocks. These rocks were soils in earlier days, and the conches and oysters are the things [living] inside water. [What happened is that] what was low has changed and become high, and what was soft has changed and become hard...[3]

It turned out that Shen Kua 沈括 (1031–1095), whom Chu Hsi read, also had similar passages:

> On Chin-hua Mountain of Wuchow there are pine[-shaped] stones (*sung-shih* 松石); there also are those similar to kernels of peaches, roots of reeds, and fishes and crabs, and so on, all turned to stones. Yet, all of these are what are native to the area, and thus are not very strange. [On the other hand,] these things are not the things that are deep underground, and also not the things that are originally in earth. This is especially strange.[4]

> A man of Tsechow digging a well in his house, saw something in the shape of a creeping dragon or snake. He was scared and did not dare touch it. Seeing that it did not move for a long time, he patted it, and [found that] it was a stone. The country people were ignorant, and smashed it. Ch'eng Po-shun, magistrate of Chin-ch'eng at the time, got hold of a piece of it. The scales and shells were all like those of the actual living thing. Probably [it was formed by] the transformation of a snake or clam, as the things like "stone-crabs" (*shih-hsieh* 石蟹) were.[5]

These passages led to some very enthusiastic evaluations — or over-evaluations — of Shen Kua's and Chu Hsi's understanding of the nature of fossils, which, it was often said, was achieved centuries ahead of the West. The explanation was that in the absence of the obstacles like belief in the creation of the world several thousand years past, which precluded the correct understanding of the nature of the fossils in the West, these Sung Chinese thinkers could see the implications of such common observations, and recognized "that some of the life-like forms discovered in the rocks were in fact remains of ancient animals".[6]

Through the work of the scholars answering Sarton's query, a long list of "forerunners", both in China and in the West, emerged.[7] In the Christian West, the list grew thickest in the Renaissance: Leonardo (c. 1508), Bernard Palissy (c. 1580), John Ray (1627–1705), and Robert Hooke (1635–1703), to mention the more famous names. Forerunners were also found in medieval Islam and ancient Greece: Avicenna (c. 1022) and Xenophanes (6th century B.C.). The idea was discerned even in Babylonian cuneiform inscriptions. Xenophanes's view, narrated by Hippolytus (3rd century A.D.), is especially interesting because of its striking similarity to the remarks of Shen Kua and Chu Hsi:

> Xenophanes thinks that a mixture of the earth with the sea is going on, and that in time the earth is dissolved by the moist. He says that he has demonstrations of the following kind: shells are found inland, and in the mountains, and in the quarries in Syracuse he says that an impression of a fish and of seaweed has been found, while an impression of a bay-leaf was found in Paros in the depth of the rock, and in Malta flat shapes of all marine objects. These, he says, were produced when everything was long ago covered with mud, and the impression was dried in the mud. All mankind is destroyed whenever the earth is carried down into the sea and becomes mud; then there is another beginning of coming-to-be, and this foundation happens for all the worlds.[8]

In China, more than 30 instances of such "forerunners" have been amassed, including the 8th-century figure Yen Chen-ch'ing 顏真卿, and extending at least as far back as *Shui-ching chu* 水經註, which quotes Lo Han's 羅含 *Hsiang-chung-chi* 湘中記 (c. 375).[9]

When Needham mentioned the names of Shen Kua and Chu Hsi in the second volume of his *Science and Civilisation in China* (SCC), published in 1956, he did not seem to be aware of these earlier "forerunners". In volume three, however, published in 1959, he had all the above information, and clarified the situation.[10] Yet, because of the overwhelming popularity of the second volume of SCC, Shen Kua and Chu Hsi are still widely thought by many people, including scholars, to have been Chinese "forerunners" who had achieved an understanding of the nature of fossils long before Westerners.[11] Later, I shall briefly look at the context of the Chu Hsi passages quoted above, but first, let me present my second example.

"Organic" World-View

Needham characterizes the neo-Confucian world-view, especially that of Chu Hsi as "organic". He uses the word "organic" in a sense opposite to "mechanical". This characterization is not unreasonable because Chu Hsi's world-view is clearly more "organic" than "mechanical". And, as Needham further suggests, Chu Hsi's ideas might indeed have inspired Leibniz's philosophy of monad, and then the organic philosophy of

Whitehead, whose congruence with modern science was emphasized by Needham:

> The great triumphs of early 'modern' natural science were possible on the assumption of a mechanical universe — perhaps this was indispensable for them — but the time was to come when the growth of knowledge necessitated the adoption of a more organic philosophy no less naturalistic than atomic materialism. That was the time of Darwin, Frazer, Pasteur, Freud, Spemann, Planck and Einstein. When it came, a line of philosophical thinkers was found to have prepared the way — from Whitehead back to Engels and Hegel, from Hegel to Leibniz — and then perhaps the inspiration was not European at all. Perhaps the theoretical foundations of the most modern 'European' natural science owe more to men such as Chuang Chou, Chou Tun-I and Chu Hsi than the world has yet realised.[12]

But what we have here is nothing more than a possible, though interesting, case of inter-cultural influence and congruence. It would be wrong to proceed from this to conclude (1) that Chu Hsi's world-view somehow has the character of modern science, which has overcome, or shed itself of, the predominantly mechanical character of the Newtonian "classical" science, or (2) that it is a desirable world-view for people to have in a modern world plagued by many problems created by the "mechanical" science.

To be sure, it is true both that Chu Hsi's world-view can be called "organic" rather than "mechanical" and that modern science has a relatively more "organic" character compared to classical science. Yet the "organic" character we see in modern science is of a level completely different from that of Chu Hsi's world-view. Although modern science does not depict the natural world in such a completely mechanical way as classical science did, it has not forsaken its mechanical character to such an extent that it has become similar to Chu Hsi's neo-Confucian world-view. Modern science remains essentially mechanical, though less so than classical science.

It is also misleading to characterize Chu Hsi's world-view as "organic", if it is interpreted as a systematic world-view in which everything finds its proper place in a harmonized whole. In fact, Chu Hsi's world-view was rather particularistic. He treated most natural phenomena, and problems involving them, basically as particulars; he did not generalize from them, as he took each phenomenon or problem as it came along without much concern to correlate it with his views about the rest of the natural world.[13]

Such a particularistic character can be seen clearly if we look at the way Chu Hsi dealt with a few problems and phenomena related to the idea of weight and compare it with what the medieval European scholastics did. Chu Hsi noted that water tends to move downward. This is similar to the views of the medieval scholastics in that whether an object falls or rises is determined by what material substance it is made of. But Chu Hsi's was not a general idea applicable to all heavy and light objects. To flow downward, for him, was an innate tendency of a particular substance, water, and not of all heavy objects. It is even unclear whether he realized that the downward flow of water is a particular case of the fall of heavy objects. Furthermore, for the medieval scholastics, the fall of heavy objects and the rise of light objects were tied with their conceptions of the structure of the world: heavy things fall down as they tend to their natural place, the center of the earth, which, for them, was the center of the world.[14]

Thus, the problem of the weight of the earth, i.e. the cause for the stability of the heavy earth in the middle of the sky, never arose for the scholastics. The concept of weight was built into their world-picture itself; the earth, made of the heaviest of the four elements, lies at its natural place, the center of the world, and thus does not move. But Chu Hsi, for whom the problem of weight was separate from that of the structure of the world, had to deal with the problem of the earth's weight. Whereas the scholastics' world-picture included the particular empirical fact of weight as an integral part of it, Chu Hsi's had to take account of the notion of weight as an additional and independent fact.

Chu Hsi's answer to the problem was that the power of the rapidly rotating *ch'i* of heaven keeps the heavy earth from falling down. To illustrate such power of moving *ch'i*, he noted that water in a revolving vessel does not spill even when the vessel is upside down. And interestingly, we find from the fragments of another pre-Socratic, Empedocles (c. 492–432 B.C.), narrated by Aristotle, a discussion of the problem of the earth's weight that is very similar to Chu Hsi's.

> ... All those who generate the heavens hold that it was for this reason that the earth came together to the center. They then seek a reason for its staying there; and some say, in the manner explained, that the reason is its size and flatness, others, like Empedocles, that the motion of the heavens, moving about it at a higher speed, prevents movement of the earth, as the water in a cup, when the cup is given a circular motion, though it is often underneath the bronze, is for this same reason prevented from moving with the downward movement which is natural to it.[15]

So again, we are faced with similarities between Chu Hsi's ideas and those of the pre-Socratics. Of course, I do not mean to say that Chu Hsi's world-view is at the level of the pre-Socratics. Rather, my point is that once we begin to judge traditional East Asian scientific achievements using the Western achievements as standards or criteria, we are very likely to end up with this kind of outcome. While we will find many East Asian "forerunners" of various Western scientific achievements in this manner, it will frequently be possible to find Western "fore-forerunners" to these East Asian forerunners. For example, we had begun the previous section by noting that Shen Kua and Chu Hsi can be seen to have hit upon the understanding of the true nature of fossils several centuries before the West. But as we started looking for forerunners, many of them were found both in the West and in the East. And in this particular case of fossils, the tracing went back farther in the West.

The Earth's Motions

My next example is a modern scientific idea whose ancient forerunners in the West are known a great deal better: the motion of the earth. Heraclides (c. 388–310 B.C.), Aristarchus (c. 310–230 B.C.), and Seleucus (fl. 150 B.C.) were all known to have had such an idea. Medieval European scholars like Oresme (c. 1320–82) also discussed the idea in some length.[16] Yet, the historians of Western science, quite rightly, have not made too much of these "forerunners".

Nor have the historians of East Asian science been much interested in finding East Asian forerunners for the idea. Instead, they have been interested in the "independent" developments — or discoveries — of the idea of the earth's motions in East Asia, assuming, of course, that such independent developments are what is to be praised. Frequently noted in this connection were the 17th- and 18th-century Korean scholars, Kim Sŏk-mun 金錫文 (1658–1735) and Hong Tae-yong 洪大容 (1731–83) in particular. Of course, there were also Chinese scholars who had touched upon the possibility: Wang Fu-chih 王夫之 (1619–92) and Huang Pai-chia 黃百家 (1643–?) for example. There have been debates about the originality of their ideas, involving the question of whether, and how, they arrived at the ideas of the earth's motions independently from the Western ideas.[17]

What is really interesting in all this, however, is that these 17th- and 18th-century "discoverers" found an East Asian forerunner of the idea of the earth's rotation, namely Chang Tsai 寂載 (1020–1077). They saw the idea in two very ambiguous passages from the "San-liang" 參兩 chapter of Chang Tsai's *Cheng-meng* 正蒙. I will first

present the two passages in full (only in Chinese characters) before
discussing them.

(Passage I)

地純陰凝聚於中，天浮陽運旋於外，此天地之常體也。恒星不
動，純繫乎天，與浮陽運旋而不窮者也。<u>日月五星，逆天而
行，— 包乎地者也，地在其中</u>。雖順天左旋，其所繫辰象隨
之，稍遲則反移徒而左爾，間有緩速不齊者，七政之性殊
也。···

(Passage II)

凡圍轉之物，動必有機，既謂之機，則動非自外也。<u>古今謂天
左旋，此直至粗之論爾，不考日月出沒，恒星昏曉之變，愚謂
在—而運者，惟七曜而已。恒星所以為晝夜者，直以地氣乘機
左旋於中，故使恒星河漢，因北為南，日月天隱見</u>，太虛無
體，則無以驗其遷動於外也。

Of modern scholars, Yamada Keiji 山田慶兒 is most enthusiastic in
seeing the idea of the earth's rotations in these passages.[18] He interprets
the underlined portion of Passage I as follows:

> The sun, the moon, and the five stars move in the direction opposite to
> the sky's rotation, and they also surround the earth. The earth is inside them,
> and though [it, i.e., the earth,] follows the sky and rotates leftward, ...

For Passage II, Yamada follows Wang Fu-chih and others in changing "*tso-
hsüan*" 左旋 (leftward rotation) to "*yu-hsüan*" 右旋 (rightward rotation),
and interprets the underlined portion as speaking of the earth's rightward
rotation instead of the sky's leftward rotation.

> People have spoken about the sky's leftward rotation. This is an extremely
> crude theory, [not knowing that it is the earth that rotates rightward rather
> than the sky rotating leftward, and] not considering the rise and setting of
> the sun and the moon, and the change of the morning and evening of the
> fixed stars. I would say that what move in the sky are only the seven
> luminaries. The reason that the fixed stars have day and night is that the
> earth's *ch'i* rotates *rightward* inside relying on the *chi* 機 [of motion]. Thus
> it makes the fixed stars and the Milky Ways [appear to be rotating] from
> north to south, and the sun and the moon appear and disappear following
> the sky.

It is not possible to determine the exact meanings of these ambiguous
passages. Yet, what is to be noted here is that such readings of the idea of

the earth's rotation in these passages make Chang Tsai appear not "creative" or "imaginative", but rather strange and difficult, if not confused and even ridiculous. Nowhere in Chang Tsai's sayings and writings does he show any awareness of the possibility of the rotation of the earth. Indeed, in the passage that immediately follows Passage II, which Yamada and others interpret as referring to the earth's rightward rotation, Chang Tsai speaks of the sky rotating leftward. Moreover, the idea of the earth's rotation would contradict the idea of the "circular heaven and square earth" (*t'ien-yüan ti-fang* 天圓地方) assumed by every one in traditional China, certainly including Chang Tsai. Even Wang Fu-chih, who read the earth's rotation from these passages, opposed the idea of a spherical earth.[19] Considering this circumstance, it is extremely unlikely that if Chang Tsai really had the idea of the earth's rotation, he would have spoken about it in such a casual manner.

Indeed, there is a perfectly reasonable way to make sense of the passages without reading the earth's rotation into them. For this new interpretation we adopt a different punctuation in one place in Passage I: making the phrase *ti tsai ch'i chung* 地在其中 part of the previous sentence, instead of beginning the next sentence with it. Then we follow Chu Hsi's commentary and take the subject of the previous sentence, *jih yüeh wu-hsing* 日月五星, to be the subject of the next sentence also: it is "the sun, the moon and the five stars" that "follow the sky and rotate leftward". The result is:

> The sun, the moon, and the five stars move in the direction opposite to the sky's rotation, and they also surround the earth, which is inside them. Although [they, i.e., the sun, the moon, and the five stars,] follow the sky and rotate leftward ...

> 日—五星，逆天而行，并包乎地者也，地在其中。〔日月五星〕雖順天左旋，…

For Passage II what we need to do is to take "*ti-ch'i*" 地氣 to be Chang Tsai's, or a scribe's, mistake for "*t'ien-ch'i*" 地氣. If we have to assume that a mistake was made, this would appear far more likely than the mistake of saying a "leftward rotation" while meaning a "rightward rotation", and at a crucial place at that. Passage II then becomes:

> People have spoken about the sky's leftward rotation. This is an extremely crude theory, [only speaking about the sky while] not considering the rise and setting of the sun and the moon, ... I would say that what moves in the sky are only the seven luminaries. The reason that the fixed stars have

day and night is that the *ch'i* of the sky rotates *leftward* inside relying on the *chi* [of the movement of the sky]. Thus it makes the fixed stars and the Milky Ways [rotate along with it] from north to south ...

古今謂天左旋，此直至粗之論爾，不考日月出沒⋯愚謂在天而運者，惟七曜而已。恒星所以為晝夜者，直以地氣乘機左旋於中—— A故使恒星河漢，因北為南⋯

In this way we can avoid the necessity of putting the idea of the earth's rotation into the thought of Chang Tsai.

It should be noted in this connection that all readings — by Wang Fu-chih, Kim Sŏk-mun, Hong Tae-yong, and Huang Pai-chia — of the idea of the earth's rotation in these passages of Chang Tsai came after the arrival of the Jesuits. While Chang Tsai's theory of the leftward rotation (*tso-hsüan-shuo* 左旋説) had been frequently discussed and praised, no one until the 17th century had read the earth's rotation from his passages. Indeed, the very fact that those post-Jesuit scholars could read such an unlikely idea of the earth's rotation from these passages suggest that they might well have known about the idea when they read the passages and were thus predisposed to see the idea in them.[20] In this sense, their readings of the earth's rotation in Chang Tsai can be better understood in the context of the so-called theory of "the Chinese origin of the Western science" (*hsi-hsueh chung-yüan-lun* 西學中源論) — namely in their eagerness to find a Chinese forerunner for a Western theory of the earth's rotation.[21]

The Roundness of the Earth

Concerning my last example, the idea of the roundness of the earth, the situation was different from the three earlier cases. For it is well known both that the idea was firmly embedded in Western natural philosophy from ancient times and that the idea was alien to the Chinese before the Jesuits came.

There were debates concerning this idea in 17th- and 18th-century China. In the spirit of the above-mentioned "Chinese origin" theory, many Chinese scholars of the time traced the origin of the idea to ancient Chinese sources. Yet, for modern historians who know well about the historical situation, the issue is not over the priority or independent development of the idea in East Asia. Freed from such issues, they can afford to exert their efforts into contextual studies. It is from these contextual studies that we learn, for example, that when the Chinese scholars like Mei Wen-ting 梅文鼎 (1633–1721) and Chiang Yung 江永 (1681–1762) accepted the sphericity of the earth, they felt the need to deal with the "cultural" or "ideological" problems caused by its conflict with the "Chung-hua" 中華 (Central Culture) theory: because if the earth is spherical, it does not make sense to speak of a center of the earth's surface.[22] In this respect, the reactions of Korean scholars of Chosŏn are interesting. Some like Yi Ik 李瀷 (1681–1763) used the idea of the roundness of the earth to counter the Chinese "Chung-hua" theory, or to support the "Korean Chung-hua" (Chosŏn chunghwa 朝鮮中華) idea, namely the notion that Korea was the true "Central Culture" now that China itself was ruled by the barbarian Manchus.[23] Other more conservative Confucian scholars like Yi Hang-no 李恒老 (1792–1868) shared the skepticism of the conservative Chinese toward the idea of the round earth.[24]

Concluding Remarks: Contexts of the Discussion of "Western" Scientific Ideas in East Asia

The above discussion of the idea of the roundness of the earth shows the importance of studying the true contexts of the occurrences of what may appear as East Asian forerunners of Western scientific ideas, and illustrates the kind of fruit one can receive from such contextual studies. I will conclude by making brief comments on the contexts of the ideas I have discussed in my three other examples.

First, Chu Hsi's assertions, interpreted as containing and understanding of the nature of fossils, were actually made while he was discussing and illustrating the cosmogony of *Huai-nan-tzu* 淮南子 and the theory of the yin-yang cyclical repetition, which together formed the basis of his cosmology. Change of sea water into mountains was mentioned as a detail of the *Huai-nan-tzu* cosmogony, according to which the primordial *ch'i* in the beginning of the world rotated, and the sediments of the *ch'i* were precipitated at the center to form the earth. But as the last sentence in the quotation indicates, Chu Hsi was also using the change to illustrate the cyclical alternation of the yin and yang characteristics: what is low (the bottom of the sea) becoming high (mountain); what is soft (mud) becoming hard (stone). Thus clearly, what he was discussing was not a paleontological theory or the nature of fossils, but simply an illustration of the yin-yang scheme and a traditional cosmogony.

The context for Chu Hsi's organic world-view, on the other hand, should be looked for in the Confucian notion of the moral order underlying the natural world of heaven and earth. This ancient notion, which gained a renewed importance in his time, in part in response to the Buddhist challenge of Confucian moral philosophy, provided what can be called the "cosmic basis of morality", and led to the convergence of the moral and the intellectual endeavors in the Confucian self-cultivation.[25]

For the idea of the earth's motions, the theory of the "Chinese origin of Western science" provides a possible context, as I have suggested. But another context that is worth a careful study is the ancient theory of the so-called "four wanderings" (*ssu-yu* 四游), the true nature of which is not clear and seems never to have been clearly understood by scholars.[26]

All these examples of the occurrences of East Asian ideas similar to those of Western science have turned out to be very interesting. It should be noted, however, that they are interesting not for the issues of "priority" or being "forerunners", but rather for the contexts — mainly intellectual, but sometimes political, ideological, or religious — of the discussion of those ideas. We began from similarities, but what we have found are differences. In the preceding pages I have touched upon only a limited number of such cases, but I expect that there are other similar cases which will be even more interesting.[27]

Notes

[1] George Sarton, "The Earliest Reference to Fossil Fishes", *Isis* 33 (1941), 56–58.

[2] L. Carrington Goodrich, "Early Mentions of Fossil Fishes", *Isis* 34 (1942), 25.

[3] Chu Hsi, *Chu-tzu yü-lei* 朱子語類 (1270 edition, reprinted in 1473, modern reprint, Taipei: Cheng-chung shu-chü 正中書局), 94.3.

[4] Shen Kua, *Meng-hsi pi-t'an* 夢溪筆談 (reprinted in 1975, Beijing: Wen-wu ch'u-pan-she 文物出版社), ch. 21, no. 373.

[5] *Ibid.*, no. 374.

[6] Joseph Needham, *Science and Civilisation in China* (Cambridge: Cambridge University Press, 1954–) (hereafter abridged as SCC), vol. 3, p. 611.

[7] A. S. Pease, "Fossil Fishes Again", *Isis* 33 (1942), 689; Robert Eisler, "Early References to Fossil Fishes", *Isis* 34 (1943), 363; Richard Rudolph, "Early Chinese References to Fossil Fish", *Isis* 36 (1946), 155.

[8] G. S. Kirk and J. E. Raven, eds., *The Presocratic Philosophers* (Cambridge: Cambridge University Press, 1957), p. 177.

[9] Needham seems to use the situation surrounding this as still another support for his belief that "from about the +2nd and to the +15th century China was much more advanced than Europe ...", SCC, vol. 3, p. 623.

[10] SCC, vol. 2, p. 487; vol. 3, pp. 611ff.

[11] Chang Chia-chü 張家駒, *Shen Kua* (Shanghai: Jen-min 上海人民, 1962), pp. 187–88; Yamada Keiji, *Shushi no shizen-gaku* (Tokyo: Iwanami, 1978), p. 163.

[12] SCC, vol. 2, p. 505.

[13] This point is discussed in more detail in Yung Sik Kim, *The Natural Philosophy of Chu Hsi (1130–1200)* (Philadelphia: American Philosophical Society, forthcoming in 2000), Sec. 14.3.

[14] For more detailed discussion of the points made in this and the following paragraphs, see *ibid.*, Sec. 13.2. For a brief account of the medieval Western world-picture, see Edward Grant, *Physical Science in the Middle Ages* (New York: Wiley, 1971), Chs. 4–5.

[15] Kirk and Raven, eds., *The Presocratic Philosophers*, p. 334.

[16] G. E. R. Lloyd, *Greek Science after Aristotle* (New York: Norton, 1973), pp. 53–55. Copernicus himself knew and mentioned certain "Pythagoreans" who had the idea of the movement of the earth, although it was mainly to show that he was not the first to come up with such a radical, apparently ridiculous, idea. See Copernicus, *De revolutionibus orbium coelestium*, Book 1, Chap. 5.

[17] Yi Yong-bŏm 李龍範, "Kim Sŏk-mun-ŭi chijŏnnon-kwa kŭ sasangjŏk paegyŏng", 金錫文의地轉論과 그思想的背景, *Chindan hakpo* 震檀學報 41 (1976), 81–107; Park Seong-Rae, "Hong Tae-yong's Idea of the Rotating Earth", *Han'guk kwahaksa hakhoeji* 한국과학사학회지 1 (1979), 39–49; Yabuuti Kiyosi 藪內清, "Richo gakusha no chikyu kaitensetsu" 李朝學者の地球回轉説, *Chōsen gakuhō* 朝鮮學報 49 (1968), 427–34; Ogawa Haruhisa 小川晴久, "Higashi Ajia ni okeru chiten (dō)setsu no seiritsu", 東アジアにおける地轉(動)説の成立, *Tongbang hakchi* 東方學志 23–24 (1980), 375–87.

[18] Yamada, *Shushi no shizen-gaku*, pp. 40–42. Ogawa Haruhisa mentions the possibility, but does not accept Yamada's interpretation: "Higashi Ajia ni okeru chiten(dō)setsu no seiritsu".

[19] *Ibid.*, p. 381.

[20] Ogawa also suggests this: *ibid.*, p. 380.

[21] John B. Henderson, "Ch'ing Scholars' Views of Western Astronomy", *Harvard Journal of Asiatic Studies* 46 (1986), 121–48, esp. pp. 139–43; Chiang Hsiao-yüan 江曉原, "Shih-lun Ch'ing-tai hsi-hsueh chung-yüan-shuo" 試論清代西學中源説, *Tzu-jan ke-hsueh-shih yen-chiu* 自然科學史研究 7 (1988), 101–108.

[22] Pingyi Chu, "When Science Crosses Its Cultural Boundaries: Debates over the Sphericity of the Earth in China, 1600–1800", *Science in Context* (forthcoming).

[23] For Yi Ik's attitude toward Western science, see, e.g., Pak Sŏng-rae 朴星來, "*Sŏngho sasŏl* sok-ŭi sŏyang kwahak" 星湖사설속의西洋科學, *Chindan hakpo* 59 (1985), 177–97.

[24] Im Chong-t'ae 임종태, "Dori'-ŭi hyŏngisanghak-kwa 'hyŏnggi'-ŭi kisul" '道理'의 형이상학과'形氣'의 기술, *Han'guk kwahaksa hakhoeji* 21 (1999), 58–91.

[25] Thomas A. Metzger, *Escape from Predicament: Neo-Confucianism and China's Evolving Political Culture* (New York: Columbia University Press, 1977), Secs. 3i and 3p; Peter K. Bol, "Chu Hsi's Redefinition of Literati Learning", in William Theodore de Bary and John W. Chaffee, eds., *Neo-Confucian Education: The Formative Stage* (Berkeley: University of California Press, 1989), pp. 151–85.

[26] The idea of "the four wanderings" appeared in various ancient sources, including the *Chou-pei suan-ching* 周髀算經 (ch. 5), and were variously interpreted as

movements of the earth, heaven or the luminaries. For brief discussions of "the four wanderings" and the movement of the earth in general, see, e.g., Yamada, *Shushi no shizen-gaku*, pp. 29–31, 171–84; Chung-kuo t'ien-wen-hsueh shih cheng-li yen-chiu hsiao-tsu 中國天文學史整理研究小組, *Chung-kuo t'ien-wen-hsueh shih* 中國天文學史 (Beijing: Ke-hsueh ch'u-pan-she 科學出版社, 1987), pp. 171–73.

[27] A particularly promising case is the ideas of Liu Hui 劉徽 (+3c), on which very sophisticated research has been done recently by many scholars. See, e.g., Fu Daiwie, "Why Did Lui Hui Fail to Derive the Volume of a Sphere?", *Historia Mathematica* 18 (1991), 212–38; Christopher Cullen, "How Can We Do the Comparative History of Mathematics? Proof in Liu Hui and the Zhou Bi", *Philosophy and the History of Science: A Taiwanese Journal* 4(1995), 59–94; Karine Chemla, "What Is At Stake in Mathematical Proofs from Third-Century China?", *Science in Context* 10 (1997), 227–51.

The Science of Eurasia: Meiji Seismology as Cultural Critique

GREGORY CLANCEY

National University of Singapore

The term 'East Asian Science' remains common and serviceable at the turn of the millennium — it invites potential readers toward this book, for example — but increasingly as a container for objects much more vital and mobile than itself. Once an invocation of essence, it can now just as well mean science *in* East Asia, *between* sites within (or linked outside of) that geography, or even practices that *signify* East Asia, as for example, acupuncture in Europe. This re-rendered East Asia is more alive with tradition-making, the forging of hyphenated identity, and the formation of pidgins across zones of curiosity or trade. Such concepts invoke yet go beyond Joseph Needham's 'great titration'. Our instinct, moreover, need not be to untangle them as 'problems' in the manner of the scientists we study, but to learn something from our inability to fully do so.

Those of us who tell stories about 'modern' science in this region — which are stories set in colonial and post-colonial worlds — are yet in need of more flexible geo-imaginaries. 'Japanese science', for example, is a set of practices, identities, and strategies deeply moulded by locationally-induced tensions. Archipelago and nation-state are merely two corners of the space which Japanese science initially sought its place within. One might call this spatial imaginary, with reference to its Meiji-period origins, 'Eurasia'. It is not to be mistaken, however, for the physical geography that goes by the same name, although both share the virtue of having no recognizable center nor clearly-delineated periphery. Its identity lies more within an instinct to transcend, re-draw, or confound boundaries it cannot, given the contradictions of its modern birth, entirely accept nor reject. I want to tease out these characteristics of Japanese science through the story of just one problematic disciplinary manifestation: seismology, the science of earthquakes.

Contemporary scientific textbook introductions tell us that seismology was 'Western' science brought to Japan in the 19th century, in the same manner as steam engines and top hats. This makes sense to the casual lay or expert reader, because seismology is a modern, instrumental, geo-physical science, and the west-to-east trajectory of such practices is well-charted. When one looks more closely, however, the description is troubling, because there was no coherent set of theories, practices, texts, and institutions, called 'seismology' in 'The West' ready, in the middle of the 19th century, to migrate to and enlighten Meiji Japan. The formative Japanese academy of this period, even had it wanted to import people credentialed as 'seismologists', would have had a difficult time finding any. In fact it never made the effort, even though it famously scoured Europe and America for many other species of foreign scientific expert.

European scientists had been interested in earthquakes from time to time prior to the 19th century, and as Perry sailed into Tokyo Bay, Italian researchers were inventing instruments to bring them under surveillance. None of these people called themselves 'seismologists', however, as most were more interested in other, more regular (and hence more observable) phenomena of nature. The term 'seismology' was coined in the 1860s by Robert Mallet, an English civil engineer then investigating an earthquake in Naples (which remained, alas, his only one).[1] Those with a serious interest in watching the earth convulse were more likely to end up with volcanoes, which had the virtue of always being in the place one had last marked them.

It was in Japan itself, more specifically Tokyo of the 1880s, that the modern science of earthquakes — the most direct ancestor of the present discipline — actually speciated. It was Tokyo University that established the world's first chair in seismology, built the best-equipped seismology lab in the world, and graduated the first students with the word 'seismology' (*jishingaku*) on their diplomas. The 'seismograph' — the direct ancestor of the device still in use — was invented in the Meiji capitol, and the first 'seismograms' recorded there. The first seismological journal and first nation-wide series of seismological stations first appeared in Japan.[2] By 1891, the year of the Great Nobi Earthquake (after Naples, the most scientifically productive seismic event of the 19th century) Tokyo was the principle center of world seismology, a science that remained for at least another decade an amateur or tangential pursuit in the academies of Europe, and did not even exist in the United States.[3]

It is true that the 'founders' of seismology in 19th-century Japan were largely foreigners, mostly Englishmen who had been hired to teach science and engineering in the formative Japanese academy. The original seismograph, for example, was named the 'Ewing-Grey-Milne' device after its three British inventors, and was physically and theoretically embedded in a long European instrumental tradition. To stop there, however, as most textbook-writers do, is to ignore the crucial matter of context. None of Japan's proto-seismologists arrived in Tokyo with prior documented interest in earthquakes. The three inventors of the seismograph were hired as lecturers in, respectively, physics, telegraphy, and geology, and turned their attention to seismometry — as they named the fashioning of earthquake-recording instruments — as an extracurricular, and inter-disciplinary pursuit. Ewing and Grey abandoned earthquakes for other interests when their contracts expired and they returned home to Britain. The 'birth' (and as importantly, the rearing) of seismology, like that of most science projects, was thus highly location- and time-dependant.[4]

The Seismograph in the Japanese Landscape

Before the emergence of 'Anglo-Japanese' seismology, the standard European method of investigating earthquakes was to survey damage to buildings, which was called 'observational seismology' by its leading practitioner Robert Mallet. Only by mapping ruins over a large area — seeing how and in what direction walls and chimneys fell — had Europeans constructed maps of an earthquake's relative strength and direction of movement. These had been the methods by which Mallet famously 'recorded' the Great Neopolitan Earthquake of 1857. Ruins, and ruins almost alone, had allowed Mallet to locate and map an 'epicenter' and surround it with concentric 'iso-seismal' lines.[5]

The trouble for geologist John Milne, eager to record an 1880 Japanese earthquake in a way that would allow it to be compared with Mallet's data from Naples, was that Japanese buildings lacked the necessary display of damage.

> Everywhere the houses are built of wood and generally speaking are so flexible that although at the time of a shock they swing violently from side to side in a manner which would result in utter destruction to a house of brick or stone, when the shock is over, by the stiffness of their joints, they return to their original position, and leave no trace which gives us any definite information about the nature of the movement which has taken place.[6]

Mallet, whose views on earthquake movement Milne had resolved to dispute, had been able to gather data like an archaeologist — a familiar role for Englishmen in southern Europe. As he described the "meizoseismal area" (the epicentral zone) of an Italian earthquake,

> The eye is bewildered by 'a city become a heap'. [The seismologist] wanders over masses of dislocated stone and mortar ... Houses seem to have been precipitated to the ground in every direction of azimuth.[7]

Earthquakes were to be 'seen' by sorting all this wreckage into patterns, and the location and depths of 'epicenters' were to be mathematically calculated from the angles of wall-cracks. For Milne, it was exceedingly difficult to measure, trace, map, 'see' the earthquake of February 1880. Because it had not sufficiently inscribed itself into the Japanese cultural landscape, it could not be fixed on the natural one.

It was this very problem of insubstantiality, however, that propelled Japanese seismology in the direction of instrumentation and Milne toward membership in the Royal Society. The lack of traces in the post-earthquake landscape of Japan meant that Anglo-Japanese seismologists devoted inordinate attention, by European standards, to inventing and using inscripting devices. "At this time", wrote Milne, "Tokyo was in reality a city of many inventions ... their name was legion."[8] At the first meeting of the Seismological Society of Japan, Milne remarked "we shall see around us a mighty forest of pendulums, springs, and delicately balanced columns". It was at the same meeting that James Ewing, professor of Mechanical Engineering and Physics at what would soon be Tokyo University, unveiled his 'seismograph' — the kernel of the later Ewing-Grey-Milne device — essentially a seismometer attached to a disc of smoked glass (and later a continuous roll of smoked paper) on which various earthquake features could be recorded.[9] The seismograph became the instrument around which the new science of seismology was built, yielding as it did a product (a 'seismogram') that was not only highly readable, but, like a telegram, vendible (between Tokyo and Europe) and reproducible in scientific reports and papers.[10]

The seismograph, later so useful everywhere, was thus at the beginning peculiarly useful in Japan. Moreover, as soon as its pendulums and tracers were properly balanced, it began to inscribe (onto paper) earthquakes that had hardly been felt, a kind that had never been important to European scientists before. Through the medium of this one instrument, Japan went from being a place where earthquakes could be felt but not traced (thus confounding European methodologies) to a site where they could be traced even if not felt. Trans-Eurasian traffic in recordable earthquakes was now set to become a Japanese specialty.

Defining 'Japan's Earthquake Problem'

Instrumental seismology began to find a home in the Imperial university and various government agencies beginning in the mid-1880s. The government had become convinced that what Milne was doing after hours was promising enough to warrant their sustained institutional support. And what was, from the standpoint of the Japanese academy, seismology's promise? Partly it was the awareness that the 'science of earthquakes', not yet fully formed in the West, was a scientific niche that Japan might readily fill and dominate. It was the recognition of *novelty*, rather than the instinct to emulate or copy, that helped give the study of earthquakes resources in Japan that it would never enjoy to the same extent elsewhere; that convinced Japan to train young people in a 'Western' discipline in which they would not (for a while longer at least) have young Western colleagues.

Sekiya Seikei and Omori Fusakichi, the first seismologists to emerge from the Japanese academy (both graduated in the 1880s), would build on the relationship with European science begun by their foreign mentors. Omori in particular would go on to make a foreign reputation that equalled or perhaps surpassed that of his teachers.[11] The Japanese scientists would have the luxury, however, at least at the beginning, of not having to mold their discipline too closely to Western scripts. They would cultivate within and around seismology a set of attitudes, stances, and arguments, political and time-dependent, which provided Japanese seismology a certain nativist identity.

The Meiji government patronized seismology not only because of its international promise, but because it offered to both define and ultimately solve 'Japan's earthquake problem'. This problem, as seen from turn of the century Japan, had a specific and intimate relation to the modernizing, westernizing project itself. As Education Minister Dairoku Kikuchi put it in 1904:

> ... on the practical side [of seismology] were considerations of houses built in foreign styles, arches, bridges, chimneys, etc., new to Japan; these were constructed without proper attention to the probable effects of the earthquake shocks peculiar to this country.[12]

During the Great Nobi Earthquake of 1891, large 'Western-style' railroad bridges, factories, and post-offices had, to the surprise of their engineers and the government alike, readily collapsed. On the other hand, certain Tokugawa-period structures like Nagoya Castle had ridden out the waves. Thus was the new foreign-derived infrastructure, which was largely the materiality of the new Japanese state itself, shown to be unexpectedly fragile before Japanese nature. This assignment of failure and success to 'foreign' and 'Japanese' buildings was of course mediated by more than 'nature'. Indeed there was a long and heated debate over the proper lessons and meaning of the Nobi earthquake, a story with too many intricacies and twists to present here. Suffice it to say that in the aftermath of the Nobi earthquake, Japanese seismology emerged not as a simple complement to other imported forms of knowledge, but as the site of a uniquely critical stance toward elements of the new foreign learning — 'facts' about nature which had, quite literally, let Japan down. Seismology in this context was both contrarian and corrective. It actively doubted, and not without reason, the ability of contemporary Western knowledge to explain and anticipate the behavior of its own objects on the Japanese earth.[13]

Seismologists — both the original foreign teachers and the Japanese they trained or mentored — began as early as the 1870s to discover 'aseismic adaptations' in 'traditional' Japanese architecture. They focused in particular on five-storey pagoda (*gojunoto*), which famously did not fall during earthquakes that easily toppled factory chimneys. The central mast of pagoda, it was claimed, were actually hanging pendula, which kept the center of gravity within the base in the same way that smaller pendula did within seismographs. Thus were ancient Japanese cultural objects and new geo-physical instruments made to quite literally converge. The pyramidal shape of Japanese castles, the joint structure of temples and shrines, and the concave stone walls of palace moats were also read by seismologists as evidence that Japanese, from a relatively ancient age, had been close students of seismic forces. Seismology was thus a 'natural' science for Japan not only because earthquakes were so common there, but because knowledge of seismic forces had a long, perhaps unique, Japanese history. Seismologists devoted a portion of their research work to historic buildings; Omori literally hooked up seismographs to pagoda, demonstrating their shared physics in his longest English-language article.[14]

This convergence between modern geo-physics and traditional carpentry was not accidental, incidental, nor casually drawn. Indeed it was bound up with seismology's uniquely strong emergence within the Meiji academy in a period of *kokusui hozon*, or "preservation of the national excellencies", the first cultural reaction against the uncritical absorption of foreign models. Seismology's instinct to tie itself closely to the Japanese landscape was also at least partly a function of its lack of close ties to a distant European one. But this 'Japanese' identity of Japanese seismology had a positive role to play as well in the arena of international science. Demonstrating even a folk understanding of seismic waves over historical time helped create deep context for modern earthquake science in Japan, adding additional, countervailing, tropes to the story of seismology's origin as a local 'Western' science founded by expatriates.

European Lines and Japanese Maps: The Omori Scale

Omori Fusakichi's published research — much of it written in English for a European scientific audience — often defines 'Japan' in the course of defining nature. Among all the seismograms, iso-seismal maps, graphs of foreshocks and aftershocks — the regular currency of a growing Eurasian seismological trade — which Omori Fusakichi tucked into the English-language *Proceedings* and *Bulletin* of the Imperial Earthquake Investigation Committee, there are numerous demonstrations of Japan's historic ability to contain destructive earthquakes, and the West's historic and ongoing difficulty at controlling the same.

Following the Great Neopolitan Earthquake of 1857, Europeans began mapping earthquakes as a series of isometric ('iso-seismal') lines, indicating degrees of intensity ('acceleration') at various distances from the epicenter. Even after the seismograph became the micro-graphic record of an earthquake, the iso-seismal map remained its macro-graphic one. In 1883 Swiss and Italian researchers had developed the Rossi-Forel Scale of Earthquake Intensity in the first effort to standardize the placement of these lines, the standard being the human and physical geography of Europe. For example, "8" on the Rossi-Forel Scale marked the area where chimneys fell and walls cracked. The zone just beyond, where the walls did not crack but plaster fell, was marked "7". Each gradation was also defined by non-physical markers. Within a magnitude 7 zone, for example, the population experienced "general panic" and church bells rang. In the zone marked "6", even further from the epicenter, there was a "general awakening of those asleep, general ringing of bells, oscillation of chandeliers, stopping of clocks", and other traces, but no falling plaster or panic. And so on.[15]

Needing to draw iso-seismal lines for Japanese earthquakes, Omori found the Rossi-Forel Scale nearly useless, given Japan's paucity of church bells, chandeliers, chimneys, etc. Omori also calculated that the scale's maximum level of intensity (expressed as the inner-most line, marked "10") marked an acceleration rate of 2,500 mm m/s 2. After that degree of ground movement the Rossi-Forel Scale simply ceased, because it presumed devastation of the built environment would be total; there would be no further markers of seismic acceleration for a seismologist to read as he approached the epicenter. But as Milne had anticipated, and Omori would confirm, the built landscape of Japan, continued to survive and sustain measurable degrees of damage after this "maximum acceleration" had been reached. Japanese seismology thus had to come up with additional categories, and key all of them to a new series of 'Japanese' markers, in what became known as the "Omori Scale".[16]

The Omori Scale maps the Japanese landscape as a collection of foreign and domestic objects with strikingly different performances. When an earthquake reaches 2,000 mm m/s 2, according to Omori's scale, "the majority of the ordinary brick [European-style] houses are partially or totally destroyed". At the higher rate of 2,500 mm m/s 2, however, only "about 3% of the wooden [Japanese-style] houses are totally destroyed". As the acceleration reaches 4,000 mm, "great iron bridges are destroyed", but 20–50% of the Japanese houses remain standing. At the maximum acceleration — above 4,000 mm m/s 2 — all buildings are completely destroyed "except for a few wooden houses". In justifying the termination of his categories before "utter devastation" — which in European terms defined the maximum acceleration rate — Omori explained that "a few wooden houses could not be totally destroyed by an earthquake, however violent".[17]

In re-writing the Rossi-Forel Scale to Japanese specifications, Omori had written the survivability of 'Japanese' buildings and the fragility of 'foreign' ones directly into the content of seismology. Their unequal relationship was now fixed in Japanese practice as scientific fact. A "2,500 mm m/s 2 earthquake" was, by definition, one in which about 3% of the Japanese houses are destroyed. The ratio (in some cases the percentage) of destroyed 'Japanese' and 'foreign' structures was now to be the normative sign of a particular acceleration of seismic waves, the principle calculation by which seismic waves were to be seen and explained over very large areas.

It is equally telling to compare Omori's to a second Italian earthquake scale, the Mercalli, developed around the turn-of-the century as a replacement for the Rossi-Forel, and adopted by most European seismologists as the discipline internationalized in the first two decades of the 20th century. While the Omori and Rossi-Forel Scales assume a seismologist reading architectural and landscape inscriptions, Mercalli's anticipates that equal stress will be placed on oral interviews. In the Mercalli Scale, an area is to be marked "2" if the earthquake is "observed only by persons in a state of perfect quiet, especially on the upper floors of houses, or by very nervous people". In the zone marked "3", "people say 'it was scarcely felt' without any apprehension, and generally without having noticed that it was an earthquake". And so on. Even after the intensity increases to the point where physical damage becomes visible, Mercalli does not abandon human seismometers. In the area of magnitude 8 (categorized as "Ruinous") the earthquake not only "collapse[s] some houses" but is "observed with great terror".[18]

In Omori's scale, not only are there no human reactions to be collected, but there are no dead bodies to be counted. Mercalli's, after a certain point, is full of victims. Relying on physical markers, Omori's scale has a greater number of categories in the range of "serious" quakes, while Mercalli's, relying on human perception, makes its finest divisions among lesser quakes, before any physical damage occurs. Yet the two scales have this much in common: each provides the seismologist with not one, but a pair of everyday seismometers. In Mercalli's case it is Italian people and Italian buildings, both of which are presented as ubiquitous and unproblematic (normative) devices for the collection of data about nature. In Omori's, the equivalents are "wooden" (i.e. Japanese) houses and the iron and masonry infrastructure recently introduced from abroad. The difference in the choice of seismometers highlights the way in which Omori and other Japanese seismologists had come to read their own landscape: as above all a collection of 'native' and 'foreign' elements. In Japan, fundamental knowledge about natural phenomenon would come not from the resonance between two 'local' seismometers, as in Italy, but the dissonance between seismometers local and foreign.

It may have been the very importance of 'foreign' seismometers that made oral interviews seem as problematic to Omori as they seemed natural to Mercalli. Would Euro-Americans resident in Japan record the same reactions as the Japanese? If they did not, what would that mean? Would the reactions of Japanese, gathered through interviews, be accepted in the West as normative — a proper data set from which to draw iso-seismal lines? Omori's teacher Milne had generally favored the foreign over the Japanese body as an accurate or normative seismometer. That is, he had not always found what ordinary Japanese said (through translators) to be useful as data. As he wrote from the Nobi Plain:

> Attempts to find out what sensations were experienced by the people at the time of the shock are unsatisfactory. People questioned will tell trivial circumstances — how they tumbled from the top to the bottom of the stairs whilst hurrying to get out of doors — girls tell how they began to cry, etc.[19]

Japanese had obviously experienced sensations, but had not experienced seismology. They had yet to be trained to notice and record those specific sensations seismologists found useful as data.

On the occasion of a rare British earthquake (in Hereford) in 1896, the geologist Charles Davison expressed surprise that "one in every five" of those interviewed in the affected zone "gave unasked his impression of the direction of the shock". What had surprised Davison was that so many Englishmen had learned the lesson that earthquakes were waves, and knowing they were waves, had expected them to move directionally. With little 'common' knowledge about earthquakes to interfere with their science-derived knowledge about wave theory (which was becoming 'common' knowledge to educated Englishmen), and without having to worry that their roofs were about to fall in, inhabitants of Hereford had unreflectingly made themselves into seismometers. The Mercalli scale and others like it were in one sense sets of instructions to the populations of 'earthquake countries' as to what should be felt, said, and remembered.[20]

The living Japanese whom Milne had most trusted as seismographs were civil servants, and then secondarily, as when he sent out "barrages" of questionnaires to meteorological stations in order to pinpoint the best locations for instruments. When Milne left Japan in the early 1890s, his students inherited the kernal of a nation-wide monitoring system that made human reaction all but superfluous.[21] Omori was thus acculturated into a discipline which from the beginning eschewed Japanese bodies as reliable recording devices. We have seen that one of the reasons seismometry found its locus among the foreign community of Tokyo, rather than in Europe, was because the Japanese landscape lacked European-style 'inscription': the wealth of damaged buildings and toppled villages that allowed Italian earthquakes to be located, measured, and chartered by the naked eye alone. It was not just the buildings, however, but the people of Japan who were not inscribed by earthquakes in ways European scientists could easily read. This distrust in Japanese witnessing carried over from Anglo-Japanese into Japanese seismology because the "correct" witnessing procedures had been embodied in instruments and practices even before the first students were matriculated.

The result was that Japanese seismology relied less on human seismometers than the same science practised in Italy, Britain, or eventually America. In his textbook *Earthquakes* of 1907, the American geologist William H. Hobbs, describes the post-earthquake investigation this way:

> Notebook and map in hand, the student traverses the wrecked district, and while memory is still fresh, gathers, sifts, and correlates the observations made by an army of non-scientific observers.[22]

These field-interviews were to be followed up by barrages of questionnaires, on which would depend the final placement of the iso-seismal lines. Japanese seismologists also depended on local information in order to locate traces (and place their instruments) but their mapping techniques relied less on "an army of non-scientific observers" than on displays of overturned gravestones, the ratios of collapsed foreign and domestic structures, and other information literally objective to a foreign scientific audience.

Objective/instrumental seismology was thus simultaneously a set of colonizing practices and a potentially powerful commentary on the colonial project itself. As long as Omori relied solely on physical markers, questions about the reliability of Japanese witnessing would never be raised. On the other hand, Omori's particular arrangement of physical markers in iso-seismal mapping raised a host of questions about the reliabilities of European engineering knowledge. The Omori Scale of Earthquake Intensity managed to isolate a fragile foreign infrastructure within a landscape of indigenous tenacity.

The cultural politics imbedded in Omori's science is more evident when one compares it to British seismology in India, which drew different cultural and political conclusions from similar circumstances. Following the "Great Indian" or Assam earthquake of 1897 the Indian Geological Survey experienced the same problems of iso-seismal mapping. The lines it eventually drew over the 160,000 square miles "effected" by the earthquake were, according to geologist Charles Davison, "purely diagrammatic", revealing only where the lines would "probably" be "if we might suppose that local conditions were uniform". Explained Davison in his later reading of the Survey's report:

> It must be remembered that one third of the area over which the shock was sensible was one from which no observations could be obtained, while another third was inhabited by ignorant or illiterate tribes.[23]

Tribes whose houses had, nonetheless, not fallen in. In the 30,000 square miles surrounding the epicenter (⅓ the area of Great Britain), only 1,542 people died, and 600 of these in a single landslide. Within the same area, according to the Survey, "all brick and stone buildings" (the buildings of the British tribe) "were practically destroyed". It was only by locating and charting these scattered examples of shattered brickwork — essentially cataloging the damage to the colonial infrastructure — that seismic mapping was able to take place at all; that "the 30,000 square miles surrounding the epicenter" became a geographical and geo-physical entity; that the earthquake was describable as "effecting" 160,000 square miles (the area within which "nearly all brick buildings were damaged"). When Davis writes that the maps "suppose that local conditions were uniform", he means they suppose that India were not India, but Britain. They map, quite literally, where general damage and destruction would have occurred had India been constructed with European materials and methods.[24]

To Davison, areas where "no observations could be obtained" map land inhabited by "ignorant or illiterate tribes". Omori maps them, in the Japanese context, as areas of indigenous skill and stability. To Davison the work of the Indian Geological Survey is a triumph of European science amid difficult local conditions. To Omori iso-seismal lines trace out the failure of European science to come to terms with unexpected local anomalies.

Japanese Scientists in Western Ruins

After the turn of the 20th century Japan would defeat a European nation (Russia) in war, and Japanese seismology would begin mounting overseas expeditions in the manner of the European sciences. Omori and his student-assistants could be found gathering data in India, Chile, and post-catastrophic San Francisco, the last site at the invitation of American geologists just beginning to awake to seismology's value. Children stoned the Japanese scientists as they walked among the ruins, injuring Omori, and drawing an official apology from California's governor. As that incident made clear, Omori did not live in a world where a "Japanese scientist" was an unproblematic identity. The Imperial Earthquake Investigation Committee was nonetheless coming to consider the whole Asia-Pacific area as its natural research base. In 1908 it would even follow the "Earthquake Belt", a trans-Eurasian feature that Omori helped chart, name, and publicize, into Europe itself.

On December 28, 1908, the most destructive earthquake in modern European history occurred in Calabria and northern Sicily, centered near the city of Messina. Within 35 seconds, about 120,000 people died or were mortally wounded, 55,000 in Messina itself and 9–10,000 in nearby Reggio. Both cities were essentially laid flat, about 98% of their houses collapsing, killing, in the case of Messina, over 50% of the urban population.[25]

Omori was already acquainted with numbers of Italian scientists. He immediately led a Japanese research team to the earthquake zone, collaborated with Italian seismologists, and filed a "Preliminary Report" on Messina in his *Bulletin* a few months later in 1909.[26] "The enormity of the destruction in Messina" , wrote Omori, "is really beyond one's imagination." It was not rare, he stated, "that 15 or more dead bodies were found buried one upon the other in the space of a single small room at the ground floor". Because the apartment houses in Messina were four to six storeys tall and the streets narrow, "it was certainly impossible for the majority of the people to save themselves, even if they had succeeded in escaping out of doors".[27] Omori compared the horrors of Messina to the situation in Nagoya following the Nobi earthquake. The rates of acceleration (intensity) in Messina and Nagoya were roughly comparable, he wrote, the Nobi quake being, if anything, slightly stronger.

> The population of Nagoya in 1891 was 165,339, which was nearly equal to that of Messina and the vicinity, and of which only 190 were killed in the earthquake. Thus, even supposing the intensity of seismic motion in Messina (1908) to be equal to that in Nagoya (1891), the number of the persons killed in the former city was about 430 times greater than that in the latter. That is

to say, about 998 out of 1000 of the number of the killed in Messina must be regarded, when spoken in comparison to a Japanese city, as having fallen victims to seismologically bad construction of houses.[28]

Omori had actually made this observation about lethality rates at least nine years earlier, and using data that had been available not only to him, but to every seismologist in the world, since 1891. Even in those areas most devastated by the Nobi earthquake, he noted in an article of 1900, only 4 to 5% of the population had died. Within the 80,000 destroyed houses in the earthquake zone, about 7,000 people were killed, or one for every eleven houses collapsed. "The comparatively small number of the killed", he wrote in 1900, "was doubtless due to the fact that common Japanese houses are built of wood."[29] This meshed with the claim of his colleague Sekiya Seikei, made as early as 1887, that when Japanese houses 'collapsed' they often did so only partially, sometimes allowing their inhabitants to walk from the ruins.[30]

Thus did Japanese seismology begin, in the early 20th century, to speak to its Western colleagues from post-catastrophic Western landscapes. It was Omori's science that first made comparative lethality rates a category of fact-creation, predicated as Japanese seismology was on the comparison of the native and the foreign. His arguments, made in perfect English prose and illustrated by superior multi-colored maps and diagrams, would not go unheard. Italian seismologist Alfredo Montel, in his book *Building Structures in Earthquake Countries*, published the year after the Messina quake, discussed the devastation and sought to draw lessons from the ruins. Much of what he claimed to have learned about earthquakes and aseismic structures he credited to the published research of Omori. There is, one suspects on reading Montel, a felt solidarity with Japan, another 'earthquake country', politically unified in the same decade as Italy, whose native seismologists had only recently taken control of their own landscape from the scientists of Britain.[31]

Conclusion

The common analogy of "transfer" — evoking a box being packed, shipped, and its contents assembled at the other end — does not well-describe the early history of Japanese seismology. Nor, for that matter, does it fit what happened to Western engineering, which was ultimately reformulated by Japanese scholars to account for an unstable earth, a bit of physics outside the training of British civil engineers. The typicality of these examples can only be tested by further studies from Meiji Japan and elsewhere in the colonial world. But we have taken it too much for granted, I suspect, that speciation in science is more easily modeled and explained than in language, religion, theatre, literature, art, and other more obviously 'cultural' activities; that 'Western' science ca. 1900 meant a something made in Europe and America, with the rest of the world providing the raw material called 'data'. This is not to obscure or deny the strong Eurocentricity of 19th- and early 20th-century 'international science'. Indeed, if I had continued this history a bit longer, we would see Britain and Germany seizing the initiative and control over 'international' seismology, despite the comparative quietude of the ground in both places. But the historic ability of northern Europe and America to seize and dominate agendas in modern science (as much else) need not invite us to read the first generations of non-Western scientists as distantly-placed laboratory assistants, complicit in their own colonization. Meiji-period Japanese had ambitions for their Western science beyond the desire (though this was there as well) to be considered helpful in London and Berlin.

Notes

[1] Robert Mallet, *The Great Neopolitan Earthquake of 1857: The First Principles of Observational Seismology* (1862).

[2] "Historical" introductions to seismology texts tend to concentrate on European origins and gloss over their discipline's late 19th-century re-organization in Japan. Italian and Swiss seismologists of the late 19th century, by contrast, often went out of their way to recognize advances made in Tokyo. See footnotes 9, 10, and 31.

[3] This and later sections encapsulate arguments made at greater length in my doctoral dissertation, "Foreign Knowledge, or Art Nation/Earthquake Nation: Architecture, Seismology, The West, and Japan, 1876–1925" (Ph.D. dissertation, MIT, 1998). On the founding of seismology in the Meiji academy, see also Graeme J. N. Gooday and Morris F. Low, "Technology Transfer and Cultural Exchange:Western Scientists and Engineers Encounter Late Tokugawa and Meiji Japan" in Morris F. Low, ed., *Beyond Joseph Needham: Science, Technology, and Medicine in East and Southeast Asia* (Osiris, 2nd Series, v. 13, 1998), pp. 121–27.

[4] *Ibid*. These were James Ewing, Thomas Grey, and John Milne.

[5] See chapter on Mallet in Charles Davison, *Founders of Seismology* (Cambridge: Cambridge University Press, 1927), pp. 65–86.

[6] John Milne, "The Earthquake in Japan of Feb. 22, 1880", *Transactions of the Seismological Society of Japan* (hereafter *TSSJ*), Part II, 1880, pp. 1–115 (quotation from page 1).

[7] Quoted in Davison, p. 76.

[8] Quoted in A. L. Herbert-Guster and P. A. Nott, *John Milne: Father of Modern Seismology* (Tenterden, Kent: Norbury, 1980), p. 80.

⁹ Milne, "Notes on the Recent Earthquake of Yedo Plain ...", p. 35; A "seismograph" invented by the Italian earthquake researcher Luigi Palmieri (who also invented the word in 1859) was actually operating in Tokyo's meteorological observatory when Milne arrived in 1876. The seismographs invented by Ewing and Grey operated on an entirely different principle, however, and represented a new level of sensitivity. The English downgraded Palmieri's device by re-classifying it as a "seismoscope" and eventually had it replaced at the Tokyo observatory with a Ewing-Grey-Milne machine.

¹⁰ Wrote the Swiss earthquake investigator F. A. Forel in 1887: "More had been learnt from the seismograph-tracer of the Anglo-Japanese observers in two years, than twenty centuries of European science had been able to show", Prof. F. A. Forel, *TSSJ*, v. XI, 1887, p. 165. "Seismometry", as the invention of seismic recording devices was called in Tokyo, was at least indirectly influenced by the strong local interest in telegraphy. The new seismographs and telegraphs were in effect siblings, as Thomas Gray, one of the seismograph's three 'fathers', and the one most responsible for the construction of a working prototype, was a professor of Telegraphic Engineering.

¹¹ As the Secretary of Japan's Imperial Earthquake Investigation Committee (hereafter IEIC) from 1897 to his death in 1923, Omori would oversee the publication of its influential foreign-language *Report* series, and, following the San Francisco earthquake of 1906, its English-language *Bulletin*. The *Bulletin* of the IEIC was the first seismological journal in the world to include a detailed report of that earthquake. Omori would author the vast majority of papers in both outlets. The English seismologist Charles Davison would pay him the posthumous complement of a chapter in his book *Founders of Seismology* [see citation 5].

¹² Kikuchi Dairoku, *Recent Seismological Investigations in Japan* (Tokyo, 1904). Kikuchi's book was a summation of the accomplishments of Japanese seismology, written in English for a foreign audience, on the eve of the founding of the International Seismological Association in Strasbourg.

¹³ See my "Foreign Knowledge ..." [citation 3] for a full discussion of the Nobi earthquake debate.

¹⁴ Omori Fusakichi, "Measurement of the Vibration of Gojunotos, or 5-story Buddhist Stupas (Pagodas)", *Bulletin of the Imperial EIC*, vol. IX, 1918–21, pp. 110–52; *ibid.*, "Note on the Form of Japanese Castle Walls", pp. 30–32.

¹⁵ I am relying on the 1883 version of the Rossi-Forel Scale printed in Charles Davison, *A Manual of Seismology* (Cambridge, UK: Cambridge University Press, 1921).

¹⁶ In this and the following paragraphs I am relying on Omori's own English translation of his scale in Omori Fusakichi, "Seismic Experiments on the Fracturing and Overturning of Columns", *Publications of the Imperial Earthquake Investigation Committee*, no. 4, 1900, pp. 138–41.

¹⁷ Omori, pp. 138–41.

¹⁸ I take these translations of the Mercalli scale from Alfredo Montel, *Building Structures in Earthquake Countries* (London: Charles Griffin & Co., 1912), pp. 9–11. Montel also publishes Omori's scale and attempts to reconcile it with Mercalli's in the form of a chart.

19 *Japan Weekly Mail*, Nov. 7, 1891, p. 559.
20 Charles Davison, *The Hereford Earthquake of Dec. 17, 1896* (Birmingham: Cornish Bros., 1899), p. 266.
21 John Milne, "Suggestions for the Systematic Observation of Earthquakes", *TSSJ*, vol. IV, Jan.–June 1882, p. 110. By 1904, Japan had a network of 71 local meteorological stations equipped with seismographs, and 1,437 others equipped with lesser data-recording devices (Kikuchi, p. 9).
22 William H. Hobbs, *Earthquakes: An Introduction to Seismic Geology* (N.Y.: D. Appleton, 1907), p. 227.
23 Charles Davison, *Great Earthquakes* (London: Thomas Murby & Co., 1936), pp. 140–41.
24 *Ibid.*
25 Various casualty figures have been published for this earthquake. My figure of 120,000 comes from Wakabayashi (1986). The percentage of collapsed houses, first reported by Omori [see citation 26], was repeated by Davison in 1936 (Charles Davison, *Great Earthquakes* [London, 1936] p. 203).
26 Omori, "Preliminary Report on the Messina-Reggio Earthquake of Dec. 28, 1908"; *Bulletin of the Imperial EIC*, v. 3, no. 1, 1909.
27 *Ibid.*, pp. 38–39.
28 *Ibid.*, pp. 39–40.
29 Omori, "Notes on the Great Mino-Owari Earthquake of Oct. 28, 1891", *Publications of the EIC in Foreign Languages*, No. 4 (Tokyo: EIC, 1900), p. 13.
30 Sekiya Seikei, "The Severe Earthquake of the 15th of Jan., 1887", *TSSJ*, vol. XI, 1887, p. 83.
31 See Montel [citation 18].

"Good Luck" in the History of Science in Japan

SUGIYAMA SHIGEO 杉山滋郎
Hokkaido University, Japan

Introduction

Historians of science generally agree that Japan made rapid progress in science after the Meiji Restoration of 1868. How and why, then, was she able to do it? Some scholars have pointed out that the cultural tradition Japan had developed in the Tokugawa era (1603–1867) contributed greatly to introducing Western science.[1] Others have indicated how effective the policy was which the Japanese government launched after the Meiji Restoration with respect to the promotion of Western science.[2] Yet others may suggest the diligent nature of Japanese people.

It seems to me, however, that one important factor heavily responsible for this rapid progress has not been properly noted in academic literature of the history of science in Japan. The factor is, I would argue, timeliness. The period when Japan began to introduce Western science energetically and tried to keep up with the Western scientific community was the latter half of the 19th century. This and the early years of the 20th century were a time when some leading disciplines of science were in a state I would characterize by the concept "region-bound". The "region-bound" character of scientific disciplines of this period greatly helped Japan to develop Western science in her own way. Japan was lucky in this sense.

Basalla's Model

To present my idea clearly, I will make use of George Basalla's three-stage model. His paper discussed how modern science, which originated in Europe, has been and will continue to be introduced into non-European countries, and proposed a model, with a graph (Fig.1) conceptually representing the model.[3] The ordinate of the graph stands for the level of scientific activity;[4] and the abscissa stands for time. With this graph Basalla claimed that modern science had diffused from Western Europe and had established itself in the rest of the world through three overlapping phases or stages.

43

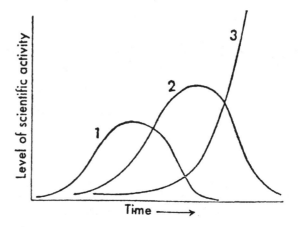

Fig. 1. Sequence of phases in the diffusion of Western science

During "phase one" nonscientific society provides a source for European science. Europeans visit new lands, survey and collect their flora and fauna, study their physical features, and then take the results of this work back to Europe. Science during this phase is an extension of geographical exploration. "Phase two" is marked by a period of colonial science. Colonial science means dependent science, and the scientific activity in this phase is based primarily upon institutions and traditions of a nation with an established scientific culture, that is, a European nation. Colonial science, as Basalla defined, can occur in situations where there is no actual colonial relationship. The dependent country may be a colony of a European nation, as in the case of India or the United States, or may not be a colony of a European nation, as in the case of Russia or Japan.

The colonial scientist, if formally trained, will have received some or all of his scientific education in a European institution or from European teachers invited to his country. The colonial scientist will have purchased his books, laboratory equipment, and scientific instruments from European suppliers. The colonial scientist's role within scientific fields and problems is delineated or determined by European scientists. Basalla states that Japan in the Meiji era, at least in its earlier years, as well as in the Tokugawa era, was dependent, as far as her science was concerned, and was therefore in the second stage.

"Phase three" is characterized by the following six elements. A scientist will (i) receive most of his training at home; (ii) gains some respect for his calling, or perhaps earn his living as a scientist, in his own country; (iii) finds intellectual stimulation within his own expanding scientific community; (iv) is able to communicate easily his ideas to his fellow scientists at home and abroad; (v) has a better opportunity to open new fields of scientific endeavor; and (vi) looks forward to the reward of national honors — bestowed by native scientific organizations or the government — when he has done superior work. We can say that in the case of Japan phase three came to override phase two around 1900 or in the 1910s according to the definition Basalla has given.

I have no objection to his general scheme as I have just outlined it above. But I believe that an important point is missed. He indeed admits that the shape of the curves can vary from country to country, or from nation to nation. He pointed to the case of Japan as an example. Japan had an unusually long, and initially slow-growing, second phase because of the policy of political, commercial, and cultural isolation practised by her rulers. This long interval quickly reached a peak after the Meiji Restoration, when Japan was fully opened to Western influence. Thus, he admits that the graph for the case of Japan is a peculiar one.

The possibility that the shape of the graph can greatly vary from time to time, however, is not explicitly taken into account in his explanation. The Japanese scientific community moved easily and swiftly from the second stage into the third by the early years of the 20th century. If the Japanese scientific community had undergone the transition process in earlier years or much later, it could not have transformed itself so easily and rapidly. Thus, if the ease and swiftness were caused by the fact that Japan experienced the transition process in that period of time — in the latter half of the 19th century — it should explicitly be acknowledged that the shape of the graphs depends on time.[5]

That timing contributed much to the Japanese scientific community's transition from phase two to phase three can effectively be shown by the practice leading to the founding of the International Latitude Observatory in 1899 at Mizusawa, Iwate prefecture, located in the northern part of the main island of Japan. Leonhard Euler had already suggested around 1785 the existence of polar motion with a period of about ten months. During the years between 1884 and 1888, about 120 years after Euler's suggestion, F. Kustner found that latitude variation was really taking place. It was also discovered in 1892 that the manner of the variation observed in Berlin was the reverse of that in Honolulu, which is nearly 180 degrees from Berlin in longitude. This proved that latitude variation was caused by polar motion. Moreover, S. C. Chandler made it clear in 1891 that the period of polar motion was actually fourteen months, not ten months as suggested by Euler, by analyzing observed data of the variation.

Thus did the problem of latitude variation become an important topic for geodesy and astronomy towards the end of the 19th century. The International Association of Geodesy proposed to carry out, as an international cooperative project, continuous observations of latitude at places on the same latitude, in order to determine precisely the way plar motion takes place. Thus, in 1898, at the IAG conference in Stuttgart, six places at latitude 39.8 north were selected for the project: Mizusawa, Japan (141.8 E); Tschardjui, Turkmenistan (63.29 E); Carloforte, Italy (8.19 E); Gaithersburg, the U.S. (77.12 W); Cincinnati, the U.S. (84.25 W); and Ukiah, the U.S. (123.13 W).

Why was Japan chosen as one of these places? It was because observation places at the same latitude needed be spread uniformly in respect to the directions East and West. Thus at least one place had to be set in the far east region. In those days Japan was the most advanced country in terms of scientific culture in the area, and was the only member in Asia of the IAG.

We can recognize here an element of luck. Luck, however, is not my concern at this juncture. What I would pay attention to is the overall timeliness that the period of the latter half of the 19th century represented in the promotion of scientific activity in Japan. The isolation of Japan from foreign countries was over in 1850s, and phase two soon built toward its peak. At this precise moment, scientific research, in this case geodetic and astronomical research, required global observation places to develop further. They needed a place for observation in the far east region, as they desperately demanded data which could not be obtained in their own lands alone. Under these circumstances, when scientists in Japan succeeded in obtaining the data and provided it to the Western scientific community, they were enthusiastically welcomed.

The acceptance of research results by the Western scientific community lent both to the appreciative recognition of the Japanese researchers who produced it and to the recognition of the scientific community to which the researchers belonged. Basalla has pointed out in his paper the importance of a scientist's becoming a member of an "Invisible College", when a country where he is working is in the process of transition from the phase two to phase three. Basalla argued that when scientists become members of the invisible college, they can exchange their latest ideas and news of the advancing frontiers of science and they can benefit from "continuing mutual education". Colonial scientists were, in general, unable to share in this informal scientific organization. However, a scientist's being appreciably recognized as a first rank researcher was extremely effective in aiding his entry into the invisible college.

For example, Kimura's finding of the Z-term in the formula of latitude change was immediately recognized as highly significant. This was an important turning point, and the academic circles of astronomy in Japan then started to play an important role in the world-wide community of astronomers. The science of geomagnetism around the turn of the century is another example. Research in the latter half of the 19th century had reached the stage where the global distribution of geomagnetism and its time variation required survey, and the cause of the phenomenon was being probed. Scholars working in the field of geomagnetism, therefore, strongly needed observational data collected at various parts of the globe. Thus, scholars in Japan were led to observe geomagnetism by themselves just after the opening of this field in foreign countries, in response to the request of Europeans. The Hydrography Department of the Ministry of the Navy began geomagnetic observations twice per month at meteorological observatories for one year from November 1882. In the following year of 1883, hourly observations were started at the temporal magnetic observatory founded on land of the Ministry of Engineering. Measurements in these observatories were carried out in the frame of international cooperation.

Scientific investigations which were planned by Western scientific communities often promoted the exchange of experimental or measurement skills. This then led to the leveling-up of skills and accelerated the transition process from phase two to phase three. The International Latitude Observatory founded at Mizusawa was provided with a German-made observation apparatus by IAG, which led to reliable data being produced by Japanese hands. On the occasion of a joint observation in the 1880s of Mars passing in the front of the Sun, a group of Japanese scientists used a high performance apparatus donated by a foreign group, which was to enhance greatly the astronomical observations later carried out in Japan. Besides, we can easily imagine that international cooperation in calibrating observational and measurement apparatus, which was inevitable in any cooperative research, helped Japanese scientists improve apparatus and develop skills.

The expectations imposed by various groups of scientists abroad made it easy to explain to the general public how necessary and important research was. Sometimes, even financial assistance was given by foreign groups of scientists, as in the case of establishing the International Latitude Observatory in Japan. In such cases it became much easier to explain to the public the necessity of founding the institution. Basalla pointed out that society's approval of scientists' labors must be ensured in order for phase two to be replaced by phase three. In this respect, expectations promoted the transition between phases.

"Region-bound" sciences

In this manner, expectations by the Western scientific community regarding contributions of scientists in other countries make those country's transition from phase two to phase three easy and swift. The reason why their contributions are expected is that they are seen as having advantages in acquiring the needed results. The advantage is caused by the fact that the Western science is in a state where it requires for its further development data or information which can only be obtained in (or are bound up with) specific geographical regions. I would characterize this state of science as being "region-bound".[6] Not a few scientific fields or disciplines were "region-bound" in the latter half of the 19th century and the early years of the 20th, which to a great extent contributed to promoting the Japanese scientific community's transition to phase three.

It is probable that disciplines within "geoscience" display, sometime in the course of their development, a "region-bound" character. But even disciplines which are usually not regarded as "geoscience" show in some cases a "region-bound" nature.

The study of ginkgo or ginkgo biloba conducted by Hirase Sakugoro in the 1890s is one of examples of "region-bound" research. In the latter half of the 19th century, the taxonomical place of ginkgo was an important problem in the community of botanists centered in Europe. Although ginkgo was classified as gymnosperm, it had at the same time great similarity with ferns in cryptogam. One European scholar guessed that ginkgo might have spermatozoid as fern did, and another tried to find it only to fail. Under these circumstances, Hirase discovered the spermatozoid of ginkgo. One factor in his successful discovery was his regional advantage; the motion of ginkgo spermatozoid are more active in Japan due to its warm climate than in the climate in Europe. There are also many more ginkgo trees in Japan than in any European country. Therefore, it was hardly possible, though not impossible, for European scholars to discover ginkgo spermatozoid in their countries.

Hirase reported his discovery of ginkgo spermatozoid in a distinguished journal of botany, *Botanisches Centralblatt* published in Germany, which greatly facilitated Japanese botanical circles' scientific exchanges with the European community.

Another example can be found in medical science. In the latter half of the 19th century, medical scientists began to search for specific bacterium as the cause of each infectious disease, triggered by the fact that Robert Koch specified bacteria as the cause of tuberculosis and cholera. Thus, as soon as the spread of the black plague was reported in Hong Kong, Kitazato

Shibasaburo and medical scientists of Tokyo Imperial University went there to try to identify a bacterium. Medical scientists of the Pasteur Institute in France also came to Hong Kong with the same purpose. These researches resulted in the discovery of plague bacillus.

Research to discover a bacterium causing an infectious disease which spread in a specific region of the Earth could be carried out only on the spot. It was meaningless to transfer patients from Hong Kong to a laboratory in Tokyo or in Paris in order to identify a bacterium. In this sense, the bacteriological study of infectious diseases was "region-bound" in the years of the late 19th century and in the earlier years of the 20th century. Nevertheless, after a bacterium was identified and came to be cultivated, it could then be transferred for the purpose of, for example, analyzing its DNA structure. It follows that as scientific research in any discipline proceeds and its research objects turn out to be more fundamental or elementary, the discipline becomes in most cases less "region-bound".

Needless to say there are disciplines that have never been or will never be "region-bound". Studies in the kinetic theory of gas or in statistical mechanics could be pursued in Europe, in Japan, or even on the Moon. Thus, the kinetic theory of gas and statistical mechanics has never been a "region-bound" research project. Pierre and Marie Curie started their study of radioactivity towards the end of the 19th century. They acquired a large amount of pitchblende for their research from abroad. The mineral was easily transferable to Paris. Therefore, the science of radioactivity was not "region-bound" either.

Furthermore, most scientific disciplines which were once "region-bound" have generally come to be less so or entirely lost that character in the course of their development. Most geosciences have experienced this in the 20th century, a change which was caused to a large extent by the emergence of telecommunication systems and satellites orbiting the earth.

It is true that not all scientific disciplines nor even most were in the state of being "region-bound" in the latter half of the 19th century. However, the number of disciplines in such a state was appreciably larger at that time and the degree of being "region-bound" was obviously higher than in any other period. It was luckily in this period that Japan began its transition from phase two to phase three.

Concluding Remarks

A Japanese astronomer Hagiwara Yusuke once alluded to a geographical advantage enjoyed by Japan in astronomical observation, referring to *Kanae* or a tripod. The historian of science Nakayama Shigeru pointed out the same thing.[7] *Kanae* has three legs, with which it stands quite stably. Hagiwara and Nakayama likened one of these three legs to Japan in Asia, and the other two legs to a European country and an American country, respectively. Thus, they pointed out that Japan could enjoy geographical advantage in the Asian region. My arguments in this paper support their contentions, but are different in two ways.

I would argue that geographical advantage, which is closely related to the concept I proposed of "region-boundness", is limited neither to astronomy among scientific disciplines nor to Japan among the countries of the globe. I would say that every country could have its own advantage if only it was endowed with the opportunity to take advantage of it. Another point is that the importance of geographical advantage varies from time to time. It was quite great during the period between the late 19th and the beginning of the 20th centuries because not a few leading sciences of those days were strongly "region-bound". And it was just at that time that Japan opened itself to foreign culture and moved from phase two to phase three. It is for this reason that Japan could make the most of "region-boundness".

Notes

[1] See for example Bartholomew, James R., *The Formation of Science in Japan: Building a Research Tradition*, New Haven and London: Yale University Press, 1993.

[2] See for example Hiroshige, Tetsu, *Kagaku no Syakai-shi (Social History of Science)*, Tokyo: Chuou-Koron-sya, 1973.

[3] Basalla, George: "The Spread of Western Science", *Science*, vol. 156 (1967), pp. 611–22.

[4] Note, however, that the level of scientific activity should not be considered in any strict quantitative way, because, as he said, qualitative factors such as the judgements of historians of science on the contributions of individual scientists are to be included in the definition of scientific activity, as well as quantitative factors such as number of scientific papers produced, manpower utilized, and honors accorded.

[5] This implies that any country now experiencing the transition process from phase two to phase three may not be able to accomplish it so easily and quickly as Japan did in the latter half of the 19th century.

[6] I have preliminarily used the concept in my paper, "Astronomy and geosciences in Japan towards the end of the 19th century"(in Japanese), *Ido-kansoku 100-nen (A Hundred Years of Latitude Observation)*, Mizusawa: Division of Earth Rotation and the Mizusawa Astrogeodynamics Observatory, 1999, pp. 24–29.

[7] Nakayama, Shigeru, *Nihon no Tenmongaku (History of Astronomy in Japan)*, Tokyo: Iwanami-syoten, 1972.

Knowledge of the Dual Character of Technology: Viewpoints on Technology in Ancient China

GAO XUAN

Institute for the History of Science and Technology & Ancient Texts, Tsinghua University

Introduction

This paper is part of a series of research projects we are working on at our institute. This serial research focuses on viewpoints concerning technology in ancient China. Let me begin with an explanation of several concepts mentioned in the text.

Concept 1: 'technology'. The word 'technology' in this paper is used contextually, corresponding to the level of technology in ancient China. Here it mainly means the technology of handicrafts mastered by the carpenters, smiths, masons and other skilled workmen at that time.

Concept 2: 'Ancient China'. This refers to the time from the beginning of the *Zhou* (周) to the end of *Qin* (秦) Dynasties, or from about B.C. 1064 to B.C. 206. Sometimes we also include the early years of the *Han* (漢) Dynasty.

About 30 titles have been identified for this project:

1. *Shi Jing* (詩經, Book of Odes)
2. *Shang Shu* (尚書, Book of Documents)
3. *Zhou Yi* (周易, Book of Changes)
4. *Zhou Li* (周禮, Record of the Rites of *Zhou*)
5. *Li Ji* (禮記, Record of Rites)
6. *Chun Qiu Zuo Zhuan* (春秋左傳, Master *Zuo*'s Commentary on the Spring and Autumn Annals)
7. *Chun Qiu Gongyang Zhuan* (春秋公羊傳, Master *Gongyang*'s Commentary on the Spring and Autumn Annals)
8. *Chun Qiu Guliang Zhuan* (春秋谷梁傳, Master *Guliang*'s Commentary on the Spring and Autumn Annals)
9. *Lun Yu* (論語, Conversations and Discourses)
10. *Xiao Jing* (孝經, Filial Piety Classic)

11. *Er Ya* (爾雅, Literary Expositor)
12. *Meng Zi* (孟子, The Book of Master *Meng*)
13. *Xun Zi* (荀子, The Book of Master *Xun*)
14. *Lao Zi Dao De Jing* (老子道德經, Canon of the *Dao* and its Virtue)
15. *Zhuang Zi* (莊子, The Book of Master *Zhuang*)
16. *Lie Zi* (列子, The Book of Master *Lie*)
17. *Mo Zi* (墨子, The Book of Master *Mo*)
18. *Yan Zi Chun Qiu* (晏子春秋, The Spring and Autumn of Master *Yan*)
19. *Guan Zi* (管子, The Book of Master *Guan*)
20. *Shang Jun Shu* (商君書, Book of the Lord *Shang*)
21. *Shen Zi* (慎子, The Book of Master *Shen*)
22. *Han Fei Zi* (韓非子, The Book of Master *Han Fei*)
23. *Sun Zi* (孫子, The Book of Master *Sun*)
24. *Wu Zi* (吳子, The Book of Master *Wu*)
25. *Lü Shi Chun Qiu* (呂氏春秋, Master *Lü*'s Spring and Autumn Annals)
26. *Guo Yu* (國語, Discourses on the States)
27. *Zhan Guo Ce* (戰國策, Records of the Warring States)
28. *Xin Yu* (新語, New Discourses)
29. *Huai Nan Zi* (淮南子, The Book of *Huai Nan*)
30. *Shan Hai Jing* (山海經, Classic of the Mountains and Rivers)

In these books we find plentiful records of many kinds of technology. They show that technology was closely related to daily life at that time, and exerted a great influence on ideologists and their theories.

We are gathering and sorting out the useful records from these ancient books, including description, analysis, analogy, and their affirmation or denial of the importance of technological change. We have divided their viewpoints on technology into several aspects:

(a) the relation between technology and nature;
(b) the relation between technology and society;
(c) the relation between technology and human beings;
(d) the relation between technology and other aspects, such as politics, economy, military, etc.;
(e) the positions and functions of technology and technologists;
(f) the influence of technology on Chinese philosophy;
(g) the positive effect of technology;
(h) the negative effect of technology.

We are trying to categorize viewpoints on technology into these eight different aspects. We have discovered that people held a very abundant and thorough set of viewpoints on technology in ancient China. Such viewpoints on technology produced more than 2000 years ago are still relevant today. This paper will mainly discuss ancient Chinese knowledge of the dual character of technology (i.e. its capacity for good and ill).

Knowledge of the Active Effects of Technology

In ancient China, the level of material culture was relatively low. People used very simple technology to ensure their existence. The relations between technology and politics, economy, military, philosophy or other aspects of society were very close. We find many references to technology in ancient texts, such as *Bai Gong* (百工) (a hundred kinds of handicrafts), although such books mainly discuss history, geography or other topics. The writers often thought that technology was helpful to power and the prosperity of the country.

These are some examples:

A. The morality and ability of a king are shown through his excellent governance of the country. The excellent governance must improve the living of people. Improvement of the living of people must develop the technology of handicrafts related to water, fire, metal, wood, soil and crop. (These are a summary of important handicrafts.) *Yu Shu Da Yu Mo* (德惟善政，政在養民，水火金木土穀惟修。《虞書•大禹謨》)

B. Ministers liked their official duties, monarch lead the country toward progress, and all kinds of craftsmen were successful in their work. *Yu Shu Yi Ji* (德惟善政，政在養民，水火金木土穀惟修。《虞書•益稷》)

C. There are 6 kinds of professions in a country. *Bai Gong* (白工) (a hundred kinds of handicrafts) is one of them. *Zhou Li Kao Gong Ji* (國有六職，百工與居一焉。《周禮•考工記》)

The above quotations speak to the perceived importance of technology.

Writers would sometimes quote rules or artificial processes of technology to support their opinions. Here are some examples:

A. The wise technologists teach people by using rules, and followers must learn how to apply such rules. *Meng Zi Gao Zi Shang* (大匠誨人必以規矩，學者亦必以規矩。《孟子•告子上》)

B. Jade will not become a useful appliance if it is not carved. *Li Ji Xue Ji* (玉不琢，不成器。《禮記 • 學記》)

C. Straight wood can meet the rope criterion. It can be curved to make a wheel, and is suitable to the standard of a circle. It cannot become straight any more even if it is dry because it is processed into curves. So the wood will be straight after correction according to the rope criterion, and metal will be sharp after grinding. Masters will be wise and will not make mistakes if they have plentiful knowledge and think of themselves everyday. *Xun Zi Quan Xue* (木直中繩，揉以為輪，其曲中規，雖有槁暴，不復挺者，揉使之然也。故木受繩則直，金就礪則 。《荀子 • 勸學》)

These examples show a popular method of discourse at that time. People liked to quote terms, rules and methods of technology to illustrate their thoughts in other realms. They believed that technological ideas were very significant in helping people to improve their minds. So technological words and ideas often appear in ideologically-oriented books of that time.

We can also see another interesting tendency in these books. In China, many important inventions were credited to people who had special abilities. They called such inventors "sages", and gave them great honor. This shows a kind of social psychology of worshipping inventors of technology.

A. What handicraftsmen do is what sages invented. *Zhou Li Kao Gong Ji* (百工之事，皆聖人之作也。《周禮 • 考工記》)

B. The ancient great sages studied astronomy, geography, provided 8 hexagrams *Gua* (卦), and determined ways of living. Shen Nong (神農) taught people to plant 5 kinds of crops. Huang Di (黃帝) made a house for people living. Hou Ji (侯記) cultivated land, and planted mulberry and sesame. Da Yu (大禹) controlled floods. *Xin Yu Dao Ji* (先聖觀天文、察地理、畫八卦、定人道；神農教民食五谷，黃帝作宮室，候稷闢土地、種桑麻，禹治水。《新語 • 道基》)

In Chinese legends, the great men and leaders, such as Fu Xi (伏羲), Shen Nong (神農), Huang Di (黃帝), Hou Ji (後禹), Da Yu (大禹), etc., were well known in ancient days as specialists in technology. Other inventors, such as Xi Zhong (奚仲), who created the wheeled vehicle in recorded legend, Chui (垂), who invented the bow and arrow, and Lu Ban (魯班), who invented the saw and was a very famous carpenter in ancient China, also held very high honour in these books.

A famous example is Da Yu, who was a specialist of water conservancy and was successful in controlling very big floods in his time. He had a crucial influence on the country. Then he won popular support and became the first king of the first dynasty of China.

During the period we are discussing, China was divided into many small states. War frequently occurred among them. So these countries faced an urgent need to rapidly develop military power. Only those countries that had developed agriculture, handicraft, and the military could survive at that time. Technology was closely related to the existence of a country. So it was very normal that technology became a popular topic. For example, the number of war vehicles became a symbol of the power of states. In these books, we can read passages like "the country possesses 10,000 chariots" (萬乘之國) or "the country possesses 1,000 chariots" (千乘之國). This number shows whether the country was strong or weak. Technology to make chariots was particularly important at that time. In the book entitled "*Kao Gong Ji* (考工記)" written in the early Warring States period, there were many lines which describe the names, sizes and forms of various parts of a chariot.

Kao Gong Ji is a special technological book. It described the technological rules and methods of many kinds of handicrafts, including the technology of woodwork, metal work, leather work, dye work, jade work and pottery work. It contains much useful information on technology in ancient China.

Knowledge of the Negative Effects of Technology

Ancient people also realized the negative effects of technology as well as the positive ones. Another tendency in these books was to emphasize the need to be vigilant about technology. Some writers tell stories to teach people technology's negative influence.

Story A: Mo Zi (墨子) had made a wooden bird for three years. But the wooden bird flew only one day before it was damaged. The followers of Mo Zi thought their teacher was skilful to make a flying bird. But Mo Zi told his followers that he was not as skilful as the workers who could make chariots in one day because the vehicles were useful in transporting heavy things. *Han Fei Zi Wai Chu Shuo* (《韓非子‧外儲説》)

Story B: A skilled worker in the Song State had made a brown leaf by using jade for 3 years. This man-made leaf was very similar to a real leaf, and people could not find it when mixed in among authentic leaves. But Lie Zi (列子) did not appreciate this skill. He said: "The man spent 3 years to make a leaf. If all those natural leaves were made in this way, we could see very few of them." *Lie Zi Shuo Fu* (《列子 • 説符》)

These two stories come to a similar conclusion: that if a technology can be used for the development of the country or the improvement of people's lives, it will be very important. But if the technology can only be used for entertainment, it is not important although it may be very skilful and complicated.

Many references in these books give the same opinion when discussing technology and its products. The ideologists consider that when the technology and skills exceed the fundamental need of people, it might have a negative influence on the country. Furthermore, the ideologists thought that negative technology and skills could affect popular morality in that it made people crafty.

A. It is enough for sages to make clothes suitable for their bodies and skins. It is unnecessary to show their clothes as beautiful to others. *Mo Zi Ci Guo* (故聖人為衣服，適身體和肌膚而足矣，非榮耳目而觀愚民也。《墨子 • 辭過》)

B. Those who have cunning devices use cunning in their affairs and those who use cunning in their affairs have cunning hearts. Such cunning means the loss of pure simplicity. Such a loss leads to restlessness of the spirit, and within such men the *Dao* (道) will not dwell. *Zhuang Zi Tian Di* (有機械者必有機事，有機事者必有機心，機心存乎心中，則純白不備，純白不備則神生不定，神生不定者，道之所不載也。《莊子 • 天地》)

C. Don't keep the machination and deceit in your heart. *Huai Nan Zi Ben Jing Xun* (機械詐偽，莫藏于心。《淮南子 • 本經訓》)

D. The more skilful apparatuses the masses hold, more easily confusion occurs in the country. The more technology and skills people hold, more easily fantastic matters take place. *Lao Zi Dao De Jing* (名多利器，國家滋昏。人多技巧，奇物滋起。《老子 • 道德經》)

Therefore, negative effects of technology and skills were those harmful for the government. Many governors made decrees prohibiting some kinds of technologies. We can read the following records in some books.

A. Don't make odd or luxurious appliances to confuse the king. Makers of appliances must think of their positive effects first.

Li Ji Yue Ling (毋或作為淫巧以蕩上心，必功致為上。《禮記•月令》)

B. Craftsmen who display their carvings to each other are called as violating a prohibition. *Guan Zi Zhong Ling* (工以雕文刻鏤相稺也，謂之逆。《管子•重令》)

C. These people who compose decadent music, or fabricate odd clothes, or make strange art and appliance will be killed. *Li Ji Wang Zhi* (作淫聲異服奇技奇器以疑眾，殺《禮記•王制》)

It is valuable for us that some ideologists analyzed the purpose of technology. If technologists only consider narrow purposes and ignore negative effects, it will lead to unlimited development of technology, which is not profitable to human beings, and perhaps harmful. We should be careful about technology.

A. The craftsmen who make vehicles wish more people to be rich, but the craftsmen who make coffins wish more people to be dead. It does not mean that vehicle craftsmen are nobler than coffin craftsmen. *Han Fei Zi San Shou* (故輿人成輿則視欲人之富貴，匠人成棺則欲人之夭死也，非輿人仁而匠仁賊也。《韓非子•三守》)

B. Professionals who make arrows worry about their products being not strong enough to kill people, but professionals who make armor worry about their products being not strong enough to protect people. Wizards and craftsmen have similar thoughts as well. So we must be careful about technology. *Meng Zi Gongsun Chou Shang* (矢仁惟恐不傷人，函人惟恐傷人，巫、匠亦然，故術不可不慎也。《孟子•公孫丑上》)

Besides the negative influence of technology on human beings, the ideologists at that time also considered the negative influence of technology on nature. Some books mentioned stipulations to protect the natural environment. These stipulations prohibited burning forests, cutting trees and catching fish excessively. Ideologies to protect the natural environment have existed in China since ancient times.

A. According to the decrees made by the wise king, it is forbidden for people to bring axes to enter forests when plants are growing, or to bring fishnets and poison to a lake or river when aquatic animals are laying eggs. Their growth must be protected. *Xun Zi Wang Zhi* (聖王之制也，草木榮華滋碩之時，則斧斤不入

山林，不夭其生，不絕其生，不絕其長也黿龜魚鱉，孕別之
時，罔罟　藥不入澤，不夭其生，不絕其長也。《荀子•王
制》)

B. When trees are growing, the local administrators are ordered
    to enter the forests to patrol and protect them. *Lie Zi Tang Wen*
    (樹木方盛，乃命虞人入山行木，毋有斬伐。《列子•湯
    問》)

Conclusions

Before the Spring and Autumn period, technology developed very slowly but
continuously, and promoted the development of society. During the Spring
and Autumn period, technology made great progress with the production of
iron. Iron represented one kind of new technology. The book *Kao Gong Ji*
described a series of tight regulations, operation programs, and divisions of
this technology. Material culture had reached a higher level than merely
satisfying the fundamental needs of life. With the development of the market
economy, wealth increased and gathered rapidly. The calm life of the people
was replaced by severe competition, and the pursuit of wealth, enjoyment
and luxury. Many small states appeared, and each state was different in
politics, economy and society. Many kinds of theories were produced and
exchanged at that time. Political form, economic form and social ideas faced
important reform through technology, which was developed over a long time.
By the time of the Warring States technological thought was mature.

Following the above analyses, we can conclude that:

A. In ancient China, technology promoted the development of
    society and the improvement of people's lives. Technology had
    a great influence on the ideologies of that time.

B. The ideologists realized there is dual character to technology.
    One is positive to the development of a country and life of a
    people. Another is negative. People must be careful in developing
    technologies which are not profitable to political stability,
    fundamental production, and human thought.

C. Technology should not be excessively developed. The
    civilization of human beings must be paid attention to when
    embarking on technological plans. It is very important, for
    example, to protect the natural environment.

Today, we can derive inspiration from ancient ideologists, as many situations we now face are similar to those in ancient China. Cautions about the development of technology by Meng Zi have realistic significance in the contemporary world. In fact, we are now realizing that the unlimited development of technology leads to negative results in human society. For example, an excessive military budget and the development of powerful weapons are threatening to human beings. The development of some industries is destroying the ecological environment. Therefore, we must pay more attention to the interaction between technology and society. It is important to develop profitable technology and confine the harmful factors. We must remember the warning given by ideologists some 2000 years ago.

Note

I would like to thank Prof. Hua Jueming (華覺明) and Prof. Wang Bo (王波) for their help on this paper.

Traditional East Asian Views of Nature Revisited

LEE SUNG KYU 李成奎
Inha University, Inchon, Korea

It is well known that the view of Nature held within a culture can deeply influence the characteristics and development of its science and technology. In this paper I would like to make an argument confirming the deep mutual relationship between the East Asian view of nature and its medical tradition.

Organic View of Nature

Traditional East Asian views of nature generally show organic characteristics. The world is a holistic one, in which every part is related to every other. Humans do not occupy a privileged position reigning over nature. On the contrary, we are to be merged and harmonized with it. The human and physical worlds cannot be essentially distinguished. Furthermore, the two worlds keep on influencing each other, so that there can be no regular and unalterable laws in the physical world. The relationship between man and nature is harmonizing and continuous rather than confrontational and discontinuous. Everything in the world makes up an organic whole, in which there is no hierarchical order. To East Asian people, nature was never the object of rules or utility.

This kind of world-view has became a matter of attention for late 20th-century mankind, when the destruction and pollution of the environment is recognized as a serious world-wide concern. People have come to think that the Western mechanical world-view must have been the main agent in the deterioration of the environment, while the East Asian view is pro-environmental. Nowadays it is generally said that, in the sense that it was the West which constructed and developed modern science, the Western view of nature, which sees nature as an object of development and conquest, is responsible for the results. For example, Lynn White expressed such an opinion in 1966 in his article, "The Historical Roots of our Ecologic Crisis".[1]

Qi and the East Asian Cosmogony

At the centre of the East Asian view of nature is the concept of *Qi*. Let us look at the universal explanation of this cosmological process, which appears in several ancient Chinese classics.

There was a time of chaos when nothing could be distinguished, and in the middle of it there was *Qi*. The *Qi* moved to separate into two spirits, *Yin* 陰 and *Yang* 陽. The chaotic state then separated into two. The pure and light material became the heaven and the turbid and heavy material became the earth. The heaven is the spirit of *Yang* and the earth is the spirit of *Yin*. The two spirits mixed together and made up all things in the universe and furthermore caused all changes. All things in heaven and earth affect each other, so if any change occurs on one side, the other side will change in response.

The theory was originally found in *Huainanzi* 《淮南子》, an encyclopaedia written early in the Han dynasty (c. 100 B.C.). Also according to the reading of *Yijing* 《易經》, *Qi*, by gathering and dispersing, produces everything in the universe. The above is virtually the only theory at the centre of East Asian cosmogony throughout its history. In a sense, the history of East Asian cosmogony is the history of the interpretation and elaboration of the original explanation. Inevitably the view of nature based on *Qi* should have been organic.

Qi and East Asian Medicine

Among the various fields of sciences which were developed in the cultural circle of East Asia, medicine reflects most dramatically the East Asian view of nature. The concept of *Qi* is the nucleus of East Asian medicine. In *Zhuangzi* 莊子 there is a passage that says when *Qi* gathers it means life, and when *Qi* disperses it means death. Traditional medicine describes *Qi* as follows:

Qi is the basic force which humans possess from birth. *Qi* forms life and gives nutrition to the body. Therefore *Qi*, flowing inside the human body, is indispensable to life. Although *Qi* goes out of and come into the body continuously, if it goes out of the body completely, one dies because it is *Qi* that manages life. *Qi* also make bodily activity possible and is the source of physical stamina. If *Qi* suffers any kind of damage the person gets disease. The length of human life is also decided by the *Qi* one received from birth.

The oldest classic of Chinese Medicine, *Huangdineijing* 《黄帝內經》 (The Yellow Emperor's Manual of Corporeal [Medicine]) deals in its aetiology with the symptoms of six kinds of diseases. According to the manual, when *Qi* flows all over the human body, by going through six types of conditions, six kinds of diseases appears. Though, with the passage of time, several different specific theories appear in the explanations of the types and conditions of change of *Qi*, there appear no different opinions in the argument that the change of *Qi* manifests itself in various characteristics and becomes the cause of all diseases. Since *Qi* was discussed systematically in the *Huangdineijing* for the first time, it remained the chief concept in the history of East Asian medicine. We see no medicinal criticism against the existence of *Qi* all through time till today. Presently, the theory of *Jingluo* 經絡, that is, the path through which *Qi* flows through the body, and in which there are 360 important points (*Jingxue* 經穴), is the central element in East Asian medicine. Naturally *Qi* and *Jingluo* theory form the foundation of pathology and physiology in this medical science. The concepts of *Jingluo* and *Jingxue* are the essential elements for acupuncture and moxibustion that are the two main therapeutic measures in Eastern medicine.[2]

Among the many fields of science, the field of medicine is one of the most practical. That is to say medicine could have secured its existence in the beginning only if it showed practical ability to heal and cure the patient and disease. I mean that medicine is not a field of science based on an empty theoretical system. East Asian medicine spread throughout East Asian cultural circles and showed excellent capability to cure disease and pain for thousands of years. Nowadays the superiority of East Asian medicine is attracting the attention of the Western world and emerges as an alternative medicine confronting and complementing modern Western medicine.

So-called 'Chinese medicine' can be said to be Han 漢 medicine, from which both the *Han** 韓 medicine of Korea and *Hanfang* 漢方 medicine of Japan derived their theoretical systems. In this sense, the term 'East Asian Medicine' which I have been using here gets its validity without any difficulty.

How different is the *Qi* which forms the basis of the East Asian view of Nature from the *Qi* used in the field of medicine? In short they are the same. In other words, the *Qi* that is indispensable to the maintenance of life and the *Qi* which is the aboriginal substance forming the universe are essentially the same. On the one hand *Qi* is full in the heaven and earth, and on the other hand, *Qi* has its fountainhead just below the navel (called *Dantian* 丹田) in the human body. Hereby we see the idea that the universe is an organism made of *Qi*.

Qi rooted itself deeply in the everyday language, custom and consciousness of East Asian peoples. According to my understanding, nobody in the traditional East Asian world would have questioned the existence of *Qi*, because *Qi* played the central role in Eastern thought for thousands of years, sometimes as the fundamental matter of the universe, sometimes as the central concept of medicine, sometimes as psychological terminology explaining mental condition, and sometimes as a kind of physical power to subdue one's opponent, *Qi* manifests itself in various guises. However there is only one essential being of *Qi*.

The concept of *Qi* appears under the name of *Qigong* 氣功 for the cure of patients or in the realm of martial art. Namely, if *Qi* in the human body becomes strong enough, it not only enhances the health of the possessor, but also could cure diseases or illnesses of another person by the stronger one sharing his *Qi* with the weaker or subduing the opponent physically by ejecting *Qi*. Now we see the establishment of *Qigongfa* 氣功法 and the appearance of *Qigongshi* 氣功師.

The Existence of *Qi*

Recently, interest in and discussion of *Qi* has become lively and spread rapidly in the Western world as well as in East Asia. In Japan, a book which asserted every disease comes from *Qi* became a best seller.[3] Someone in Korea has claimed that the coming 21st century will be the age of *Qi*. One report says that in New York City alone there are more than one hundred *Qi* therapeutic institutes. In Korea and Japan, according to my information, there are quite a few industrial companies, without mentioning universities, which have opened *Qi* research centres.

As the discussion of *Qi* increases, the search for 'scientific proof' of the existence of *Qi* is being actively undertaken. To take an example, some scientists have attempt to explain *Qi* as a kind of electromagnetic wave. However, to the regret of innumerable believers in *Qi*, the route of finding 'scientific proof' has usually ended in failure, because *Qi* is not to be seen, touched and measured. A logical consequence of that failure is that the Western world does not admit the existence of *Qi*. A dramatic example of this rejection was the international symposium titled "Science and the Mental World — A Conversation between the West and East", held in November, 1984 and sponsored by Tsukuba University of Japan and the National Broadcasting Company of France, with about 40 of physicists, astronomers, psychologists, philosophers and martial artists in attendance.[4] The main topic was *Qi*. For this symposium there were martial arts demonstrations by *Qigongshi* who showed a kind of super-ability by using *Qi*. After a fierce discussion over several days the conclusion of the French side was that they could not admit the existence of *Qi*. They, who were totally ignorant of East Asian medicine and its *Jingluo* theory, consequently rejected the concept.

Two Different Paradigms

I would like to propose a problem. When we call for the demonstration of the existence of *Qi* by the scientific method, what kind of science are we discussing? Is it Western modern science, which obtained its dominant position in the 17th century? Is this Western modern science the most perfect system of thought created by mankind? Standing at the end of the 20th century, the limited nature of the rationalistic and mechanistic way of thinking of Western modern science is being widely discussed. Western rationalism has dominated natural science since its foundation in the Scientific Revolution, its main method being analytical reductionism based on the mechanistic world-view. This mechanistic world-view which is the paradigm of Western science had its starting point in the Cartesian dualism of mind and matter, where except for the human mind, anything that man could not touch, see, or measure could not be admitted as existing.

Isn't it only moderate to say that such an 'over-simplified' and deterministic way of thinking is not enough in order to grasp the real image of intricately intertwined reality? If so, a new system of thought should be sought that can overcome the limits of modern science's explanatory principle. Some who criticize the Western rationalism of modern science have observed the East Asian view of nature and looked for a way to overcome modern science from an Eastern perspective. For example, Carl Gustav Jung (1875–1961), offered a new creative concept of collective unconsciousness, claiming that the Western world should admit and accept the traditional Eastern way of thought which relies mainly on intuition, while pointing out the shortcomings of modern Western science and its way of thinking. For him the main error which the Western world committed was the way of reasoning which sees every event only in terms of causality, the necessary sequence of cause and effect.[5]

The typical characteristics of the Eastern paradigm of thought are: first, the reliance on intuition rather than analytical understanding, and second, the monistic world-view which holds to a sense of oneness between humans and Nature, and whose holism is centred on *Qi*. In the Eastern paradigm, there is no clear distinction between mind and body, no dichotomy between psychological world and physical world. Also there is no discontinuity between matter and life, and between matter and spirit. Herewith we meet a totally different and independent tradition of thought from the Western tradition of dualism.

Conclusion

Here are the questions I pose: Is it justifiable to judge an Eastern paradigm using the standard of a Western paradigm? Is it impossible for the two paradigms to coexist? Can it be expected in the future that a new paradigm unifying the two thought systems could appear?

It would not be enough if we only tried to give conceptual definition to *Qi* or its translation into Western language, which has been attempted by many scholars.[6] A very important point is whether or not we choose to admit the existence of *Qi*. To me *Qi* exists between mind and body, and connects the psychological and physical worlds. I think that the existence of *Qi* was demonstrated above all in the realm of Eastern medicine. Once we accept the East Asian view of Nature and its way of thinking, the limits of the Western way of thinking, which is now dominant worldwide, will be overcome and 21st-century science will jump to a new stage.

Notes

[1] Lynn White, Jr., "The Historical Roots of Our Ecologic Crisis", *Science*, March 10, 1967, pp. 1203–1207.

[2] As for the importance of *Qi* 氣 in the East Asian medicine, refer to Yamada Keiji 山田 慶兒, *Chugokuigaku wa Ikani Tsukuraretanoka* 中國醫學はいかにつくられたのか (How was the Chinese Medicine established?) (Tokyo: Iwanami Shoten, 1999) and Manfred Porkert, *The Theoretical Foundations of Chinese Medicine: Systems of Correspondence* (Cambridge: MIT Press, 1974).

[3] Takada Akikazu 高田 明和, *'Yamai wa Ki kara' no Kagaku*「病は氣から」の科學 (Science of 'Disease comes from *Qi*') (Tokyo: Kodansha, 1989).

[4] As for the details of the symposium, refer to Yuasa Yasuo 湯淺 泰雄, *Shukyo to Kagaku no Aida* 宗教と科學の間 (Between Religion and Science) (Tokyo: Meichokankokai, 1993), pp. 146–79.

[5] As for Jung's ideas, refer to *ibid.*, pp. 27–118.

[6] For example see Kim Yung Sik 金永植, "Chu Hsi esoui *Qi* Gaenyum ui myokaji Chukmyun" 朱熹에서의氣개념의 몇가지 측면 (On several aspects of the Concept *Qi* of Chu Hsi), in Kim Yung Sik 金永植, ed., *Chungkuk Jeontongmunwha wa Kwahak* 중국 전통문화와 과학 (Traditional Chinese Culture and Science) (Seoul: Changbisa, 1986), pp. 166–82.

PART II

Science & Technology Policy

On the Mistakes of Emperor Kangxi's Scientific Policy

XI ZEZONG 席泽宗

*Institute for the History of Natural Sciences,
Chinese Academy of Sciences, Beijing*

During the 67 years before the destruction of the Ming Dynasty in 1644, there had appeared in China seven scientific monographs of world level academic standards, such as *Bencao Gangmu* (Compendium of Material Medicine, 1578) and *Luxue Xinshuo* (A New Account of Musical Acoustics, 1584), as well as a crop of brilliant scientists, such as Xu Guangqi, Li Shizhen and Xu Xiake *et al.*[1] After the foundation of the Qing Dynasty in China, however, not only were few important contributions to science made by Chinese, but the backwardness of Chinese science, compared with the Western science, also became more and more obvious. This paper argues that Kangxi — the second emperor of the Qing Dynasty (ruled 1662–1722), should bear the blame to some extent. In comparison with Emperor Kangxi's contemporaries in the West such as Louis XIV of France (ruled 1661–1715) and Peter the Great of Russia (ruled 1689–1725), we see that he badly trailed in science policy making. And what is more, his science policy was even backward than that of Xu Guangqi (1562–1633), a premier as well as a scientist 100 years before. When he was in charge of the work of the calendar reform, Xu Guangqi proposed a set of methods for promoting astronomy as follows:

> To develop astronomy, we should first make a thorough and careful study in theory and mathematics, choose cautiously the ablest person to take up the work, produce proper instruments for observing and calculating, and then make observation at all times so as to make the results tally with the realities of the sky ... there are otherwise no other ways (*Chongzhen Lishu. Hengxing Lizhi. Xumu*) (Chongzhen Reign-Period Treatise on Calendrical Science, Introduction to the Treatise on Fixed Stars, 1631).

The above methods can be regarded as policy. I introduced this quotation once in a symposium on astronomy held in Beijing in 1996, and called the passage "a key policy". The attendants were so impressed by Xu Guangqi that some even proposed to have these words engraved on "LAMOST" (Large Area Multi-Object Spectroscopic Telescope), which is being built by them and will be the largest Schimidt telescope in the world. Checking what Kangxi did with Xu Guangqi's words, we can find how far he was from this policy.

On the Selection and Employment of Capable Person

Xu Guangqi knew how to choose the right persons for the right job. For example, before his death, Xu invited Li Tianjin, who was proficient in astronomy, from Shandong Province to Beijing to take charge of the work of the Bureau of Calendar. However, Kangxi treated the Han scholars in quite a different way. When Kangxi ascended the throne at the age of 8, he encountered the lawsuit of Yang Guangxian vs. J. A. Schall von Bell, a German Jesuit missionary. This conflict, tangled by problems of learning and politics as well as religion, ended with the failure of Yang Guangxian. On April 1, 1669 Kangxi appointed Belgian Ferdinand Verbiest (1623–88) as the vice-director of the Imperial Bureau of Astronomy. Declining this position Ferdinand Verbiest was reappointed to *Zhili Lifa* (the general editor of the calendar) with equivalent status to that of vice-director. The vice-director was in fact the top leader in vocational work, while the director was a Manchu official. This pattern was maintained unchanged until 1826, when the Portuguese Vervissimo da Serra returned to his homeland because of sickness, and the Imperial Bureau of Astronomy ceased to employ European missionaries as heads of the management. Needless to say, these European missionaries had done some beneficial things. However, they were not professional astronomers. There is a distinction between missionaries with scientific background and the professional scientists with religious beliefs. The purpose of the former was to do missionary work for which science was only a tool or stepping stone. There was no need of any improvement in exploration when apparatus production or calendar calculation satisfied the requirements of the imperial palace.

To see the difference, compare Kangxi's engagement of Ferdinand Verbiest to that of Louis XIV's engagement of the Italian astronomer G. D. Cassini (1625–1712). Could there not have been Chinese directors of the observatory during the more than 150 years? In the early years of the Kangxi period there had been two outstanding Chinese astronomers widely known as "Southern Wang and Northern Xue". The former was Wang Xishan (1628–82) of Jiangsu Province while the latter was Xue Fengzhuo (1600–1680) of Shandong Province. Both were not only well versed in mathematics and astronomy, but also learned Chinese and Western Science. Especially Wang Xishan, on whom Professor N. Sivin of the USA was invited to present a 10-page biography for "the Dictionary of Scientific Biography" edited by

American G. G. Gillispe,[2] in which only nine Chinese scientists were included. Can Kangxi, who paid no attention to such a prominent young scientist, who was around him and just over 40, be considered an Emperor who respected talent? Even if Wang Xishan was declined for his dissident political outlook, should not Xue Fenzuo be employed? In fact, Kangxi harbored such a deep mistrust of the Han Chinese scholars that he treated Mei Wending (1633–1721) with only ostensible courtesy later on. Even more, when the Mengyang Zhai (An institution for compiling astronomical and mathematical books) was built in 1713, in which Li Guangdi (1642–1718) took a substantial role, its policy-making was still controlled by the court with Yinzhi (1677–1732), the third son of Kangxi, as the full-power deputy.

On Training Persons of Ability and Collective Research

During Emperor Kangxi's reign period, in the West, the Royal Society of London was founded in 1662, the French Academy of Sciences launched in 1666, and the German Academy of Science built in 1700 at Berlin with G. W. Leibniz as president. Kangxi was surely not kept unknowing as to what was happening in Europe regarding academic systems. According to Dr. Han Qi's research, both Joach Bouvet and J. F. Foucquet had introduced to Kangxi "Gewu Qiongli Yuan" of France (i.e. the French Academy of Sciences) as well as the "Astronomic Palace" (i.e. Paris Observatory built in 1667). The building of Mengyang Zhai and the undertaking of the land survey in the scope of the whole country was relevant to this.[3] The Mengyang Zhai, however, later became an institution for merely compiling books with little research undertaken and ending its mission after compiling the 100 volumes of *Luli Yuanyuan* (Encyclopedia of Astronomy, Mathematics and Musical Acoustics). The *Huangyu Quanlan Tu* (The Kangxi Jesuit Atlas), completed in 1718 as an important achievement of the land survey in the scope of the whole country, had been kept in the Palace and never opened. No mapping methods were ever written down, either. As a consequence, the Court had to ask the Jesuit missionaries to direct the work when land surveys were undertaken again (1756–59) in the Qianlong period.[4]

The land survey carried out in the Kangxi period was actually part of a scientific research project of the French Academy of Sciences, for which Kangxi became unconsciously the organizer. While kept in strict secrecy in China, the Atlas was published in Paris in 1735 and was known widely after that. French scholar C. Jami pointed out correctly:

> Strictly speaking, it would be improper to use the words 'scientific exchange'. According to the Jesuit missionary literature, Emperor Kangxi

sent Joach Bouvet to Europe in 1693 for a diplomatic mission of bringing back to China some other scholars, and he intended to make this trip as an act of the diplomatic corps. As a matter of fact, however, Kangxi never formulated a policy on the scientific exchanges with France. It was only made by the French Jesuit missionaries who described all this as a kind of academic exchange between the two countries.[5]

The only channel of exchange with Europe in the Kangxi period was that of the Jesuit missionary. The first Chinese who went abroad was Fan Shouyi (Liru), who stayed in Europe for 28 years (1682–1709) and mastered Latin as well as Italian. After returning home, however, he was only interviewed by Emperor Kangxi once, at Bi Shu Shan Zhuang (Summer Mountain Villa) at Chengde and left without any designation.[6] As an emperor of a country, Kangxi need not have to learn a foreign language himself. But it was such a great pity that neither did he ever consider setting up a school to educate the children of the Royal family in foreign languages, which might easily have been done as so many foreign missionaries were staying in China by then. Nor did he organize Chinese scholars to translate foreign scientific works. Consequently, such a great chance for development as was nearly granted was lost!

On the Instrument Making and Observation

Confucius said: "*Gong yu shan qi shi, bi xian li qi qi*" (A craftsman who wants to do good work must first sharpen his tools). In *Chongzhen Lishu* (Chongzhen Reign-Period Treatise on Calendrical Science) translated and compiled under the direction of Xu Guangqi, there were 10 volumes of *Celiang Quanyi* (the Complete Treatise on the Measurement of the Heavens, 1631). It was the first time that the word "instrument" emerged in *Yiqi Tushuo* (Illustration of the Instruments), volume 10 of *Celiang Quanyi*,[7] which indicates that the making of and research on scientific instruments was being put into practice by then. The invention of the telescope was a great leap in the history of development of astronomical instruments. The news that Galileo observed the celestial phenomena by telescope was heard in China soon after it took place in 1609. As early as 1618, Johann Schrek (1576–1630) already took a small telescope to China. In the same year, J. A. Schall von Bell and Li Zubai introduced Galilean discovery and invention by their translation of *Yuan Jing Shuo* (Account of the Telescope). In 1629 Xu Guangqi proposed that a telescope should be made for observing the planets and the solar eclipse that would appear on September 9 of the year, which was only 20 years after Galileo's first observation of celestial phenomena by telescope. Forty years later in 1669, however, when Kangxi ordered Ferdinand Verbiest to make astronomical instruments, the telescope was not on the list. Far from not knowing of telescopes, Ferdinand Verbiest wrote in volume two of the *Lingtai Yixiang Zhi* (Record of Newly-made Instruments in the Observatory) the words "glass telescope, microscope", but he never made them. All this, I think, is mainly due to Kangxi. Dominique Parrenin had said in a letter written from Beijing on August 13, 1730 to Dortous de Mairan, president of French Academy of Science, that "Many telescopes, clocks and watches here in the Palace were produced by the most capable craftsman of Europe. Although Emperor Kangxi knows more clearly than anyone else that the telescope as well as pendular clock are indispensable for accurately observing the astronomical phenomena, he did not order his mathematicians to make any use of these instruments."[8]

What was even worse was that none of the instruments made by Ferdinand Verbiest were really employed in the observation. Prof. Pan Nai found out that all the data in the catalogue of stars of *Lingtai Yixiang Zhi* were reckoned rather than observed: the data for longitude came out by adding those data from *Chongzhen Lishu* with 37', the accumulative data for the precession of the equinoxes; the data for latitude came out through the same way as that of longitude; and those data for right ascension and declination were on the whole from the calculation of longitude and latitude.[9] Not only this, all the catalogues of stars compiled in the Qing Dynasty were formed on the basis of those of the older generation or European's with the addition of the precession of the equinoxes, in which little data came from actual observation at the observatory.[10] We do not see the least of what Xu Guangqi had requested: "produce proper instruments for observing and calculating, and then make observation at all times so as to make the results tally with the realities of the sky".

It is unfortunate that instead of sharing the ingenious instruments with the observatory where they could be put to good use, the Qing rulers regarded all these as royal gifts that should be hoarded in the imperial palace to be employed only by the Emperor. In consequence, apparatus set at the observatory were much less than those kept in the Imperial Court. According to Professors Li Di and Bai Shangshu, there are nearly one thousand pieces of scientific apparatus stored in the Palace Museum. Among these about one or two hundred pieces were telescopes, most of which were produced in the Kangxi-Qianlong reign period.[11] What punishment should Kangxi deserve for keeping so many scientific instruments court and far away from research!

On Theory

In Xu Guangqi's policy statement, the most important point was "making a thorough and careful study in theory and mathematics". One of the defects in traditional Chinese science is the lack of theory and system of which *Kangxi Jixia Gewubian* (Studies in the Natural Phenomena by the Kangxi Emperor in His Leisure Hours) is a case in point. For example, all the things that Kangxi focused on in astronomy were general knowledge. He never laid any stress on or studied the theoretical matters introduced from Europe, whether it was the Ptolemaic system, the Tycho System or Copernican theory. There are 53 volumes in total in *Shuli Jingyun* (Collected Basic Principles of Mathematics, Part III of *Lu Li Yuan Yuan*) which was compiled in two sections. It seems that the work strongly stressed system and theory in section A "Li Gang Ming Ti" (Establishment of the Principles and Elucidation of the Bases), and section B "Fen Tiao Zhi Yong" (Applications in Different Cases). Section A consists of 5 volumes: *Shu Li Ben Yuan* (Foundations of Mathematics, Vol. 1), *Ji He Yuan Ben* (Elements of Geometry, Vols. 2–4) and *Suan Fa Yuan Ben* (Elements of Arithmetic, Vol. 5). The first volume relates the "founding myths" according to Chinese tradition, discussing the *Hetu* (River Diagram), *Luoshu* (Writing from the Luo River), and the *Zhou Bi Suan Jing* (Zhou Mathematical Classic of the Gnomon). The *Jihe Yuanben* (Elements of Geometry) mentioned here was not the one translated by Matteo Ricci and Xu Guangqi, but the lecture notes that the missionaries had written for Kangxi when they taught him mathematics. In this book Euclid's geometry was broken up without its axiom and deductive system remaining. This was Kangxi's attitude towards mathematics and theory.[12]

On "Western Learning Originated from China"

Though the theory that "Western learning originated from China" (*Xixue Zhongyuan*) was not initiated by Kangxi, his advocacy, however, doubtlessly promoted its spread. On November 21, 1704, when holding court, Emperor Kangxi delivered a paper "Sanjiaoxing Tuisuanfalun" (On the Triangle, about 600 words or so), in which he said:

> Those who do research on this (triangle calculation) always considered the differences between the old (ancient) and new (nowadays) (western) methods, but they do not know that the method of calculating the calendar originated from China, then spread to the west, where it was well kept and revised and enlarged through observations year after year; hence the different results.[13]

In 1711, when discussing with Zhao Hongxie, Governor of Zhi Li, now Hebei Province, about mathematics problems, Kangxi said: "Principles of calculation were all from *Yi Jing* (the Classic of Changes). Though the Western methods of calculation are nice, they are from China, so they have been called 'algebra' which means 'imported from the East'."[14]

One of the most important steps adopted by Kangxi was calling in Mei Wending on May 11, 1705, talking with him for three days, and granting him the four characters "Ji Xue Can Wei" (Unrivalled in width and depth of understanding). Mei Wending thereupon was overwhelmed by the Emperor's flattery and honor, and was grateful for such special favors. He said once after returning home:

> The Emperor's 'On the Triangle' pointed out that Western learning originated from China. What a great idea it is! Never has a writer said such a thing before." (in *Jixuetang Shichao*, The Collection of Mei Wending's Poems, Vol. 4)

> I read with respect 'On the Triangle', in which the Emperor said that our calendar of ancients spread to the West, where it was studied and revised to preciseness by the people there. Such a great statement can surely put down the debates coming from all schools." (*ibid.*)

> I have the honour to read 'On the Triangle', which said that the method of calculating radian by radii crossing the centre must have existed in ancient China and then spread to the west. While unable to be handed down on this land, it was kept and developed there. So you see that what the sage said can be the guiding principle in our calendar compiling ("*Lixue Yiwen Bu*", Vol. 1).

"*Lixue Yiwen Bu*" (Addenda for the Queries on Astronomy) was Mei Wending's representative work on the theory of Western learning originating from China. With its publication in the first year of the Yongzhen period (1723) as a part of *Meishi Lisuan Quanshu* (Complete Work of Mei Wending on Astronomy and Mathematics), the theory was spread throughout China and learnt by scholars in a wide scope. In the same year, *Shuli Jingyun* was also formally published, among which including "Zhoubi Suanjin Jie" (Explanation on the Zhou mathematical classic of the gnomon), which said:

> When J. A. Schall von Bell, Ferdinand Verbiest, A. Thomas and C.-F. Grimaldi successively took charge of calendar administration, they all had good command of mathematics. The theory of mathematics, therefore, had been gradually full and clear. Whenever asked where their theory is from, they would all say that it was originally from China and spread to the West thereafter.

Now that had been advocated by "the saint and kind ancestor of Qing Emperors", demonstrated by Mei Wending, authority on the calendar and mathematics in the Qing dynasty, and approved uniformly by Jesuit missionaries, the theory of Western learning originating in China became the main trend of ideology in the Qianlong-Jiaqing periods.[15] To be introspective about this painful national subjugation, old adherents of the Ming Dynasty returned to "Liu Jing" (the Six Classics), whereas the early Qing rulers in seeking their own way of domination, reached the same goal by a different way. They believed that according to "the theory that Western learning originated from China" they could find out from "Liu Jing" not only the way of "cultivating their moral qualities, building up their families fortune, and administering the country well and making the world peaceful", but also advanced science and technology. This caused a situation in which Chinese academic research took the way of returning to the ancients, and scholars did not study nature but buried themselves in outdated writings. Thus we had Ruan Yuan's edition of *Chouren Zhuan* (Biographies of Mathematicians & Astronomers); Dai Zhen's writing of *Kaogongji Tuzhu* (Illustrations for the Artificer's Record); and Chen Maolin's compiling of *Jingshu Suanxue Tianwen Kao* (Investigations on Astronomy and Mathematics of the Classics).

Just when our ancestors indulged themselves in the theory of Western learning originating from China, and looked upon returning to "Liu Jing" as the objective of their effort, science and technology in Western countries, made greater progress than ever before. Not until our gate was opened in the Opium War by powerful Western weapons, one of the fruits of industrial revolution (1770–1830) of Britain, did Chinese intellectuals begin to realize that this way was wrong and that we were falling far behind.

To sum up, we can come to the following conclusion: in the light of a developing tendency in the late Ming Dynasty, Chinese science should have blossomed and probably shifted to modern science. This process was broken off by the Qing intrusion which resulted in war and damage. Up to the Kangxi period, the country had been recovering from the damage and on the whole it had been united. By then political reform was being undertaken, the economy was quite developed, and what's more, with so many Jesuit missionaries who knew sciences staying in China and serving the imperial court, there was a great chance for China to develop in science to a level similar to that of Europe. Unfortunately, such an opportunity was missed because of Emperor Kangxi's bad policy.

Postscript

On October 11, 1999 *Kexue Shibao — Dushu Zhoukan* (Science Times — Reading Weekly) published on page B4 an essay of Zhao Xinshe, "Viewing Kangxi from a Different Angle", which introduced the new work of Prof. Tian Shitang *et al.*, *Emperor Kangxi and Peter the Great of Russia — the Pities behind the Prosperity of the Kangxi-Qianlong Reign-Period.*

The authors in this book compared Kangxi with his contemporary Peter the Great of Russia and analyzed what effect Kangxi's thought and action had brought to China afterwards. They pointed out that in 1700 the GNP of China was 23.1% of the whole world while that of Russia was only 3.2%. However, the two countries developed very differently later on. The reason is that the two governments adopted quite a different way and system for each countries' development. Since the way adopted by Kangxi caused China to miss the chance at an Industrial Revolution and then to fall from feudal society into semi-colonial and semi-feudal society, which brought China a century of mortification, Kangxi should bear the blame. Such a conclusion happens to agree so well with what I said in my above paper that I thought it worth mentioning here.

Notes

[1] Chen Meidong, "The revival of science and technology in the Ming Dynasty and the solid learning ideology", in *Yazhou Keji Yu Wenming* (Science and Civilization in Asia), ed. Zhao Lingyang and Feng Jingrong, pp. 64–84, Hong Kong: Ming Boa Press, 1995.

[2] N. Sivin, Wang Hsi-shan (1628–82), in *Dictionary of Scientific Biography*. XIV, pp. 159–68, New York: Charles Scribner's Sons, 1976.

[3] Han Qi, "The Scientific Activities of French Jesuit Missionaries in China during the Kangxi Reign Period", *Gugong Bowuyuan Yuankan* (Bulletin of the Palace Museum), 1998, No. 2, pp. 68–75.

[4] Du Shiran *et al.*, *Zhongguo Kexue Jishu Shigao* (A Draft of the History of Science and Technology in China) (II), p. 213, Beijing: Science Press, 1982.

[5] C. Jami, "The Contact of Sino-French Science in 18 Century", translated by Geng Sheng, *Qingshi Yanjiu* (The Study of the History of Qing Dynasty), 1996, No. 22, pp. 56–60.

[6] Fang Hao, *Zhongxi Jiaotong Shi* (A History of China-West Communication) (IV), p. 187, Taibei: Huagang Press, Inc., 1997.

[7] Zhang Baichun, *The Europeanization of Astronomical Instruments during the Ming and Qing Dynasties* (doctoral dissertation), p.82, Beijing, 1999.

[8] Zhu Jing (translated and edited), *Yangjiaoshi Kan Zhongguo Chaoting* (The Chinese Imperial Court Seeing by the Foreign Missionaries), Shanghai: People's Publishing House, 1995.

[9] Pan Nai, *Zhongguo Hengxing Guance Shi* (A History of Star Observations in China), p. 377, Shanghai: Xuelin Publishing House, 1989.

[10] Zhang Baichun, *The Europeanization of Astronomical Instruments during the Ming and Qing Dynasties* (doctoral dissertation), pp. 253–55, Beijing, 1999.

[11] Li Di, Bai Shangshu, "An Outline of the Collections of Scientific and Technological Relics in the Palace Museum", *Zhongguo Keji Shiliao* (Chinese Historical Materials of Science and Technology), 1981, No. 1, pp. 95–100.

[12] Fan Hongye, *Yesu Huishi Yu Zhongguo Kexue* (The Jesuit Missionaries and the Chinese Science), p.236, Beijing: China People's University Press, 1992.

[13] There are two versions on "Yuzhi Sanjiaoxing Tuisuanfa", one is of Manchu-Chinese bilingual version collected in *"Man-Han Qi Ben Tou"*, printed around 1707; another one is the vol. 19 of part 3 in *"Kangxi Yuzhi Wenji"*. Wang Yangzong and Han Qi have a different point of view on the writing time of "Yuzhi Sanjiaoxing Tuisuanfa". In this paper I adopted Wang's idea, see Wang Yangzong, "The Hypothesis of the Sinological Origin of Western Learning: Emperor Kangxi and Mei Wending", in *Chuantong Wenhua Yu Xiandaihua* (Chinese Culture: Tradition and Modernization), 1995, No. 3, pp. 77–84; Han Qi: "Joachim Bouvet's Study of the Yi Jing and the Theory of Chinese Origin of Western learning during the Kangxi Era", in *Hanxue Yanjiu* (The Chinese Studies), 1998, Vol. 16(1), pp. 185–201.

[14] *Shenzhu Shilu* (The Veritable History of Kangxi), Vol. 245, p.431, Beijing: Zhonghua Press, 1985.

[15] Jiang Xiaoyuan, A Tentative Discussion on "the Theory of Western Learning Being of Chinese Origin" of the Qing Dynasty, *Ziran Kexueshi Yanjiu* (Studies in the History of Natural Sciences), 1988, Vol. 7(2), pp. 101–108.

The Late Qing Government's Policy Towards Science and Technology: The Case of Jiangnan Manufactory

LI DI 李迪

Institute for the History of Science
Inner Mongolia Normal University, Huhhot, China

Qin Tian Jian (欽天監 the Imperial Boards of Astronomy) of various dynasties were characterized by their scientific research, but were mainly organs of administrations for calendar compilation and of astronomical studies relevant to calendars. Their functions were multifarious.[1] In the Qing dynasty, nothing in this system was changed. Although Frenchman Joachin Bouvet (1656–1730) mentioned in his report to Louis XIV (1638–1715) that Emperor Kangxi wanted to establish an academy of science,[2] no Chinese material about it has been found. In the early Qing, Chinese science had fallen behind its Western counterpart, and the gap between China and Europe became larger and larger until, by 1840, China was behind by about 200 years.[3]

The Opium war between China and Britain astonished the Chinese, who realized the priority of Britain and the backwardness of China in military equipment, and therefore in science and technology. Some Chinese intellectuals began to promote the study of foreign science and technology. Among others, Wei Yuan (魏源) raised the clarion call, "*Shi yi changji yi zhiyi*" (师夷長技以制夷, Study foreign advanced science and technology to control foreigners); Li Shanlan (李善兰) advocated the development of mathematics as a step toward studying Western science and technology; Ding Gongchen (丁拱辰), Ding Shoucun (丁守存) and Zheng Fuguang (郑复光) were active in the practice of studying Western knowledge.[4] Facing the upset of *Taiping Tianguo* (太平天國 Taiping Heavenly Kingdom) and invasion by the allied forces of Britain and France, the Qing government, beset with difficulties both at home and abroad, had to pay attention to developing science and technology. In the early Tongzhi (同治) reign the Qing government established schools for foreign language in Beijing, Shanghai, and Guangzhou. Mathematics and astronomy were afterward added to the curriculums in some schools. Manufactories were also set up in Shanghai and other places. In addition, Zuo Zongtang (左宗堂, 1812–85) set up an

79

engineering school, training shipbuilding engineers and workers. Thus, a movement toward developing military industry and studying western science and technology was underway. This is the so-called "Westernization Movement" of China.

Because of the urgent need for military equipment, the late Qing government began to study the manufacturing techniques of foreign weapons, and therefore to study Western science and technology. When the war between the Qing court and the *Taiping Tianguo* was at the critical moment in 1863, Li Hongzhang (李鴻章) set forth a proposal for studying the foreign technology of firearms, and set up manufactories to make weapons, thus solving the urgent need for their supply.[5] The *Jiangnan* Manufactory was thus established in Shanghai, and thirteen manufactories and one firearms mill were afterward established throughout China. Their names and the dates of their establishment are listed in Table 1.

Table 1   The Names and Dates of Manufactories Established in the Late Qing

Name of Manufactory	Date	Name of Manufactory	Date
Jiangnan Manufactory (江南制造厂)	1863	Jilin Manufactory (吉林制造厂)	1881
Tianjin Manufactory (天津制造厂)	1866	Jinling Manufactory (金陵制造厂)	1883
Fujian Manufactory (福建制造厂)	1870	Shanxi Manufactory (陕西制造厂)	1884
Shandong Manufactory (山东制造厂)	1875	Zhejiang Manufactory (浙江制造厂)	1885
Hunan Manufactory (湖南制造厂)	1876	Taiwan Manufactory (台湾制造厂)	1887
Sichuan Manufactory (四川制造厂)	1877	Yunnan Manufactory (云南制造厂)	1890
Guangdong Manufactory (广东制造厂)	1878	Hubei Fierarms Mill (湖北制造厂)	1890

The manufactories were set up for making weapons and some of them also made warships. These kinds of factories were also set up later in other provinces, for instance, in Fentian (today's Liaoning), Xinjiang and Shaanxi. Among others, Jiangnan manufactory was the largest and best equipped, and reflected the government's policy towards science and technology. In *Jiangnan*, subfactories and institutes were also established (see Table 2).

It can be seen from Table 2 that most of the subfactories aimed at making military equipments or served that task. But three of them, that is the Translation Section, the Planetarium and Observatory, and the Technology

Table 2 The Subfactories and Institutes in Jiangnan Manufactory

Names	Founded Time	Names	Founded Time
Machine-building Works (机器厂)	1865	Bullet Mill (枪子厂)	1875
Shipyard (造船厂)	1867	Powder Magazine (火药库)	1876
Translation Section (翻译馆)	1868	Cannon Mill* (改汽锤厂为炮厂)	1878
Planetarium and Observatory (天文馆与天文台)	1868	Mine Mill (水雷厂)	1881
Pneumatic Hammer Plant (汽锤厂)	1869	Steelworks (汽锤厂)	1890
New Gun Mill (另建枪厂)	1869	Chestnut Powder Mill (栗色火药厂)	1892
Artilleryman school (操炮学堂)	1874	Smokeless Powder Mill (无烟火药厂)	1893
Black Powder Mill (黑火药厂)	1874	Technological School (工艺学堂)	1898

Note: *The Cannon Mill was transformed from the Pneumatic Hammer Plant.

School, were not connected directly with military equipment. It is necessary to discuss these.

First, the Translation Section. Why was this Section set up in a manufactory? After the establishment of *Jiangnan*, Li Hongzhang, who was one of the most important Qing Court officials and was in charge of foreign affairs, realized that the task of manufacturing quality military equipment could not be fulfilled without studying the necessary theory and principles of machinery building. He thought that it was necessary to translate and study foreign works on science and technology. In a memorial presented to the emperor, he, together with Zeng Guofan (曾国藩) wrote: "Translation (of foreign books) is the foundation of the manufacturing industry. Foreign manufacture is based on mathematics, and all its mysteries can be drawn out by illustrations." Their proposal was approved, and the Translation Section was set up. Some foreigners such as Englishmen Alexander Wylie (1815–87), John Fryer (1830–1928), and American Daniel Jerome MacGowan (1814–93), along with some native scientists such as Hua Hengfang (华蘅芳, 1833–1902), Xu Shou (徐寿, 1818–84), Xu Jianyin (徐建寅, 1854–1901), Li Fengbao (李凤苞, 1834–87), Zhao Yuanyi (赵元益, 1840–1902), and Wang Zhensheng 汪振声), were recruited to translate foreign books. The first group of translated works were *Qiji fachuang* (汽机发创, Rudiments of Manufacturing Steam Engines), *Qiji wenda* (汽机问答, Questions and Answers on Steam Engines), *Yungui yuezhi* (运规越旨, Outline of Applications of the Compasses), and *Taixi caimei tushuo* (泰西采煤图说, Illustrative Coal Mining Engineering of the Western Countries), the first two of which are about the manufacture and principle of steam engines, the third of which was a book on drafting for machinery, and the last of which is easily explained by its title. Scientific and technological works ranging from mathematics, physics, mineralogy, and meteorology were translated in succession. By the turn of the century the number of works translated into Chinese and published in China was very large. Fryer himself took part in translating 75 works.[6] According to Li Hongzhang's idea, Chinese people should study Western science and technology in three steps: translating foreign books firstly, then learning them, and finally writing books themselves.[7] The idea itself was appropriate, but it did not emphasize doing research work, so that the latter task was not fulfilled and few high level works were written by Chinese scholars.

Second, the Planetarium and Observatory. They were set up in the Engineering section of *Jiangnan*, In another paper, I have shown that the Planetarium and Observatory mainly served to compile *Hanghai Tongshu* (航海通书, Nautical Almanac). Therefore, their tasks were different from those of Beijing Observatory. The nautical almanac was prepared as a guidebook for ships. At first, it was mainly translated from English books. But because all the computations in foreign books were completed according to the latitude of Greenwich instead of that of Beijing, and the starting point of astronomical computation was at the Spring Equinox instead of the Winter Solstice (used in Chinese astronomy) these had to be reedited to satisfy the needs of Chinese ship's crews. Jia Buwei (贾步纬, 1827–1902), a famous astronomer and mathematician, was appointed to take charge of the work. To do this a planetarium was established. Perhaps the Qing Government wanted to compile it independently of foreign books, so that necessary observation data were demanded, and a new observatory thus was set up and equipped with telescopes and other devices. Besides Jia Buwei, Li Dian (李佃), an astronomer from Hubei, was also engaged in the research work in the Observatory. Jia imported a Thomas calculation machine from England and used it in compiling various tables for astronomical calculations. However, the original target of compiling independently of Western nautical almanac was not hit, because it was a difficult and heavy work, and could not easily be completed by a few people.[8]

Third, the Technology School. The aim of the school was training technicians and engineers. This can be seen from the curriculum and the identity of teachers. The courses belonged to two categories. The first were common courses, including Chinese, English, mathematics and drafting, and the second were special courses which followed those of Osaka Technology School in Japan and belonged to two subjects: chemistry and machinery. Among the school staff, Hua Hengfang taught mathematics; Xu Huafeng (徐华封) taught chemistry; Wang Shiwan (王世缓) taught technology; Yang Jiankui (杨建逵) taught drafting; and Hua Beijue (华备钰) taught machinery.

The above measures, however, did not meet the needs of high level technicians and engineers. The situation became worse and worse, so a proposal for sending students abroad was made in 1904 by Wei Guangtao (魏光涛) and Zhang Zhidong (张之洞, 1837–1909). They pointed out in their memorial to the Emperor that because of a lack of high level administrators, technicians and engineers, the Chinese machinery industry had to rely on foreigners. They asked to select 20 officials, 40 students and 40 craftsmen and send them to Germany, Belgium, Japan and other countries, studying respectively administration, technology, and engineering administration for five years. Although the proposal was appropriate, it was too late to develop Chinese industry and to change the passive attitude of the Qing government.

As a leading manufactory and a military industry group, *Jiangnan* played an important role in the development of the arms industry of modern China. But it should be pointed out that the Qing government's policy towards it had its limitation. All measures were taken for improvement of the arms industry, but not of the development of science and technology. In fact, representative figures such as Li Hongzhang and Zeng Guofan, had realized that military manufacturing technology could be engaged in civil uses. It is lamentable that they were not given consideration.

It is necessary to also mention *Tongwenguan* (同文馆) in Beijing which was established in 1862 as a college of translation. At first, only students of foreign languages such as English, French, and Russian were trained there. Yi Xin (奕欣, 1833–98), who was in charge of foreign affairs for the Qing court, realized that students of mathematics and sciences were more needed. In 1866 he set forth a proposal to set up a department of mathematics and astronomy in *Tongwenguan* and to invite foreigners to teach science there.[9] His proposal was approved by the Emperor. It was apparent that students in *Tongwenguan* had studied much more scientific knowledge than those in *Jiangnan*.

The establishment of manufactories and institutes such as *Tongwenguan* were the core feature of the Westernization Movement. Here we have taken *Jiangnan* Manufactory as a typical case to discuss the Qing government's policy towards science and technology. This should be considered a supplement to a more general discussion of the Qing government's policy towards science and technology during the Westernization Movement.[10]

We also would like to point out that the Qing government missed many opportunities to learn science and technology from Western countries. Many officials of the Qing, including Li Hongzhang himself, went abroad to survey industry or do diplomatic work. But they visited neither any important scientific society, nor academy of science, nor famous university. There was no place for such visits in the officials' minds. The establishment of *Jiangnan* Manufactory, together with the Qing government's emphasis on the arms industry, was only an expedient measure toward solving problems of domestic trouble and foreign invasion. The Qing government did not develop a comprehensive policy towards development of science and technology. In fact, this was also a case of the prevalent argument "*zhongxue weiti, xixue weiyong*" (中学为体, 西学为用, Take Chinese learning as essential, meanwhile load Western learning in application).

In conclusion, the establishment of *Jiangnan* Manufactory in Shanghai and its later development reflected the Qing Government's policy towards science and technology. At first, the manufactory was set up only for military purposes. There was not an exclusive plan for the introduction of foreign science and technology. But soon afterwards it was realized that a concentration on military equipment only was useless. Therefore, the Manufactory was developed as a scientific and technological center where foreign military technology was mainly studied and introduced. Meanwhile general scientific and technological knowledge was also studied and foreign books were translated. The manufactory was also an educational center. A similar story unfolded in *Tongwenguan* in Beijing. These facts show that the

Qing government changed its policy on science and technology. The reason of the change and its influence on the modernization of science and technology in China are worth studying in detail.

Notes

[1] Guo Shirong (郭世荣), *Zhongguo gudai tianwen jigou de zhineng* (中国古代天文机构的职能, The Function of Astronomical Organs in Ancient China), in *Tongxiang xiandai kexue zhilu de tansuo* (通向现代科学之路的探索, Exploration of the Way of Science from Ancient Times to Today), Huhhot: Inner Mongolia University Press, 1993, pp. 6–9.

[2] Bouvet F. J., *The Emperor Kangxi*, tran. by Zhao Chen (赵晨), Harbin: Heilongjiang People's Press, 1981, pp. 50–51.

[3] Li Di, *Zhounguo kexue fazhan de yige gaofeng shiqi—shilun mingqing erbai nian jian zhongguo kexue de fazhan qingkuang* (中国科学发展的一个高峰时期 — 试论明清二百年间中国科学的发展情况, A High Leap in the Development of Sciences in China — A Study of the Development of Science in China During 200 Years Between the Ming and the Qing), in *Zhongguo kexue jishu shi lunwen ji* (中国科学技术史论文技集, Collected Papers on History of Science and Technology in China), Huhhot: Inner Mongolia Education Press, 1991, pp. 84–105.

[4] Li Di, *Diyici yapian zhanzhen qianhou chuanru woguo de xifang kexue jishu* (第一次鸦片战争前后传入我国的西方科学技术, Western Science and Technology Introduced into China around the Period of the First Opium War), *ibid.*, pp. 106–152.

[5] Li Hong Zhang (李鸿章), A Memorial to the Emperor on the 4th of Ninth Month (of Chinese Calendar) in the Second Year of Tongzhi (同治) Reign, in *Chou ban yiwu shimo* (筹办夷务始末, The Whole Story of Management of Foreign Affairs) in Tongzhi (同治) Reign.

[6] *Jiangnan* Manufactory, *Jiannan zhizao ju ji*, ed. (江南制造局记, The Story of *Jiangnan* Manufactory), Vol. 2.

[7] Zeng Guofan (曾国藩), A Memorial to the Emperor on the 2nd of Ninth Month (of Chinese Calendar) in the Seventh Year of Tongzhi (同治) Reign, in *Chou ban yiwu shimo* (筹办夷务始末, The Whole Story of Management of Foreign Affairs) in Tongzhi (同治) Reign.

[8] Li Di 李迪, *Jian shu jiangnan zhizao ju tianwentai* (简述江南制造局天文台, A Brief Description of the Observatory at Jiangnan Manufactory), in *China Historical Materials of Science and Technology*, Vol. 16, No. 4, pp. 59–63.

[9] Yixin (奕欣), A Memorial to the Emperor on the 5th of Eleventh Month (of Chinese Calendar) in the Fifth Year of Tongzhi (同治) Reign, in *Chou ban yiwu shimo* (筹办夷务始末, The Whole Story of Management of Foreign Affairs) in Tongzhi (同治) Reign.

[10] Lao Hansheng 劳汉生 and Li Di, *Yangwu yundong yu zhongguo jindai keji wenhua zhence de bianqian* (洋务运动与中国近代科技文化政策的变迁, The

Westernization Movement in China and the Change of Modern Chinese Policy Towards Science, Technology and Culture), in *Zhongguo kexue jishu shi lunwen ji* (中国科学技术史论文集, Collected papers on History of Science and Technology in China), Huhhot: Inner Mongolia Education Press, 1991, pp. 47–72.

The Institutionalization of Higher Science Education in China: A Case Study of Higher Chemistry Training before 1937

ZHANG LI

*Institute for the History of Natural Science,
Chinese Academy of Sciences, Beijing*

Brief Review

T'ung Wen Kuan (同文馆) was the first government school to begin teaching chemistry in modern China. In 1871, a Frenchman Anatole Billequin (1837–94) was invited there as a chemistry instructor. A few years later, John Fryer (傅兰雅, 1839–1928) and Hsü Shou (徐寿, 1818–84) founded *Ko-chih shu-yüan* (格致书院, Shanghai Polytechnic Institute) to promote public education in modern science and technology. They made many chemical demonstrations, beginning in 1895, and always greatly surprised the public. They also invented a system of nomenclature that has survived down to the present day. At the same time, many mission schools also offered various levels of chemistry training.

During this period (the second half of 19th century) translation was the main channel through which knowledge of chemistry entered China. Meanwhile, a lot of young Chinese had been sent to Europe and America by the government or local administrations for advanced training. Though almost all of them had been assigned to learn telegraph operation, shipbuilding, navigation, mining, and so on, they had learned much about chemistry in various European countries.

Soon more Chinese people chose to go to Japan instead of to Europe and America for various reasons. And Chinese translators looked increasingly to Japanese texts. The first science periodical published by the Chinese themselves, the *Ya-chüan Tsa-chih* (《雅泉杂志》, Ya-chüan Magazine), which appeared between 1900 and 1901, was composed almost exclusively of articles translated from Japanese. The law of periodicity and news of the discovery of 10 more chemical elements, including radium and polonium, reached China through this magazine.

In 1903, the first law for modern education (also called *Kui-mao hsüeh-chih* 癸卯学制) was promulgated by the Emperor. This made it a rule that scientific training must be a necessary part of education. Therefore this reform became very important in the history of Chinese education. According to the law, chemistry instruction ought to be offered in all schools (primary and secondary) as well as colleges, and departments of chemistry should be founded. However, there was no one to teach until the *hua-hsüeh men* (化学门, similar to department of chemistry) at the Imperial University of Peking was founded in 1910.

Institutionalization

After reviewing the above, we know that in China higher education in chemistry began around 1910. In my opinion, its early history should be divided into three stages:

> 1910–19: higher chemistry training emerged;

> 1919–28: the system of higher chemistry education gradually completed;

> 1928–37: more and more young students being trained professionally, and real academic research began.

Historians of science always have paid more attention to the third period (1928–37), in which modern science finally sank its roots in China and then was developing rapidly in all relevant aspects. Compared with this, the second period has been ignored. I think, however, there are some interesting issues to raise about this period, such as how departments of chemistry were founded, how social and academic conditions were at that time, and how chemistry training had influenced the public. Therefore, my paper will focus on this period.

Emergence of Higher Chemistry Training, 1910–19

Though systematic chemistry education began formally after *Kui-mao hsüeh-chih* (癸卯学制), there had been little progress until 1910. Along with other scientific courses, chemistry was offered in the first institutions under Chinese management established around the turn of the century: Peiyang University (北洋大学, founded in 1895), the Imperial University of Peking (京师大学堂, founded in 1898), and Shansi Imperial University (山西京师大学堂, founded in 1901).

In 1910 the Imperial University of Peking's four-year course was divided into seven *ko* (科, faculties) including law, engineering, agriculture and science and so on. Each *ko* had several *men* (门, specialties), for example, *ko-chih ko* (格致科, science faculty) had two specialties named *hua-hsüeh men* (化学门) and *ti-chih men* (地质门, specialty in geology). In the same year, *hua-hsüeh men* began its enrolment and became the first proto-department of chemistry in China. All students had to take inorganic chemistry, organic chemistry, analytic chemistry, applied chemistry, theoretical chemistry, chemical equilibrium theories, and chemical laboratory. Textbooks were translated from European books originally published in 1850–90. Most teachers were Japanese and a few Westerners; the only Chinese chemistry teacher was Yu Tong-kui (1876–1962), who had gotten his master's degree in chemistry from Liverpool University. Meanwhile students in *hua-hsüeh men* had to learn German as a preparatory course because science in Germany was regarded the best. In 1918, there were seven students as the first graduating class from the university. At the same time, there were no such departments in other universities. That was why many students had to go abroad for advanced training.

Chinese higher education improved after the establishment of the Republic in 1912. Many institutions of higher learning were founded. Among those well known were Tsinghua College (founded in 1911) and Nanking Higher Normal College (founded in 1915, later becoming National Southeastern University). The Tsinghua College was set up in 1911 to prepare Chinese students for their future studies in America. In the early years, over half the teachers were Americans. Later, most were Tsinghua graduates who had completed advanced study in America. All instruction was in English and followed the American model. The chemistry department at Nanking Higher Normal College was founded by Wang Chin, one of the first notable Chinese teachers of science. He graduated from an American university in 1914, with a bachelor of science degree in chemistry, and then returned to China. However, there were few chemistry departments like these at the time. Therefore, in the late Ch'ing and early Republican periods, chemistry advanced slowly and haltingly in colleges and universities.

Institutionalization of Higher Chemistry Education: 1919–28

During this period a lot of well-known chemistry departments were founded in succession, led by the department at Peking University, which was reorganized from previous *hua-hsüeh men* in 1919 as one of the results of the educational reformation prompted by the May Fourth Movement.

I. Chemistry Departments

In 1922, the Administration of Education of the Republic issued new regulations for education, which combined models of America and Europe with actual Chinese conditions. It was an important reform in public education and we call it *Ren-xu xue-zhi* (壬戌学制). One of its purposes was to strengthen scientific training, so that it encouraged a lot of institutions of higher learning to be founded.

Table 1   Departments of Chemistry or Chemical Engineering Founded in the 1920s and Their First Deans

Universities	Time of Related Departments Founded	The First Deans	Univ. Graduated	Academic Degree
National Peking Univ.	Chem., 1910	Yu Tong-kui	Liverpool	M.S., 1907
Nankai Univ.	Chem., 1921	Qiu Zong-yue	Clark	Ph.D., 1920
Xiamen Univ.	Chem., 1921	Leo Shoo-tze	Columbia	Doc., 1919
Nanking Univ.	Chem., 1921			
National Southeastern Univ.	Chem., 1922	Wang Chin	Lehigh	Bach., 1914
Yenching Univ.	Chem., 1923			
Peping Normal College	Chem., 1923	Liu Tuo		
National Chungshan Univ.	Chem., 1924 Chem. Eng., 1931	Whang Siar-hong Chen Zong-nan	Berlin	Ph.D., 1930
Tsinghua Univ.	Chem., 1926	Chang Tzsu-kao	M.I.T.	M.S., 1915
Szechwan Univ.	Chem., 1926	Lin Chao-zong		
National Chekiang Univ.	Chem., 1928 Chem. Eng., 1928	Lih Kun-hou Li Sheo-hen	Munich Illinois	Ph.D., 1932 Ph.D., 1925
Wuhan Univ.	Chem., 1928	Wang Xing-gong		
Lingnan Univ.	Chem., 1928			
Fujen Univ.	Chem., 1929			
Shantung Christian Univ.	Chem., 1930	Tang Teng-han		

All universities mentioned in Table 1 could be divided into two categories, one derived from previous lower colleges like Tsinghua College and Nanking Higher Normal College, and another founded newly following the regulation for education.

Compared with the universities mentioned above, chemistry in missionary higher education was better. Around 1920, mission colleges and universities emerged as the leading centres of Chinese higher education, particularly in the sciences. For example, by 1926 the Protestants operated 16 colleges and universities in China and all of them had departments of chemistry and physics.

II. Chemistry Professors

From Table 1, we can also see that most founders of chemistry departments in universities under Chinese control were the first generation who had returned from America or Europe, some of them having lower academic degrees such as master's or bachelor's. They generally returned to China as soon as they got their degrees, so they had no real experience of research. They built the chemistry departments of the 1920s, gradually turning them over to their students, who returned with advanced degrees from American and European universities in the 1930s.

In the late 1920s, chemistry was doing better in Tsinghua College, in part because of following the University of Wisconsin as a model. Laboratory facilities were in plentiful supply, and several excellent men were appointed as chemistry professors (refer to Table 2). The chemistry courses were said to be the equivalent of first-class university courses in America.

Table 2 shows faculties of chemistry departments in leading mission colleges, which were largely composed of foreigners in the early years but of Chinese later on. With more money, the mission schools improved rapidly and produced notable research results.

Table 2   Faculties in Chemistry Departments in the 1920s

Nanking Higher Normal College around 1920		
Wang Chin	bachelor, Lehigh, 1914	taught analytic chemistry
Chang Tzsu-kao	M.S., M.I.T., 1915	taught physical chemistry
Sun Hung-fen	not clear	taught organic chemistry
Tsinghua College before 1926		
Chang Tzsu-kao	M.S., M.I.T., 1915	chairman of the chemistry department
C. A. Pierle	Ph.D., Wisconsin, 1919	
Yang Kuang-pi	M.S., Wisconsin, 1917	Pierle's assistant

(*cont'd overleaf*)

Table 2   (*continued*)

Tsinghua University in the early 1930s		
Kao Ch'ung-hsi	Ph.D., Wisconsin, 1926	chairman of the chemistry department
Chang Ta-yü	Ph.D., Dresden, 1933	
Chang Tzu-kao	M.S., M.I.T., 1915	taught general and quality analysis
Huang Tzu-ch'ing	Sc.D., M.I.T., 1935	taught analytic chemistry
Li Yün-hua	Ph.D., Columbia, 1927	
Sah Pen-t'ieh	Ph.D., Wisconsin, 1926	taught organic chemistry
Leading mission colleges in the late 1920s		
Yenching University		
William Adolph	Ph.D., Pennsylvania, 1915	
Stanley Wilson	Ph.D., Chicago, 1916	
Earl Wilson	M.S., M.I.T, 1928	
Nanking University		
James C.Thomson	M.S., Columbia, 1917; Ph.D., Columbia, 1932	
Lingnan University		
Henry Frank	M.S., Pittsburgh, 1922	

III. Curriculum

Table 3 is the curriculum of the chemistry department in Peking University during 1923–24. Among the courses are several, such as electric chemistry and colloid chemistry, offered to seniors as the result of new specialty-formation.

Table 3   Curriculum of the Department of Chemistry at Peking University during 1923–24

1. General Chemistry and Laboratory
2. Organic Chemistry and Laboratory
3. Qualitative Analysis and Laboratory
4. Quantitative Analysis and Laboratory
5. Physical Chemistry and Laboratory
6. Advanced Inorganic Chemistry
7. Advanced Analytical Chemistry
8. Advanced Organic Chemistry
9. Advanced Physical Chemistry
10. Applied Chemistry

Table 3 (*continued*)

11. Chemical Engineering
12. Colloid Chemistry
13. Electrochemistry
14. Metal Examine
15. Metallography
16. Metallurgical Chemistry
17. Chemistry for Fuels and
    Fire-resistive Materials
18. Chemistry for Medical Jurisprudence
19. Pharmaceutical Chemistry
20. History of Chemistry
21. Progress in Chemistry made recently

As another example, by 1922 Southeastern University enrolled twenty-five students in fourteen chemistry courses, including academic specialties and applied specialties such as industrial, agricultural and so on. Texts and reference materials were in English, instruction in some combination of English and Chinese.

During 1919–28, the government of the Republic attempted to establish a nationwide system of modern education, and Chinese who had had advanced chemistry training in America and Europe worked hard as educators as well as organizers. Therefore the system of higher chemistry education was being gradually formed.

Prosperity of Higher Chemistry Education: 1928–37

After 1928, the new Ministry of Education adopted several important measures to reorganize the education system. The basic law on higher education, promulgated in 1929, required that each university include at least one college of science, engineering, agriculture, or medicine. At the same time, all private schools were required to register themselves with the government. Therefore the chaos of higher education during 1920s was ended and the system was finally standardized.

On the other hand, the second generation of scientists returning from abroad dominated all academic activities in China. They provided the leadership in teaching, research, and the application of new techniques. All of them had better education backgrounds than their predecessors did. Every one had gotten his doctorate from the leading universities in Europe or

America, published scientific treatises and performed work to the standard
he had learned in the West.

Owing to these political and social developments, Chinese chemistry
advanced rapidly during the 1930s. Table 4 gives clear evidence of this
improvement.

**Table 4   Numbers of Students in Colleges and Universities, by Department and
College, 1931**

College	Number	%
Colleges of science		
Chemistry	1, 062	24.2
Math	676	15.3
Physics	640	14.5
Biology	414	9.4
Geography	151	3.4
Geology	85	1.9
Psychology	59	1.3
Astronomy	36	0.8
Other	825	18.8
N.A.	459	10.4
Total	4, 407	100.0
Colleges of engineering		
Civil	1, 246	38.2
Electrical	690	21.1
Mechanical	500	15.3
Mining	282	8.6
Chemical	176	5.4
Architecture	61	1.9
Other	196	6.0
N.A.	116	3.5
Total	3, 267	100.0

I.   High Quality in Chemistry Training

Tsinghua became a national university, and chemistry was the largest of its
science departments. There were eighty chemistry students and a dozen full
and assistant professors, including several of China's most prominent chemists
(refer to Table 2), all younger men who returned to China in the late 1920s
or the early 1930s.

The Tsinghua curriculum combined a core of required courses in each branch of chemistry with a wide variety of electives. All courses included lab work. The standard ratio was three hours of lab for each two hours of lecture. A new chemistry building, containing fifteen laboratories, opened in 1933. The library held over seven hundred reference books and subscribed to more than forty journals in this field. The university published two journals.

II. Post-graduate Programs in Chemistry

Post-graduate programs were one of the most important characteristics of development in Chinese higher science education. It would keep the better students and teachers at home, and stimulate the development of education generally. Yenching was the first to award the master of science degree, beginning in 1922. And after 1934, six public and five private universities were authorized to grant master's degrees.

A program of post-graduate training in chemistry began in 1931 at Tsinghua University. The institute of chemistry enrolled eleven students and produced two notable chemists: Ma Tsu-sheng (Ph.D., Chicago 1938) and Chang Tsing-lien (Ph.D., Berlin, 1936) before 1937. Higher chemistry education got to a new stage with the establishment of post-graduate training.

III. Prosperity of Chemistry Research

Chinese chemists of the second generation concentrated not only on the fundamental problems of chemistry, but also on special topics with Chinese characteristics. For example, research on analyzing and extracting Chinese herbal medicine has always been one of foci for Chinese chemists. Table 5 shows a distribution of Chinese chemists' interests in 1930s.

Table 5   The Numbers of Papers in Various Branches of Chemistry, Published in *Chemistry* (化学) during 1934–37

Years	1934	1935	1936	1937
Volumes	1	2	3	5
Issues	1–4	1, 2, 4	1–6	1–3
General and inorganic chemistry	53	84	109	36
Physical chemistry	40	62	105	46
Analytical chemistry	44	57	58	2
Organic chemistry	114	57	118	40
Biological and physiological chemistry	55	116	176	68
Medical chemistry	40	50	118	43

During this period, the leading centers of chemistry research were still the universities. Table 6 demonstrates the superiority of the universities. *Journal of the Chinese Chemical Society* (中国化学会志) was the leading organ in this field during the years 1933 to 1937, showing that university-based scholars published the largest number of articles.

College seniors were required to do research theses and research was expected to educate the younger generation of chemists. In universities, education and research improved each other and then Chinese chemistry achieved a great deal in both aspects.

Table 6   Institutional Affiliation of Authors in the *Journal of the Chinese Chemical Society*, 1933–37

Institution	Number of Articles	%
Chinese universities		
Tsinghua	64	
Peking	44	
Central	19	
Nankai	6	
Other	14	
Total	147	66
Mission universities		
Yenching	22	
Other	16	
Total	38	17
Research institutions		
Academia Sinica	13	
Henry Lester Institute	5	
National Geological	4	
National Academy of Peiping	3	
Other	12	
Total	37	17
Total	222	100

Conclusion

Higher chemistry training was a miniature of Chinese higher education in science. Based on analyses made above, we get a clear result: the system of higher science education was formed in 1920s and standardized before 1937. Obviously, its success was due to three factors as follows:

Chinese People had Gradually Developed a Psychological Ability to Accept the Western Sciences by this Period

In Chinese history, education reform that encouraged scientific training began in the late Ch'ing dynasty, when the traditional civil service examination was abolished. From then on, the attitude to science of ordinary Chinese people had experienced a quite complex progress with the passage of time: knowing nothing — rejecting — getting in touch passively – understanding — researching. By the 1920s, Chinese had the ability to accept science as a part of the content of their daily lives. This situation fostered the higher science training engaged in by some Chinese.

Policies that Favored Science Education were Made in Succession

With the improvement of society and transformation of political power, more and more administrators realized the importance of science for the country and nation and adopted policies that favored utilitarian goals in the first quarter of 20th century, particularly the late 1920s. As a result, the scope and the quality of science training received a great boost throughout China. For example, the government encouraged students to learn science and engineering, while limiting access to literature, politics, humanities and the arts. In this period, education addressed national needs more than before.

Many Chinese Students Returned from Abroad and Carried Out Academic and Related Activities as Advocates, Organizers and Educators

From its first appearance to its period of maturity, higher science education in China developed rapidly, mostly because two generations of scientists returned to China. Back in China, they offered an increasingly academic curriculum, which provided a broad foundation of learning that would equip students to deal with the needs of the future. Meanwhile, they continued the work that they had begun as graduate students and designed new research projects according to actual Chinese conditions, and the best of such work was published in the world's leading scientific journals and cited by colleagues abroad. They had brought not only the model of the European and American universities, but the latest scientific achievements, ideas, experiences and methods of both teaching and research to China.

References

The department of chemistry in Peking University, 1910–95, edited by Faculty of Chemistry in Peking University, 1995.

The department of chemistry in Tsinghua University, edited by Department of Chemistry in Tsinghua University, 1996.

J. Reardon-Anderson, *The Study of Change, Chemistry in China 1840–1949*, Cambridge University Press, 1991.

A Shift in Academia Sinica's Mission in 1935

DUAN YIBING & FAN HONGYE

Institute of Policy and Management,
Chinese Academy of Sciences, Beijing

Introduction

President Cai Yuanpei (蔡元培) proclaimed the founding of Academia Sinica on 9 June 1928 in Shanghai, and stated its mission as probing the mysteries of the universe, making new discoveries, and creating inventions. He also said it would pursue sustained and quintessential, not hasty, progress.[1] Actually, in the early days of Academia Sinica, scientists selected suitable problems to research independently. However, President Cai issued a new statement on Academia Sinica's mission in April 1936, and scientists were called to study the important practical problems of China.[2] This paper discusses the background, process, and impact of the shift.

Ting's Advocacy of Laying Particular Stress on Practical Studies

In tracing this shift, we should begin with V. K. Ting (丁文江), the secretary-general of Academia Sinica in 1934–36.

Japan attacked the Manchurian capital, Mukden (奉天), on 18 September 1931, and subsequently conquered Manchuria. Scholars in Peking felt it their duty to do more for national security than research and teaching. Ting and other professors started to publish *Review of Independence* (獨立評論) in May 1932. Many topics were discussed in the journal, including northeast questions, prospects of Sino-Japan relations, and problems of China's security. Ting wrote sixty-four articles for *Review of Independence* in three years. He paid close attention to China's industrialization, studied Japanese internal affairs and diplomacy, and advocated a new-style autocracy in China. His main thought was to develop China's comprehensive strength and prepare in anticipation of an anti-Japanese war.[3]

In Nanjing, Qian Changzhao (錢昌照) proposed to set up a national security plan committee (國防設計委員會).[4] About 30 scholars, industrialists and experts, including Ting and Wong Wen-hao (翁文灏), were invited to discuss the proposal with Jiang Jieshi (蔣介石). The committee was set up on 1 November 1932 and Jiang was the chairman, Wong the secretary-general. The Committee of National Security Plan was renamed the Resources Committee (資源委員會) in April 1934.

V. K. Ting was an active member of the Committee. When Jiang presided over a meeting of all committee members in summer of 1933 at Lushan Mountain (盧山), some problems related to the anti-Japanese war were discussed and Ting offered a lot of suggestions, which made a good impression on Jiang ([4], p. 40).

After accepting President Cai's invitation to become the secretary-general of Academia Sinica in June 1934, Ting had a chance to show his management talent and patriotic ideals. Three reforms were soon put into effect ([3], pp. 74–76). The first was to change the research budget of Academia Sinica. Before Ting was the secretary-general, every institute in the academy was given ten thousand yuan per month over which the Academia itself had no special control. Ting asked the institutes to make smaller budgets and used the leftover funds to promote proposals, especially for programs with economic profit and related to national security. The second was to organize a committee to manage the Foundation of Academia Sinica. Since the Nanjing Government had allocated a sum of 500 thousand yuan to the Foundation of Academia Sinica in 1928, a lot of profit had been made. Also, some products such as steel, porcelain, and glass in the Institute of Engineering,[5] and teaching instruments in the Institute of Physics[6] made good profits, although these were not being handled in the Academia. Ting thought it necessary to organize a committee to manage Foundation revenue and all the income of the institutes. A part of the interest was put aside to be used for lectures, scholarships, investments, and buildings, which were accounted to the Academia's benefit. The third was to establish the First National Research Council (to be discussed further).

The purpose of these reforms, according to Ting in a paper of January 1935, *The Mission of Academia Sinica* (中央研究院的使命),[7] was "Academia Sinica certainly should lay particular stress on practical studies". On Ting's promotion, Academia Sinica began to shift from pure science to practical studies. The shift was carried out more completely after the First National Research Council (首屆評論會) was established in 1935.

Solutions to Important Problems Requested by the First National Research Council

The Nanjing Government adopted the Organic Rules of the National Research Council in May 1935. Thirteen presidents from national universities, Cai, and

Ting attended the election meeting on 20 June in Nanjing. Thirty members were elected from sixty candidates. Adding directors of ten institutes and the president of Academia Sinica, who were certainly members, forty-one members made up the First National Research Council. Ting took up a concurrent post as the secretary in the Council.

The first meeting of the Council was held on 7 September and received 7 motions. Ting's motion on promoting research on the principle of mutual aid was the keynote. The second meeting was on 16 April 1936 and Wong Wen-hao submitted a new keynote motion that Chinese scientists should attach importance to practical and important problems of the country and society (中國科學研究應對于國家及社會實際急需之問題特為注重案) ([2], p. 43). This was the first time that practical and important problems were put on Academia's agenda. What should or should not constitute an important problem and how to study them were considered over in the following three meetings. All these efforts aimed to link up the government, society, and academy, and compel Academia Sinica to shift studies from pure science to practical problems.

At the second meeting of the First National Research Council, Wong Wen-hao's motion was discussed and its draft resolution was that ([2], pp. 43–56):

(a) Special attention should be paid to important problems for the nation and society, including all subjects of science.

(b) Academia Sinica informs its institutes and other important academic units in China that important problems should be considered when they distribute research funds.

(c) The National Research Council gives a list of the most important problems in contemporary China, which are, if possible, to be solved in two or three years. Academia Sinica to distribute the list to related institutes after receiving them.

(d) Departments in the government may consult with Academia Sinica if they have problems needing to be solved.

(e) Every academic unit should report the results of the above-mentioned problems to the Council.

Accounting to the resolution, the Academia sent letters to some important research units, famous universities and government departments. They replied with active agreement. However, they wanted Academia to cooperate to research routine work and Academia replied that it could only do this partly. In fact, the proposals from members of the Council were more important and useful. Every group within the Council proposed important problems with research significance, content, and method. For example, the group of engineers proposed to study optical glass, an important material in telescopes, microscopes, and lenses. The proposal said that if war broke out, China would not be able to import optical glass, which might interfere with the Chinese military. All these problems involved many departments and were complicated in content. But the problems related to national security had priority.

Impacts of the Shift

After 1935, all the activities of institutes with Academia Sinica put national interests first. Research close to nation building was approved and obtained more support. Pure science was in a difficult position.

Another impact was increasing cooperation among different subjects, different institutes in Academia, and between Academia and the external world, especially the close relationship between the academy and government. A new and greater degree of contact and even control became one of the characteristics of modern Chinese science.

Conclusion

In conclusion, Academia Sinica had a mission shift in 1935. Practical studies and important problems were paid more attention to and close relations between the academy and government became a general tendency in Chinese scientific society.

Notes

[1] Cai Yuanpei, Foreword in Monthly Bulletin of Academia Sinica, January 1929. Also in Gao Pingshu, A Detailed Chronology of Cai Yuanpei (*Cai Yuanpei nianpu changbian*), Beijing: People's Education Press, 1998. III (1), pp. 357–58.

[2] Cai Yuanpei, Outline of Academia Sinica's Research. The First Report of the First National Research Council (*guoli zhongyang yanjiuyuan shoujie pingyihui diyici baogao*), 1937, p. 108.

[3] Lei Qili (ed.), An Impression on V. K. Ting (*dingwenjiang yinxiang*). Shanghai: Xuelin Press, 1997, pp. 74–76.

[4] Qian Changzhao, The Recollections of Qian Changzhao (*qian changzhao huiyilu*). Beijing China Culture and History Press, 1998, pp. 36–37.

[5] Zhou Peide, A Biographical Sketch of Zhou Ren. Historical Material of Chinese Metallurgy (*zhongguo yejin shiliao*), 1987 (2), pp. 71–82.

[6] The General Report of Academia Sinica in the 23rd year of ROC (*guoli zhongyang yanjiuyuan ershisan niandu zongbaogao*), 1934, p. 23.

[7] Seen Tao Yinghui. The Draft History of Academia Sinica (*zhongyang yanjiuyuan shi chugao*). Taipei: Academia Sinica, 1988, p. 32.

Enhancing China's International Prestige: The 1948 Sino-American Scientific Expedition to Mount Amne Machin

CHEN SHIWEI

Lake Forest College, Lake Forest, Illinois, U.S.A.

This paper investigates a previously unknown story, the Sino-American scientific expedition to Mount Amne Machin (*Jishishan*) in northwest China in 1948. This story is extremely interesting because it highlights the significance of Academia Sinica, Republican China's leading institution of advanced research, and its post-war scientific cooperation with the international scientific community at a critical juncture in the Chinese Revolution. The consequence of the 1948 Sino-American scientific expedition was also extraordinarily important, because it not only politically discouraged China's postwar national aspiration to enhance its prestige as a scientific Great Power in East Asia, but also served as a catalytic agent for the growth of Chinese nationalism, which eventually reached its peak after the 1948 event.

Initiation of the Expedition

For centuries, the barriers of Mount Amne Machin, the "Great Ancestor of the Yellow River" in Qinghai Province, had remained one of Asia's great exploratory challenges.[1] Located in remote Western China near the fountainhead of the Yellow River, the Amne Machin Range was celebrated for its harsh weather, perilous peaks, and perennial snow. Since the 1920s, scientists, adventurers and mountaineers from Europe and the U.S. had attempted to conquer the mountain, but none of them had achieved the goal with success. In 1929, Dr. Joseph Rock, noted botanist exploring for the American National Geographic Society, reached a point about 50 miles east of the great peak after crossing the Yellow River at Radja Gomba. As the only foreigner ever to have seen Amne Machin at close range, he made a hypothesis that "[the peak of Mount Amne Machin] is one of the world's greatest peaks, probably more than 25,000 feet in height".[2] Since the world's highest peak, Everest, at that time was calculated to be only 29,141 feet, Dr. Joseph Rock's narrative on the height of Mount Amne Machin became a challenge that fascinated subsequent adventurers.

In the 1940s, *Life* magazine decided to initiate an expedition to Amne Machin and made an agreement with Dr. Bradford Washburn, a world-renowned expert on mountains working at the Boston Museum of Science.[3] Celebrated for his explorations to Alaskan mountains, Dr. Washburn was a leading mountain photographer, who pioneered the use of ultra-high-frequency radio for field communications in the 1940s. Questioning the theory that Amne Machin was as high as Everest, Dr. Washburn personally agreed that an expedition "to scour the unmapped region from Minya Konka to the big bend of the Yellow River" would be worthwhile to study the glaciers, geology, and rivers in the vicinity of the great mountain.[4] Thus, finding out the true altitudes of the major peaks in the Amne Machin range would only be a small part of the overall project.

In late 1947, *Life* succeeded in interesting Milton Reynolds, a millionaire ballpoint pen manufacturer from Chicago, in becoming a sponsor of the expedition. Born in Prussia in 1892 and having immigrated to the United States with his parents, Milton Reynolds was celebrated for his introduction into the American market of the ballpoint pen, a new writing tool invented by Hungarians Laszlo Biro and his brother, George. When the first ballpoint pen went on sale on October 29, 1945, at Gimbel's Department Store in New York City, its price was $12.50. By the time he walked away from the pen business in early 1948, Reynolds had accumulated over $5 million despite federal income taxes of 70% which existed at that time![5] Reynolds, like most American businessmen during that time, had an old dream of opening the China market. The exploration of Mount Amne Machin, inevitably, would provide him with a golden opportunity to make publicity. It was because of his ambition to develop ballpoint pen manufacture in China that Mr. Reynolds decided to sponsor the expedition.

In January 1948, Dr. Washburn and Reynolds flew to China to negotiate an agreement with the Chinese Government permitting the expedition team to bring a huge reconditioned B-24 bomber to China and to make a series of reconnaissance flights over Western China and the Northeastern part of Tibet.[6] When they landed in Shanghai, they discovered that this country was deeply distressed by two serious problems: the Communist movement and inflation.[7] In 1948, the Nationalist government was suffering from unprecedented political and financial problems. The renewal and extension of the civil war between the Nationalists and the Communists had been going on for several years, producing a disastrous political situation. On the other hand, due to the gravity and complexity of social problems which were already rampant during the Sino-Japanese War and to mistakes made by the government, the galloping inflation had become completely uncontrollable. Inflation not only did permanent damage to Nationalist prestige by creating speculation and corruption, but also threatened all those whose incomes did not keep up with rising costs, especially the professional classes such as the scientists.

On January 23, Dr. Washburn and Reynolds moved to Nanjing to visit their future expedition partner, Academia Sinica, the highest scientific research institution in China.[8] At Academia Sinica, they were warmly received by Dr. Sa Bendong, eminent physicist and the Secretary-General of Academia Sinica. In 1948, Dr. Sa was the Director of the Chinese National Research Council, in charge of the nation's scientific research and coordination. As a technocrat, Sa had a vision that the Amne Machin expedition would serve as an opportunity for Academia Sinica to develop its geographic research and establish a scientific relationship with the American scientific community.[9] This vision was shared by the Nationalist Government, who observed the expedition as a great chance to enhance China's international prestige and to gain, more importantly, military and political support from the U.S. in its war against the Communists. It was under these auspices that the Sino-American scientific expedition to Mount Amne Machin was initiated.

On January 19, 1948, a joint committee named the "China Committee of the Amne Machin Expedition" was officially organized under the auspices of Academia Sinica. Participants of this meeting consisted of the representatives from various government, military, and scientific organizations, including the Ministry of Defence, the Air Force, the Ministry of Foreign Affairs, the Ministry of Education, Academia Sinica, the Institute of Geographical Research, and the National Geological Survey.[10] Between January and February, the China Committee held a series of meetings with the American team, represented by Dr. Washburn, at Academia Sinica. As a result, a detailed agreement stipulating the expedition's objective and obligations was made. According to this agreement, the main objective of this Joint Expedition was to explore from the air the Amne Machin Mountains and the region between these mountains and Minya Konka located between longitudes 96° and 103° east and latitudes 29° and 36° north. To accomplish this objective, the Chinese Government should: (1) allow the American team to bring 30,000 gallons of gasoline tax-free into China for the expedition's use; (2) permit all the expedition team's equipment to pass through customs without examination; (3) allow the expedition's airplane to make all necessary landings at Chinese airports, taking Chengdu as its primary base and Lanzhou as the secondary base; landings en route also might be made at Shanghai, Nanjing, Hankou, and Chongqing; (4) in exchange, the American team would provide a converted B-24 bomber, which would be equipped with state-of-the-art radar and aerial cameras to take photographs of Amne Machin from above and to measure the mountain's height, and Chinese scientists would be allowed to board the airplane for each trip.[11] On the February 11th, the Executive Yuan of the Nationalist Government officially ratified the agreement.

Once the Agreement became effective, the Amne Machin Expedition was a front-page story in all the leading Chinese newspapers. On January 26, the *China Press* of Shanghai published a picture of Madame Chiang Kai-shek receiving Reynolds and Dr. Washburn at her home.[12] This picture further ensured the Government's support for the expedition and thereby aroused a hot debate. There were basically three different positions in the Chinese media. The first opinion affirmed that the expedition was a "distinguished pioneering work" which would eventually benefit China's scientific research by revealing the secrets of Mount Amne Machin.[13] Expressed by the China Committee of the Amne Machin Expedition and sustained by the government, this opinion reflected the most optimistic view. A second opinion, however, was completely different from the first one, and strongly questioned the expedition's scientific value and the priority given it. This was put forward by a group of Chinese scientists who were outsiders to the expedition, but followed the development of the situation with close interest. Their primary argument was that "the isolated fact of the height of Amne Machin, though alluring to some millionaire, does not promise adequate scientific information so as to warrant the diversion of the hard-pressed Chinese geologists to its exploration at the present time when they are carrying out their own projects such as the study of the glaciation of the lower Yangtze basin, the completion of a geological map for China, ... etc."[14]

The third opinion was the most peculiar one. Given by some officials from the Ministry of National Defense, it was deeply concerned that the Sino-American expedition might reveal "China's national defense secrets" by allowing foreigners to explore China's Northwest.[15] In 1948, there was a rumor that the real motive of the Reynolds' expedition to Amne Machin was to survey China's uranium deposits from a high altitude, on behalf of an American secret military agency, for purposes of making atomic weapons.[16] Interestingly enough, this view also was shared by Radio Moscow and the Chinese Communists.[17] The Chinese Communists, for example, charged that the expedition was a "deep penetration of the aggressive arms of American Imperialism into the Western parts of our country".[18] It was later announced by the Communist broadcast that "the death sentence is recommended at once for any of the imperialists and traitors aboard the expedition's plane if they chance to fall into Communist hands".[19] It was under these circumstances that the expedition finally raised its curtain.

The Expedition Aborted

On March 7 at 3:24 p.m., Reynolds and the American expedition mission arrived in Shanghai in a used B-24 bomber. Members of the expedition team included Dr. Bradford Washburn, Milton Reynolds, Dr. Walter McKay (radar expert of the Massachusetts Institute of Technology), Dr. Richard Goldthwait (geologist of Ohio State University), Grant Ross (photo lab expert from Boston University), Bill Odom (Reynolds' private pilot), Tex Sallee (co-pilot), Julian Levi (Vice President of Reynolds' Pen Company in Chicago), and Philip Wootton (representative of *Life* magazine).[20] An eye-catching title, "Reynolds Boston Museum China Expedition", was prominently painted on the B-24 bomber. With the arrival of the American team in Shanghai, the publicity for the expedition reached its peak. In the following days, Reynolds was obsessed by his ballpoint pen advertising. He gave away free pens on a Shanghai street and spent much time in his hotel "setting up a 10,000-pen assembly line".[21] He even held a press conference in Peiping, announcing that he would soon establish a branch of his ballpoint pen factory in China. "Managed and worked by Chinese", Reynolds proclaimed, "this factory will manufacture 400 million pens and the majority of its earnings will be donated to Madame Jiang Jieshi for her charities and the New Life Movement".[22]

While Reynolds was concentrating on making publicity for his pen business, Dr. Washburn disappointingly discovered that these was "plenty of trouble ahead", when he took his first glimpse of the inside of the airplane:

> The camera hatch is in the side of the nose with NO HEAT at all. There is no vertical hatch at all. So all mapping is automatically impossible. Only 5 "demand" oxygen masks (all in front). The cabin has only 4 outlets of the old free-flow type and the whole O system leak badly. No throat microphone. No electric flying suits, no negative envelopes — BUT 750 POUNDS of pens which M.R. is already distributing freely to everyone (not the poor children!).[23]

To make things worse, people discovered that the B-24 bomber was heavily leaking oil and had to be sent back to Los Angeles for repairs. At that time, a team of Chinese scientists, together with two American scientists, had gone ahead to Lanzhou to make preparations for the expedition. On March 31, the B-24 returned to Peiping again after being repaired in Los Angeles and was finally ready for takeoff for Amne Machin. At 8:40 a.m., the plane started its engine and began slowly to move toward the runway.

But at this moment, an accident happened that eventually aborted the whole expedition. Dr. Washburn gave a vivid picture of this accident in his diary:

> Well, this should have been April Fool's Day! At 8:56 AM this plane's right wheel collapsed and she came to a rest almost on the tip of the right wing. Odom had just started for Lanchow which gave us a perfect weather report. He turned too sharply taxying from the brick apron (where we've been parked) over to the concrete runway. He let his right wheel get off the bricks. It sank a bit into the dirt [and] that was the end.[24]

The accident was a big frustration to everybody, but what happened in the following days was even worse. After estimating that it would take weeks to repair the damaged plane, Reynolds, without consulting with his Chinese partners and American colleagues, immediately announced that the expedition was terminated. On April 1, Reynolds and his "Explorer" landed in Shanghai on three engines, leaving behind him in Peiping an unpaid hotel bill of over 270 million Chinese *yuan*.[25] On April 2, the "Explorer" mysteriously and illegally left Shanghai at 6:00 a.m., without advance notice of its intention and destination. But it was an even bigger surprise when the airplane returned to Shanghai at 6:45 p.m the same day. Asked where the plane had flown during the thirteen missing hours, Reynolds explained that he and his crew were on their way to the United States via Calcutta, but after a few hours flight he suddenly discovered that his visa for Calcutta, India, "had expired".[26] The mysterious flight of the "Explorer" immediately caused the China Committee and the American scientists to suspect that Reynolds had already flown over Amne Machin by himself.[27]

On the evening of April 2, Zhu Jiahua, Acting President of Academia Sinica, on behalf of Academia Sinica, sent a telegram to the National Bureau of Civil Aviation, asking for the authority to detain the "Explorer".[28] Meanwhile, in a press conference held in Nanjing, Sa Bendong declared that "this matter has now become a diplomatic incident".[29] Following Academia Sinica's petition, the "Explorer", stationed at the Longhua Airport in Shanghai, was under the custody of armed guards. Under these circumstances, the ballpoint pen millionaire realized that he had to hurry to Nanjing to explain to the China Committee of the Amne Machin Expedition "why he started back to the United States without the formality of bidding them farewell".[30]

In the morning of April 3, an "inquest" into the demise of the Amne Machin expedition took place at Academia Sinica in Nanjing. "Chain-smoking cigarettes in an amber holder", Reynolds was asked to relate in detail to the China Committee all the circumstances of the "Explorer"'s rapid and unpredictable movements during the past week, and in particular to describe his "flight to Calcutta" yesterday.[31] Reynolds admitted that he was guilty of "negligence, discourtesy and lack of consideration" for his Chinese co-sponsors, but he insisted that he had made no attempt to fly to Amne Machin the previous day.[32] Reynolds' testimony, however, was an absolute lie. A few months later on October 7, 1948, Washburn ran into T. Carroll Sallee, Reynolds' flight engineer, in Boston. Sallee proved to Dr. Washburn that Reynolds did attempt to conduct the "sneak" flight over Amne Machin on April 2; it was only because of the bad weather around Amne Machin that Reynolds had to give up his plan and turn back to Shanghai.[33]

On the evening of April 3, after promising that his plane would not leave China until all the Chinese members of the expedition still in Lanzhou had returned to Nanjing, Reynolds was allowed to return to Shanghai.[34] The morning of the following day, Reynolds summoned a taxi and picked up the three flight crew members. The driver was instructed to proceed to Longhua Airport. At 1:45 p.m., Reynolds and his flight crew reached the Airport. At customs, the three flight crew were allowed to board the plane but Reynolds himself was blocked. At that time, there were three armed Chinese guards keeping sentry beside the airplane. Reynolds told the airport manager that he wanted to start the engine to test whether it was still working well. In addition, he said he needed to collect a batch of pens from the airplane for some friends. Since nobody suspected that Reynolds would take off, the millionaire was allowed to enter the airplane without delay. A few minutes later, the engine started hurriedly. Then, Reynolds went to the door of the B-24 bomber and threw several handfuls of ballpoint pens to the guards. "While the guards rushed to get the pens, the airplane took off!"[35] Later on, according to a detailed report from Academia Sinica to the U.S. Consulate General, "the plane, without contacting the [control] tower rushed from its parking area and entered the N-S main runway at the middle of its length and, ignoring all safety rules, took-off downwind without bothering [to taxi] to the end of the strip".[36]

Realizing Reynolds' venture to escape, the Chinese Air Traffic Control Center (CATC) immediately instructed the airport tower to contact the plane, but the tower got no response. At 4:55 p.m., when the intentions of Reynolds' party became obvious, the CATC radioed the American Far East Air Force Headquarters in Tokyo for information. The CATC was informed that the "Explorer" had landed in Tokyo.[37] In Shanghai, both Dr. Washburn and the U.S. Consulate General were horrified by what had happened. The U.S. Consulate General telegrammed General MacArthur in Tokyo, asking him to force the airplane to return to China. The Chinese government also requested Reynolds' return to Shanghai. MacArthur, however, declined all requests and "let Reynolds get off unscathed and without a passport ... to Midway".[38]

Reynolds' escape from Shanghai greatly embarrassed his American team members. On April 6, the Boston Museum of Science made a public statement on behalf of Dr. Washburn to Academia Sinica, announcing its repudiation of "any connection with the recent actions of Milton Reynolds in the Orient", and extending "its deep regrets to the Academia Sinica Committee which has worked so long and diligently to make the expedition a success".[39] But in Tokyo, Reynolds told the press a different story. The millionaire described his escape from Shanghai as "a melodramatic escape by flinging gifts of ball-point pens at gun-brandishing Chinese guards".[40] Reynolds related: "I got in [to the airplane] and threw out the pens — fifty of them — and stemmed the door ... I lay on the floor, holding the door shut ... We heard the Shanghai airport alerting a fighter squadron, so for the first 350 miles we stayed about twenty feet above the ground and water".[41] When informed of the anger of Sa Bendong over his breaking of the expedition agreement, Reynolds unblushingly bragged: "Of course I broke every promise I made to him in China. I always will break promises I make under duress and the shadow of tommy-guns at my back."[42]

Conclusion

In his study of China's foreign relations in the Republican era, William C. Kirby argues that the Republican era was a "high tide of internationalism", during which the Nationalist government achieved a series of accomplishments in its political prototypes, military persuasions, cultural and economic internationalism, and educational establishments through an "independent diplomacy". "By 1945," Kirby maintains, China "had become an important factor in the global balance of power and in the victory of the Allied coalition that it had joined ... as a partner more than a supplicant. Indeed China was formally now a 'great power', a status attained by performance in war and diplomacy, and confirmed by a permanent seat on the Security Council of the new United Nations."[43] While few would dispute China's achievement of what Kirby called "great power" in many fields of the postwar international relations, this conclusion, however, can hardly be employed to characterize the field of Chinese science and technology. Indeed the Chinese government's international status still remained in a relatively unfortunate situation in the post-war period, one in which China had no power to maintain sophisticated foreign relations as a nation of equal stature and depended heavily on the political and military assistance of the U.S. in its war against the Communists. The failure of the Mount Amne Machin Expedition in 1948 is an excellent example of how hollow China's "great power" status was and of the frustration felt by China's scientific community in the post-war era.

To be sure, the Chinese government did not lack the willingness or determination to pursue international scientific cooperation. Since its establishment in 1927, the Nationalist government had made a series of efforts to enhance China's political status and prestige in both internal and external affairs by promoting science and technology. These efforts can be clearly seen from the founding of Academia Sinica in 1927, the establishment of the National Research Council in 1935, the initiation of atomic research in the post-war era, the promotion of the Sino-American Amne Machin Expedition in 1948, and the organization of the Assembly of the Academician Council in the same year.[44] The stress on international solidarity and cooperation fitted appropriately into the Nationalist regime's post-war development framework, for it not only implied that other powers, especially the U.S., would assist the Nationalist regime in its task of postwar reconstruction, but also that otherwise dissident elements in the country might adopt discordant policies toward China to the detriment of political unity within the country.[45] As Jiang Jieshi claimed, the nation's "goal of international scientific cooperation" was an enduring theme of the Nationalist government's state building program.[46] The paradox, however, lay in the fact that the Nationalist regime lacked both the capacity to regulate and direct foreigners' activities within China and the ability to use international laws in the service of China's science and technology. Moreover, the unfavorable domestic situation made the Nationalist government so impotent that it was incapable of pursuing postwar international cooperation.

In 1948, China was at a turning point in its bloody internal strife as the Civil War between the Nationalists and the Communists entered the last stage after the failure of the Marshall Mission. The Nationalist government suffered from a series of disastrous military defeats, rapidly accelerating inflation, and the strong criticism of disheartened intellectuals who hoped for greater reform and freedom. Washington D.C., on the other hand, had decided in its foreign policy to rebuild Japan as an alternative base for American power in East Asia since it would entail far less commitment of resources and offer more assurance of success.[47] This situation made it extremely difficult for the Nationalist regime to pursue its international cooperation and national reconstruction projects. It was under these circumstances that the Sino-American Mount Amne Machin Expedition was initiated, which eventually provided the Nationalist government with a dear opportunity to enhance its international prestige, redeem its domestic errors, and, more importantly, demonstrate its "goodwill" to the U.S. in order to gain more political and military support for its war against the Communists.

During the preparation and organization of the expedition, Academia Sinica played a key role in cooperating with the American scientists. Sa Bendong, as the Secretary-General of Academia Sinica and the Chairman of China Committee, devoted a great deal of time and energy to planning and executing the expedition's scientific exploratory program. Unpublished data known only to Chinese geologists and geographers were made available to the expedition. A dozen meetings and conferences were held to discuss the details of the expedition. A number of government agencies, including the Ministry of Defence, the Air Force, and the Ministry of Foreign Relations, were involved in the operation. Five Chinese scientists, three geologists, and two geographers, each a specialist in his field, were requested to put aside their immediate work to accompany the expedition. Even the salary of all Chinese personnel was paid by the affiliated institutions as usual, while all the expenses of additional Chinese participants above the number agreed upon were borne by the China Committee.[48] All of the goodwill made by the Chinese scientists, however, failed to prevent Reynolds from his unlawful practice. The expedition, therefore, eventually deteriorated after Reynolds' ridiculous escape from China, which made a huge impact on the Chinese scientific community.

First, it deeply wounded the Chinese scientists in their search for "goodwill" from the U.S. and gave the Communists the leverage to manipulate the political agenda. For a long time following the founding of the Nationalist government in 1927, there were the beginnings of contacts and exchanges with Western scientists and the emergence of a small number of intellectuals educated in the universities who were trained in Western science.[49] Associated with Academia Sinica, this group of scientists played a crucial role in portraying a beautiful image of Western society and promoting international scientific cooperation. In their search for international scientific cooperation, these scientists perceived the U.S. as the Number One partner and friend because of its advanced scientific development and generous assistance to China in World War II. Equally vital was the Chinese scientists' belief that the U.S. could play a leading role in helping China's pursuit of modern science and technology in the postwar era. This beautiful image, however, was completely broken by the 1948 event, in which a total fraud, Reynolds, insinuated himself into an unsuspecting Chinese scientific community.

After the 1948 incident, Chinese scientists were unwillingly thrown into a self-imposed dilemma. On the one hand, they wanted to protect China from external humiliation and, on the other hand, they sought to accommodate the U.S. In this unprecedented situation, many scientists, even some of those who shared Western political values and graduated from Western institutions, began to change their attitude toward the U.S., while a nationalist sentiment of Anti-Imperialism was generated by the Communists in China. Moreover, those who had been affected by inflation and the Civil War started to take this opportunity to vent their anger and discontent with the Nationalist government. This new situation not only slowed down the Nationalists' political reform process, but also snowballed into a social and psychological sentiment against the U.S. An article published by *Dagongbao* after Reynolds' escape in April 1948, for example, indignantly remarked: "Can we still believe that every person in America, the land of prosperous industry and commerce, is a man of generosity and willing to spend money without calculated aims?"[50] The answer was surely "No".

Second, while the misfortune of the Amne Machin Expedition in 1948 won the goodwill of neither Chinese scientists nor public opinion, it ironically deepened the enduring theme of Chinese Nationalism, which eventually reached its peak after the Communists took power. Nationalism, as the "moving force", was a constant factor in China's long revolutionary history.[51] It played an important role in modern China's advocacy of state sovereignty and international egalitarianism. It also initially displayed a strong anti-foreign resentment and xenophobic strain in opposing imperialism. In 1948, China was experiencing a rejuvenation of nationalism after a series of accidents occurred between Chinese and the U.S. Marines in China. First, on September 22, 1946, a Chinese rickshaw puller named Zang Dayaozi was beaten by an American naval enlisted man "in a street altercation" in Shanghai, and subsequently died of the injuries.[52] Three months later, a U.S. Marine was reported to have raped a Beijing University student, Shen Chong, on Christmas eve in Peiping.[53] On March 30, 1947, yet a third incident occurred in Qingdao, where Su Mingcheng, a Chinese rickshaw puller, was murdered by an American sailor outside a bar, possibly due to a dispute over the fare for the rickshaw ride. The Su Mingcheng incident, together with the previous two affairs, inspired a great wave of Chinese nationalism. Angered by the acquittal, the Chinese local government and the Chinese Foreign Ministry contended that the U.S. government did not fulfill its obligations in carrying out justice, while tens of thousands of Chinese students, together with pullers, factory workers, and office clerks, poured out to the streets to launch a prolonged protest against the U.S. troops in China.

It was under these circumstances that the Mount Amne Machin incident, a case showing China's sense of inferiority and resentment *vis-à-vis* foreign expertise, occurred. It inevitably caused already intensified Sino-American relations to deteriorate. Since the American military presence was the most visible foreign influence in post-war China, any outburst of nationalist feelings was, inevitably, directed against American troops and the U.S. policy of which they were the most obvious executors. The tension in the relationship between China and the U.S. was therefore unavoidable, like a power keg waiting for a match. The stupidity of Reynolds and his expedition drama had eventually provided the spark. This time the immense weight of historical grievances against Western imperialism was borne by the Chinese scientists and intellectuals. It was more rational and more powerful. As an article in *Dagongbao* argued:

> China has always been the paradise of adventurers. The lowliest foreigners coming to this country became first-class gentlemen and messenger boys would turn into ambassadors. This is entirely due to the Chinese worship for foreigners, and so we cannot blame Mr. Reynolds, who has made money from his exploration in China, for his sudden departure ... The Chinese should by this time awaken from their dream and realize the true aim not only of Mr. Reynolds but of most American businessmen who worship gold ... If China has proper ambitions, she must not only help her scientists to complete this exploration work with success, but will have to build up a prosperous and powerful nation so that she will not permit such avaricious and money-loving merchants to enter this country and play such a ridiculous comic act here ... Henceforth the Chinese must awaken from their superstition of worshipping foreigners; they must cultivate self-respect and confidence in themselves.[54]

The Sino-American Mount Amne Machin Expedition started with the aim of enhancing China's prestige and ended up by harming China's dignity. But this incredible exploit succeeded in stirring up a lot of new interest in Amne Machin. On April 16, 1948, with the dust from Reynolds' antics barely settled, an airplane sponsored by the Chinese Central Air Transport Corporation flew over the mountain range and proved that the highest peak of the Amne Machin was no higher than 19,000 feet.[55] This report put a temporary stop to the expedition of Amne Machin. Since *Life* magazine, the Boston Museum of Science, and the Chinese Authorities soon inflicted the worst possible punishment on Reynolds by deciding never to publish a single word on the expedition till after Reynolds' death, the episode of the Amne Machin expedition eventually became a forgotten story.[56]

Notes

[1] Bradford Washburn, "Farewell speech at Academia Sinica, April 10, 1948". Academia Sinica Archives. Nanjing: The Second National Historical Archives. Catalog No. 393, Vol. 28. Henceforth referred to as: *ASA: 393/28*.

[2] "Amnyi Machin Expedition", manuscript provided by Bradford Washburn.

3 Bradford Washburn was Director of Boston's Museum of Science from 1939 to 1980 and was Chairman of the Museum's Corporation from 1980 to 1985. He holds both A. B. and A. M. from Harvard University and was an instructor at Harvard's Institute of Geographical Exploration from 1935 to 1942. Although his life's work has been focused primarily on the founding and building of Boston's Museum of Science into one of the Nation's leading Museums, Dr. Washburn is also a noted cartographer, photographer and leading expert on Mt. Everest as well as on Alaska mountains and glaciers. He pioneered in the use of ultra-high-frequency radio for field communications in the 1940s. He has led numerous Alaskan exploratory expeditions and has published a large-scale map of Mt. McKinley. At the end of World War II, he received the U.S.A. Distinguished Civilian Award for his services in connection with the development and testing of cold-climate and high-altitude equipment for the United States Army Air Forces. On September 2, 1994 at St. Moritz, Switzerland, Dr. Washburn received the first King Albert Gold Medal for lives of "Outstanding Achievement in the Mountain World".

4 Bradford Washburn, "Amnyi Machin-1947–48", manuscript provided by Bradford Washburn.

5 *Ibid.*

6 Bradford Washburn, *Diary of 1948*, January 5, 1948. Provided by Bradford Washburn.

7 According to Dr. Washburn's diary, the exchange rate of currency on January 20, 1948 "is $217,000 Chinese to $1.00 US". Bradford Washburn, *Diary of 1947*, January 20, 1948.

8 Bradford Washburn, *Diary of 1947*, January 23, 1948.

9 Dr. Sa Bendong once told Dr. Washburn: "this is the first time in history that Chinese and American scientists have joined together on an expedition in China — this will really be a memorable deal if it can be pulled off". Bradford Washburn, *Diary of 1948*, January 23, 1948.

10 "*Shangtao meiguo leinuohua gongsi deng tuanti laihua tanxian rujing wenti huiyi jilu* (Minutes of the Meeting for the Matters of Reynolds' Expedition, January 19, 1948)", *ASA*: 393/29.

11 "*Shangtao meiguo leinuohua gongsi deng tuanti laihua tanxian rujing wenti huiyi jilu* (Minutes of the Meeting for the Matters of Reynolds' Expedition, January 23, 1948)", *ASA*: 393/28.

12 Bradford Washburn, *Diary of 1948*, January 26, 1948.

13 Wang Mingyang, "*You Lei Nuo tanxian shuoqi* (On Reynolds' expedition)", *Dagongbao*, February 3, 1948.

14 "Letter from Pen-Tung Sah (Sa Bendong) to the Editor of the North-China Daily News", *ASA*: 393/28.

15 "*Shangtao meiguo leinuohua gongsi deng tuanti laihua tanxian rujing wenti huiyi jilu* (Minutes of the Meeting the Matters regarding Reynolds' Expedition, January 19, 1948)", *ASA*: 393/28.

16 Liu Xian, "*Lun Leinuo tanxiandui zhi laihuan* (On Reynolds Expedition in China)", *Shenbao*, February 2, 1948, *ASA*: 393/150.

[17] Li Lie, "*Zhongguo jishishan tancetuan Li Lie baogaoshu* (Concluding report of the China Committee of the Amne Machin Expedition)", *ASA*: 393/26.

[18] "Special News Dispatched over the Communist Radio from South Shensi, China, Feb. 15, 1948", *Washburn Archives*.

[19] *Ibid.*

[20] Bradford Washburn, *Diary of 1947*, December 14, 1947.

[21] Bradford Washburn, *Diary of 1948*, March 12, 1948.

[22] "*Lei Nuo zai Peiping tan tanxian dongji* (Reynolds explains his expedition motives in Peiping, March 16, 1948)", *ASA*: 393/150.

[23] Bradford Washburn, *Diary of 1948,* March 7, 1948.

[24] Bradford Washburn, *Diary of 1948,* March 31, 1948.

[25] "Letter from Washburn to Reynolds, April 4, 1948", *ASA*: 393/28.

[26] "Odom Explains Flight Mystery, April 3, 1948", *ASA*: 393/28.

[27] "Academia Sinica Asks Detention of Reynolds", *ASA*: 393/28.

[28] "*Guoli zhongyang yanjiuyuan siyue erri mizi siliusan hao daidian* (Confidential Telegram from Academia Sinica to the National Bureau of Civil Aviation)", *ASA*: 393/27.

[29] "Request to Detain Reynolds and Party, April 2, 1948", *ASA*: 393/28.

[30] "Reynolds Admits Guilt, April 3, 1848", *ASA*: 393/28.

[31] *Ibid.*

[32] *Ibid.*

[33] "Amnyi Machin-1947–48", *Bradford Washburn Archives*.

[34] "Reynolds Admits Guilt, April 3, 1848", *ASA*: 393/28.

[35] *Ibid.*

[36] "Report from Academia Sinica to the U. S. Consulate General, April 10th, 1948", *ASA*: 393/27.

[37] *Ibid.*

[38] Bradford Washburn, *Diary of 1948,* April 5, 1948.

[39] "Statement to the Academia Sinica from the Boston Museum of Science", *ASA*: 393/28.

[40] "Reynolds Tells Story of Shanghai Escape, April 5, 1948", *ASA*: 393/28.

[41] "Reynolds Escapes to Tokyo by Ruse, April 5, 1948", *ASA*: 393/28.

[42] "Won't Reply Question on 'Sneak' Trip", *ASA*: 393/28.

[43] William C. Kirby, "The Internationalization of China: Foreign Relations at Home and Abroad in the Republican Era", *The China Quarterly*, No. 150, June 1997, pp. 433–58.

[44] Shiwei Chen, *Government and Academy in Republican China: History of Academia Sinica, 1927–1949*, Ph. D. Dissertation, Harvard University, 1998.

[45] Akira Iriye, "Japanese aggression and China's international position, 1931–1949", in John K. Fairbank and Albert Feuerwerker, eds., *The Cambridge History of China*, Vol. 13. Cambridge: Cambridge University Press, 1986, p. 541.

[46] "*Zongtong xunci* " (Speech of Jiang Jieshi at the First Assembly of the Academician Council, Academia Sinica, September 23, 1948), *ASA*: 393/136.

47 Immanuel C. Y. Hsu, *The Rise of Modern China*. Oxford: Oxford University Press, 1983, p. 638.

48 "Letter from Pen-Tung Sah (Sa Bendong) to the Editor of the North-China Daily News", *ASA*: 393/28.

49 Merle Goldman and Denis Fred Simon, "Introduction", in Denis Fred Simon and Merle Goldman, eds., *Science and Technology in Post-Mao China*. Cambridge: Harvard University Press, 1989, p. 6.

50 W. H. Chang, ed., "Through Chinese eyes", *Dagongbao*, *ASA*: 393/28.

51 Mary C. Wright, "Introduction: The Rising Tide of Change", in Mary C. Wright, ed., *China in Revolution: The First Phase, 1900–1913*. New Haven: Yale University Press, 1968, p. 3.

52 A memorandum from W.T. Kenny, U.S. Naval Attache to Commander Naval Forces Western Pacific, "Embassy Letter to Ministry of Foreign Affairs regarding the Trial of Edward Roderick, Coxwain, United States Navy, by General Court Marital", January 27, 1948, Box 6140, RG 313, NACP. Cited from Yang Zhiguo, *U.S. Marines in Qingdao: Military-Civilian Interaction, Nationalism, and China's Civil War 1945–1949*, Ph. D. Dissertation, University of Maryland, 1998.

53 Record of proceedings of a General Court Marshal regarding the case of William G. Pierson, Corporal, U.S. Marine Corps, January 17, 1947, Office of Judge Advocate General of the Navy, Washington Navy Yard. Cited from Yang Zhiguo, *U.S. Marines in Qingdao: Military-Civilian Interaction, Nationalism, and China's Civil War 1945–1949*, Ph. D. Dissertation, University of Maryland, 1998.

54 "Through Chinese Eyes, April 8, 1948", *ASA*: 393/28.

55 "*Jishishan tance jingguo* (The story of the Amne Machin expedition, April 16, 1948)", *ASA*: 393/150.

56 Pursued by the Internal Revenue Service, Reynolds soon fled to Mexico where he lived luxuriously until his death in 1976. Bradford Washburn, "Amnyi Machin-1947–48", provided by Dr. Washburn.

Glossary

Chengdu	成都
Chongqing	重慶
Dagongbao	大公報
Hankou	漢口
Jishishan	積石山
Lanzhou	蘭州
Liu Xian	劉咸
Longhua	龍華
Qinghai	青海
Sa Bendong	薩本棟
Shenbao	申報
Shen Chong	沈崇

Su Mingcheng	蘇明成
Xiamen	廈門
Zang Dayaozi	臧大咬子
Zhu Jiahua	朱家驊

References

Academia Sinica Archives, Nanjing: The Second National Historical Archives, hereafter referred to as *ASA*.

"Academia Sinica Asks Detention of Reynolds", *ASA*: 393/28.

"Agreement Between the Reynolds-Boston Museum of Science China Expedition, 1948, and The China Committee of the Amne Machin Expedition, 1948", *Bradford Washburn Archives*. Provided by Bradford Washburn.

Akira Iriye, "Japanese aggression and China's international position, 1931–1949", in John K. Fairbank and Albert Feuerwerker, eds., *The Cambridge History of China*, Vol. 13. Cambridge: Cambridge University Press, 1986, p. 541.

"Amnyi Machin Expedition", manuscript provided by Bradford Washburn.

Bianco, Licien, *Origins of the Chinese Revolution, 1914–1949*. Stanford: Stanford University Press, 1971, p. 161.

Chang, W. H., ed., "Through Chinese eyes", *Dagongbao*, *ASA*: 393/28.

Chen Shiwei, *Government and Academy in Republican China: History of Academia Sinica, 1927–1949*, Ph. D. Dissertation, Harvard University, 1998.

————, "Interviews with Dr. Bradford Washburn, July 2, 1998, March 12, 1999". Boston Museum of Science.

Goldman, Merle and Simon, Denis Fred, "Introduction", in Denis Fred Simon and Merle Goldman, eds., *Science and Technology in Post-Mao China*. Cambridge: Harvard University Press, 1989, p. 6.

Gostony, Henry B., "The Incredible History of Milton Reynolds and His Ballpoint Pens", *Pen World*, Vol. 8, No. 1, September/October, 1994, p. 18.

"*Guoli zhongyang yanjiuyuan siyue erri mizi siliusan hao daidian* 國立中央研究院四月二日秘字四六三號代電 (Confidential Telegram from Academia Sinica to the National Bureau of Civil Aviation)", *ASA*: 393/27.

Hsu, Immanuel C. Y., *The Rise of Modern China*. Oxford: Oxford University Press, 1983, p. 638.

Huang Jiqing 黃汲清 and Wang Chaojun 王朝鈞, "*Jishishanqu tanxian lueshi* 積石山區探險略史 (Brief history of the Mount Amne Machin expeditions)", *ASA*: 393/150.

"*Jishishan tance jingguo* 積石山探測經過 (The story of the Amne Machin expedition, April 16, 1948)", *ASA*: 393/150.

"*Jishishan tacetuan zhongguo weiyanhui huiyi jilu* 積石山探測團中國委員會會議記錄 (Meeting Minutes of the China Committee of the Amne Machin Expedition, January 29, February 3, 7, 11, 24, March 4, 1948)", *ASA*: 393/28.

Kim, Samuel S., *China, the United Nations, and World Order*. Princeton: Princeton University, 1979, p. 47.

Kirby, William C., "The Internationalization of China: Foreign Relations at Home and Abroad in the Republican Era", *The China Quarterly*, No. 150, June 1997, pp. 433–58.

"*Lei Nuo zai Peiping tan tanxian dongji* 雷諾在北京談探險動機 (Reynolds explains his expedition motives in Beiping, March 16, 1948)", *ASA*: 393/150.

"Letter from Pen-Tung Sah (Sa Bendong) to the Editor of the North-China Daily News", *ASA*: 393/28.

"Letter from Washburn to Reynolds, April 4, 1948", *ASA*: 393/28.

Lin Hongxi 林鴻禧, "Sa Bendong 薩本棟 (Biography of Sa Bendong)", *Zhongguo xiandai kexuejia zhuanji* 中國現代科學家傳記 (Biographies of Contemporary Chinese Scientists), Vol. 1. Beijing: Kexue chubanshe, 1991, pp. 126–31.

Li Lie 里烈, "*Zhongguo jishishan tancetuan Li Lie baogaoshu* 中國積石山探測團里烈報告書 (Concluding report of the China Committee of the Amne Machin Expedition)", *ASA*: 393/26.

Liu Xian 劉咸, "*Lun Leinuo tanxiandui zhi laihuan* 論雷諾探險隊之來華 (On Reynolds Expedition in China)", *Shenbao* 申報, February 2, 1948, *ASA*: 393/150.

Pye, Lucian W, "How China's Nationalism was Shanghaied", in Jonathan Unger, ed., *Chinese Nationalism*. New York: M. E. Sharpe, 1996, p. 110.

"Report from Academia Sinica to the U. S. Consulate General, April 10th, 1948", *ASA*: 393/27.

"Request to Detain Reynolds and Party, April 2, 1948", *ASA*: 393/28.

"Reynolds Admits Guilt, April 3, 1848", *ASA*: 393/28.

"Reynolds Escapes to Tokyo by Ruse, April 5, 1948", *ASA*: 393/28.

"Reynolds Saw Peak, Ohio Prof. Suspects", *ASA*: 393/28.

"Reynolds Tells Story of Shanghai Escape, April 5th, 1948", *ASA*: 393/28.

Rowan, Roy, "Chinese Mountain High", *Smithsonian*, Vol. 28, No. 12, March 1998, pp. 121–34.

"*Shangtao meiguo leinuohua gongsi deng tuanti laihua tanxian rujing wenti huiyi jilu* 商討美國雷諾華公司等團體來華探險入境問題會議 (Minutes of the Meeting for the Matters of Reynolds's Expedition, January 19, 23, 1948)", *ASA*: 393/29.

"Special News Dispatched over the Communist Radio from South Shensi, China, Feb. 15, 1948", *Washburn Archives*, provided by Dr. Washburn.

"Statement to the Academia Sinica from the Boston Museum of Science", *ASA*: 393/28.

"Through Chinese Eyes, April 8, 1948", *ASA*: 393/28.

Unger, Jonathan, "Introduction", in Jonathan Unger, ed., *Chinese Nationalism*. New York: M. E. Sharpe, 1996, p. xiii.

Wang Mingyang 汪名揚, "*You Lei Nuo tanxian shuoqi* 由雷諾探險說起(On Reynolds's expedition)", *Dagongbao* 大公報, February 3, 1948.

Washburn, Bradford, "Farewell speech at Academia Sinica, April 10, 1948", *ASA*: 393/28.

_____ , "Amnyi Machin-1947–48", Private Archives of Bradford Washburn.

_____ , *Diary of 1947*. Provided by Dr Bradford Washburn.

"Won't Reply Question on 'Sneak' Trip", *ASA*: 393/28.

Wright, Mary C., "Introduction: The Rising Tide of Change", in Mary C. Wright, ed., *China in Revolution: The First Phase, 1900–1913*. New Haven: Yale University Press, 1968, p. 3.

Yang Zhiguo, *U.S. Marines in Qingdao: Military-Civilian Interaction, Nationalism, and China's Civil War 1945–1949*, Ph. D. Dissertation, University of Maryland, 1998.

Yang Zhongjian 楊鐘鍵, "*Jishishan tancetuan zhi zhanwang* 積石山探測團之展望 (Prospects for the Mount Amne Machin Expedition)", *ASA*: 393/150.

Zeng Shiying 曾世英, "*You tankan jishishan tandao cetu, cangtu yu dutu* 由探勘積石山談到測圖，藏圖與讀圖 (Discussions about the Mount Amne Machin Expedition and maps)", *ASA*: 393/150.

"*Zenyang tance jishishan* 怎樣探測積石山 (How to measure the Mount Amne Machin)", *Dagongbao* 大公報, March 3, 1948, saved in *ASA*: 393/150.

"*Zhongmei jishishan tacetuan hezuo yuewen* 中美積石山探測團合作約文 (Agreement of the Sino-American Amne Machin Expedition)", *ASA*: 393/28.

"*Zongtong xunci* 總統訓詞" (Speech of Jiang Jieshi at the First Assembly of the Academician Council, Academia Sinica, September 23, 1948), *ASA*: 393/136.

When State and Policies Reproduce Each Other: Making Taiwan a Population Control Policy; Making a Population Control Policy for Taiwan[*]

KUO WEN-HUA
Program in Science, Technology and Society
Massachusetts Institute of Technology (M.I.T.)

Problematizing Taiwan's Population Control

Over the past three decades, rapid fertility transitions have been observed in many developing Asian countries. Researchers have credited the state's role in this process. When economic policy makers noted an upsurge in births, they considered this population explosion as a potential obstacle and designed programs to deal with the threat. Although all the programs had to be carried out by health administrations, they were considered a part of the states' strategy for development, and had a clear economic goal. The degree to which they could be imposed depended, of course, on the degree of authority the state could exercise (for examples, see Freymann:14–16 and Kwon 1997: 3–4, 20–21).

The above process, which I will call the "Asian Population Control Model", echoes previous studies on the relationship between states and scientific policy. It can be as complicated as what William H. McNeill calls the "Military Industrial Complex", in which the state leads in the making of industrial policies. Using competition between the strong powers of 19th-century Europe as his examples, McNeill argues that technologies could not be considered as entities in themselves. They were parts of an interactive system in which the state comprised the dominant part (Chapter 8).[1] This is clear in the case of Third World countries, where science and technology are perceived to be tools for catching up and for helping to solve domestic problems (Clarke 1985). Although the strategy applied may vary from that of First World countries, the state still plays an active role in this process (for example, see Stepan 1981).[2]

Taiwan's program of population reduction has been singled out as exemplary. The effective reduction of the birth rate, combined with the country's rapid development, appear to give proof of its success; as Sun and Soong argue, the shift from high to low fertility was most rapid during 1965–69, when the Family Planning for All of Taiwan Project was instituted and applied (Sun 1990 and Sun & Soong 1979). However, I will argue that two major problems concerning Taiwan's political status and population situation complicate these statements, making Taiwan an exception from the start. First, in the context of population policy, Taiwan's problem had a different orientation. Its over-population was not the result of a natural increase but due to political immigrants. Second, in the context of global politics, Taiwan was exceptional among Asian countries due to the ambiguity of its political status. The *Kuomintang* (Nationalist party, KMT) did not perceive Taiwan's over-population as a problem. Since Taiwan was not considered a well-defined demographic territory, over-population held no significance for the KMT. In its mind, Taiwan with Mainland China made up an ideological Republic of China (ROC); the claim of "fighting back to the Mainland" was the KMT's ideological tenet.[3]

This paper will critically review the formation of Taiwan's population control policy since the 1950s. Given the facts of Taiwan's unique population problem orientation and of its political status, it will attempt to make historical sense of why Taiwan was forced to "fit" itself to this standard solution. By discussing related international and domestic issues, I will argue that Taiwan's population problem and that of its political status were interlocked; they had to be "solved" together. Taiwan could not have been considered as a state until the advent of the Cold War period. The logic of the Cold War gave all countries a reference by which the terms "enemies" and "friends" were made clear. In this sense, Taiwan and its communist counterpart, the PRC, were put in opposing categories, and functioned independently as modern states. Only with this development did Taiwan's over-population became "visible" to the KMT regime. A state population control agenda, which matched US interests in foreign policy and its belief in rigorous calculations and models, could be introduced as the only way to save the state from its economic difficulties. In the end, the KMT not only made a population control policy for itself. Inadvertently, it made Taiwan a clear political territory in which practical statehood would be inevitable.

Taiwan's Over-Population in the Early 1950s

I will begin by examining the reasons why the Asian population control model is not adequate in Taiwan's case. This model presumes a stable population composition in which no mass migrations occurred during the years under discussion. In Taiwan, however, there were extreme demographic changes. The most important change was the arrival of political immigrants from Mainland China.[4] Between 1949 and 1954, an additional population of about 1.5 million, the major part of which consisted of young male soldiers led by the defeated KMT, entered this island. These people, who accounted for about 15% of Taiwan's inhabitants, soon became a heavy burden, and they gave Taiwan's over-population a different orientation (Table 1).

Table 1   Adjusted Demographic Data on 1950s Taiwan[5]

Year	Registered Taiwanese (1)	Estimated Immigrants from the Mainland = (2) – (1)	Adjusted Population in Total (2)	Immigrants in Percentage = (2) – (1)/(2)	Birth Rate (per thousand)	Population Growth Rate (per thousand)
1951	7,268,557	1,187,411	8,455,968	14.04	47.97	30.44
1952	7,478,544	1,234,820	8,713,664	14.17	45.82	32.00
1953	7,724,000	1,268,210	8,992,210	14.10	43.97	33.00
1954	7,983,087	1,308,596	9,291,683	14.08	43.00	36.41
1955	8,224,955	1,405,028	9,629,983	14.59	43.35	34.44
1956	8,444,965	1,516,716	9,961,681	15.22	41.70	34.65
1957	8,676,022	1,630,830	10,306,852	15.82	40.37	33.34
1958	8,943,399	1,707,105	10,650,504	16.02	39.22	33.19
1959	9240,465	1,763,508	11,003,973	16.02	38.99	31.40
1960	9,512,776	1,836,670	11,349,446	16.18	37.54	30.51

Source: Adopted from Executive Yuan, ROC 1976: 16 and 66, and Lee Dong-Ming 1969: 223.

According to the above data, the rise in the population and birth rates was not as great as was claimed by population experts (see Sun *et al.* 1979: 24–25). These numbers roughly match those of Korea in the late 1950s and early 1960s (Kwon 1977: 149–53 and Kwon 1993: 43–44). They were high, but not severely so. This increasing population rate certainly resulted from the incorporation of these immigrants into the denominator; however, these were the same people whose arrival made Taiwan's population in 1951 reach the same level it would have reached in 1956 if this unusual immigration had not occurred. Since Taiwan's over-population was not the result of a natural increase but due to these immigrants, it could not be technically solved by simply reducing the birth rate as did other Asian countries.

Here a comparison between Taiwan and Korea can help. While Korea experienced the same kind of political migrations — 646,000 North Koreans moved south while 286,000 remained held by the North — this imbalance was not felt profoundly, since the country had suffered from a loss of approximately 1.64 million of its original population due to the war (for details, see Kwon 1977: 203–204). While over-population was not a severe problem for Korea, Taiwan's situation, as a result of the enormous influx of people from Mainland China within a much more limited area, was much more serious. Although Taiwan and Korea had almost the same birth rate in the mid-1950s, Korea did not really feel the same impact because of its bigger territory (for Korea's total population at the time, see Kwon 1977: 12). Without any increase in births, Taiwan was already over-populated.

Yet the KMT refused to discuss the issues of over-population openly. For the KMT, any attempt to control Taiwan's population not only presented practical problems, but also created an ideological crisis: if Taiwan were to decrease its population, the KMT would not have enough manpower to fight against the PRC in the future; and, more importantly, it might give the impression of abandoning its hope to reclaim Mainland China. As observed by Hsu Chih-Chu (許世鉅), the early promoter of birth control in Taiwan and the Chief of the Rural Health Division of the Sino-American Joint Commission on Rural Reconstruction (JCRR), birth control was taboo because it badly contradicted the teaching of Dr. Sun Yat-sen, National Father of the ROC. Thus, when a booklet that suggested some traditional methods of contraception was published, he pointed out, it was soon prohibited by the KMT because it was "a communist plot to weaken the military forces" (Hsu: 7–8).

Scholars observed that, on the other hand, if Taiwan were to ignore its immediate over-population problem, it might risk immediate economic difficulties (Chen and Yin: 20–23). However, their worries were simply buried beneath the ideology of the ROC. If a way were to be found to ease Taiwan's over-population, the KMT would have insisted upon waiting "until we re-gain the Mainland" rather than immediately introducing an appropriate policy. How to identify Taiwan's political status and how to solve its urgent over-population problem became two facets of the same question and created a complicated deadlock for the KMT regime.

Breaking the Deadlock: Changing Ideas of Taiwan and Its Population

The above situation started to change when the Cold War came. Before the outbreak of the Korean War, America had a negative "hands-off" attitude toward the KMT on Taiwan. This policy changed dramatically when the North Korean forces attacked South Korea. In 1950, John F. Dulles, who was in charge of US negotiations in Asia, asked the KMT to "freeze" Taiwan's political status temporarily; thus, if a world war broke out again, the US could incorporate Taiwan in its defense line controlling the Western Pacific. Just a few months earlier, US Secretary of State Dean G. Acheson outlined his Asian policy in a speech before the Press Club in Washington, DC. He included Japan, Okinawa, and the Philippines within the US's *cordon sanitaire* but excluded Taiwan and Korea. Now, Dulles assured Wellington Koo, the ROC Ambassador to the US, that "Taiwan was just *within* this defense line" (Chiu 1973: 236–37, emphasis mine).

However, this placed the KMT in a political dilemma. If it were openly conceded that Taiwan was an integral part of China, the US would have no right to intervene, since the "Taiwan question" would be a purely Chinese affair. Taiwan badly needed US forces to prevent the Communists' attack. However, if the KMT were to accept the detachment of Taiwan from the PRC, it would risk losing its legitimacy to rule the island. The KMT finally decided to declare that it accepted *in principle* the US proposal but denied its interpretation of Taiwan's international legal status. Under the KMT's ambiguous response, the US resumed its aid to Taiwan.[6]

Taiwan was thus saved and became an "unsinkable aircraft carrier" in the West Pacific in General Douglas MacArthur's terms, and based on this view, US experts became aware of Taiwan's over-population problem. Their awareness coincided with changing ideas concerning population in the post WWII era. Population was no longer the source of a state's "human power", these experts claimed, since resources in the world were limited. In order to achieve stability, "a world authority is needed which has the power of making, interpreting, and enforcing, within specified spheres, laws which are directly applicable to the individual" (Brown 1954, in Young: 426). The US started to look at its own role in Taiwan's over-population problem through this lens.

In 1953, within the context of the Sino-American Joint Commission on Rural Reconstruction, policy experts turned their attention to the first professional demographic investigation in Taiwan. The statements and reports by the JCRR were not intended to be arguments for a decrease in Taiwan's population or a decrease in its birth rate. Its conclusions make only the following points: (1) since such a big immigration like the one in 1949 seemed *not* likely to happen in the near future, the government had to pay attention to the increasing population in Taiwan; (2) the Taiwanese government did *not* prefer a birth rate reduction policy, therefore a consensus had to be reached before considering the way to solve Taiwan's over-population; and

(3) if the productivity could be raised to a reasonable level, reducing the birth rate was unnecessary. In order to improve understanding of population so that a more reliable decision could be achieved, more detailed demographic research was called for (Barclay: 58–59). Because of political and ideological uncertainty, these experts chose to make their suggestions ambiguous until Taiwan's political status became clear.

In 1954, the PRC made an air raid in the Taiwan Strait just as the Southeast Asia Treaty Organization was being founded in Manila. The US, the leader of the free world, was pressured by public opinion to do something in order to cease further conflicts in Asia. The KMT, on its side, eagerly urged a guarantee of its own safety. The suspended mutual defense treaty between the US and the ROC was thus signed. This treaty, which was basically the same as the one the US signed with the Philippines, declares the US's determination to defend Taiwan and the Pescadores with the assumption that Taiwan is an independent unit. The other KMT-held islands, although then not legally conceived as parts of Taiwan's territory, were accepted *de facto* after two major PRC strategic attacks on these islands during 1955 and 1958 (for details, see Chiu 1979: 157–72). However, during the time, the KMT also recognized that the US would not support its wish to regain the Mainland. The US promised the KMT to guarantee the security of Taiwan, but, in return, asked for the KMT's pledge not to use military force against Mainland China (Chiu 1973: 250–53).

The PRC expressed its anger concerning this agreement; it declared in public that the US had created "two Chinas", something it would never accept. But one might point out that when the above argument was made, the PRC had, in fact, accepted the presumption of "two Chinas". Defense Minister Peng Teh-Huai (彭德懷), for example, warned the KMT in 1958 that the US had started to call Taiwan "small China" and to indicate the PRC as "big China". "Thank heaven!" he added, "our country is seen by an American lord" Taiwan would thus be treated by the US as a "*de facto* political unit" but not as a country (Chiu 1973: 288–89). Peng's message should be read as the PRC's understanding that the US had begun to think of the PRC, or "our country" in his words, as a "*de facto* political unit", which did not include Taiwan or other KMT-held islands. Obviously, the PRC, which began to suffer conflicts with the Soviet Union but had not yet begun a formal relationship with the US, appreciated this subtle approval (for details of the PRC-Soviet rift in the late 1950s, see Lawrance:Part II). Armed conflicts between the PRC and Taiwan ceased at the time, and have not resumed for over forty years. However, before President Nixon's formal visit to the PRC in 1972, secret political negotiations continued between the US and the PRC, and between the US and Taiwan (for details, see Chiu 1979: 172–79). Indeed, an unarticulated consensus of "one Taiwan-one China" or "two Chinas" was formed among the PRC, the KMT, and the US. As the PRC foreign Minister Chen Yi (陳毅) commented later in November 1958, "Quemoy, Matsu, Formosa and the Pescadores must be liberated *as a whole*" (Yim: 112, emphasis mine). The future of Taiwan and of the Pescadores was closely bound with that of Quemoy and Matsu.

The Introduction of State Population Control: From Military Conflicts to Population Control Games

The process of shaping Taiwan as a state, and one receiving US aid, corresponds to the introduction of a new understanding of population linked to the state development strategy. Let me reiterate, the political concern of a state and its population were again interlocked; however, this time it was scientific policies that guided the direction of the state's development.

This can first be seen in Taiwan's military manpower readjustment program. Jacoby pointed out that during 1950–65 Taiwan's military forces were among the world's largest in proportion to the state's population (Jacoby: 118). These career soldiers from the Mainland, who made up a substantial fraction of the military personnel, were still not shown on any demographic record (Barclay: 7–8). When the Korean War came to an end, the "superfluous military force" in Taiwan became a real burden for the US. As Jacoby indicated, all of Taiwan's domestic production, plus the US Aid injection, functioned as a pool which kept the Taiwanese government consumption to a minimum level while most of it went to military expenses. This, from the viewpoint of the US, would probably aggravate Taiwan's economy. One of "the critical issues", Jacoby pointed out, was "whether military outlays could have been reduced under their *actual* amounts" (Jacoby: 221, emphasis mine).

Chiang Kai-shek hesitated to reduce his military power; however, and through subtle negotiations the US and the KMT finally fixed Taiwan's force at a size which could be accepted by both sides. In the fall of 1953, a program which consisted of a new institute for training veterans was drafted by the Mutual Security Agency (MSA) and the Taiwan Department of Defense to eliminate "unqualified soldiers in order to strengthen the military power". The MSA was concerned that Taiwan did not have enough new positions to absorb these dismissed soldiers, most of whom might not have been qualified for jobs in industries. The agency suggested that a Veterans Affairs Commission (VAC, or formally called Vocational Assistance Commission for Retired Servicemen, VACRS), a cabinet-level organization for settling retired soldiers, should be founded, and this was accomplished in November 1954. The MAC reciprocated with a larger four-year disbandment plan which would require 42 million US dollars from the 1955 FY Economic Aid Budget (Chao: 224–25). The above process of negotiation, as Wen writes, well reflects two promises the US made to the KMT if it were to reduce its military power and expenses — that the KMT would retain the full control of its military force, and a promise of extra aid for retired veterans (Wen: 168).

In the following ten years, the VAC helped to shift over 130 thousand soldiers into Taiwan's society as *Jungmin* (榮民, honorable national). The demographic effect of these originally non-registered people was first seen in KMT's first Taiwan Population Census on 1956. Although these veterans were no longer a military burden, their appearance, marriages, and fathering of new births, brought into visibility Taiwan's over-population and its impact on Taiwan's economy to a degree. More importantly, their existence was articulated in the form of statistical numbers that nobody could ignore.

The above situation became tougher when it related to the new logic of population — population could hinder the development of a state if it were not useful to its economy. This logic was introduced by the US and closely linked to its Cold War foreign policy. The financial support given by the US to developing countries was not always disinterested; it sought to prevent the expansion of Communism. The US felt that it could claim that the achievement of a stable government and rapid economic development would result from the acceptance of a democratic political ideology. Most countries which received aid from the US had fragile economic structures. Since they could not export competitive goods to accumulate national capital, most of their income depended on this aid. Obviously, the US did not have enough money to feed so many people, so the only way to solve this problem was to warn these countries to control their populations.

In this sense Taiwan was treated by the US in the same way. When Taiwan's Second Four-Year Plan was about to start in 1956, the US removed population from positive factors in evaluating economic development while making land, or agricultural productivity, as a constant. The economic development thus only depended on the ratio between population and capital accumulation. In 1959, the director of the Agency for International Development Westley C. Haraldson, gave a talk titled "The Economical Development of Taiwan", pointing out that in the last five years Taiwan had received over 500 million dollars from the US, but still could not achieve rapid development. Most of this money was used for consumption by the growing population in Taiwan; the Taiwanese government could not invest more money in basic construction, or further accumulate state capital. Therefore, Haraldson concluded that Taiwan had to decrease its population in order to control its general consumption, implying that if Taiwan did not take this advice, the US might decrease, even cease, its financial support.

Haraldson's harsh comment pressured the Taiwanese government. Based on his 8-point suggestion, Taiwan announced the 19-point program of Economical and Financial Reform in 1960 and added this to its Third Four-Year Program for "accelerated economic development". Taiwanese economic experts started to persuade the public of the "new logic" of making rapid development. In this Plan they set a rigid target of 8% annual growth of per capita GNP; and, in order to achieve this, K. Y. Yin (尹仲容), the minister of the Economy Department, claimed in 1962 that what Taiwan wanted was not the Keynesian model of expanding market by increasing consumption, but decreasing new consumption by decreasing the new birth rate. Although introducing new technologies and foreign investment was also very important, the key to rapid growth of per capita GNP, as Yin repeatedly emphasized, was absolute population control. "If the increasing rate of population goes down to 1.5%," he once argued, "we will not only save more US aid money to invest in industries, but also, more importantly, get a higher number of per capita GNP", which, in his terms, was the golden index for evaluating the development of a country (Yin: 219–28).

To reach a higher per capita GNP was not the US request only for Taiwan; it was a rule for competition among Asian countries receiving the US Aid. In Haraldson's talk he compared the consumption rates and the investment rates of these countries, warning that Taiwan was the second to last country in the competition for economical development; and the economic crisis of the Philippines, the country behind Taiwan, demonstrated what would happen if Taiwan still had high birth rate. The "race" metaphor was thus introduced with the new logic of development by K. T. Li (李國鼎), the successor to K. Y. Yin. He described the 1965 Far East Economy Conference "like a game in which every country tried its best to display its economical development.... Led by our great leader Chiang, Taiwan and its 1.3 million people placed well among these countries" (Li: 27–29). A newspaper even reported: "a 'Family Planning' game, even more important than the Olympic Games, is being held among Asian countries now. Any country which can cut its population increase rate in half will win this game" (*Chungyang Jihpao* 中央日報 [Central News], 7 July 1966). This was the game of scientific policy; everything was judged by the standard of numbers set by the US. Taiwan had no choice but to join the others.

When Things became "Normal": The State and Policy in Reproduction

In August 1965, the Taiwanese Health Department announced its five-year birth control plan, which became Taiwan's biggest public health maneouvre of the 1960s. The program, included in Taiwan's Fourth Four-Year Economic Plan, explained why Taiwan needed a population reduction: the population in 1963 was increasing at the rate of 3.2%; if Taiwan's birth rate continued to rise, its population would reach 16.3 million in 1973 and 26.2 million by the 1990s. This program was proposed in order to eliminate this "superfluous" population, and reduce the rate of increase to an appropriate 1.86% before 1968. To achieve this goal, 600 thousand Taiwanese women would have to have intrauterine devises implanted within five years (Taiwan Health Department 1965). A conference on manpower regulation was then arranged in July 1966, during which the population control issue was first announced in public. The whole policy was settled on in November 1967 when Chiang Kai-shek met with economic experts in a national security meeting. Chiang announced afterwards that Taiwan should have a clear state population policy in order to achieve rapid economic development (for details, see Kuo 1997).

As discussed in the second section, Taiwan's over-population resulted not from severely high birth rate but from huge immigration; simply controlling new births still could not appropriately solve this problem. However, under the logic of state development, no decision could be questioned. Chiang Kai-shek, still a strong man in the mid-1960s, realized that the KMT would never replace the PRC without US help. The US was glad to back up Taiwan's population control as long as the KMT accepted Taiwan's political status. Chiang made his decision. Upon this unarticulated commitment, these new births became urgent and real, and a birth control program became inevitable — just as if Taiwan's problem were like that of any other developing countries. Critics of this program, some of whom academically questioned the vague Malthusian theory on which this program was grounded, were simply mis-interpreted as conservative and ideology-oriented (for an analysis, see Kuo 1997: Chapter 1). With selective financial support, the KMT government improved Taiwan's hygiene, further achieving a more perfect control of reproductive bodies. When the health experts could proudly *predict the long-term* change of Taiwan's population, Taiwan's statehood and the KMT's bio-power were mutually born.[7]

The formation of Taiwan's population control policy reached its closure when it was internalized and normalized as self-evident logic for the Taiwanese government and health experts. Chou Lien-Pin (周聯彬), a renowned scholar and policy maker, recalls with enthusiasm that Taiwan's birth control is "the triumph of public health in the 1960s", for by the 1970s it had earned Taiwan an international reputation in public health policy (Chou 1993). Obviously, these experts enjoyed this triumph approved by, in Pierre Bourdieu's terms, the state's nobility.[8] When Taiwan's "exceptions" disappeared into her "glorious history", namely, the "Taiwan Miracle" narrative created by experts and scientists, the marriage of the state with policy was completed.

Conclusion

This paper demonstrated a salient example of the complicated relationship between the state and policy. As we have seen, by ignoring Taiwan's exceptional problem orientation, the logic of state population control became a taken-for-granted reference in the development strategy of Asian countries.

I want to highlight again two sets of relationships in this story. First, when the KMT government and its military force moved their main body into Taiwan, they created Taiwan's over-population. This fact was simply ignored because the KMT held the ideology of an integral ROC which consisted of Taiwan and Mainland China. This state ideology denied the actual problem of economy and public health. Second, the state population control was introduced during the Cold War period when the US concern over Taiwan's role in global politic was at its height. In this sense, the population control project helped Taiwan, perhaps unintentionally, to identify itself as a territory in which a practical statehood could be established. This was a situation of scientific policy functioning backward to shape a state.[9]

We are compelled to ask how the characteristics of the KMT regime on Taiwan and its relationship with population control policy can be described. They seem to be so deeply interlocked that we cannot locate them in any traditional sense of political science or history of science and technology; no clear boundary, or cause-and-result relationship between the state and policy, can be drawn. It is proper to bring Benedict Anderson's remarkable work on nationalism in here. He points out that the last wave of nationalism that followed both world wars gave birth to new countries in Asia and Africa, but that these new countries were established on the languages and bureaucratic systems imposed by colonialists (Chapter 7). Taiwan's situation was exceptional. After taking over, the KMT introduced Mandarin Chinese, a language which was not understood by the Taiwanese, especially by those who were well-educated; although the KMT kept the local bureaucratic system established by Japan, it added a detached "central" government of the ROC. As I have insisted, Taiwan's statehood could not be established until the settlement of a state/policy complex in which the state and scientific policies could define each other.[10]

However, trickily, Taiwan has become a state with no legal recognition by most of the countries in the world. The US gradually "normalized" its relationship with the PRC, and withdrew its formal recognition of Taiwan in 1978. As the Cold War lost its energy, Taiwan's recently acquired statehood became an oddly dissonant note in the post Cold War harmony which marked the "new world order". Since it was expelled from the WHO in 1973, Taiwan has not been able to participate in any global public health projects, or in other international activities, as a result of the PRC's political intervention. The scientific policy seemed to pave the way for the ideology of the PRC's plan for Taiwan where it never governed for even a single day. In regard to the contemporary question of Taiwan's statehood and that of its people's right to health,[11] I hope this paper can provide new insights in the complicated historical landscape of global politics and that of public health.

Notes

* Parts of this paper were presented at the 2nd Conference on East Asian Studies in RSEA at Harvard University, February 20, 1999. My thanks to the many people with whom I worked in Taiwan and am working at MIT; they inspire my ideas on this topic, and help to make them concrete. I am particularly grateful for the fruitful discussions with Yang Tsui-Hua (楊翠華), Liu Shih-Yung (劉士永), Joseph Wong (黃一莊), Joseph Dumit, Chu Pingyi (祝平一), Lee Dong-Ming (李棟明) and Christopher Kelty. My presentation at the 9th ICHSEA conference is made possible by the financial help from the CCK Foundation.

[1] For other studies regarding state policy and the development of big science, mainly military science, see Everett Mandelsohn, Merritt Roe Smith and Peter Weigart, ed., 1988.

[2] Some research has been done that critically addresses interactions between countries on scientific issues. Sharon Traweek brought attention to the cultural and political issues in the practice of modern science in her study on high-energy physics in Japan, which neatly describes the cultural practices that have shaped the Japanese scientific community's strategy to establish its own big science. See Traweek 1992.

[3] Many scholars have remarked on the KMT authoritarian control of Taiwan since the end of W.W.II; the control became more severe when the KMT was defeated by the Chinese Communists and moved onto the island. For a general understanding of Taiwan's domestic politics under the KMT regime, see Tien: Chapter 1.

[4] Taiwan experienced three major changes in its demographics. The first of the other two changes occurred during Taiwan's involvement in W.W.II. The Japanese government conscripted young Taiwanese males to send to war fields overseas; meanwhile, the US attacked cities in Taiwan by planned bombing. Both Taiwan's resident registration system and population statistical information system broke down. The February 28 massacre in 1947 marked the second major demographic

change. Thousands of Taiwanese citizens and intellectuals, most of whom were innocent, were arrested and slaughtered for demonstrating against the corrupt take-over by KMT bureaucrats. This event made Taiwanese people flee and seek shelter everywhere on the island. As a result, the KMT government stopped recording population statistics for the next eight years.

5 In 1950 the KMT imposed two years of military service on all 18-year-old males in Taiwan; as a result, the numbers of estimated immigrants from the Mainland that we see in this table should be a little lower and the actual number of Taiwanese at the time should be a little higher, because soldiers were not registered in any formal statistical data. In fact, lacking solid data, no exact number can be given for these political immigrants. Long Kwan-Hai (龍冠海) writes that including the soldiers, it must be over one million. Chang Chin-Yuan (張境原) points out that with the addition of the defeated soldiers, it should be over 1.25 million.

6 Before the Korean War, no special aid plans were settled for Taiwan. In 1949, as a result of the US's negative attitude towards the KMT regime, China Aid was suspended. Funds under the China Act were diverted to other areas in East Asia. Although the JCRR moved to Taiwan in August 1949, the Economic Corporation Administration (ECA) asked its surrogate chairman Raymond T. Moyer to change the method of payment from annual to biannual, and empowered him to end the JCRR, and even the ECA in Taiwan as well, if Taiwan were to fall (Chao: 1–11). When the US decided to resume the aid, it extended the deadline of 1949 FY Aid to the end of June, 1950, quickly sending $8.5 million worth of essential goods from the US (Chao: 11); following that was $50 million in military aid, $42 million in economic aid in early 1951. Later, in October, the Mutual Security Act of 1951 was passed in the US Congress. This act, which incorporated previous foreign aid plans, provided a legal basis for giving aid in order to prevent the spread of communism. The ECA under the Marshall Plan was replaced by the Mutual Security Agency (MSA) and the Department of Defense took charge of all military aid plans (Wen: 50–51). Three billion dollars in aid to Taiwan was issued under this Act.

7 In his *History of Sexuality* volume I, Michel Foucault argues that, along with the development of modern states, the discourse and agenda for regulating population should be understood as a part of the new power mechanism, which he called bio-power (1980: 138–46). For details concerning the ways in which the KMT's biopower was established by imposing Intrauterine Devices on Taiwanese Women, see Kuo 1998.

8 See Boudieu 1994, especially pp. 12–18. In fact, the attitude of these experts toward numbers can be traced back to the time when the mode of biopower and that of the modern state were merged. Ian Hacking shows how the rise of biopower coincided with that rise of statistics, and how both nurtured the authority of individual professions, such as physicians, in the late 1880s. See Hacking 1982.

9 Scholars of political science also notice this transformation in terms of "technocratic power". Those people who specialized in management and climbed up through the governmental bureaucratic ladder, they point out, began to appear in the decision

core of the KMT in the late 1960s and became the mainstream in the mid-1970s. In contrast with the original ideology-oriented party cadre and security-conscious generals, these technocrats were development oriented. Backed by the US Aid institutions and responsible for Taiwan's economy, they were pragmatic in mentality and oriented toward efficiency. For further discussion, see Wen: 231, Tien: 125–29, and Jiang & Wu: 82–83.

[10] Lucian W. Pye's excellent analysis addresses a bit about the relationship between Taiwan's state building and its economic development (1986). Pye argues that up to the early 1980s, Taiwan had been developing a complex social and political system based on the fact that Taiwan is an independent political unit, although one badly influenced by the US. This fact, of which Pye reminds the leaders of the PRC and the US, can help readers to clarify the terms the state/policy complex and "practical statehood".

[11] Taiwan has sought to rejoin the WHO since the early 1990s. Brought to world attention by the Medical Professionals Alliance in Taiwan (MPAT), Taiwan's campaign to join the WHO continues every year when the WHO annual congress is held. In May, 1999, the WHO again decided to leave Taiwan's request for a role in the UN-affiliated body pending at its current annual congress. Two months later, when interviewed by the German Sound about Taiwan's relationship with the PRC, Taiwan's President and KMT leader Lee Teng-Hui made a special "state-to-state" claim. This is the first time the KMT has admitted that the PRC is a state rather than a "political entity" or a "rebel group"; and it is the first time the KMT has clearly defined the ROC as a state only consisting of Taiwan and other KMT-held islands. For a brief introduction to Taiwan's political debate and that of Taiwan's participation in the WHO, see Time Magazine, 26 July, and the website at http://www.mofa.gov.tw/88who/88who.htm.

References

Anderson, Benedict. 1983. *Imagined Communities: Reflections on the Origin and Spread of Nationalism*. London: Verso.

Barclay, George W. 巴克萊 1955 [1954]. *Taiwan Jenko Yenjo Baogao* 臺灣人口報告 [A report on Taiwan's population]. Translated by the Sino-American Joint Commission on Rural Reconstruction (JCRR). Taipei: The JCRR.

Bourdieu, Pierre. 1994. "Rethinking the State: Genesis and Structure of the Bureaucratic Field". *Sociological Theory*, vol. 12 no.2 (March 1994), pp. 1–18.

Chao, Chi-Chang. 1985. *Meiyuan di Yunyung* 美援的運用 [An analysis of the US Aid to Taiwan]. Taipei: Union Publisher.

Chen, Chih-Chi. 陳志奇. 1980. *Meiguo due Hua Jengtse Sanshi Nien* 美國對華政策三十年 [A study of US foreign policy regarding the ROC, 1948–1978]. Taipei: Chung-Hua Daily News.

Chen, Tien-Si 鎮天錫 and Yin Cheng-Chung 尹建中. 1983. *Jenko Jentse di Shinchen yu Jientao* 人口政策的形成與檢討 [The formation of the population control policy in Taiwan: a historical review with critical notes]. Taipei: Union Publish Co. Ltd.

Chiu, Hungdah, ed. 1973. *China and the Question of Taiwan: Documents and Analysis.* New York and London: Praeger Publisher.

_____ , ed. 1979. *China and the Taiwan Issue.* New York and London: Praeger Publisher.

Chou, Lien-Pin 周聯彬. 1993. "Taiwan Gonggong Weishen Fajan Shi Fantan Jilu: Chou Lien-Pin (III)" 臺灣公共衛生發展史訪談記錄: 周聯彬 [The interview of Chiu Lien-Pin for the History of Taiwan's Public Health]. Interviewed and Recorded by Lin Hui-Fang 林慧芳 and Hsieh Ya-Hsu 謝亞栩. Unpublished manuscript.

Clarke, Robin. 1985. *Science and Technology in World Development.* Oxford and New York: Oxford University Press/UNESCO.

Executive Yuan, ROC. 1976. *Taiwan Dichu Huji Jenko Tongji Chi TiaoCheng* 臺灣地區戶籍人口之調整 [Adjusted Data of Taiwan's Population, 1951–1973]. Taipei, Executive Yuan of ROC.

Feldman, Harvey. "The Development of US-Taiwan Relations, 1948–1987", in Harvey Feldman, Michael Y.M. Kau and Ilpying J. Kim, ed. *Taiwan in a Time of Transition.* New York: Paragon House.

Foucault, Michel. 1980. *The History of Sexuality, volume I Introduction.* New York: Vintage Books.

Freymann, Moye W. 1965. "India's Family Planning Program: Some Lessons Learned", in Minoru Muramatsu and Paul A. Harper, ed. *Population Dynamics: International Action and Training Programs — proceedings of the international conference on population, May, 1964, The John Hopkins School of Hygiene and Public Health.* Baltimore, Maryland: The John Hopkins Press.

Hacking, Ian. 1982. "Biopower and the Avalanche of Printed Numbers". *Humanities in Society*, vol. 5 nos. 3 & 4 (summer and fall 1982), pp. 279–94.

Hsu, Chih-Chu. 1969. *From Taboo Ton National Policy: the Taiwan Family Planning Program up to 1970.* Taichung: Chinese Center for International Training in Family Planning.

Jacoby, Neil H. 1966. *US Aid to Taiwan A Study of Foreign Aid, Self-Help, and Development.* New York, Washington, London: Frederick A Praeger.

Jiang, Ping-Ling and Wen-Cheng Wu. 1992. "The Changing Role of the KMT in Taiwan's Political System", in Tun-jen Cheng and Stephan Haggard, eds. *Political Change in Taiwan.* Boulder and London: Lynne Rienner Publishers.

Kuo, Wen-Hua 郭文華. 1997. *Yijiowulin Ji Yijiochilin Niendai Taiwan Jiating Jihua: Yiliao Chengtse yu Nusinshi de Tantao* 一九五〇年代至一九七〇年代臺灣接頭計劃: 醫療政策與女性史的探討: [Politizing medical policies and medicizing Taiwanese women's bodies: a politico-cultural study of Taiwan's population/birth control from the 1950s to the 1970s]. Master thesis for History of Science Department, National Tsing-Hua University, Hsin-Chu, Taiwan.

_____ . 1998. "(Re)producing Technologies for Making (Non-)reproductive Bodies/Lives: an Anthro-Historical Observation of the Birth Control Program in Taiwan". Paper presented in the 97th American Anthropology Association annual Conference. December 6, 1998, Philadelphia.

Kwon, Tai Hwan. 1997. "The National Family Planning Program and Fertility Transition in Korea", *The East-West Working Papers, August 1997*. Honolulu: the East-West Center.

————. 1977. *Demography of Korea: Population Change and Its Components 1925–1966*. Seoul: Seoul National University Press.

————. 1993. "Exploring Socio-Cultural Explanations of Fertility Transition in South Korea", in Richard Leete and Iqbal Alam, ed. *The Revolution in Asia Fertility: dimensions, causes and implications*. Oxford: Clarendon Press.

Lawrance, Alan. 1975. *China's Foreign Relations Since 1949*. London and Boston: Routledge & Kegan Paul.

Lee, Dong-Ming 李棟明. 1995. *Taiwan dichu tsaochi jating chihua fajan chihsian* 臺灣地區 早期家庭計劃發展 [Family Planning Program in Taiwan: the history of its early period]. Taichung: The Family Planning Institute.

————. 1969. "Kwanfu ho taiwan jenko shehui tsenja ji tantao" 光復後臺灣人口社會增加之探討 [an investigation on the social increase of Taiwan's population after W.W.II]. *Taibei wenhsien* 臺北文獻 [Archieves on Taipei], Nos. 9 & 10.

Li, K. T. 李國鼎. 1966. "Jinji Fajan yu Jenli Tsuyuan chih Kaifa yu Yunyung: Wushiwu nien Chiyue Chiji tsai Jenli Tsuyuan Yentaohui Jiantsu" (經濟發展與人力資源之開發與運用五十五年七月七日在人力資源研討會講詞 [The relationship between the economical development and human resources: A speech at the Conference of Human Resource, 1966.7.7.], in *Jenli Tsuyuan Yentaohui Tsuliao Techi* 人力資源研討會資料特集 [Proceedings of the Conference of Manpower Resource Regulation, 1966.7.7], Taipei.

Liu, Ta Jen. 1978. *A History of Sino-American Diplomatic Relations, 1840–1974*. Taipei: China Academy.

Mandelsohn, Everett, Merritt Roe Smith and Peter Weigart, ed. 1988. *Science, Technology and the Military*. Dordrecht, Boston and London: Kluwer Academic Publishers.

McNeill, William H. 1982. *The Pursuit of Power: Technology, Armed Force, and Society since AD 1000*. Chicago: University of Chicago Press.

Pye, Lucian W. 1986. "Taiwan's Development and Its Implications for Beijing and Washington", *Asian Survey*, vol. 26 no. 6 (June 1986), pp. 611–29.

Stepan, Nancy. 1981. *Beginning of Brazilian Science*. New York: NY Science History Publication.

Sun, Teh-Hsiung 孫得雄, *et al.* 1979. *Woguo Jenko Jentse yu Jenko Jihua chi Tantao* 我國人口政策與人口計劃之探討 [A research on population policy and population control in Taiwan]. Taipei: Executive Yuan.

————. 1990. "Sunyulu Dadao Tidaishuijun Ji ho" 生育率達到代替水準之後 [When the birth rate is going down to its replacement level] in the Chinese Association of Demography, ed. *Taiwan Juanshin Hochi di Jenko Siensian Yentaohui Lunwenchi* 臺灣轉型後期的人口現象研討會論文集 [Proceedings of the Conference on the Population Phenomenon after Taiwan's Demographic Transition]. Taipei: Chinese Association of Demography.

Sun, Teh-Hsiung and Y. L. Soong. 1979. " On Its Way to Zero Growth: Fertility Transition in Taiwan, Republic of China", in Cho Lee-Jay and K. Kobayashi, ed. *Fertility Transition of the East Asian Populations*. Honolulu: The University Press of Hawaii.

Tien, Hung-mao. 1989. *The Great Transition: political and social change in the Republic of China*. Stanford, California: Hoover Institution Press, Stanford University.

Traweek, Sharon. 1992. "Big Science and Colonialist Discourse: Building High-Energy Physics in Japan" , in Peter Galison and Bruce Hevly, ed. *Big Science: The Growth of Large-Scale Research*. Stanford, California: Stanford University Press.

VACRS, the (Vocational Assistance Commission For Retired Servicemen). 1964. *Ten-Year Operation of the VACRS*. Taipei: the VACRS.

Wen, Hsin-ying 文馨瑩. 1990. *Jingji Chiji di Beiho: Taiwan Meiyuan Chingyen di Chengjing Fenhsi, 1951–1965* 經濟奇蹟的背後: 臺灣美援經驗的政經分析 [The politico-economical analysis of US Aid to Taiwan, 1951–1965: a critical note on Taiwan's Economic Miracle]. Taipei: Independence Daily Inc.

Yim, Kwan Ha, ed. 1973. *China & the US 1955–63: Relations During the Eisenhower & Kennedy Administrations*. New York: Facts on File, Inc.

Yin, K. Y. 尹仲容. 1963. *Wo Dui Taiwan Jinji di Kanfa Shubein* 我對臺灣經濟的看法續編 [My view of Taiwan's economy, Volume II]. Taipei: Council on US Aid.

Young, Louise B., ed. 1968. *Population in Perspective*. New York, London and Toronto: Oxford University Press.

The Internet in China: Policy and Problems

LIU YIDONG

*Institute for the History of Natural Science,
Chinese Academy of Sciences*

Introduction

Worldwide National Information Infrastructure (NII) construction has been developed with unprecedented concern and support both socially and governmentally.[1] Under this most significant and influential circumstance, China (which here means mainland China) has become the focus of world attention for its fast development and tremendous potential. Although the internet, a rudiment of the NII, was introduced into China only five years ago, it has since undergone tremendous development, an important process yet seldom a case for study. What I am going to discuss in this paper is the internet and information policy in China including its history and problems.

Brief History of the Internet of China

The computer network in China has achieved remarkable progress ever since the internet entered in 1987 and become an essential element in the development of the CNII (China National Information Infrastructure). As statistics show, by the end of June 1999, there were 1.46 million computers in China connected to the internet. The number of users here hit 4 million, 29045 domain names were registered in the domain "CN", and about 9906 www net sites were set. The total capacity of Bandwidth was 241Mbps. The U.S., Canada, Britain, Germany, France, Japan as well as Hong Kong etc. were reachable via the internet.

The expansion of the internet in China has proceeded in two phases. From 1987 till 1994 was the "sprout period", during which Chinese academic organizations initiated study and cooperation on networking, then succeeded to connect with e-mail. This period witnessed the first Chinese e-mail "Reaching out to the world over the Great wall" sent at five twenty-three on September 20, 1987. Domestic users from scientific and educational circles then were able to email via the transformative system of the internet.[2] In 1994 the internet entered its prosperity phase in China. In April of that year, The China Science

138

and Technology Network opened an international line of 64kpbs to the U.S., entered the internet, established TCP/IP, and initiated overall service for CSTNET's members. So far, CSTNET, CHINANET, CERNET as well as CHINAGBN are the national computer networks in China, each directly on the internet. In June 1997 the four domestic giants joined together. Data concerning the four networks in China are listed in Tables 1 and 2.[3]

Table 1   Four Main Chinese Networks

Networks	Establishment Dates	Administrators	Types
CSTnet	1994/4	中国科学院	ac
Chinanet	1995/5	邮电部	com
CERnet	1995/11	国家教委	edu
ChinaGBN	1996/9	电子部	com

Table 2   Development of Bandwidth of Four Main Chinese Networks (Mbps)

	1996	1997	1998	1999.6
CSTnet	0.128	2.128	4	8
Chinanet	5.512	20	123	195
CERnet	2.32	2.256	8	8
ChinaGBN	0.256	2.256	8.256	18
Total	8.216	26.64	143.256	229

Table 3   Statistics of Development of China Internet

	1995	1996	1997.10	1998.6	1998.12	1999.6
Chinese Internet Users	$4*10^4$	$10.8*10^4$	$62*10^4$	$117.5*10^4$	$210*10^4$	$400*10^4$
Computers on Internet			$29.9*10^4$	$54.2*10^4$	$74.7*10^4$	$146*10^4$
Domain Names Registered in "cn"			4066	9415	18396	29045
Number of www servers in China			1500	3700	5300	9906
Total Capacity of Bandwidth (Mbps)			18.64	84.64	143.256	241

The internet is growing rapidly in China. Table 3 shows its development during recent years.[4] The number of people surfing the internet jumps at an annual growth rate of over 300%, dwarfing the average world increase. Meanwhile, the Chinese internet-population only accounts for 3% of the total in the world, which is fairly unbalance compared with the great population of China. Hence, great effort still needs to be made in the future.

Brief History of Internet Policy of China

Being an important component of CNII, the internet in China has been expanding following state info-policy instruction. The policies concerned here are mainly those that are relevant to the internet. After the NII plan was started in the U.S. in September 1993, Chinese scientists made an immediate response. Joining concerned others, Professors Ye Peida, Chen Junliang and Zhong Yixin initiated the prospect of an "Information express network for China" which gained great attention in the government. The State Council set up a State Economy Informatization joint conference which was made up of representatives from over 20 government organizations as well as enterprises, and was chaired by Deputy Prime Minister Zou Jiahua. Afterwards the Chinese government formulated the strategy of "China National Informatization Infrastructure".[5] Establishment of the joint conference made it possible to manage informatization procedure integrally over the range of the whole country. 1994 marked the start of comprehensive furthering of CNII through the setting up of the joint conference, success in overall linkage to the internet, the note about Three-"Golden"-Projects-Related Subjects issued by the State Council, as well as the overall spread of the project of the "Golden Bridge". At the 5th meeting of the Central Committee of the 14th Communist Party of China held in October 1995, "The strategy of accelerating the national economy informatization procedure" was formulated. Later the 4th session of the 8th National Congress enveloped advancing informatization into the "Extract of the 9th 5-year Framework for National Economy and Social Development & 2010's Outline", which was approved in March of 1996. The Extract for the first time openly listed informatization as a strategy of priority in a long-term framework, which confirmed the electronic industry as one of the five basic industries of national economy development.[6]

The State Economy Informatization joint conference was reformed into the State Council Steering Committee on National Information Infrastructure in January 1996, headed still by Deputy Prime Minister Zhou Jiahua. Eighteen administrations participated in the committee.

Consequently, some departments and governments at the lower level set up similar independent organizations. A basic national administration system for informatization was forming.[7] The major response of the State Council Steering Committee was to study developing strategy, principals and policy for economic and social informatization, set schemes and technical criteria for informatization, and organize major multi-regional and multi-organizational informatization projects and scientific research. Meanwhile the Committee carried out control over the computer information network and internet at the macro-level.

After its establishment, and the joining of the concerned departments, the Committee drafted the "Extract of the 9th 5-year Framework for National Economy and Social Development & 2010's Outline", the "Focus of Informatization over 1997–1998", and revised the "Tentative Administration Stipulation on Computer Network & Internet of People's Republic of China", issued in February 1996. It also published the "Tentative Administration Regulation on Internet Domain Name Registration in China" in May 1997. CNNIC was formed in June 1997. Meanwhile a CNNIC Committee was set up. Organizing of CNNIC and issuing of a set of law and regulations regarding the internet indicated the beginning of scientific systematization of the internet administration in China. The first national informatization conference held in April 1997 officially formulated the concept of a national system of informatization, which covered information resources, information networks, application of information technology, information technology and industry, human resources for informatization, and regulation and criteria for informatization. Stressing development and application of information resource as a core priority, it tried to achieve a comparatively integrated national information system on a large scale by the year 2000.[8] An orderly period of prosperity for the internet in China began.

Problems and Discussion

In the process of the development of China's internet, there were some problems such as a small transmission capacity, too much cost, little information in Chinese, invasion of unwholesome information, hidden trouble on security, and the existences of ISPs, etc. As we know, there are many discussions already held on those topics, therefore, what I wish to focus in this paper is China's information policy and the key role it played in the process of developing China's internet.

I am going to discuss two aspects of the information policy of China's internet. First, according to the dominant ideas of policy makers, I will touch on whether the policy remained the same after being implemented. Secondly, I will try to give some new criteria with which to evaluate the policy.

1.   Both vacillation and self-contradiction are prohibited in the process of policy formulation and implementation

One thing the Chinese government was always concerned about was to strengthen the important role it played in macro-economic regulation within the process of economic construction and informational development. The first national working conference on information had added further "national domination" in the article on guiding principles. The joint conference has been replaced by the informational leading group which embodies the importance of national domination systematically. The IDC China Report had critically stated at the time, "Perhaps, it is a sign of the State Council that Chinese leaders have already made the decision to give the right to the State Council on the plan of developing the national long-range communication data, but not to all departments directly under the State Council."[9] Recently, Chinese academic circles have called on the Council to establish an authoritative organ of power to represent it in order to carry out unified leadership without any departmental discrimination. Early in 1993, at the first high-level academic conference on information, the chairman of the conference pointed out that the reason why Koreans could be so successful in the development of their information technology industry were their strong state leadership and practical economic plans, as some Korean scholars had pointed out.[10] A number of conferences were held by the Center of Social Development Studies, Chinese Academy of Social Sciences on the problem of how to make a wholesome environment for the development of China's information technological industry. The conference had emphasized that China should plan for establishing an organ of power directly under the leadership of the President instead of the present associating authority.[11] In 1998, the Chinese government implemented an adjustment of the organizational structure. Since then, the leading informational working organization has been assigned to the Ministry of Informational Industry. In this particular area, some scholars have pointed out, "It is doubted whether the Department of the Ministry of Informational Industry will be able to perform its duty properly like the trans-departmental associating authority usually did under the circumstance of the present unchanged structure, especially the non-separation of the enterprise from the state in the departments of post and telecommunications."[12] It is hard to understand why the government tries to strengthen national domination on one hand, and to reduce the power of the associating authority on the other.

For this, there is also no other explanation or remedy declared by the government.

However, the government pays more attention to the problem of some unwholesome information on the internet and makes strict regulations for internet management. The regulation was made not only for stipulating production, publication, and transmission of this unwholesome information, but also for referencing and reviewing such programs.[13] The investigation usually takes place strictly in the domestic net service companies. For instance, a famous network service company named Sohu once was nearly suspended after being carelessly connected with some unwholesome sex information.[14] However, leaving one side of the net open to foreign companies is generally practised in the country. For example, "Yahoo Chinese" can transmit any kinds of websites no matter if it is regarded as unwholesome by the Chinese government for our domestic network users. As we know, there are no national boundaries on the network. This kind of keeping inside information from outsiders reduces not only the limitation of this unwholesome information, but also weakens the prestige of the department. Because of this, some domestic network service companies have simply established two kinds of networks, both inside and outside respectively. Some information cannot be referred within the domestic network but can be easily received from the outside. Those inside and outside networks appear on the computer screen less than a few centimeters apart.

Scholars have pointed out many times that this unwholesome information cannot be kept out of the country with the fast development of information technology.[15] Therefore, we have to cast away illusions and look reality in the face — the unwholesome sex information certainly cannot be expunged from people's lives. The only thing we can do today is to stress the importance of relieving congestion and education in accordance with different people.[16] So far, this kind of proposal has still not been accepted and implemented by the government. This is the only example of information policy which is obviously self-contradicting. We know that a lack of strong and powerful unified leadership is not only occurring in informational construction, but also in policy formulation.

2. That the nucleus of informatization is to heighten the organizational level as well as to renew ideas of management

The formulation of information policy mainly depends on the understanding of informatization. The understanding of informatization is inseparable from the understanding of the present difficulty which the government faces. After a reform opening to the outside, the Chinese people have realized that mistakes in macro-policy decisions were always the biggest faults of the past and that backwardness of ideas of management is the biggest cause of falling behind in the present.[17] It is regretful that this sense of the intellectuals could not be adopted in the process of development of information technology and policy design. The development and popularization of information technology could lead to a reform of management structure and innovation in ideas of management. The scientific function of society has been unprecedentedly strengthened by the support of commonly used informational resources and a special knowledge of information disposal by specialists. It will cause a tremendous change in the power of the social structure if all duties of study and design on policy, regulation, and settlement of all problems are transferred to academic circles.

We see clearly that the policy of information in China focuses only on the development of information technology and the information industry. It does not devote much attention to how to make use of the technology in order to improve the standard for systematic management. It hardly touches upon the problem of renovating people's concepts and ideas. There is little plan made for encouraging scholars to play important roles in design and study of important cases, policy, and regulations. It should be clearer that China's basic problem is that a group of scholars who are able to make and provide good suggestions and ideas to some big problems are lacking. At present, China also lacks a dynamic management system and operating mechanism as well as the social environment to guarantee that scholars will play important roles. Therefore, China has not yet come into the virtuous circle of mutual promotion between science and society.

The standard of China's technology involving the internet is quite advanced. For example, China's network of science and technology has already reached the international advanced level after applying group-net technology and equipment.[18] But in the case of usage, there is no innovation. After having investigated nineteen examples of application from the network of science and technology, we found that all of them belong to the conventional realm, and there are almost no new ideas brought forth. The so called "electric government" has been given priority in the development of all aspects of NII proposed by all countries. However, the project of using the internet to service governmental departments began from January 1999. China plans to have 80% of the governmental departments using the internet by 2000. In the present situation, China has not yet made a new kind of process and operating mechanism, even though the steps made for the governmental departments to use the internet have been accelerated.

The situation results mainly from three reasons: (1) Stress on technology; overlooking the academy; lacking of acknowledge that serving the country and society is a serious goal. Recently, the application of science and technology in industries has been strengthened in China, emphasizing the contribution of technology to economic increase. However, this idea is crippling. It is acknowledged that direct benefit for a country and society is needed for China. Not only does acknowledge of his kind create social effects, but also remarkable economic ones. We may put knowledge-producing values into two groups. The first one (or Knowledge I) may directly help enterprises or individuals. In another words, it is realized through enterprises and evaluated by the market. Examples on W. Shockley's study of transistors; Bill Gates' study of software and sales; Wang Xuan's study of the Chinese electronic publishing system. The more common ones are those that are deduced through R&D; the second type of knowledge (or Knowledge II) mainly benefits a country or society. It is appreciated by a country or society, and its value can hardly be assessed by market. Examples include J. M. Keynes' economics; V. Bush's study of scientific research management and science-technology-related policy; and Ma Yinchu's study about population. The numerous achievements in this group are relevant to study social and multi-lateral problems. Compared with Knowledge I, Knowledge II may create as much economic effect, or even more than the former, yet the latter does not gain deserved investment and attention, since the efficiency brought about by this highly and quickly rewarding domain cannot be straight-forwardly reflected.

Stressing technology, overlooking the academy means to emphasize Knowledge I and ignore Knowledge II. For example: while the cost for subscribing to international academic magazines is shrinking, expenditure for importing technological products is jumping.

To complete anything but instinct requires a plan. The more complex the thing is, the more options are available to solve it; consequently, the more remarkable the economic and social contrast aroused by suggestions of different quality is. Making a plan differs from doing algebraic exercises, since everybody can make a plan. Because of this very reason, scholars have not been considered participates in this job. Yet, a feature of making a plan is that programmers of poor quality cannot even imagine what a first-class plan should be like, as one already quite close to success when everyone thinks alike. We cannot even know what we are losing if we cannot create a first-class plan. In one word, the essence of planning is that nothing may be achieved without it; deviation results from where an advanced or a defective plan is accepted. The core of Knowledge II is scientific research for planning as well as developing solutions to various major problems and it is of unparalleled importance. However, investment for research of the second type is limited, and the income of researchers is little in China. It is impossible to obtain a great number of first-class researchers. "Progress in science depends on the number of first-class human resources" (Harvard's President J. B. Conant said). Shortage of material resources retards development of a domain which should have created considerable economical and social effects, impeding its great power.

(2) Inadequate realization of the nature of scientific research. Planning for construction of bridge, power station or any other important projects completely relies on experts. Yet, why shouldn't planning of policies or solutions to major problems proceed the same way? It results from ignorance of scientific research. After professional research came into being, a relevant huge social system needed to be formed. A feature of a professional researcher's work is that his or her research is done within a SRSS (scientific research support system). SRSS has three layers. The first one is knowledge support that consists of data, theories, methods, criteria and cases. The second layer is organizational support, which is composed of science associations, academic organizations, research institutions, and universities. It offers exchange, conferences, cooperation, competition, and evaluation among scholars. The third layer is social support, which is made up of governmental, military, industrial, and social departments requiring the knowledge. Generally speaking, the difference between the professional researcher and non-professional ones is the completeness of the SRSS, which make it impossible for the non-professional to compete with professional researcher when plans for solutions to major problems are required. Science cannot solve all puzzles. However, all problems need plans for their solution provided by scientists. Today, as the information monopoly is falling apart, and the public sharing of data resource is prospering, with more and more advantage given information-selecting as well as information-analyzing, professional scholars who are versed in managing but occupy no administrative posts will be the uncrowned kings.

(3) Everybody enjoys equal opportunity evaluating truth: a plan-evaluating and selecting mechanism should be set up immediately, promising the full participation of science. Examining the progress of information policy in China, we lack both an effective cooperation system between the government and academic circles, and a feedback system for the research and suggestions of scholars. An essential reason that the U.S. became the world's leading power was equal cooperation between the government and academic associations rather than the dictation of superior and inferior. The initiator

and promoter of this idea, V. Bush, praised as an Engineer of the American Century,[19] may have had a stronger impact on the development of science and engineering in the 20th century than any other person.[20] In my humble opinion, it can never be over-appreciated that to allow science to work fully as brain of society rather than a simple tool. Rule of equal individuality in front of truth should be erected to guarantee genuine participation of scholars in social activities. Respect towards acknowledge means obey the whoever as long as he or she is right. To set up an effective operational system of plan-evaluation and prompt improvement for acceptation and application of qualitative solutions to major problems by borrowing distinctive assessing and self-improving mechanism, experiences, computer network of scientific activities is imminent for China.

References

[1,11,15] 刘吉、金吾伦等：《千年警醒：信息化与知识经济》，社会科学文献出版社，1998年，第10–13页，第486页，486页.

[2,3,4] http://www.cnnic.net.cn

[5] 鲁品越等：《中国未来之路－信息化进程在中国》，南京大学出版社，1998年，第65页.

[6] 中华人民共和国国民经济和社会发展"九五"计划和2010年远景目标纲要.

[7,12] 汪向东：《信息化：中国21世纪的选择》，社会科学文献出版社，1998年，第88页，第265页.

[8] 蒋亚平：全国信息化工作会议在深圳举行，《人民日报》，1997年4月19日.

[9] IDC China Report：《计算机世界》，1996年第29期.

[10] 李京文、郑友敬、小松崎清介、G•鲁赛尔•派普主编：《信息化与经济发展－国际议论文问萃》，社会科学文献出版社，1994年，第17页.

[13] http://www.cnnic.net.cn/cnnic/info/chinapolicy.htm

[14] 白水：网上黄毒闪不及，《中国计算机报》，1999年1月25日.

[16] 江晓原：电脑时代的色情文艺，《方法》1997年，第9期，第42–43页.

[17] 惠益民、邬宽明：《科学管理和管理科学》(《现代科学技术基础知识》参考丛书之一科学技术文献出版社，1994年，前言.

[18] 中国科学院计算机网络信息中心：中国科技网简介，1999年4月.

[19] 社克：我国图书馆文献资源的现状不容乐观，中国科学报，科普周刊，1997年7月25日.

[20] Harvey Brooks: Endless Frontier: Vannevar Bush, Engineer of the American Century, *Nature*, 6 November 1997, p. 34.

The Impact of Local Government Policies on the Development of Traditional Technology: A Case Study of the Daning Salt Factory

LIAO BOQIN
Southwest China Normal University
Chongqing, China
WANG LIMING & JOHN DAVIS
The Queen's University of Belfast, U.K.
JIA BOSHENG
Wuxi Middle School,
Wuxi County, Chongqing, China

Introduction

The Daning River, once named Wuxi (巫溪) and Changjiang (昌江), originates from Dongxi (东溪) and (西溪) in the south of Dabashan (大巴山), passes through Wuxi and Wushan (巫山) county, and finally flows into Changjiang (长江). Along the Daning River, there have been mysteries which people today cannot clearly explain, such as how in ancient times their ancestors put hundreds of coffins on a steep cliff (岩棺); when and how they made the one-hundred-kilometres-plus *zhandao* (栈道) (a plank road built along the face of a cliff); and so on.

The Daning Salt Factory is situated at the north bank of the Houxi (后溪) which is a branch of the Daning River and flows into it at Ningchang town (宁长镇) in Wuxi county, bordering on Hubei (湖北) and Shanxi (陕西). The most famous character of Danang is that there is, in the factory, a natural white salt spring which was used from the earliest times in China.[1] It is said that the original inhabitants used bamboo to lead the salt water into pots and cooked it.[2] Other records tell us that: "There was a goddess in Yanshui (盐水) who stopped Linjun (廪君) from leaving and told him: 'it is so wide, with many kinds of fish and salt, so I wish that we could live here together'";[3] "It was said that this spring could produce salt because there is salt gas in the water";[4] "In the Song dynasty, there was a salt spring flowing from the south of the mountains, 69 *li* from the Dachang county (Daning county)".[5] These ancient records tell us that using the salt spring to produce salt has a long history, dating back to 316 BC.

This paper attempts to examine the impact of local government policies on the development of Daning Salt Factory, especially its traditional technology of salt production. Key features of salt production and policy interaction are outlined in the next section. An empirical analysis on the relationship between government policy and the performance of the factory is carried out in the third section and the reasons for the factory's bankruptcy are also discussed. The final section contains conclusions and draws out some suggestions on promoting traditional technology in China.

The Nature of Salt Production and Policy Interaction

From the Chronicle of Factory Events in Appendix A,[6] we can see that there are four different kinds of policies which affected the development of technologies in salt production. Firstly, local policies related to the national policies of China affected salt production. For example, the factory was affected in 1958 by the Great Leap Forward. People in the factory had to pay attention to iron and steel smelting, so that some aspects of production, such as profit, total output value, salt production, earnings from sales, and total salaries, decreased compared with other years. Also, the Cultural Revolution impacted policy making in the factory so that salt production was also affected. Figure 1 shows that the profits of the factory were clearly impacted after changes in national policies. For example, the profit was very low between 1963–66 after the Great Leap Forward in China, and between 1977–80 after the Cultural Revolution.

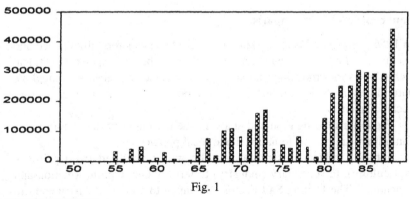

Fig. 1

PROFIT

Secondly, the policy of changing seasonal production promoted the development of technology. The salt density of water is changeable, being low in summer but high in winter. In order to increase salt production, the factory improved technology. For example, an electrodialysis machine was built to enrich salt water in 1976.

Thirdly, in order to modify production efficiency the factory took different actions to reform its technologies. Figure 2 shows that the total output value changed along with the development of technology. For example, after 1972, the factory paid more attention to developing technology and spent more money on investment, so that the total output value was higher after 1972 than before.

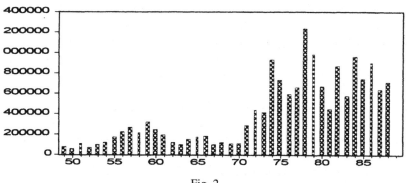

Fig. 2

TOV

Bankruptcy and a Comparison

In 1956, Wuxi Salt Factory (Daning Salt Factory) was formally opened and Daning salt dominated the markets in Sichuan, Hubei and Shanxi. Traditional technologies for producing salt were used in the factory because it suited the natural geographical and human environments till 1994, when the factory became bankrupt.

From our investigation and analysis, the reasons why the factory faced bankruptcy are related to national and local policies:

Firstly, industrial structure in the world changed from primary (agriculture), to secondary (industry) to tertiary (service, information and education). The Daning Salt Factory belonged to cottage industry and did not develop to the second stage. When China was facing change in its industrial structure, the Daning Salt Factory could not undergo that change, even though some big factories such as Zigong Salt Factory already had successfully faced economic reformation.

Secondly, the government had a policy of protecting big industries and controlling the salt market. The Daning Salt Factory was a small one and could not meet government standards. For example, the salt had no iodine and a high content of sulphur because the coal contained about 5.5–7.5% sulphur and its smoke and dust polluted the salt. The salt in China should have iodine added when it is sold at market, but the government limits the ability of small salt factories (like Daning) to add iodine, so the Daning Factory could not sell its salt in formal markets. Zigong Factory, on the other hand, got the right from the government to sell its iodized salt at market.

Thirdly, the factory's technology was traditional and not exploited in the most economical way. The production costs in the factory were higher than that in other factories. Furthermore the density of the Daning salt spring was not high enough (in winter, 5–6% salt content, in summer, 2–3% salt content) and the factory could not produce salt for 6 month of the year (in winter). But in the Zigong Salt Factory, by contrast, the technologies were quite developed and the production much higher than that in Daning.

Fourthly, because of its geographical situation it was difficult for the Daning salt factory to communicate with outside suppliers. There were no other chemical industries nearby to buy its salt. However, near Zigong Factory there were several chemical factories needing salt from the factory. So Zigong Salt Factory not only sold salt for people's daily life but also sold it to different factories in the same city.

Concluding Comments

Industrial structure sometimes evolves from a primary (agriculture and cottage), to a secondary (big machines and flow-line production), to a tertiary (tourist economy) stage. But Daning Salt Factory, as a cottage industry, goes directly towards the third stage, skipping over the second. It is reported that businessmen from Japan or Hong Kong have visited the factory several times recently. They want it to remain as it is, with its traditional technologies, and exploit it as a sightseeing place together with the Three Gorges.

Development from primary to tertiary industry is the way for small factories such as Daning to free themselves from economic difficulty.

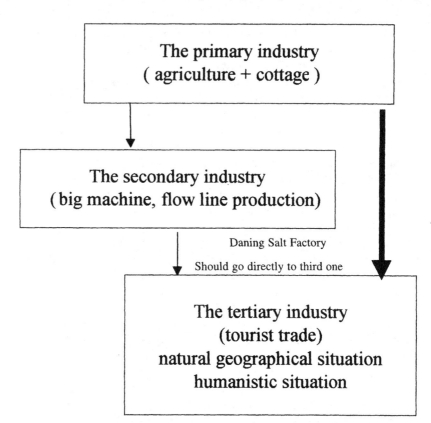

Fig. 3

Notes

[1]   Lin Yuanxiong etc. *Zhongguo jingyan kejishi* (History of science and technology of well salt in China), *Sichuan Kexue jishu chubanshe* (Science and technology publishing house in Sichuan), 1987, p. 277.

[2]   (Song dynasty), Leshi, *Taiping huan yu ji*: "shan nan dong dao qi. Kui zhou lu" （宋）乐史：《太平寰宇记》："山南东道七。夔州路".

[3]   *Hou han shu*, "nan man zhuan" 后汉书・南蛮传.

[4]   (Northern Wei dynasty), Li Daoyuan, *shui jing zhu*, vol. 14 (北魏) 郦道元：《水经注》卷十四.

[5]   Leshi, *ibid*.

[6]   Records of Wuxi Salt Factory, 1989《巫溪县盐厂志》.

Appendix A: Chronicle of Factory Events

Before 316 BC:	started to produce salt.
991:	a stone dragon built to protect salt production.
1195–1224:	a bamboo bridge was built to transfer salt water to the south bank of the Houxi.
1665–1772:	there were 336 stoves using wood. It was the heyday of the factory.
1803–1881:	flooded; 72 coal stoves and 34 wood stoves left. All stoves had become private since the Qin dynasty because of salt water supporting problem.
1942:	stoves changed from 'long' to 'tian' so that high quality salt could be produced.
1951:	salt stoves became publicly owned, later all belonged to the state because all private business was distributed to the public at that time.
1952:	The factory became private again because government wanted to develop private business and concentrate financial resources.
1954:	two wood-fire stoves left; April–May, state owned again.
1957:	official structure of the factory simplified and management personnel decreased.
1958:	movement of iron and steel smelting, locals destroyed the temple near the salt spring.
1959:	the level of salt water was raised 5 meters high, and the pipe for transferring salt water was changed from bamboo to pottery material. In this year the factory received medals respectively from prefecture, province, and ministry because of high salt production.
1966:	introduction of electrical motor, changing the shape of cauldrons by help of Zigong salt water so that working intensity was decreased.
1967–71:	a power station was built, affected by the cultural revolution.
1972–73:	attempt to build new modern stoves failed because of several technical problems.
1976:	new technology for enriching salt water was successfully invented.
1982:	power station expanded to enrich salt water.
1983:	production was contracted by workshops in the factory.
1987:	success at getting rid of fluorine.

PART III

Medicine and the Life Sciences

Traditional Medicine and Pharmacopoeia in the Philippines, 16th and 17th Centuries[1]

MERCEDES G. PLANTA

University of the Philippines, Diliman

> They are not accustomed here to give syrups, nor purgatives, and still less
> to practice bloodletting, their only medicine being commonly known herbs
> and their juices. When they are in a fever they immerse themselves in cold
> water up to the neck... They do not sew up cuts but heal them with banana
> buds roasted, and when still hot, soaked in oil.[2]
>
> Antonio Galvao, *A Treatise on the Moluccas*, 1544

Traditional medicine in Southeast Asia has been influenced by the medical
systems of the other regions in and outside of Southeast Asia. The Indian
Ayurvedic theory had its students in Burma, Thailand and Java. Chinese
medical theory was influential in Vietnam and to a lesser extent in Thailand.[3]

The Roman physician Galen elaborated the Greek humoral pathology
based on the Hippocratic doctrine of the four humors.[4] According to this
medical system, the basic functions of the body were regulated by blood,
phlegm, yellow bile and black bile. The humors had "complexions" with
characteristics or qualities of fire, earth, water and air or vapor. Whereas
blood was hot or warm, phlegm was cold and wet; yellow bile was hot and
dry, black bile was cold and dry. Phlegm had no specific location in the
body, but blood was centered in the liver, yellow bile in the gall bladder
and black bile in the spleen. Each quality's intensity was ranked on a scale
of one to four. For example, "cold" and "dry" vinegar was described as
F_1S_2 or *frio* in the first (and least) degree and *seco* (dry) in the second
degree. Humans as well as medicines, food, and most natural objects had
complexions, which were based on pairs of the qualities of temperature and
degree of moisture.[5]

Famous Arab physicians such as as-Razi (sometimes referred to as
Rhazes, 841/50–903/06 A.D.) added to this doctrine. Although the medicine
practised by the Arabs was chiefly of Greek origin, it had also absorbed some
Indian and Old Persian elements. In fact, it is even doubtful if Hippocrates
"fully expounded" the doctrine of humors that is regarded as an essential part
of his medical system.[6]

Humoral pathology was later transmitted to Spain when the Moslems from the 8th to the 11th centuries dominated the country. This medical complex later diffused from Spain to Latin America during the conquest of the "New World". The Moslems were not only responsible for the diffusion of humoral pathology to Spain, they also carried this disease concept and associated curing techniques to Southeast Asia.[7]

Filipino Humoral Doctrine of Medicine

Filipinos were also aware of the four humors, as suggested in a 19th-century novel, *Sin Titulo*. This novel narrates the traditional beliefs of the people and elaborates on their medical beliefs and practices. In this novel, a Chinese physician Tiang-Song, is asked to examine an ailing young *mestiza,* Charing. When her father, Don Anselmo, requests Tiang-Song's diagnosis, he replies:[8]

> "Heat!!"
>
> "How?" said Don Anselmo... "Do you feel hot? Bah! The windows shall be opened in this house!"
>
> "No... I say what the girl has is... heat! Do you see Charing? It is necessary that you become cool... that you not eat spiced foods... and above all that you not eat eggs, nor *pansit* (noodles usually cooked with diced meat), nor piquant fruits... What you have is heat! And it is understood..."
>
> "What do I have?" asked Don Anselmo at last...
>
> "You... cold!"
>
> "The young are hot... the old are cold... the earth is hot... the air is cold... the heat of the earth gives heat... cold of the air, cold... What is good for you, for me is bad... The human body is not constant... Always different... This one has blood... that one none... here, yes... there, no... this one strong... this one weak!"
>
> "Ah, understood, understood... Señor Tiang-Song... therefore there are two... One of the fright (*susto*) and another of the heat and of cold... It is clear... That is why the Spaniards, when they experience fright, say: *no se calor usted...* "
>
> "And please tell me, Señor Tiang-Song, can Charing take chocolate?"
>
> "Ah, no, hot."
>
> "And mangoes?"
>
> "Also hot."
>
> "And eggs?"
>
> "Ah, no... hot."
>
> "And ice cream?"
>
> "Ice cream... yes... because it is cold!... "
>
> "Chicken, hen?... That cannot harm her?" asked Don Anselmo.

"Why not? Hot! *Lechon* (roast suckling pig) is good, *tajuri* (a white gelatinous food sold by the Chinese), the *atole* (a ground corn gruel), the *nido* (bird nest), squash... are good for her."

The preceding vignette suggests that the concept of humoral pathology existed in the Philippines. Good health, according to this system, required a person to maintain his/her individual balance of the four humors and their qualities. Sickness, on the other hand, was the result of a humoral imbalance and extremes of hot and cold, dry and wet. The seasons, wind, mode of living (e.g. diet), age and climate could influence one's humoral equilibrium.[9]

Filipino belief in humoral pathology could also be seen in statements/ descriptions made by the native *hilot* when making certain diagnoses, such as *may lamig* (cold) or *hangin sa likod* (air in the back), when referring to people who take a bath before drying their backs of perspiration. Sometimes, the *herbolarios* would say that a person is not well because *mainit siya*, meaning, the person has too much heat in the body.[10]

In this connection, Rizal made the comment: *El aire, el calor, el frio, el vapor de tierra y la indigestion, son las unicas causas patogenas que se admiten en el pais* ("Air, heat, cold, mist, and indigestion are the sole pathogenic causes in the country [Philippines]").[11] Jean Mallat, who made references to this subject, associated these beliefs with the Chinese, not Filipinos. He said:

> The Indios maintain that air plays a great role in all diseases, and the Chinese add that it is the lack of a balance between cold and hot and the perpetual struggle between these two principles that cause all disturbances that afflict the harmony of the human body.[12]

Such Chinese practices could also be seen in the Filipino practice of forcing a person with fever to perspire profusely by wrapping him in a blanket in order to release the heat from the body so as to cleanse it of impurities. These beliefs and practices may also be traced to the idea of establishing balance of hot and cold in the body as well as to cleanse out impurities.

Thus it was for extracting air that the early Filipinos employed a great number of topical remedies and repercussives. The *tandoc*, an instrument of their invention, was used for this purpose. It was a piece of buffalo horn pierced at the end and which, applied by suction, had the effect of a cupping glass. A leaf or a bamboo skin placed at the suitable moment to support the vacuum closed it. Applied for instance on the back and along the posterior part of the neck, it was moved from top to bottom (*astiran tandoc*), producing an effect stronger than ordinary cupping glasses.[13]

To re-establish harmony between hot and cold, the early Filipinos administered certain remedies possessing one of these two qualities. When they talked about a fruit or a plant, they said that one was hot and the other cold. They also practised *pisil* or massage, which almost immediately relaxes the fatigued members and disposes one to fitful sleep. It was a kind of magnetism that brought out the most beneficial effects.[14]

The emphasis on cooling rituals and medicines, in this context, seems to belong as much to Filipino as to Chinese practitioners. These beliefs, however, were not alien to the Southeast Asians. Long before these medical practices were brought to Southeast Asia, people already had these concepts in their belief systems. According to Reid, "the most practical effect of these beliefs was to strengthen what may have been an indigenous Southeast Asian belief that illness arose from excessive heat brought about by fever or pregnancy, from dangerous loss of heat due to childbirth, or from the entry of excessive dry or moist air".[15]

Local practitioners relying on local folk remedies and herbs did the majority of the healing at that time in relation to the experimental "science", primarily that of anatomy and surgery, which began to make progress in Europe starting in the 16th century.[16] This gave an impression that:

> Medicine cannot merit the name of a Science amongst these people... They trouble not to have any principle of medicine, which they have learnt from their ancestors, and in which they never alter anything. They have no regard to the particular symptoms of a disease; and yet they fail not to cure a great many.[17]

In the 16th and 17th centuries European surgeons specializing in amputation, the setting of serious fractures, the removal of growths, and even bleeding were already in demand in Asian cities. However, the interventionist style of early European "scientific" medicine probably killed more patients than it cured. Europeans in the East initially had more to learn than to teach. They found out from experience that for most complaints, it was safer to trust an Asian practitioner than a European. The reason for this was: "Asian medicines do not alter nature, but assist her in her ordinary functions, drying up the peccant humours without any trouble to the sick person at all."[18]

Hygiene

The use of herbs, baths and massage formed the medical cures of the early Filipinos and Southeast Asians. A feature of Filipino life which always attracted the Spaniards was the Filipinos' propensity to take a bath. Dr. Antonio Morga, in the early 17th century, notes:

The young and the old ordinarily bathe their entire bodies in the rivers and streams without regard to whether this may be injurious to their health, because they find it to be one of the best remedies to be healthy; and when a child is born, they immediately bathe him and likewise the mother.[19]

Morga's contemporary, Father Chirino, observed that the early Filipinos were a very clean and hygienic people. He said:

From the day they are born[,] these islanders are raised in the water, and so from childhood both men and women swim like fish and have no need of a bridge to cross rivers. They bathe at all hours indiscriminately, for pleasure and cleanliness, and not even women who have just delivered avoid bathing or fail to immerse a newly born infant in the river itself or in the cold springs.[20]

"The Indios, men as well as women, take the greatest care of everything that has to do with the cleanliness of the body, and perhaps the former still more than the latter; they take a bath and wash themselves everyday in the river, and at least twice a week at home; for them it is a pleasure as well as a need."[21] Chirino further states:

They also take a bath for medicinal purposes, for which God our Lord has given them a number of hot springs. During these past few years, those of Bay, on the shores of the lake of the same name, have become very popular, and many Spaniards, men and women, ecclesiastics and religious, suffering from various illnesses have had recourse to them and have recovered their health.[22]

Whereas the conventional wisdom in Europe was that bathing was voluptuous or dangerous, the Filipinos associated it with purification and "cooling", without which the body could not be healthy. The Filipinos' desire for cleanliness, as manifested in their propensity to take a bath, almost amounted to fastidiousness. Moreover, the care for the body, and neatness and elegance in dress were all matters of great importance. Morga further relates:

Both men and women, particularly the prominent people, are very clean and neat in their persons, and dress gracefully, and are of good demeanor. They dye their hair and pride themselves with keeping it quite black. They shampoo it with the boiled bark of a tree called *gogo*[23] and anoint it with oil of sesame, perfumed with musk and other sweet-smelling substances. They are all careful of their teeth, and from their early age, they file and even up their teeth with grinders and other implements of stone, and give them a permanent black color which is preserved until their old age, even if it be unpleasant to the eyes.[24]

It could be interpreted then that the Filipinos' desire for cleanliness was a function of their belief that being clean corresponded to being healthy.

The Betel

The abundance of medicinal drugs and herbs in the Philippines, called *tambales*, constituted the traditional medicinal cures of the early Filipinos. The word *tambal* refer to all plants, roots, leaves, or bark with medicinal properties. For this reason the Spaniards called the *tambalan* an *herbolario*, "one who has knowledge of plants (*hierbas*), their powers and properties". *Tambales* were taken internally or externally, with plasters or poultices distinguished as *haklup* or *tampus,* respectively.[25] Many of these herbs were effective against various digestive and intestinal disorders as well as infection. Of all the herbs consumed, however, particular attention should be made to the betel leaves because the early Filipinos of all ages, backgrounds and social classes constantly chewed these.[26]

Betel chewing was introduced to the Philippines two thousand years ago when sailors from the Malay Archipelago spread the practice throughout Southeast Asia. Since then, betel has been chewed for social, medicinal, magical and symbolic purposes.[27]

The natives reported marvelous cures obtained from the betel nut. Whatever juice entered the stomach was found to be advantageous. It tasted good and cured certain ailments. It strengthened and fortified the teeth and preserved the gums from rheum.[28]

According to local tradition, betel chewing also aided in digestion and prevented dysentery. The juice of the betel leaf was used against eye infections, infections in wounds and sores, and various menstrual and other ailments.[29] With the juice produced by mastication the early Filipinos rubbed the bodies of children to strengthen them. The residue, which is called *zapa*, was ordinarily thrown away as useless; sometimes, however, they swallowed it or applied it as a topical remedy in the epigastrium of sick children. This remedy was often effective.[30]

It seems likely that the betel-chewing habit alone may have protected the Filipinos, as well as the Southeast Asians in general, from many waterborne diseases, as well as contributed to the remarkable freedom from infection noted by several observers.[31]

Local and Herbal Medicines

One major source of information for traditional medicine and pharmacopoeia during the 17th and 18th centuries was the indefatigable Jesuit priest, Father

Francisco Ignacio Alcina. He produced a manuscript in 1668 that listed the plants and herbs that the Filipinos of that period used for medicinal purposes. Alcina believed that evaluating the benefits of these medicinal roots, which he had seen himself and/or were brought to his attention by the native experts who became his informants, would be useful to those who would live on the islands.[32] I have limited the list to those plants and herbs that have been provided with scientific names.

From Alcina's catalogue the Filipinos were known to have cured mouth blisters with the juice of the *anonang* (*Cordia sebestena*; Eng. sebestian plum, soapberry leaves). So powerful was the *anonang* that its juice immediately revived an unconscious person. An infusion of its bark also soothed ulcers and stomachaches (see Fig. A).

Fig. A. *anonang* (*Cordia sebestena*: Eng. sebestian plum, soapberry leaves)

Fibers from the fruit called *salacsalac* (*Luffa cylindrica*; Tag. *patolang bilog*; Eng. vegetable sponge) were an effective purgative. The Spaniards called it "cucumbers of Saint Gregory". They were supposedly effective in relieving constipation and gastric problems. It was also used as antidote for poison and venom. According to Fr. Alcina, he himself has cured many sick persons who were almost beyond remedy with its seeds mixed with *tierra de San Pablo*.[33] It enabled him to "extract" worms, of more than three palms in length and hair-like in texture, as well as other harmful matters from the human body.

To the Visayans, the little tree called *salibutbut* (*Tabernaemontana laurifolia*), known to the Tagalogs as *pandacaqui*, had numerous curative powers. Alcina observed two kinds of *salibutbut*: one, with the white bark, which he considered more effective as remedy and the other a black-barked *salibutbut*. The juice of the *salibutbut* leaves, which was squeezed into fresh wounds, and its roots when made into charcoal, were effective in preventing convulsions and drying out ulcers in a few hours. A poultice of *salibutbut* leaves was considered effective even for treating old wounds (see Fig. B).

Fig. B. *salibutbut* (*Tabernaemontana pandacaqui*)

The *agonoy* (*Wedelia biflora*) herb, which the Spaniards called "herb from the Moluccas", was common in the Islands and was considered medicinal by Alcina. It was used for ulcers and old wounds. Its leaf when ground and boiled in coconut oil cleanses sores, draws out the matter from the body and restored the flesh quickly. The root relieves stomach pains. When boiled in water, it stopped the swelling of the legs and other joints. Distilled, the fluid of the leaves dissolved kidney stones and relieved difficulties in urination.

Two different little trees, the *balocas* and *ronas* (*Smilax fistulosa*; Tag. *hampás-tigbalang*) were valuable for their ability to relieve constipation and other similar illnesses. They both prevented estitequez and other stomach ailments. Their roots, however, could affect defecation in extreme use.

A variety of herbs, leaves and roots that the Visayans called *pamughat*, were used to heal *sinat* or relapses in sickness. The boiled water from the leaves of the *argao* (*Premna odorata*; Tag. *alagau*) tree was an effective remedy for colds. The hot leaves when placed on top of the head, down to the eyebrows and neck, stopped bad air from penetrating into the sick person. It must be pointed out that the natives, in the 15th and 16th centuries, made a reference to "bad air/wind" that caused illness or discomfort. This affirms the Filipinos' belief in the concept of humoral pathology as cited previously.

The *tangantangan* (*Ricinus communis*; Tag. *liñgang-sina*; castor oil plant) referred to by the Spaniards as "little fig trees of hell", was another variety of *pamughat*. They were considered to be very effective for cold stomachs, illnesses of the abdomen, chest and other parts of the body. The natives considered the variety of the *tangantangan* with reddish stems or twigs to be more effective. The *tangantangan* oil, which we call castor oil today, was said to be effective for malignant fevers and dropsy. It even killed tapeworms and relieved earaches and fluxion. Women used the oil from its fruit to treat their hair and scalp. The Tagalogs called this process *liñgansina*.

There were also various herbs, shrubs and trees that had special value to women. Visayan women who suffered from menstrual cramps were advised to drink the water of the boiled root of the *arogangan* (*Hibiscus rosasinensis*; Tag. *gumamela*; Eng. hibiscus, China rose). The buds mixed with aloe vera, were pounded and applied to the forehead for the same ailment. Similarly, mixed with lime, it could be applied as plaster to burst growths and inflammations. When the buds finally bloomed, they were boiled to induce sweat (see Fig. C).

Fig. C. *arongangan* (*Hibiscus rosasinensis*; Tag. *gumamela*; Eng. hibiscus, China rose)

The roots of the *marol* (*Nyctanthes sambac*; Tag. *sampagita*; Eng. Arabian jasmine) which the Tagalogs called *sampaga*,[34] aided women in childbirth because of its ability to heal wounds. Women who have just given birth, however, were cautioned to use this in small quantities because a large amount could be harmful (see Fig. D).

Fig. D. *marol* (*Jasminum sambac*; Tag. *sampagita*; Eng. Arabian jasmine)

The *coloncogon* (*Ocimum virgatum*; Tag. *sulasi*; Eng. sacred or holy basil) a rough variety of basil, for example, was not only used for pickling and other stews because of its fragrant and aromatic mint. It was said to be effective for stomach ailments. Newly-born babies were also rested and wrapped on its leaves. The Tagalogs sometimes called it *locoloco*, while the Spaniards referred to it as "mountain basil" (see Fig. E).

Fig. E. *coloncogon* (*Ocimum virgatum*; Tag. *sulasi*; Eng. sacred/holy basil)

A tree of wide and long trunks called "white canary" or *pile* (*Canarium album*) was said to be effective for swellings on the legs and stomachaches. Placed at the back, it induced perspiration and relieved cough. The water in which the *pile* root was boiled was given to women who had just given birth. The oil from its fruit was considered to be an effective medicine to stop bleeding, making it ideal for childbirth. It was also considered effective for rheumatism when smeared on the affected part (see Fig. F).

Fig. F. *pile* (*Canarium album*)

As part of their hygiene, Visayan women used the oil from the fruit of the tree called *balocanar* (*Reutealis trisperma*) because it killed the "little animals" on their heads. They also anointed the heads of their children with its oil, presumably, to prevent them from getting lice.

There were also medicines of animal or mineral origin. The soil of the abandoned termite mounds was good for fumigating; calculous stones found in a shark's head were a cure for stones in the bladder or kidney; and powdered *duyong*[35] were a panacea, good for bloody stools, diarrhoea, colds and fevers. A kind of jet-black coral was not only taken internally, but also worn on the wrist as jewelry and as a preventative for rheumatic pains.[36]

Roles of Traditional Medical Practitioners

Because of the wide range of skills demanded of them, many of the traditional medical practitioners were not limited to a single role. The *hilot* and *herbolario* may have had primary roles but they also possessed other complementary functions. The *hilot* as the traditional birth attendant is someone who may also be knowledgeable in the use of certain medicinal plants.[37]

The practitioner's role, therefore, may be divided into two categories: (a) the empirical; and (b) the magico-religious. Empirical roles include bone setting and massage, the use of medicinal plants, midwifery, treatment of wounds and dentistry. Magico-religious roles include the shamanic (acting as medium for the spirits), the use of prayers and sacrifical offerings, magical methods of divination and specialties such as recovering lost or wandering souls.[38]

Thus, by the 17th century, the Filipinos already had a defined medical tradition. Their practices, though, were not based on the European practitioners. Practitioners were a select group. According to Tan, "the acquisition of skills, especially the empirical ones such as bone setting, birth delivery and the use of the *tambales* were associated with apprenticeship".[39] These healers played important roles in their communities and were highly respected; generally, only a member of the *maginoo* or *principalia* (elite) class could occupy these positions.

Like patients of contemporary times, patients then were highly selective. They chose the practitioner who had the best reputation for getting results and every successful practitioner mixed elements of magic with medicine.[40] The Spaniards, therefore, in the period when they formally established their colonial state in the Philippines, believed that magic had a larger place than "science" in our country's belief system.

Notes

[1] Paper read at the *9th International Conference for the History of Science in East Asia*, August 23–27, 1999, Singapore. This research paper is part of my thesis for the degree of Masters of Arts in History, Department of History, College of Social Sciences and Philosophy (CSSP), University of the Philippines, Diliman, 1101, Quezon City, Philippines. The thesis is entitled, *Traditional Medicine and Pharmacopoeia in the Colonial Philippines, 16th to the 19th Centuries.*

[2] Anthony Reid, *Southeast Asia in the Age of Commerce, 1450–1680*. Volume One: *The Lands below the Winds* (New Haven: Yale University Press, 1988), p. 52.

[3] *Ibid.*

[4] Donn V. Hart, *Bisayan Filipino and Malayan Humoral Pathologies: Folk Medicine and Ethnohistory in Southeast Asia*. Data Paper no. 76, Southeast Asia Program (New York: Cornell University, November 1969), p. 3.

[5] *Ibid.*, p. 4.

[6] *Ibid.*, p. 3.

[7] *Ibid.*, pp. 4–5.

[8] *Ibid.*, pp. 11–12. I have not myself read the novel but Donn V. Hart quoted this. See also Francisco de Paula Entrala, *Sin Título* (Manila: Ramirez y Giraudior, 1881), pp. 40–45. One purpose of Entrala in writing *Sin Título* was to ridicule the popular

mediquillo of 19th century Manila. Ee Also wenceslao E. Retana, Noticías histórico-bibliográficas de el teatro en Filipinas desde sus orígenes hasta 1898 (Madrid: Libreria General de Victoriano Suarez, 1909), pp. 108–110. A *mediquillo* was a "simple practitioner with some notions of medical science; a little or petty physician", in José Nuñez, "Present Beliefs and Superstitions in Luzon", Emma Blair and Alexander Robertson, *The Philippine Islands, 1493–1803* (Cleveland: Arthur H. Clark, 1903–1909), Vol. 43, p. 314.

9   *Ibid.*, pp. 4–5.

10 From my own personal experience because I have often times consulted *herbolarios* for relief of certain bodily discomforts.

11 Jose P. Bantug, *A Short History of Medicine in the Philippines During the Spanish Regime, 1565–1898* (Manila: Colegío Médico Farmacéutico de Filipinas, Inc., 1953), p. 12. See also Jose P. Rizal, "La curacíon de los hechos hechizados", Bantug's Edition, *Revista Filipina de Medicína y Farmacía*, Vol. 22 (Manila: n.p., December 1931).

12 Jean Mallat, *The Philippines: History, Geography, Customs, Agriculture, Industry and Commerce of the Spanish Colonies*. Translated from the original French by Pura Santillan-Castrence in collaboration with Lina S. Castrence (Manila: National Historical Institute, 1983), p.281.

13 *Ibid.*, p. 285.

14 *Ibid.*

15 Reid, p. 52.

16 *Ibid.*, p. 53.

17 *Ibid.*

18 *Ibid.*, pp. 53–54.

19 Antonio de Morga, *Sucesos de las Islas Filipinas*. With an annotation by Jose P. Rizal (Manila: Jose Rizal National Centennial Commission, 1962), p. 246.

20 Pedro Chirino, *The Philippines* in 1600. Translated from the original Spanish by Ramon Echevarria (Manila: Historical Conservation Society, 1969), p. 258.

21 Mallat, p. 29.

22 Chirino, pp. 258–59.

23 According to Rizal, "rather than the bark it is the body of the shrub that is crushed but not cooked". Morga, p. 246.

24 Morga, p. 246.

25 Scott, p. 62.

26 Reid, p. 54.

27 Paulo K. Tirol, "When Betel-Chewing Was In", *Philippine Daily Inquirer* (PDI), Sept. 6, 1999.

28 Morga, pp. 257–58.

29 Reid, p. 54.

30 Mallat, p. 284.

31 Reid, p. 54.

32 Francisco Ignacio Alcina, "Breve resumen de las raices, hojas o plantas medicinales mas conocidas" (A Brief Summary of the Best-Known Medicinal Roots, Leaves

and Plants, etc.), Parte I, Capitulo 26, *Historia de las islas e indios de bisayas*. Translated from the original Spanish by Cantius Koback and Lucio Gutierrez, *Philippiniana Sacra*, Vol. 32, No. 94, January–April 1997, p. 96. Alcina's list contains the following plants, pp. 96–135. Pictures of plants are taken from Blanco's *Flora de Filipinas*. Reissue of the Grand Edition, 1877–83. Intramuros, Manila: San Agustin Convent, 1993.

[33] A kind of oil.

[34] The old Tagalog dictionaries like Noceda-Sanlucar, have only the entry *sampaga, Flor como el jazmin*, p.283. Jose Villa Panganiban's *Diksyunaryo-Tesauro Pilipino-Ingles*, 1st ed., differentiates *sampaga* from *sampagita*. *Sampaga*, jasmine-type shrub with white flowers are bigger than those of the *sampagita*. *Sampagita*, jasmine-type with white flowers are smaller than those of *sampaga*, referred to often as "tropical rose", adopted as the National Flower of the Philippines. The term *sampaguita*, spelled with an "u", has come into Philippine languages through Spanish. *Sampaguita* comes from the Arabic *zanbag* and means jasmine or lily.

[35] Panganiban, 1st ed., s.v. "*isdang laot* (Tag); sea cow (Eng.)" .

[36] Scott, p.120.

[37] Michael Lim Tan, *Traditional Medical Practitioners in the Philippines* (University of the Philippines, Diliman: College of Social Sciences and Philosophy Publications, 1996), p. 8.

[38] *Ibid.*

[39] *Ibid.*, p. 11.

[40] Reid, p. 57.

Biology in Song China

LUO GUIHUAN (罗桂环)
Institute for the History of Natural Sciences
Chinese Academy of Sciences

The period of the Song dynasty was a very prosperous one for lore about organisms or biology in China's history. As more information about plants and animals was obtained there was tremendous incentive for the development of biology. The various biological writing rapidly increased. Some were developed from previous more extensive works such as "Yi Bu Fang Wu Lüe Ji" (益部方物略记 A Short Record of Products in Yi Zhou 益州) and "Nan Fang Cao Mu Zhuang" (南方草木状 The Herbs and Trees in the South of China). Some were derived from former Ru (儒) classics, such as "Pi Ya" (埤雅) and "Er Ya Yi" (尔雅翼) [both names mean the extension of Er Ya (尔雅)]. Some were more substantial in content than former works of the same kind, like "Tong Pu" (桐谱 Records on Paulownias), "Ju Lu" (橘录 The Systematic Records of Citrus), and so on. Meanwhile natural histories (herbals) made notable advance as well. Some very important monographs with many illustrations were written, which helped people greatly in recognizing plants and animals as well as spurring the development of botanical writings such as the "Tu Jing Ben Cao" (图经本草 The Illustration Classic on Herbals). Most importantly, however, Zheng Qiao (郑樵) wrote the treatises including "Kun Chong Cao Mu Lüe" (昆虫草木略 The Monograph on Animals and Plants), which not only dealt with lore about plants and animals (biology), but also set this as one of the 20 ancient important subjects. Furthermore such biological classics as "Qin Jing" (禽经 Classic on Birds) spread biological knowledge. Yet another scholar edited a book on botanical information named "Qüan Fang Bei Zu" (全芳备祖 Historical Records on Beautiful Plant). These and other works illustrate that there were many scholars endeavoring to set up the study of plants and animals as a subject during the Song dynasty.

As in many other nations, the Chinese study of organisms was closely concerned with ancient agriculture, medicine, and horticulture. But this lore also had features uniquely Chinese, particularly relating to the research of Ru (儒) classics. One of the greatest of China's philosophers, Confucius, had said: "to study Shi Jing (诗经 one of the Ru's classics) — you can know many names of birds, mammals, herbs and trees".[1]

During the North Song dynasty (960–1127), some famous officials made important contribution to the lore on organisms. They loved natural history and described the new plants and animals with enthusiasm when they worked in the south of China. This greatly increased people's knowledge of the plant and animal world in the south of China. It is well known that by the middle period of the Tang dynasty, the south of China was less developed than north China in both economy and culture. The wealthy of plants and animals in the south of China was little known by outsiders before the Eastern Han (东汉) dynasty. From the Eastern Han dynasty to the Tang dynasty (25–618), many natural history books about the south of China were written. These writings described a great many organisms there, actively helping "ancient biology" forward. Scholars in the Song dynasty developed this tradition in a more detailed way, some of them paying more attention to the records in plants and animals. There appeared a new kind of biological work, the most famous of which was Song Qi's (宋祁 998–1061) "Yi Bu Fang Wu Lü Ji" (益部方物略记 Short Notes on Products in Yi Zhou).

Song's was the first book on plants and animals in Si Chuan (四川). He wrote the "Yi Bu Fang Wu Lü Ji" when working as a prefecture official in Si Chuan. His writing principle was to note the things which could not be seen in Eastern China, or which could be seen but were different in South China. Song's was different from previous natural history works in that it described all plants and animals there. Song recorded 65 forms of plants and animals with drawings. Those organisms were divided into seven categories: herbs, trees, medicine, birds, mammals, insects and fishes.

Song's book was really able to record some distinctive characters and the nature of plants and animals distributed in Si Chuan and even south-west China, such as Nanmu (楠木 *Phoebe nanmu*), Alder (*Alnus cremastogyne*), Alocasia (*Alocasia macrorrhiza*), Hosie Ormosia (*Ormosia hosiei*); Snub-nose monkey (*Pygathrix roxellanae*), giant salamander (*Megalobatrachus davidianus*) and so on. His descriptions were also very remarkable.

Progress in horticulture during the Song dynasty contributed to many special treatises on ornamental plants, especially famous flowers and fruits plants. The content of these writings included much historical information and observations by the authors themselves. There was a long history of describing single sorts of famous useful plants going back to middle of the 4th century, when "Zhu Pu" (竹谱 Notes on Bamboo) was written. Down to the Song dynasty, much more of this kind of writing appeared. The most famous of these were Ou Yang Xiu's (欧阳修 1007–1072) "Luo Yang Mu Dan Ji" (洛阳牡丹记 Records of Tree Peony in Luo Yan), Liu Fen's (刘攽 1022–1088) "Shao Yao Pu" (芍药谱 Notes of Peony), Zhou Shi Hou's

(周师厚) "Luo Yang Hua Mu Ji" (洛阳花木记 Notes on Flowers and Decorated Trees in Luo Yang), Chen Zhu's (陈翥) "Tong Pu" (桐谱), Cai Xian's (蔡襄) "Li Zhi Pu" (荔枝谱 Notes of Litchi), Liu Meng's (刘蒙) "Ju Pu" (菊谱 Notes of Chrysanthemum), and "Xie Pu" (蟹谱 Notes of Crab, written about 1059), etc. A few treatises described ornamental plants, such as "Luo Yang Hua Mu Ji". However most of these described the form of a famous flower or fruit, such as "Shao Yao Pu" and "Li Zhi Pu". Some works were of more botanical interest: Chen Zhu's "Tong Pu" was a good example.

About 1054, the country scholar Chen Zhu wrote a treatise on Tong Chen's book, whose writing method differed from the former treatises. He first planted some Tong trees on a plantation for observation, then collected the information to write his work. In the treatise the author made distinctions among the several trees which had the name Tong (included the modern Phoenix tree *Firmiana simples*, Paulownia *Paulownia spp.*, and Tungoiltree *Vernicia fordii*). Those trees are very similar in the shape of their leaves and trunk. Chen accurately recorded their characteristics, summarized the methods of their culture and control of harmful insects. As a piece of botanical workmanship, it is remarkable for the accuracy of its description of shapes of flower and form of fruit, as well as spread of seed.

In addition to the new natural history books and the writings on ornamental plants, some scholars still wrote other kinds of treatises about organisms. There are many sorts of writings about plants and animals with the name Jing (经 classics) written in the Northern Song dynasty (960–1127), such as "Cao Jin" (草经 Classic on Herbs), "Hua Jing" (花经 Classic on Flowers), "Qin Jing" (禽经 Classic on Birds), etc., which introduced biological knowledge to the population. Although "Cao Jing" has not survived, "Qin Jing" has fortunately remained, which allows us to understand this sort of writing. The author of "Qin Jing" was written as Shi Kuang (师旷), a famous scholar in the Spring and Autumn dynasty (ca. B.C. 500). However, its first appearance was in Lu Dian's (陆佃) "Pi Ya" (埤雅) (ca. 1078–1085), and a lot of its information came from writings produced between the Jin (晋) dynasty and Tang dynasty (265–618), so it generally was considered to have been formed in the Tang or Song dynasties. Considering that this book had never been recorded in writings before the Song dynasty, and that other biological monographs were edited in the Song time, such as Nan Fang Cao Mu Zhuang (南方草木状), I think it most plausible that it was produced in the Song dynasty. "Qin Jing" was named "Jing", and assumed Shi Kuang's name. Perhaps the real author wanted to give the readers the idea that the history of the subject of plants and animals was very old.

"Qin Jing" was the first treatise about birds in China's history. It recorded a number of birds including osprey, eagle, red kite, great cormorant, kingfisher, crane, woodcock, pelicans and dove as well as various pheasants such as golden pheasant, etc. The accounts included various names of bird, their appearance, their habit and cries, as well as adaptability to environment and migration. For example, the author wrote: "Wild gooses association each other by calling, they live in groups, one of which was guardian for safety of a mass of birds". The author also pointed out how the various shapes of the mouth of the birds adapt to the way they live. "The water birds which eat small fishes in shallow water have a long beak, the birds which eat seeds and herbs have a short beak, the birds of prey have a sharp beak". The author also points out that the feathers change their colours with the change of season and environment.

The progress of herbals was also evident in the north Song period. While the herbals are mainly pharmacopoeia, most of their contents deal with the plants and animals. In the beginning of the Song period, because governments paid so much attention to the medicine, some herbals were edited. Among which the "Tu Jing Ben Cao" (图经本草) edited by Su Song (苏颂) was the most important in history of biological development. It contains an extensive summary of previous studies on creatures.

Su had deep historical knowledge, and he published "Tu Jing Ben Cao" in 1061, which contains some 600 kinds of medical plant and animals and included more than 900 illustrations. The work follows the tradition of adding illustrations to the herbal since the Tang period. But Su also made some changes to the method. Both statement and illustration of each plant or animal were set together, not as in former herbals where all of the illustrations were set together as a part of book. The layout of Su's book could help reader better understand the creatures recorded in the book, thus he went beyond his predecessors by the more elaborate treatment. This method influenced later eminent plant monographs such as "Jiu Huang Ben Cao" (救荒本草) in the Ming period and "Zhi Wu Ming Shi Tu Kao" (植物名实图考) in the Qing period. "Tu Jing Ben Cao" is an indication of the early development of the natural history in China. Su's monograph not only contained many illustrations, but also contained a lot of accurate description about organisms. It is why the monograph were praised highly by successors and its descriptions about plants and animals always were cited gladly by later herbals.

Moreover, the development of new Ru philosophy also encouraged the development of studies about plants and animals in the Song dynasty, since the new Ru philosophy made many scholars pay more attention to research in the classics. The classics "Shi Jing" (诗经 Book of Songs or Classics of Poetry) and "Er Ya" (尔雅 Names of Thing and Its Explanation), both recorded many animals and plants. For this reason, many scholars in China endeavored to study plants and animals in the former works and so expanded the knowledge about organism continually. Down to the Northern Song dynasty, Lu Dian (1042–1102) wrote "Pi Ya" (The Expansion of Er Ya), which was an excellent writing of this sort.

Lu Dian was a pretty orthodox scholar. His book was divided into 20 chapters, two of which were for the explanation of fishes, three for mammals, four for birds, two for insects (including worms), two for horses, two for trees, and four for herbs. Besides these, there were two for the explanation of nature. The book explained a total of 185 kinds of animals and 92 kinds of plants.

As more information about organisms was obtained in the Northern Song dynasty, there was tremendous incentive for the development of biology in the Southern Song dynasty. In the beginning of the Southern Song dynasty, the importance of plants and animals as a subject had been discussed by eminent scholar Zheng Qiao (郑樵 1103–1162).

Zheng Qiao was a famous scholar with profundity of learning. He thought that studying various plants and animals was basic. For example, if someone wanted to study "Shi Jing" he must have some knowledge about plants and animals to understand the poet's descriptions. Since people in the Han dynasty (187 B.C.–220) did not understand this, it made the "Cao Mu Niao Shou Zhi Xue" (草木鸟兽之学 The study of plants, birds and beasts) stagnant. In the Three Kingdoms Period (220–265), Lu Ji (陆机) studied these questions and wrote his "Explanation of Herbs Trees Birds Mammals Insects and Fishes in the Shi Jing Edited by Mao". But Lu's Explanation was not systematic. Even later, the subject made little progress, the reason for which was, as he said:

> In general, most of the Confucian scholars did not recognize plants and animals as a field. On the other hand, the farmers and gardeners (who knew the plants and animals well) did not write books. The two could not combine, which made the study of birds, mammals, herbs, and trees impossible to develop. Only the herbals related to the lives of human beings (through) medicine; the other book merely relayed the records of other books.[2]

Obviously he understood the obstacles to progress in China's biology, as well as the contributions of herbals to the accumulation of biological knowledge.

Realizing the importance of biology, Zheng decided to do something for the subject's development. He had lived in a mountain region and observed the various plants and animals there for thirty years. Meanwhile he often posed biological questions to farmers and gardeners, eventually writing his important monograph "Qun Chong Cao Mu Lüe" (昆虫草木略 Treatise on Animals and Plants), which is one of the twenty Lüe (treatises or monographs) within his magnum opus "Tong Zhi" (通志 A Systematic History of China). It is Zheng's contribution to utilize a biological treatise as a part of historical work. Before Zheng, there was never any historical work that had included in its content biology.

Zheng's monograph cited many predecessors, included herbals, various classics, and works about agricultural technology. This indicated that he had systematically studied the literature of natural history, and showed his great efforts to develop the "Cao Mu Niao Shao Zhi Xue" (草木鸟兽之学 The study of plants, birds and beasts). Through his writing we can see that Zhen really tried to set up a new subject by uniting antecedent results obtained by herbalists and scholars who had studied the classics, and he achieved some degree of success. His book was cited by successors, and his methods of naming organisms greatly influenced later scholars such as Li Shi Zhen (李时珍).

In "Qun Chong Cao Mu Lüe", Zhen Qiao combined the methods of classification of both "Er Ya" and "Ben Cao Jin Ji Zhu" (草木经集注 Annotation of Herbals written by the Shen Nong 神农 Emperor) by Tao Hong Jing (陶弘景 456–536). He divided his various items into herbs, vegetables, cereals, trees, fruits, insects and fishes, birds, and mammals. He collected information about different names for flora and fauna to explain 130 animals and 340 plants.

Zheng also edited other biological book like "Dong Zhi Wu Zhi" (动植物志 Notes on Plants and Animals).[3] This was perhaps influenced by the writings and illustrations of Song Qi (宋祁) and others. Zheng Qiao considered that illustrations were very important for biological works. He wrote: "For understanding the various things in nature, you have to know plants and animals. One would not be able to recognize animals and plants well without their illustrations."[4] From Zheng's work, we see that some scholars paid great attention to illustrations in biological works in the Song dynasty; this advanced tendency was a symbol of progress in biology then.

Zheng had written in his "Tong Zhi": "To summarize all the lores of the world, dividing them into different sections, named Lüe, gives a total of twenty Lüe (略). All the contents of history and of what scholars can study are included in these Lüe."[5] These Lüe, including "Qun Chong Cao Mu Lü" (昆虫草木略), recognized the importance of biology.

To consider biology as an important subject was certainly not only Zheng's idea. Another famous scholar Cheng Dachan (程大昌 1123–1195)

wrote in his "Yan Fan Lu" (演繁露): "As Da Xue[6] (大学) had said, when you want to learn, you must study the cause and effect of various things at first. According to the instruction of the sages, the pupils have to study as many names about bird, mammal, grass and tree as they can." It is obvious that to study organisms was considered an element for other learning in that time. It gradually made many scholars endeavor to enrich the knowledge of organisms.

Because of the many scholars' attention, more pure biological writings appeared in the Southern Song dynasty (1127–1279). As the economic center was now located in the south of China, so the biological writings about that place developed. One of these was "Nan Fang Cao Mu Zhuang" (南方草木状 The Herbs and Trees in Southern China) which is assumed to be written by Ji Han (嵇含 263–306). This book, which defined the record of plants in the south of China, appeared about the beginning of the 12th century[7]. The contents of the book were somewhat similar to "Qin Jing" (禽经 The Classic of Birds); both of them took much information from writings produced between the east Han dynasty and Tang dynasty. Its editor named Ji Han as the author for two main reasons. One was that Ji Han was appointed as prefecture official in Guang Zhou (though he had never been there); another was that many writings recording the products of southern China were written with various names such as "Yi Wu Zhi" (异物志 Description of Special Things) in Ji Han's time. Of course, the most important reason was to make the time of the writing appear to be earlier than it was; to make people think the origin of this kind of work was very old so as to attract more attention.

"Nan Fang Cao Mu Zhuang" has been thought of as the first botanical book on southern China. It was divided into three chapters. Chapter one describes 29 forms of herbs. Chapter two, 28 forms of trees, and chapter three, 23 forms of fruit and bamboo. It is a famous botanical monograph. Although "Nan Fang Cao Mu Zhuang" was not really written in the Jin Dynasty (265–316), it really did a very good job summarizing the botany of southern China.

Another influential writing was Fan Cheng Da's (范成大 1126–1193) "Gui Hai Yu Heng Zhi" (桂海虞衡志 Resources Noted from the Mountains and Waters in Guang Xi), which was the first book recording many plants and animals in Guang Xi. Fan Cheng Da was a famous pastoral poet in the Song dynasty. When he worked in Guang Xi he paid attention to surveying the peculiar spices, plants, animals, and minerals there, later writing a book in which he record various products not noted by the local records. Since both Yu (虞) and Heng (衡) were the separate administrators for resources from mountains and lakes in ancient China, it was very clear what the contents of the book would be.

Fan's book was divided into six chapters describing plants and animals in Guang Xi, which were birds, mammals, insects and fishes, flowers, fruits, herbs and trees. His classification consists of six parts, among which, there were 13 kinds of birds, 18 kinds of mammals, 14 kinds of insects and fishes, 16 kinds of flowers, 55 kinds of fruits, 26 kinds of herbs and trees; a total of 142 kinds of organisms were described. Fang's description of organisms also noted those of a local character, for example, peacock, parrot, white parrot, elephant, horse under fruits (tree), snuted-nose monkey, francoi's monkey, python, green turtle, and parrot snail were all distributed only in southern China. Some animals recorded in the books like the peacock, elephant and snuted-nose monkey, are extinct in Guang Xi today. As with the animals, some plants in the book also showed local characteristics, such as glanga gallangal, camellia, rangooncreeper, citrus, diversileaf artocarpus, truestar anisetree and so on.

Writings about ornamental plants and famous fruits were also remarkable in the Southern Song dynasty. For example, there were Fan Cheng Da's "Fan Cun Mei Pu" (范村梅谱 Notes on Plum in Fan's Estate) and "Fan Cun Ju Pu" (范村菊谱 Notes on Chrysanthemum in Fan's Estate), Lu Yiu's (陆游 1125–1209) "Tian Peng Mu Dan Pu" (天彭牡丹谱 Notes on Tree Peony in Tian Peng), Zhao Shi Geng (赵时庚) and Wan Gui Xue's (王贵学) "Lan Pu" (兰谱 Notes on Orchids), Chen Si's (陈思) "Hai Tan Pu" (海棠谱 Notes on Crabapple), Han Yan Zhi's (韩彦直) "Ju Lu" (橘录 Systematic Record of Citrus), Chen Ren Yu's (陈仁玉) "Jun Pu" (菌谱 Notes on Fungus) and so on. Among these Han Yan Zhi's "Ju Lu" was one of the most outstanding.

In 1178, Han Yan Zhi wrote a monograph on the orange of Yong Jia (永嘉 i.e., Wen Zhou 温州) when working as a prefect there. This writing is of great interest because of its accurate botanical descriptions. Han's book records 27 kinds of citrus fruit (sweet oranges, mandarin oranges, kumquat, etc.), discussing their characteristics, adaptability to soils, and pointing out how to control harmful insects and fungi in citrus trees. Its special excellence is in the description of the fruits of each variety, for which he gives the shape, texture of the rind, color, juice content, flavor, aroma and keeping quality. The treatise compares favorably with those of modern times.[8]

There were also monographs on pet animals, the most famous of which may be the treacherous court official Jia Si Dao's (贾似道) "Cu Zhi Jing" (促识经 Classic of the Cricket). As is well known, the cricket is a famous game insect in China. According to historical records, since the Tang dynasty down to the Song dynasty the game of cricket-fighting had been very popular at court. The game was much loved by officials and rich men. Jia was notorious for his indulgence in crickets. The book contained various descriptions of the insect, including its many local names, living environment, habits, relation between appearance of its body and fighting ability, methods of raising and taking care of crickets, etc. It contained many important facts for the development of entomology in ancient China.

In the beginning of the Southern Song dynasty, besides the descriptions of plants and animals in the south of China, as well as gardening plants and pets, some scholars in the countryside displayed immense enthusiasm in recording organisms. Wan Zhi's (王质) "Lin Chuan Jie Qi" (林泉结契 Friends in Forest and Spring) was a good example. The book recorded 73 kinds of plants and animals observed by the author when living in a mountain region. It is very interesting that Wan classified them as "mountain friends" and "water friends". The total of these two sorts were 43 forms of birds and more than 20 forms of edible plants, as well as a few mammals, amphibians, reptiles, and insects. The plants and animals included bamboo-partridge, woodpecker, cuckoo, Hwa-Mei (画眉 *Garrulax canorus*), Chinese bulbul, francolin, mandarin-duck, wild duck, red crane (Japanese crested ibis), pond-heron, and Chinese wolfberry, Xianggu mushroom, Rorippa, etc. The author's account included the color, size and living environment of each plant and animal. Wan also added a song after the description of each creature.

Meanwhile some scholars created biological writings by editing information from older treatises, adding in their own observation as well. One eminent general work of this kind was Luo Yuan's (罗愿) "Er Ya Yi" (尔雅翼 The Expansion of Er Ya). According to records, Luo Yuan (1166 Jin Shi 进士 candidate) favored history, studied hard and was an excellent scholar. His writing is similar to Lu Dian's in content, but in extent they were a little different. Luo's book is known for its account of plants and animals, so it is a more "pure" biological writing. "Er Ya Yi" is divided into 32 chapters, 12 of which are for the explanation of plants, the total consisting of 180 forms of herbs and trees, 20 of these chapters were for animals, describing a total of 235 forms. The number is larger than in Lu's writing. Luo divided all animals and plants into herbs, trees, birds, mammals, insects, and fishes (including snakes and turtles). The plants and animals which he recorded had often appeared in the classics or were common in everyday life. His book also include some animals from legend. In general Luo paid more attention to the habits of animals. He thought that the study of animals and plants should not only follow that of previous scholars, but "(the researcher should) observe fruits in autumn, look at flowers in spring, and survey fishes in the pool".[9] Because of this, he was able to accumulate more knowledge about animals and their behavior, and plants than previously possible. These purely biological monographs (derived from a Ru philosopher), indicate that biology as an independent subject was gradually being formed at that time.

With the accumulation of knowledge of various animals and plants, and especially ornamental flowers, by the end of the Southern Song dynasty, Chen Jing Yi (陈景沂) even edited a collection of information on garden flowers named "Quan Fang Bei Zu" (全芳备祖 Historical Records on Beautiful Plants). Though he collected too much literary information into the book, it proved that scholars then were endeavoring to systematically study herbs, trees, birds and animals.

The limitations of space forbid the discussion of further records. But as we have shown in the above discussions, the subject of herbs, trees, birds and mammals (or biology) was gradually forming in the Song dynasty. Not only can we see that many biological works appeared, we can also see different sorts of writing. There are some treatises such as "Yi Bu Fang Wu Lüe Ji" which recorded much new knowledge about plants and animals by survey. Some scholars began to plant trees in order to practise observation and look for real biological knowledge, such as Chen Zhu; moreover some scholars began to create classics (Jing) to suggest that their subject originated at a very early date. Some scholars, in perfecting the system of editing biological books, used the names of former scholars, such as in the "Nang Fang Cao Mu Zhuang". Meanwhile some scholars, including Zheng Qiao, clearly stated the importance of the subject for the development of culture. Zheng not only emphasized the importance of biology, but set biology as one of the twenty most important subjects. Influenced by predecessors' ideas, there are some books such as "Er Ya Yi" which mainly summarized previous biological lore. Eventually, the author of "Quan Fang Bei Zu" considered it of enormous importance to edit books of information on plants.

There are various causes for the remarkable development of biology in the Song dynasty. Some of the following may be the most important: first, the economic center transferred from the north to the south of China, making people pay more attention to the plants and animals there, especially those animals and plants with valuables both for decorative and economic reasons. For this reason many local fruits, various economic plants and animals, new pretty flowers and trees were recorded, which made people obtain more knowledge of biology and produce biological monographs.

Second, the gardening arts also developed very rapidly at that time. Whether emperors, officials or poorer scholars, men began to favor gardening. They enjoyed facing brilliant flowers to write poems and draw pictures. These conditions encouraged the writing of monographs that recorded the appearance of flowers, their forms, cultivation, etc. Furthermore, many scholars considered that beautiful trees, delicious fruits, fungus and lovely animals must be recorded as well as flowers, which doubtless promoted greatly the progress of ancient biology.

Third, the same herbals which included much new information and former lore of natural history, such as "Tu Jing Ben Cao", also greatly promoted the development of biology at that time.

Finally, the development of the new Ru philosophy (新儒家 or 理学) pushed the research in old classics such as Shi Jing, Er Ya, etc. into the study of the plants and animals recorded in those books. It was because of the above causes that those scholar brought ancient Chinese biology to a new level in the Song dynasty.

Notes

[1] Lun Yu (论语), Vol. 17.
[2] Zhen Qiao (郑樵): Tong Zhi (通志) Vol. 75, "Qun Chong Cao Mu Lüe" (昆虫草本略), Chapter 1.
[3] The book had not been reserved.
[4] Zhen Qiao (郑樵): "Qun Chong Cao Mu Lüe" (昆虫草本略), Vol. 72.
[5] Tong Zhi (通志), *Preface*.
[6] One of the Ru's classics.
[7] Luo Guihuan (罗桂环), 1990, Analysis of Current Nan Fan Cao Mu Zhuan, *Studies in the History of Natural Sciences*, 9(2): 165–70.
[8] Reed, H. S., *A Short History of the Plant Sciences*, New York, 1942.
[9] "Er Ya Yi", *Preface*.

Diseases in the *Huang Di Neijing Suwen*: Facts and Constructs[1]

PAUL U. UNSCHULD

Institute for the History of Medicine
University of Munich

Preliminary Remarks

It has been accepted for some time now that the *Huang Di Neijing Suwen* is a rather heterogenous literary source.[2] The more than 350 textual fragments compiled by unknown editors in the second or third century A.D.[3] were written by numerous authors, possibly beginning in the second century B.C., with some comments added even later.[4] The internal conceptual heterogeneity of the *Suwen* applies to its discourses on illness or disease as well.

This paper shall focus on some of the consequences that the acceptance of a new vessel theory, combined with the doctrine of systematic correspondence, entailed for the conceptualization of disease.[5] I wish to point out the range of solutions reached between an implicit request (voiced in the *Shiji* 史記 biography of Chunyu Yi 淳于意[6]) to abolish disease names altogether and interpret illness individually on the basis of the yin-yang and five-phases doctrines and vessel theory, and an obvious acknowledgment, on the other hand, that there existed specific diseases that were known since time immemorial. These carried distinctive labels because they were observed to exist trans-individually, which nevertheless required their integration into the new conceptual framework.

Disease names were not abolished by the *Suwen* authors; on the contrary, one could extract from the *Suwen* a long list of ancient and newly designed labels assigned to states of illness that were seen to affect different individuals in similar or identical ways. Hence it is from these many labels that I shall point out a continuum ranging from nosological facts to nosological constructs.

We are confronted here with a situation that has come up in the history of Chinese and European medicine alike several times. That is, traditional knowledge was met by a new style of thought. Any innovative style of thought entails its own specific phrasing of the issues it is meant to explain. One question is what to do with traditional knowledge that at first glance appears incompatible with new ways of seeing things.

In the *Suwen* we witness attempts at incorporating the new style of thought with what are well-established nosological facts. We also witness the appearance of new nosological constructs whose only *raison-d'etre* was the rise of the new vessel theory. From the many examples the *Suwen* offers, I have selected five to elucidate the cognitive dynamics at work here. These include *nüe*-malaria, cough, limpness, blockage, and recession. This selection represents a continuum ranging from what is without doubt a fact of trans-cultural validity to a health problem that is with equal certainty a cultural construct. Between these two end points it would be difficult to draw a sharp border line. The transition from fact to construct is a smooth one.

Nüe-Malaria

Nüe, from a hermeneutical perspective, is a rather unproblematic disease. *Nüe*, or malaria, has been observed in different cultures for millennia, and its symptoms have been described in similar terms, regardless of time or place. The differences in the names used to label this disease do not really matter. We can be sure that all the terms like *Nüe*, *malaria*, and *Wechselfieber*, as well as others in additional languages, referred to a disease that was and still is transmitted by anopheles mosquitoes.

The account of malaria in the two *Suwen* treatises 35 and 36 mirrors attempts at interpreting both the cause of the disease and the reasons for the different courses it may take. The contents of *Suwen* 35 are based on data gathered from a multitude of patients, possibly over extended periods of time and by more than one observer.

Suwen 35, like so many other treatises in the *Suwen*, quotes the contents of several originally separate texts on malaria but in the form of a dialogue between Huang Di and Qi Bo. While this editing may account for some inconsistencies between the questions asked and the original texts reshuffled and reassembled to answer them, the entire dialogue is nevertheless quite systematically structured. The questions pertaining to the typical appearances of the disease are answered within the conceptual framework of wind and *qi* etiology as well as wind and yin-yang-*qi* pathology.

What distinguishes malaria from any other disease is its periodicity of activity and dormancy. Hence the initial question addresses the very issue of periodicity. After elucidating that symptoms such as strong headache, trembling with cold, fits of heat, yearning for cold beverages, and others are caused by specific movements and partial abundance of yin and yang *qi*, Qi Bo's answer explains the activity of the disease as a manifestation of daily clashes between the protective *qi* moving through the organism as a guardian against intruders on the one hand and wind or water *qi* that has been able to enter the body on the other. The former is able to intrude through the open pores when a person sweats; the latter is able to enter when a person with open pores takes a bath.

The dialogue next takes up the issues of extended periods of dormancy between the outbreaks of the disease. Why does it skip one day, sometimes two days, or even several days, before it is active again? Why does its activity occur progressively later every day for some time; that is, why do the intervals between outbreaks increase for some time, before this process is reversed and the periods of inactivity are shortened a bit from day to day? Why do some patients experience cold first and heat afterwards, why do others experience heat first and cold afterwards, and why do still others experience only heat, in which case the disease is called solitary heat malaria?

The ontological perspective in the mixed ontological-functional explanatory model of the *Suwen* requires data on the location of the disease in the body. Interestingly, in *Suwen* 35, the core depots are not mentioned as possible locations of the disease. The evil *qi* moves along or inside the spine, or it passes through conduits and vessels affecting yin or yang sections of the organism in general.

Suwen 36, not structured as a dialogue and most likely dating from another historical level, is very different. Its author or authors left the notion of a single malaria; rather, they assigned a list of twelve *Nüe* diseases to the conduits associated with the bladder, the gall, the stomach, the spleen, the kidneys, and the liver, and to the six core organs, i.e., lung, heart, liver, spleen, kidneys, and stomach.

Two examples may illustrate how malaria was incorporated into the framework of the conduits and core organs. The treatment appears to have been based on a rather simple, mechanistic rationale. If the disease was diagnosed to be located in a specific conduit, the healer was advised to open this particular conduit and — in some cases — remove its blood and, hence, the disease:

> Malaria of the foot major yang [conduit]: it lets a person have lower back pain and a heavy head. Cold rises from the back. [Patients] are cold first and afterwards hot. The heat is intense [as in] harm caused by summer heat. When the heat stops, sweat leaves [the body]. [This disease] is difficult to cure. Pierce [the foot major yang conduit] into the cleft to let blood.[7]

Lung malaria: it lets a person's heart be cold. When it is very cold [it changes to] heat. While it is hot, [patients] tend to be frightened as if they had seen something [frightening]. Pierce the hand major yin and yang brilliance [conduits].[8]

In other words, the human organism is seen in *Suwen* 35 as being affected as a whole while it is interpreted in *Suwen* 36 as an assembly of different units each of which may be affected individually. This way it was easier to explain variances in the appearance of malaria in different patients. To conclude, both *Suwen* 35 and 36 apply vessel theory and the doctrines of systematic correspondence to the phenomenon of malaria. Also, both offer a mix of ontological and functional views. Nevertheless, even within this common ground we encounter two quite different perspectives whose ideological backgrounds may require further thought.

Cough

While *Nüe*, malaria, is a label denoting the entirety of a disease phenomenon including causation by wind or other evil *qi*, the notion of clashes between the protective *qi* and the intruder, as well as the periodicity of the disease's outbreaks and the various symptoms these outbreaks entail, cough — the second example on my list — denotes, at least in a modern view, a bodily reaction to a deeper pathological condition.

When the initial question in the "Discourse on Cough" in *Suwen* 38 asks "When the lung lets a person cough, how is that?" Huang Di simply alludes to the fact that the lung was known as the main organ linked to such activities as inhaling and exhaling, and, in abnormal situations, to coughing. The response by Qi Bo, however, corrects Huang Di's simple equation of cough with the lung by immediately pointing out that "all the five depots and six palaces [may] let a person cough, not only the lung".[9] This is nothing less than an introduction to a discourse integrating a simple phenomenon like cough into the five-phases and depot-palaces doctrines.

Once Huang Di has declared his interest in hearing what Qi Bo may have to say on the many types of cough, the latter opens his account by pointing out the etiology of lung cough, possibly still the primordial cough. The lung, he declares, is the location where two streams of cold may meet. One has entered the lung through the skin and the body hair, both of which are associated with the lung. The other has entered the stomach through cold food and beverages, the cold of which is transmitted via a specific vessel from the stomach to the lung. The two evils meet and settle in the lung and cause it to cough.

The text continues to explain that each of the core organs receives the disease directly from the outside during the corresponding season. Cough originating in the palaces is a problem resulting from a transmission there of uncured cough in the depots.[10] In all these cases, the text asserts, [cold] accumulates in the stomach and is sent through certain "ties" to the lung.[11] Hence all cough leaves via the lung, even though it originates in the heart, the spleen, the liver, or the kidneys.

Limpness

With the discussion of limpness, this survey of examples from the broad range of diseases named in the *Suwen* leaves those diseases and illnesses that are easily identifiable because of their trans-cultural and diachronic reality. Limpness is different in that, at first glance, it does not make much sense in a modern medical context. Limpness is not a label for a disease nor is it a standard symptom. Limpness means a lack of strength or firmness; it means drooping and exhaustion. While these are descriptions of states that can be observed in man, in animals, and in plants, they do not signify any specific underlying disease. Limpness may simply be a sign of being tired or of getting old.

The Chinese term which I have chosen to translate as limpness is *wei* 痿. We do not know whether, by the time the word *wei* was chosen to denote a certain condition, its character was written with the radical categorizing it as a name of a disease or illness, or with another radical, or even without any radical at all. In the latter case, the character *wei* 委 would offer many meanings, including that of "weak" and "decline". These are the meanings given it in the *Zhouli*, where it is equated with another term for dilapidation and decline, i.e., *tui* 頹.[12] The character *wei* 痿 may, however, also have been designed to raise the meaning of *wei* 萎 to a more abstract level.

The original meaning of *wei* 萎, attested to in the *Shijing*, denotes the wilting of trees.[13] By replacing the radical "grass" with the radical "disease/ illness", the creators of the character 痿 may have intended to move from the notion of wilting of trees to a notion of limpness of the human organism not in total but in some of its parts. The *Shiji* has 如痿人不忘起, 盲者不忘 視, "this is like a person with [the disease] *wei*, who cannot forget how [once he was able] to get up, and it is like a person who is blind and cannot forget how [once he was able] to see".[14] Obviously, *wei* refers here to a lameness of the feet, hence a "person with *wei*" is unable to get up. Similarly, the dynastic history of the Han states: 疾痿, 行步不便, "he developed the disease *wei*, [and hence] was unable to walk easily".[15]

Suwen 44, the "Discourse on Limpness" in dialogue form, opens with Qi Bo's reminder of the associations between the five depots (lung, heart, liver, spleen, and kidneys) and their correlates (skin and body hair, blood and vessels, sinews and membranes, muscles and flesh, and bones and marrow).[16] Subsequently, in a tabular listing, Huang Di was told that excessive heat in one of the depots causes its correlates to deteriorate.

A comparison of the pathology of limpness with that of cough reveals some interesting differences. In the case of cough, once the twofold intrusion of cold had entered a depot, it caused a specific type of cough. If the disease was not cured in time, it was transmitted further to the respective palace associated with the depot first affected. The correlates of the depots, i.e., skin and body hair, blood and vessels, etc., were not mentioned in this context; obviously, they were not affected by a twofold intrusion of cold. In contrast, extreme heat, for example in the liver, directly affects the gallbladder, i.e., the palace associated with the liver, and in addition manifests itself in the respective correlates, i.e., the sinews and the membranes.

The tabular listing in *Suwen* 44 of the effects of excessive heat in the five depots is not entirely consistent. Only in the cases of liver and spleen are the palaces named as being affected too. Also, given that heat in the heart results in vessel limpness, in the liver results in sinew limpness, in the spleen results in flesh limpness, and in the kidneys results in bone limpness, one might expect that heat in the lung results in skin or body hair limpness. The text, however, says something different: "When the lung is hot and when the lobes burn, then the skin and the body hair are depleted and weak. [The skin is] tense [and the body hair is] thin. When [the heat] stays, then [this] causes *wei bi* 痿躄."[17]

Giving credence to their individual meanings, one may read these two characters together as "limpness with an inability to walk". However, "limpness", *wei* 痿, and "inability to walk", *bi* 躄, may have referred to the same thing, i.e., lameness, in Han China. At one point, the *Lü shi chun qiu* states: "If much yin is present, then this causes *jue* 蹶; if much yang is present, then this causes *wei*."[18] Yang is heat. Apparently, the *Lü shi chun qiu* statement conveyed the same understanding of the etiology of *wei* as the *Suwen*. The Han-dynasty commentator Gao You 高诱 (fl. 200) commented on the *Lü shi chun qiu* passage as follows: "*Wei* 痿 [is] *bi* 躄, is: inability to walk".[19] That is, Gao You equated *wei* to *bi* to elucidate the former's meaning of "lameness".[20] Even though the term *jue*, which according to the *Lü shi chun qiu* stood for a disease caused by too much yin, will be discussed further down, it may be pointed out here that its original meaning of "to fall"[21] would parallel as a yin ailment the meaning of *wei*, "lameness", as a yang disease.

The use of the term *wei* in the *Suwen* suggests that the concept of *wei*, originally "lameness", has undergone a development similar to that of the concept of cough. The latter originally referred to a phenomenon clearly and solely tied to the lung. In a secondary development, it was subsumed under the five-phases doctrine, and henceforth cough was tied not only to the lung but also to all the remaining depots and even palaces. The meaning of *wei* may originally have been restricted an inability to walk. In a subsequent development, the term appears to have been subsumed under the five-phases doctrine, too, and its meaning of lameness was extended to a more general meaning of a limpness that could affect all the depots and their correlates.

In a second part of *Suwen* 44, another, possibly even older, explanatory model of the various types of *wei* is introduced to answer Huang Di's question of how the various forms of limpness are contracted. At first glance, this section appears not as systematic as the first part. Its conceptual basis mostly fails to parallel the clear-cut heat-limpness causality mentioned before. In the case of the lung, a relationship between a hot lung and burning lobes, on the one hand, and an inability to walk, on the other, is traced to psychological causes. The reader learns two reasons why a lung can turn hot, i.e., being upset over a loss or over the fact that one longs for something but does not get it. Why the heat caused by these emotions in the lung makes a person unable to walk is left unanswered.

Vessel limpness, sinew limpness, and flesh limpness is explained without reference to a hot heart, liver, or spleen; in the spleen, dampness rather than heat causes the problem.[22]

The only version of limpness that is explained entirely within the causal framework introduced in the first part of *Suwen* 44 is bone limpness. Interestingly, bone limpness has the same consequences as a hot lung: patients are unable to continue to walk.[23]

Basically, the conceptualization of bone limpness is an attempt to establish a theoretical link between a long-distance walk and severe exhaustion. The actual consequences of vessel limpness, sinew limpness, and flesh limpness are not specified in the text; one may assume they, too, lead to an inability to walk. This, at least, is suggested by a statement quoted in the second section of *Suwen* 44:

> "Hence, when it is said: 'Because the lung is hot and the lobes burn, the five depots develop limpness with an inability to walk,' then this is explained [by what was said above]."[24]

In the first section of *Suwen* 44, the causal chain is different. The heated lung first affects skin and body hair and only if the heat stays it causes "limpness with an inability to walk". In the second section, the statement just quoted informs us, the heated lung causes limpness with an inability to walk in all the depots. Hence, the actual consequences of vessel limpness, sinew limpness, and flesh limpness need not be mentioned again; they include — as is the case with bone limpness — an inability to walk.

Hence, it may not be too farfetched to seek the origin of the concept of *wei* as limpness in an original notion of lameness that was broadened to cover the five depots.

A third section of *Suwen* 44 is rather short and follows a perfectly systematic structure again. Huang Di asks, "how to distinguish these [states]",[25] and there is no doubt that the text continues the clear-cut heat-limpness causality mentioned above in that it simply enumerates those signs that permit a diagnosis of lung heat, heart heat, etc.

With this I leave the brief discussion of limpness, moving one step further in my list of ailments reflecting the transition from the undisputed existence of nosological facts requiring a theoretical underpinning to an undisputed plausibility of conceptual constructs requiring substantiation by pathological phenomena. In other words, with malaria and cough we have seen a situation where the question is "if such diseases exist, how could they be explained". With limpness, we have encountered an example where a nosological fact, lameness, was reinterpreted to affect not only a person's feet or limbs but his entire organism as well. The original term for "lameness" appears to have been left in exchange for a more abstract term denoting the construct of "limpness".

The subsequent diseases on my list appear increasingly to require questions such as: "if health depends on the free flow of *qi* and blood in the vessels, etc., what health problems do occur if the free flow is impeded?" In the case of malaria, a clear-cut nosological fact at the one end of the continuum, theory serves to explain pathology; in the case of recession, a clear-cut construct at the other end of the continuum, pathology serves to substantiate theory.

Blockage

Blockage is an ailment close to but not quite at the end of this continuum. Like limpness, it is an example of a development from a rather mechanical problem, in this case the closure of the urethra, to an abstract concept. In contrast to limpness, whose original Chinese name *wei* was clearly meant to signify an observable phenomenon, an inability to walk, *bi*, the term used for blockage, is a conceptual construct already in that it serves not to describe but to explain an observable phenomenon, the failure to pass urine. Similar to the development from lameness to limpness a new term was introduced that signalled the development from a concept of a concrete tubular blockage to a more abstract understanding. However, while in the case of limpness the original character was modified to express the more abstract level of conceptualization, in the case of blockage, the original term *bi* 閉 was left with the original meaning, while a completely new term, the homophone *bi* 痹, was created to denote the abstract concept.

Various appearances of the term *bi* suggest that it was borrowed from vernacular usage to indicate what was considered the reason behind an inability to urinate. At least, this is the definition given by the Maishu excavated at Zhangjiashan.[26] The vernacular term for closure, blocking, shutting in, *bi* 閉, was used in the Maishu, the Mawangdui scripts, and also in the *Suwen* when it speaks of "blocked urination".[27]

In numerous instances, the *Suwen* uses the vernacular term *bi* 閉, sometimes together with *sai* 塞, to denote not only an inability to urinate, but also various other health problems, where it was believed that a passage, which should be free, was blocked.[28]

In various contexts, the same meaning of "blocked opening" or "blocked passageway" was also expressed by the character *bi* 痹. This term appears in the received version of the *Shiji* of 90 B.C., where Bian Que is said to have acted as a physician specializing in the "blockage of ears and eyes", *er mu bi yi* 耳目痹醫, when he arrived in the state of Zhou.[29] In the *Suwen*, a blockage of the throat as a passageway of food is repeatedly termed *bi* 痹.[30]

More often, however, the *Suwen* authors employed the term *bi* 痹 to signify instances of blockage that indicate a move beyond a simple notion of a tubular plug. For example, the skin was known to have holes. When these holes were closed, sweat was unable to leave the body and this could have various unwelcome consequences. However, the skin was also conceptualized as one of the passageways of blood. In this case, the blood was not thought to move through the skin from inside out; rather it was assumed to course within the skin. It was believed that the passageways of the blood inside the skin could be blocked as much as the passage of sweat through the skin from inside out. Hence, one statement points out: "when the blood congeals in the skin, this is blockage (*bi* 痹)".[31]

Because of this blockage, the context informs us, the blood coursing in the skin cannot return to its point of departure. This is a serious disease. It is called blockage. At this level of conceptualization, blockage is both a causal factor in the development of the disease (the blockage itself is generated by a more basic causal factor, wind) and the label given to the disease. Ailments such as pain or local distension were assumed to be manifestations of *qi* accumulations. These accumulations, in turn, were thought to occur because the passage of the respective *qi* through a certain region was blocked, mostly by external agents such as cold, dampness, or wind.[32]

These statements do not yet represent the highest level of conceptualization reached in this development. From a statement on lung blockage it is obvious that the term "blockage" was dissociated from the original meaning of closure. Lung blockage, like liver blockage or kidney blockage, is a disease label conferred upon certain states of ill health. These states manifest themselves in syndromes, which are composed of specific movements in the vessels, a specific complexion, a specific emotion, and feelings that are interpreted as *qi* accumulations. The text no longer spoke of a specific passageway presumed to be blocked and, therefore, responsible for these symptoms.[33]

Of interest is an answer given by Qi Bo to Huang Di when the latter enquired about the causes of blockage. Qi Bo stated in what is at first glance a perfectly ontological conceptualization:

> When the three *qi* wind, cold, and dampness arrive together, they merge and cause a blockage. In case the wind *qi* dominates, this causes 'moving blockage'. In case the cold *qi* dominates, this causes 'painful blockage'. In case the dampness *qi* dominates, this causes 'attached blockage'.[34]

And yet, Qi Bo's reply includes a secondary construct, too. The metaphorical value of wind as something that moves all the time and does not exist if it fails to move was employed to explain a moving blockage. The experience of biting cold was transferred to an explanation of a painful blockage. Dampness can cling to anything; this may have been the image employed in the concept of an attached blockage. It is hard to imagine that the original meaning of an opening or a passageway being blocked is still retained here. Moving blockage, painful blockage, and attached blockage may, of course, evoke a notion of a plug, but to think that this plug moves through the organism, causes pain, or is attached somewhere, might be extending the original metaphor too far.

Elsewhere, Qi Bo informs Huang Di of five types of blockage, subsuming the concept of blockage under the five-phases doctrine. Accordingly, wind, cold, and dampness are the *qi* that can cause a blockage. If encountered in winter, they cause bone blockage, in spring, sinew blockage, in late summer muscle blockage, etc.[35]

Just like the *qi* causing them, the blocks, too, do not have to stay in the bones, the sinews, the vessels, the muscles, or the skin for good. If not treated successfully, the disease may move deeper into the organism.[36]

It is difficult to know which morphological image of, for examples, sinews and depots the author of these lines had in mind. Apparently, there was some underlying notion of a tubular structure which could be blocked by the *qi* of wind, cold, or dampness on their course into the body's interior. Nevertheless, how, for example, the lung or the sinews could be blocked mechanically was never spelled out in detail. Obviously, at this level of conceptualization, blockage was understood as a presence of wind, cold, and/ or dampness, but no longer a concrete closure of a tubular passageway or of an opening.

Obviously, the changing conceptualization of "blockage" was common knowledge by the time of the later Han dynasty. The *Shuo wen jie zi* 説文解字 reads *wei* 痿 as *bi* 痺, "blockage".[37] In his commentary on the *Hanshu* 漢書, the Tang author Yan Shigu 顏師古 (581–645) defined the term *wei* 痿, "limpness", as 風痺, feng *bi*, "wind[-type] blockage",[38] or simply 痺, *bi*, "blockage".[39] That is, limpness and blockage were considered now to be ailment and underlying disease. The notion of a causation of limpness by heat was replaced, at least in the understanding of the authors just quoted, by a notion of blockage as the cause of limpness.

Recession

Let us move on to the final disease to be covered by this short survey of a continuum that began with a disease of the highest cross-cultural and diachronic factuality and ends with a disease that constitutes a purely theoretical construct. In contrast to malaria, *jue* is a disease whose recognition and conceptualization is restricted to vessel theory and the doctrines of systematic correspondence in classical Chinese medicine.[40]

The term *jue* 厥, as employed, for example, in *Suwen* 05[41] and *Suwen* 45,[42] is a label that must have been in common usage by the time these treatises were written. It appears with varying radicals in a medical context in the *Yinyang shiyi mai jiujing* 陰陽十一脈灸經 of the Mawangdui-texts of the early second century B.C., in the Zhangjiashan-texts *Maishu* 脈書 and *Yinshu* 引書 of the mid-second century B.C., and in the Bian Que biography in Sima Qian 司馬遷 *Shiji* 史記 of the early first century B.C. Its pathological connotations were specified in the etymological dictionaries *Shuowen jiezi* 説文解字 of around 100 B.C. and Liu Xi's 劉熙 *Shiming* 釋名 of A.D. 200.[43]

A first question to be addressed in clarifying the meaning of *jue* pertains to its vernacular usage. Is there any non-medical appearance of the term which would allow us to trace a medical derivation, such as *wei* 痿 from *wei* 萎, and *bi* 痺 from *bi* 閉? Or, is there any evidence suggesting that the term *jue* has been the subject of an increasing abstraction, in the same way as the two terms just mentioned, from a more mechanical illness terminology to a theoretical construct?

Mencius has *jue* 蹶, read by James Legge as "to fall".[44] The *Shiji* has *gu bao jue er si* 故暴蹶而死,[45] which could be read as "hence [that person] suddenly falls and dies". In fact, most of the meanings associated with *jue* 蹶 include notions such as "to move quickly", "suddenness", "to excite". Maybe, then, *jue* was chosen to denote a rapid breakdown as is suggested by the usage of the term in the *Shiji*.

There is, however, a second layer of meanings associated with defeat,[46] exhaustion,[47] and abandonment.[48] This meaning of "ceasing" apparently was also the basis for naming a conduit *jue* yin, thereby associating it with the lowest yin category.

It may well have been that both these layers were brought together when the term *jue* 厥, eventually, was chosen to denote, in keeping with vessel theory, a pathological construct that was to express two things. First, a movement of yin or yang *qi* out of a section of the body where it should be present but ceases to be. Second, a violent reaction resulting in a sudden fall and immediate death (with this "death" being either unconsciousness or real). Hence the interpretation of the vernacular term *jue* 蹶 in the *Shuowen jiezi* is *jiang* 僵, which is attested in Han texts with the meanings *yan* 偃 "to fall" and *bi* 斃, "to perish", "to die",[49] while the interpretation of *jue* 厥, i.e., of the term designed to denote a disease, is *niqi* 逆气, i.e., "*qi* moving contrary to its normal direction". This, of course, is a purely theoretical construct. No evidence whatsoever could be observed that indicates, first, what might be considered a flow of *qi* in a normal direction and, second, a flow contrary to a normal direction.

We are able, now, to suggest the following hypothesis concerning a development in the use of the term *jue*. First, there may have been a label for an illness of "sudden fall"; this was *jue* 蹶. Once vessel theory was applied to explain the diseases of the human organism, the terms *jue* 厥 and *jue* 厥 came to be used to designate conditions thought to be caused by moving *qi* in reverse.

In *Suwen* 45, Qi Bo explains both cold recession and heat recession as consequences of an uneven presence of yin and yang *qi*.[50] This uneven presence, in turn, is seen as a result of "violent" actions, of "fighting" among yin and yang *qi* in the organism, and it is here where we may see the link with the military term of defeat and abandonment, i.e., retreat. Hence I have chosen to translate *jue* 厥 as "recession" because recession includes notions of withdrawal, of ceding back, i.e., of leaving a terrain once occupied by oneself to someone else — and this is exactly what was thought to happen: yin or yang *qi* withdraws and leaves the terrain to yang or yin *qi*, respectively.

In the *Suwen*, cold recession is a weakening of yang *qi* "below", while heat recession is a weakening of the yin *qi* "below". That is, recession has been abstracted in this context beyond its immediate literal significance. Cold recession or heat recession is not to say that cold withdraws, in the first case, or heat withdraws, in the second. Cold recession is to say that certain body parts, in this case the feet, turn cold, because the yang *qi* that has warmed them has left the feet. In contrast to malaria whose periodicity is obvious, to cough, whose emanation from the lung is obvious, and to blockage whose origin in a failure to pass urine is equally obvious, neither the etiology nor the symptomatology of recession has anything to offer requiring a conceptualization as recession.

Overtaxing oneself in autumn and winter, when yang *qi* is weak anyway, weakens yang *qi* in its function to pour into the four extremities and keep them warm. Yin *qi* is present instead and causes cold. Similarly, to have sexual intercourse while one is drunk or after one has dined to repletion causes the kidney *qi* to weaken and accumulates yang *qi* in the stomach. The kidney *qi* can no longer "nourish the four limbs", the yang *qi* dominates alone, and "it is therefore that the hands and the feet are hot". [51]

Theory dictated a further development in the usage of the term *jue*. Accordingly, recession may occur in each conduit. Hence, six different sets of signs inform a knowledgeable practitioner whether recession has occurred in the great yang, the yang brilliance, the minor yang, the major yin, the minor yin, or the ceasing yin conduits. [52]

As was the case with blockage, the notion of recession appears to have dissociated itself from its original mechanical background. It culminated in the notion of *jue ni*, recession with counter-movement, possibly diseases considered to be extreme cases of recession. Numerous ailments were traced to a presumed recession in one or another of the conduits, but what this really meant in terms of the actual mechanics of the flow of *qi* was left open. [53]

In the list of ailments discussed here, recession with counter-movement symbolizes the peak of the development towards the generation of nosological constructs in ancient Chinese medicine. A concept like recession with counter-movement makes sense only on an abstract level. I find it difficult to believe that the ancient Chinese authors who wrote about recession with counter-movement had in mind both the original mechanism of a *qi* moving back from a territory and the actual mode of passage of *qi* through the conduits implied by the term counter-movement. Ancient Chinese medical authors, like many of their contemporaries writing in other fields, must have possessed more than enough technical expertise to realize all the mechanical problems resulting from taking a notion of recession with counter-movement too literally.

One final, albeit most interesting, point remains to be noted. Let us look once again at the *Lü shi chun qiu* statement "If much yin is present, then this causes *jue* 蹷; if much yang is present, then this causes *wei* 痿." Presumably, this was originally a nosological dichotomy not uncommon in the early days of an application of the yin-yang doctrine to the realm of medicine and morphology. If there was lameness or inability to walk, apparently there had to be a yin type and a yang type. Hence two similar terms were chosen to signify the former and the latter. In this case, *jue* stood for the yin variant and *wei* stood for the yang variant of lameness. When Gao You commented on the *Lü shi chun qiu*, several centuries had passed and the terms *jue* and *wei* had acquired different meanings, dissociating them from the original yin-yang dichotomy. Obviously, *jue* had lost its original sense of "to fall" entirely, while a reading of *wei* could be based both on the earlier and on more recent historical layers. Hence, Gao You interpreted *jue* as 逆寒疾, *ni han ji*, i.e., "suffering from cold moving contrary [to its normal direction]", while he interpreted *wei* as *bi*, i.e., as an inability to walk.[54] This meaning of *wei*, enriched with a theoretical explanation as being caused by dampness, was also perpetuated by the *Yupian*, a dictionary completed in A.D. 543 and received in a version edited during the Tang dynasty. Here we read: "*Wei* 痿 is inability to walk; it is *bi* 痺, a dampness disease".[55]

To conclude, from a wide range of data on disease in the *Suwen*, I have selected a few that demonstrate: first, the conceptual flexibility of ancient Chinese naturalists *vis-à-vis* cognitive challenges presented by the emergence of an innovative style of thought associated with the foundation of the unified Chinese empire more than two millennia ago; and secondly, the dynamics of historical change in the understanding of the causes and the nature of disease during the Han era.

Notes

[1] This paper is based on the annotated translation of the *Huang Di Neijing Suwen* which I have prepared in collaboration with Dr. Hermann Tessenow and Prof. Zheng Jinsheng over the past years. The contents of this paper are an abbreviated version of a more detailed discussion of *Suwen* concepts of disease in my "Nature, Knowledge, Imagery in the *Huang Di Neijing Suwen*", which is scheduled to appear as an introductory volume to the *Huang Di Neijing Suwen* project and which I was able to write during my year as a fellow at the Institute for Advanced Study Berlin, 1998/99. I am most grateful to Wolf Lepenies, Joachim Nettelbeck, and their

extraordinarily helpful staff for an opportunity to enjoy a most stimulating academic environment. A shortened version of this paper was presented at the *9th International Congress on the History of Science in East Asia*, Singapore, August 25, 1999.

² For the most detailed study of the subject of the internal heterogeneity of the *Suwen* available to date, see David Keegan, *The Huang-ti Nei-ching: The Structure of the Compilations; The Significance of the Structure*. Dissertation. University of California Berkeley (Ann Arbor: University Microfilms International. 9816728).

³ Based on a counting by Hermann Tessenow in his forthcoming analysis of the textual structure of the *Suwen*.

⁴ For a more detailed discussion of the dating of the *Suwen*, see this author's forthcoming *Nature, Knowledge, Imagery in an Ancient Chinese Medical Text. The Huang Di Neijing Suwen*.

⁵ See Donald Harper, *Early Chinese Medical Literature* (London: Kegan Paul International, 1997), pp. 77 ff., for the early history of vessel theory in Chinese medical manuscripts from the third to second centuries B.C.

⁶ Sima Qian 司馬遷, *Shiji* 史記 (Taibei: Zhonghua shuju 台北中華書局, 1989), 11th printing, vol. 9, ch. 105, p. 2813; Harper (1997), p. 70.

⁷ *Suwen* 36-206-14. This reference and all further references to the *Suwen* refer to *Huang Di Neijing Suwen* 黄帝內經素問 (Beijing: Renmin weisheng chubanshe 北京人民衛生出版社, 1983), with the first number referring to the *Suwen* treatise, the second to the page of the 1983 edition, and the third to the line on that page.

⁸ *Suwen* 36-208-5.

⁹ *Suwen* 38-214-5.

¹⁰ *Suwen* 38-216-1.

¹¹ *Suwen* 38-216-7.

¹² *Zhouli*, Kao gong ji, zi ren 周禮, 考工記, 梓人. *Shisanjing zhushu* 十三經注疏 (Beijing, 1987), vol. 1, p. 925 (center).

¹³ *Shijing*, Xiaoya, gu feng, 詩經, 小雅, 谷風, *Shisanjing zhushu* 十三經注疏 (Beijing, 1987), vol. 1, p. 459 (center).

¹⁴ *Shiji*, 1989, ch. 93, p. 2635. Elsewhere, the *Shiji* uses the phrase yin *wei* 陰瘻 "lameness of the yin" in the sense of lameness of the male member. See ch. 59, p. 2097.

¹⁵ *Hanshu* 漢書 (Taibei: Zhonghua shuju 台北中華書局, 1987), vol. 9, ch. 93, p. 2767.

¹⁶ *Suwen* 44-246-8.

¹⁷ *Suwen* 44-246-11.

¹⁸ *Lü shi chun qiu* ch. 1.3: zhong ji 重己. Xu Weiyu 許維遹 (ed.), *Lü shi chun qiu* ji shi 呂氏春秋集釋, vol. 1, ch. 1 Chong ji 重己 (Taibei: Shijie shuju 世界書局, 1969), p. 13.

¹⁹ *Ershierzi* 二十二子 (Shanghai: Shanghai guji chubanshe 上海古籍出版社, 1986), p. 630 (bottom).

²⁰ *Ibid.*

²¹ For *jue*, "to fall", see below p. 15.

22 *Suwen* 44-248-3.

23 *Suwen* 44-248-5.

24 *Suwen* 44-247-5.

25 *Suwen* 44-248-9.

26 Harper transl., Harper (1997), p. 208.

27 *Suwen* 65-357-10; 65-359-2ff.

28 For example, *Suwen* 16-96-2; 19-126-4; 62-341-10.

29 Sima Qian, *Shiji*, 1989, p. 2794.

30 *Suwen* 38-215-6.

31 *Suwen* 10-73-3.

32 *Suwen* 10-76-3; 10-76-6.

33 *Suwen* 10-76-1.

34 *Suwen* 43-240-7.

35 *Suwen* 43-241-1.

36 *Suwen* 43-241-6.

37 *Shuowen jiezi gulin zhengbu hebian* 説問解字詁林正補合編 (Taipei: Dingwen shuju 鼎文書局, 1983), vol. 6, p. 875. No. 4729.

38 瘺, 風痹疾也, "*Wei* is a wind-blockage disease". *Hanshu* 1987, 63, p. 2767.

39 瘺亦痹病也, "*Wei*, too, is a blockage-disease". *Hanshu* 1987, 11, p. 345.

40 See also Vivienne Lo, Tracking the Pain. *Jue* and the formation of a theory of circulating *qi* through the channels, Sudhoffs Archiv 83 (1999), pp. 60ff.

41 *Suwen* 5-35-1.

42 This is the *Jue* lun 厥論, "Discourse on Recession". See the entire treatise.

43 Lo (1999), pp. 65ff. Harper (1997), p. 203.

44 *Mengzi*, Book II, pt. 1, ch. 11.

45 Sima Qian, *Shiji*, 1989, 2788.

46 *Sunzi*, junzheng 孫子軍爭. Ershierzi (1986), p. 434 (bottom).

47 D. C. Lau, Chen Fong Ching (eds.), *A Concordance to the Xunzi* (Hongkong: Commercial Press, 1996), 121, line 4 (Xunzi, chengxiang 荀子成相).

48 *Wenxuan, Ban Gu dabinxi* 文選班固答賓戲. ch. 45 (Hongkong: Shangwu yinshuguan 商務印書館, 1960), p. 990. The reading of *jue* 蹶 as "abandonment" was suggested by *Lü Yanji* 呂延濟: 棄也. See *Hanyu dacidian* 漢語大辭典 (Shanghai, 1992), Vol. 10, p. 549.

49 Xu Weiyu 許維遹 (ed.), *Lü shi chun qiu* ji shi 呂氏春秋集釋, vol. 3, ch. 21 guizu 貴卒 (Taibei: Shijie shuju 世界書局, 1969), p. 15.

50 *Suwen* 45-250-2.

51 *Suwen* 45-251-9.

52 *Suwen* 45-252-1.

53 For example, *Suwen* 40-226-3.

54 *Ershierzi* 二十二子, 1986, 630 (bottom).

55 *Daguang yihui yupian* 大廣益會玉篇, *bu* 部 148, *Sibu beiyao* 四部備要 (Shanghai: Zhonghua shuju 中華書局), 1935, vol. 2, juan 11, p. 7a.

Regulating the Medical Profession in China: Health Policies of the Nationalist Government

YE XIAOQING 葉曉青
Macquarie University, Sydney, Australia

Health policies became an important part of nation building very soon after the establishment of the Nationalist government in China.[1] Between 1929 and 1940, the government issued 44 separate regulations on medical administration, including medical qualification requirements for practising doctors, pharmacists, midwives, nurses and other paramedics. There were 24 regulations issued on public health, the prevention of epidemic disease, water and food hygiene standards, meat inspection and the like. There were a similar number of regulations on medical education and organizations.[2] Among the new health policies, regulation of the medical profession was a priority.

The regulations only tell us what the government hoped to achieve, not what was actually done. In this paper, I will examine what the government actually achieved in terms of regulating the medical profession.

The Position of Traditional Chinese Medicine (TCM)

The consensus among educated Chinese of the 19th century, including reformers like Kang Youwei and Liang Qichao, was that both traditional and Western medicine had advantages which could supplement the inadequacies of the other. This consensus broke down in the 20th century, especially after the establishment of the Nationalist government. The new Chinese elite, which was mainly Western-educated and deeply influenced by lay scientism, saw traditional medicine as part of an antiquated non-scientific tradition, and thought it should be abolished.

In pre-modern China, "medical knowledge was always considered to be a necessary part of the general education of a Confucian".[3] Confucian scholars were often able to prescribe medicines for their friends and family members. Professional practitioners of traditional medicine, on the whole, however, were not highly respected in traditional society. The government had no regulations over the qualifications or practices of the traditional Chinese doctor. Anyone could claim this title, and the quality of practitioners varied greatly. Chinese physicians were often the butt of jokes in traditional Chinese theatre. From the *zaju* 雜劇 of the Yuan, to the *kunqu* 昆曲 of the Ming, to the Peking Opera of the late Qing, physicians were portrayed in the role of *chou* 丑, a clown-like character, as a sort of comic relief. The sarcasm was aimed at the low level of their medical skills and even lower level of their medical ethics. The only exceptions were Sun Simiao and Hua Tuo, who were portrayed as Daoist priests, but that was because they had become members of the Chinese pantheon.[4] By the 20th century, not only the practitioners, but the whole corpus and practice of traditional medicine itself had come under attack.

One of the first priorities of the new Republican Government was to establish a modern education system, and this excluded most aspects of the "old culture", including Chinese medicine. The first signal of this negative attitude was that schools teaching Chinese medicine were not included in the restructured modern education system of 1912. These policies were greatly influenced by the students who had returned from Japan, as Japan had abolished traditional medicine during the modernizing Meiji period.[5]

The practitioners of TCM realized their livelihood was being threatened and in 1913 the Shanghai Shenzhou Medical and Pharmaceutical Association organized a petition, supported in 19 provinces and cities, entitled "Save Chinese Medicine From Extinction". A delegation went to Beijing to present the petition to the government, demanding that Chinese medicine be included in the educational curriculum. They were not successful.[6]

In 1914 the traditional practitioners in Beiping sought an appointment with the Minister of Education, Wang Daxie 汪大燮, applying for registration for their medical society. Wang gave a clear answer: "I have decided to prohibit their native practice and do away with their crude herbs."[7] For the next ten years the traditional practitioners continued to try to gain official recognition, without success.

In 1927 the Nationalists established themselves in Nanjing, and traditional practitioners hoped for support from the new government. In May 1928 the Nanjing Government held its first National Educational Conference. Wang Qizhang 汪企張 (d. 1955), a returned student from Japan, put forward a proposal that Chinese medicine be abolished. In the same year the Ministry of Health was established, and held its first national health conference in February 1929. The conference was chaired by the Vice Minister, J. Heng Liu 劉瑞恒 (1863–1961) a graduate of Harvard Medical School and a former researcher at Johns Hopkins University.[8] Yu Yunxiu 余雲岫, the President of the Shanghai Branch of the Chinese Medical and Pharmaceutical

Association, put forward a proposal "to abolish the old style practice in order to remove the obstacles to medicine and public health", which was accepted by the conference. It proposed restructuring the licensing of traditional medical practitioners, that all traditional medical associations be abolished, and that newspapers and magazines not be allowed to publish "reactionary views" or "unscientific propaganda" (advertisements for traditional medicine).

The Shanghai traditional practitioners were the first to react. Apart from publishing a protest in the newspapers, and sending telegrams to the government, they organized a meeting of the National Federation of Traditional Medicine Practitioners in Shanghai on March 17. The delegates numbered 262, representing 131 organizations from 13 provinces. This meeting passed a resolution designating 17 March as National Medicine Day. Three days afterwards five representatives brought a petition to Nanjing. The government agreed not to implement the resolution abolishing Chinese medicine for the time being.[9] But one month later, the Ministry of Education again issued an order that the Schools of Chinese Medicine have their names changed to "training centres", and stressed that "training centres" were not considered part of the formal education system. The traditional practitioners again sent a delegation in December 1929 with a petition to the Nanjing Government. Chiang Kai-shek ordered that the decisions of the Ministry of Education and the Ministry of Health be overturned, and Chiang himself took on the position of Minister of Education in addition to his other duties.[10]

The traditional practitioners claimed they were being suppressed by forces from outside China, and accused the opposition of being "slaves of the foreigners". Chinese medicine was part of the "national essence" and for this reason had a right to survive. After the 1929 resolution, traditional Chinese medicine was to be called "National Medicine", in parallel with the National Language and the other trappings of the modern nation state. March 17 was designated National Medicine Day.

In 1929 a high tide of nationalism swept China, especially Shanghai. When the traditional practitioners held their conference in Shanghai that year, the Shanghai Chamber of Commerce provided them with conference facilities, and they were financially supported by the Shanghai Chamber, the Chinese Products Protection Society, The All China Federated Chambers of Commerce and similar groups. Discrimination against Chinese medicine was seen by the nationalism of the commercial world to be yet another example of imperialist oppression. Slogans for National Medical Day included "To oppose the Yu Yunxiu–Wang Qizhang Proposal is to oppose imperialism" and

"To support Chinese medicine is to protect the national economy."[11] During the conference a rumour spread to the effect that "the imperialist medicine and pharmaceutical companies had subsided their pharmaceutical products to the tune of six million dollars, to try to eliminate Chinese medicine and Chinese herbs", which intensified public anger. The journal *Yijie chunqiu* 醫界春秋, which represented the interests of TCM, published a special issue with the title "The Struggle in Chinese Medicine and Pharmacology" which proposed the slogan "Support Chinese Medicine, Oppose Cultural Invasion! Support Chinese Medicine, Oppose Economic Invasion!"[12]

Public opposition to abolishing Chinese medicine meant that this proposal never became official government policy. In 1929, when the *Yijie chunqiu* demanded that the government rescind the resolution on the abolition of TCM, the response of the Ministry of Health "confirmed that the Central Committee on Health has not passed any resolutions on the abolition of Chinese medicine and Chinese herbal medicine".[13]

Although the abolition of TCM never became a formal Nationalist government policy, Chinese medicine never received any formal support or recognition from the government either, and the TCM practitioners felt under continuous threat. This was the case for the whole of the Nationalist period. In 1929 the *Provisional Regulations for Doctors* required that medical practitioners have formal qualifications from conventional government recognized medical colleges. This was organizationally impossible for the TCM practitioners, as the colleges of Chinese medicine were not part of the conventional education system. After this, there were endless protests and petitions from the TCM camp, which overshadowed the new health policies of the Nationalist government.

Resistance from the New Medical Profession

Since it was the intention of the Nationalist government to abolish TCM, resistance from the TCM practitioners was to be expected. However, if we read the newsletters and journals of the new medical associations, we can see that the new health policies were also resisted by the so-called new medical profession. As far as formal medical qualifications were concerned, a considerable number of the Western medical practitioners were in much the same position as the TCM practitioners. Many of them had not been formally educated in conventional medical colleges. Some had accumulated a lot of experience through working as paramedics in missionary hospitals, and regarded themselves as qualified to practise. One of the most famous Chinese Western-style doctors of the late 19th and early 20th centuries,

Huang Chunfu 黃春甫, learned his skills working 40 years as a medical assistant in missionary hospitals in Shanghai.[14]

What is more, most missionary hospitals had never registered with the Chinese government. Even if they had provided some conventional medical training, they were still not qualified under the new regulations. So this group also joined the protests. The Suzhou, Hangzhou, Ningbo, Nanjing, Tianjin and Shanghai branches of the National Association of Medical Practitioners expressed their opposition. Eventually they requested the Ministry of Health to modify the new regulations. In a petition to the Nationalist Party Central Committee, the Administrative Yuan and the Ministry of Health, they said: "... the regulations are not in accordance with the present situation in the country. The Health Regulations in China have just begun. We should learn from the experience of advanced countries. We need not be very strict in the beginning. What is more, there is a lack of medical personnel. The present numbers cannot meet the needs of society. If we were to register people on the basis of the present regulations, there will be only a small number of people in the large cities who may satisfy the requirements. In remote areas it would be very difficult to meet the conditions. So if these regulations are really carried out, the majority of the medical practitioners in China will be excluded."[15]

As many of the practitioners of Western medicine had no recognized formal qualifications, in 1931 the Examination Yuan was established, and published *Regulations on Examinations in Western Medicine*. These regulations also met with immediate opposition from the relevant interest groups. The leading figure in this opposition was Wang Qizhang, who had proposed the regulations to abolish Chinese medicine in 1929. Wang was especially critical of the content of the proposed examinations. Apart from questions on medicine, the questions also included "forms and formulae of official documents", Party Ideology, the Three Principles of the People, Outline of National Construction, Policies of National Construction, Important Proclamations and Resolutions of the Chinese Nationalist Party, Organizational Law of the National Government, the National Constitution and so on. Wang pointed out that doctors were not government officials, and did not need to know such matters.[16]

At the same time the regulations stipulated that the examinations would be in Chinese. Since the Western-trained doctors had studied in various countries, and even those who learnt their medicine in missionary hospitals in China used medical terminology from various different languages, this posed almost insoluble difficulties for would-be candidates.[17] On the language question, some people agreed that in the interests of national unity, and the unity of the medical profession, the examinations should be in Chinese. Adoption of a National Language and indoctrination in Guomindang principles were part of the nation building strategy of the Nationalist government.

The lack of a common language amongst the practitioners of Western medicine in China was not only a concern of the Chinese government. In 1929 Dr. Ludwig Rajchman, Director of the Health Section of the League of Nations, led an investigative delegation to China. He submitted his report in early 1930, in which he raised the matter of the factionalism and the lack of a single language amongst the Western medical practitioners. His views were clearly accepted by the Ministry of Health, and his report was made part of the work report of the Ministry.[18]

The ideological sections of the examination were not an innovation on the part of the Nationalist government. Similar regulations in the 12th century demanded that career medical officers pass examinations in classical, non-medical literature.[19]

Clearly the opposition of sections of the western medicine practitioners was also a factor in the eventual decision not to implement these examination regulations.

In response, in 1932 the Ministry of the Interior issued *Modifications to the Certification of Medical Practitioners*. This document stipulated (1) Those who have studied for four years and have graduated from an unregistered school which subsequently passed the inspection of the Ministry of the Interior, may be issued with a Certificate of Practice, (2) Those who have practised more than five years in a hospital which has passed the inspection of the Ministry of the Interior, and those who have established their own clinics before 1929 and hold a license issued by the local authorities may also be considered qualified to practise. It also mentioned that there would be no further modifications to the regulations.[20]

The next year the Shanghai Municipal Health Bureau issued rules for the registration of Western medicine practitioners. They were based on the regulations of the Ministry of the Interior, with some minor changes to make them easier to implement. For example, such evidence as advertisements in the newspapers, written prescriptions and telephone numbers could be used to demonstrate that a clinic had been in existence prior to 1929.[21]

These compromised rules also met with opposition. Some new medical practitioners argued that in their view the modified regulations were no improvement on the earlier demand of the National Association of Medical Practitioners (1929) that all Western medical practitioners be unconditionally registered.[22]

It appears that neither the proposed examinations nor the more flexible modified regulations could be implemented. In 1936, the Executive Yuan issued *Supplementary Rules on the Grading of Medical Practitioners*, which would remain in force for the time being, until the Examination Yuan had held the formal medical examinations. Grading was to be determined by a commission made up of nine appointees.[23] Whatever the qualities of this method may have been, China was on the brink of war. During the war the government moved to Sichuan and was not in control of the large cities in eastern and northern China, such as Shanghai and Peking, where Western medical practitioners were concentrated. Although regulations on medical practitioners were issued by the Central Government in 1943 and reprinted in the *Yixue daobao* in 1946, it is unlikely they were ever implemented.[24]

In retrospect, it is clear that after 1929, every time the Ministry of Health issued regulations on the compulsory regulation of medical practitioners, the government had to compromise with opposition from the practitioners of Western medicine, and issue a different set of regulations some time later. These new regulations in turn would be resisted. Obviously the Nationalist government did not have the support of Western medical circles in its attempts to regulate the profession. The Western-trained doctors, for their part, expected the protection and support of the government, something the government was not prepared, or not able, to do.

The dispute between Chinese and Western medicine was by no means only an ideological one. For most of the physicians involved, it was a matter of competition for very limited resources. In this battle, the real competition for the Chinese practitioners of Western medicine were the non-Chinese doctors resident in China, and not the TCM practitioners in the countryside or the traditionalists in the cities. The battleground was the cities, with high concentrations of educated and relatively affluent people. These people generally tended to seek treatment from foreign doctors. From an examination of the resolutions of various meetings reported in the newsletters of the National Association of Medical Practitioners, it can be seen that the issue which concerned them the most was the limitation and control of foreign, Western-trained doctors. Several petitions to the government along these lines were submitted. In the journals of the Western-style medical practitioners there are many articles critical of foreign doctors, even the missionary hospitals. According to these articles, either their medical skills were low or their medical practices ethically doubtful.

When in 1929 Wang Qizhang advocated the abolition of traditional medicine, he was denounced as a "slave of the foreigners" by the TCM practitioners. Chinese practitioners of Western medicine were also accused of being "slaves of the foreigners" and "worshipping foreign countries" by the TCM camp. Now it was the turn of the Chinese practitioners of Western medicine to use the same expressions against the foreign doctors practising in China. Their attitude towards Chinese nurses and orderlies was said to be "imperialist suppression", and Chinese who chose to seek the advice of foreign doctors were declared to be "worshipping foreign countries". And so on.[25]

Wang Qizhang divided foreign doctors into several categories: those with political connections, those with connections with various religious groups, those in both public and private practice, economic invaders, and those who had to come to China as political refugees. The reasons these people came to China were not necessarily noble and their skills not necessarily high. Thus they should also have to submit to examinations, he thought, before being permitted to practise.[26]

In the same year the National Association of Medical Practitioners held its third conference. The delegate from Shanghai proposed a resolution to send a *Request to the Government to Regulate the Practice of Foreign Physicians,* which was accepted unanimously. It stipulated that foreign doctors must (a) have been resident in China for more than two years, (b) be fluent in Chinese, (c) have formal qualifications, (d) or if lacking formal qualifications, required to take strict examinations, and (e) those possessing the above qualifications be registered by the Ministry of Health and become members of the local medical practitioners association.[27]

The Chinese practitioners of Western medicine were not in a position to demand that the government totally forbid foreign doctors to practise in China, but they certainly expected that the government would impose limitations on them. The requirements they wished to impose on foreign doctors were much the same as what the Ministry of Health was demanding of themselves, which they strongly resisted — a clear case of double standards.

Another issue which preoccupied the practitioners of Western medicine was that the government protect their rights. If a patient brought a case of alleged medical malpractice to court it would be treated as a criminal case and the doctor would be arrested. In 1929 the National Association of Medical Practitioners passed a resolution requesting the government "determine the responsibilities doctors have to the nation and to society, and what rights they should enjoy, to protect their standing as professionals".[28]

In 1934, it was proposed that the National Association of Medical Practitioners establish a committee to protect the rights of the medical profession to help doctors who became involved in medical and legal disputes.[29]

The situation had not improved by 1947. Yu Yunxiu, a leading figure in Western medical circles, wrote:

> We are now facing a crisis — disputes involving medicine. Should a patient unfortunately die after treatment, the family will take the doctor to court and try to extort something from him. For the past several months, there have been many such examples ... In Nanjing, the local court which tried Dr Qian did not adopt the standards of forensic medicine and the evidence of medical specialists, nor did it use firm and reliable rules of evidence to determine guilt. It relied on baseless rumour and anonymous evidence, and sentenced Dr. Qian to jail for one year and six months. Another newspaper reported a strange case from Suzhou in which the patient did not die, but the family tried to extort something from the doctor, claiming that the patient did indeed die.[30]

In the case in Nanjing mentioned above, the national association and the local medical associations sent telegrams to the Legislative Yuan, the Administration Yuan, the Ministry of Health and the Supreme Court in protest.[31] Chinese practitioners of Western medicine felt they had very little standing in society, no more than the practitioners of traditional medicine had in traditional society. Contemporary society regarded medicine as a second rate profession.[32] Legal disputes over medical matters actually stirred up by other doctors was another aspect of the struggle for resources and recognition.[33]

Difficulties of Policy Implementation

In only ten years, the Nationalist government issued many decrees on so many aspects of the question. But the implementation of these regulations met with quite a few difficulties.

Administratively, each province should have had a provincial level Bureau of Health to administer the policies of the central Ministry of Health. Theoretically, each province did have a Bureau of Health, sometimes as a subsection of the Bureau of Police. In reality, however, there were many places in which they did not exist, or were abolished not long after being set up. The Bureau of Health of Beiping, and that of Qingdao, were abolished in 1930 for economic reasons.[34] In places where there was no Bureau of Health, even if the practitioners had wanted to register, there was nowhere for them to do so. In many cases, the central government lacked the bureaucratic structure to implement its decisions. In Beiping, the Bureau of Health was subsumed under the Bureau of Public Security. In those places where there was a Bureau of Health, it often had its own policies distinct from those determined by the central ministry. Sometimes these local policies clashed with those in other areas. In 1927 the Shanghai Bureau of Health decreed that "Practitioners of traditional Chinese medicine must not use the methods of western doctors, or use western medicine (drugs). Practitioners of western medicine must not use Chinese herbal medicine."[35] But almost at the same time the Beiping Bureau of Health issued a document saying:

> Since the introduction of Western medicine into China, Chinese have been falling over each other to learn from the West. As a result of this, thousands of years of Chinese medical knowledge is being neglected. This is not the way to promote Chinese culture. Both Chinese and Western medicine each have their advantages ... So we should try to ensure that each complement the other. On the one hand, we should collect the classics of traditional Chinese medicine, and collect indigenous herbal medicine. At the same time, we should study Western medicine to complement Chinese medicine.[36]

The *Yijie chunqiu* sent a telegram to the Nationalist government Minister of Health, Xue Dubi 薛篤弼, in which they said "Your Excellency has always placed great stress on the reform of medicine and medical facilities. We think you certainly follow our President's [Sun Yat-sen] posthumous instruction, that we should promote national skills. So you will surely implement the policy on the complementarity of Chinese and Western medicine which has been issued by the Bureau of Health in Beiping."[37]

The minister did not express an opinion on this matter. In a terse reply, he merely said: "Medical administration is a very primitive stage in our country. As for myself, my knowledge is shallow and I lack vision. Your complimentary words make me feel even more inadequate. Thank you for your concern."[38]

The doctors trained overseas had studied in different countries and in different languages, and followed different schools of "western medicine". Even within China it was not possible to decide on one medium of instruction. The National Central University Medical College used only textbooks in English. Zhongshan University used German. Only National Beiping University used Chinese.[39] Politically, the leaders of some factions relied on different international forces. Tang Erhe 湯爾和 (1871–1940), who had studied medicine in Japan, was the founder and president of the Chinese Medical and Pharmaceutical Association, whose members were basically doctors who had studied in Japan. In 1922 Tang was a minister in the Peking Government, and in 1926–27 was Minister of the Interior and Finance. The Ministry of Health belonged to the Ministry of the Interior. During the Second World War Tang Erhe occupied an important position in the puppet Peking Government, and so was regarded as a traitor by the Nationalist government.[40]

After the establishment of the Nationalist government in 1928, the Ministry of Health came under the control of doctors who had returned to China after studies in England or the United States, and who formed a new medical elite. The problem of factionalism among various groups of Western medicine practitioners was noted in the report by Dr Ludwig Rajchman. He also pointed out that when the government wanted to unify the various factions in 1929, this impinged on the interests of all concerned, and so the problem could not be resolved.[41]

An even greater difficulty was in the cities, where the practitioners of Western medicine were concentrated. For example, the Nationalist government had no power in the Shanghai Concessions. Shanghai was divided into the International Settlement, controlled by the British and the Americans, the French Concession and Greater Shanghai. These three sections came under three different administrations. Greater Shanghai (under the control of the Nationalist government) occupied a relatively small part of the whole. It was impossible to implement the policies of the Nationalist government in most of Shanghai. In 1929, the League of Nations Report revealed that there were several so-called medical schools in the concessions, but the quality of teaching was extremely bad. The Ministry of Education considered ordering them to be closed down, but the Chinese government had no control there.[42]

In 1931 the International Settlement established its own Medical Board, which approved the rules for the registration of medical practitioners, dentists and veterinary surgeons, both Chinese and foreign.[43] So when the Shanghai doctors of Western medicine demanded that limitations be put on foreign doctors, all the government could do was to issue *Rules of Registration of Foreign Doctors* without being able to implement them.[44]

Another difficulty was the gap between the ideal situation and actual reality. The series of regulations issued by the Nationalist government were demanded by the returned students from England and America, and were modelled on similar regulations in the United States. Chinese realities, however, were far removed from Western conditions. The practicability of many of these regulations was a problem. Some of the government's policies were no more than a reaction to the report from the League of Nations. For example, the report pointed out that the sanitation system even in cities like Nanjing, Hangzhou and Changsha was not up to standard. The majority of Chinese still drank unclean water, and the Ministry of Health hoped to get the financial support of the League of Nations to ameliorate the situation. The intention was to choose three or four places to establish facilities for purifying water and effective sewerage systems. The Ministry's regulations on the use of running water and the administration of wells[45] were also reactions to the League of Nations' report. Any understanding of the actual conditions in China's villages would have revealed that it was unlikely that the regulations on the use of wells could be implemented.

In regulating the medical profession, the government also faced a dilemma. On the one hand the government hoped to control strictly the standard of the qualifications of the medical practitioners. On the other hand there were really very few new-style Western-trained doctors in China. In 1925,

> It has been estimated that there were 4000 to 5000 doctors trained in modern medicine in China, although only 918 had registered with the new Ministry of Health that year. With a population of about 450 million, the ratio of doctors to population was 1 to somewhere between 90,000 and 112,000. This may be compared to the figures of 1 to 800 in the United States in 1930, or 1 to 1490 in England and Wales in 1927. Even by 1937, "even in the most developed rural districts more than 65 percent of the population continued to rely on practitioners of native medicine, while about 26 percent died without receiving any treatment at all".[46]

The practitioners of Western medicine also used this argument. Since there were so few doctors, to disqualify the majority of them was not reasonable, and would make the situation much worse. The medical schools were the same. The schools which met the required standards were proud of their high entrance requirements. Students who did not pass the examinations would not be able to continue their courses, and only the better students could graduate. The Peking Union Medical College, for example, accepted a maximum of 30 students for its medical and nursing courses every year. During the eight-year training period, one-third of these dropped out. Clearly, it provided very few qualified medical personnel.[47]

Another factor was that people in the countryside could not afford to buy Western medicines, nor did they trust it. Some idealistic young graduates of medical schools who went to work in the villages met with numerous obstacles. In 1935 the Medical School of the Henan Provincial Normal University established a clinic in a village called Baiquan 百泉. They tried to introduce courses on concepts of hygiene, vaccination against cholera and other contagious diseases, with little success. Their free course for midwives did not attract a single enrolment.[48] Even fully qualified midwives found it difficult to find work, even in the cities. The villagers did not trust the new-style midwives. It was said that midwives wishing to work in the villages would have to develop a spirit of self-sacrifice like that of the missionaries.[49]

The poverty and the low educational level of the population in most areas of China proved major obstacles to the implementation of the new health policies. The central government could issue regulations, but changing or improving the actual situation on the ground in China's cities and villages was another matter.

Conclusion

Since its inception, the Nationalist government intended to abolish traditional Chinese medicine. For a variety of reasons it was not able to. Most of the Chinese population, particularly in the villages, still relied on Chinese medicine. After the Second World War there were 830,000 practitioners of traditional medicine in China, whereas those trained in Western medicine numbered less than 30,000.[50] As the Ministry of Health was not willing to recognize traditional medicine as appropriate for a modern state, the practitioners of TCM were excluded from the regulating process. At the same time, practitioners of Western medicine continuously objected to the new regulatory regulations of the government, one of their most common objections being the fact that the TCM practitioners did not have to undergo formal examinations as a condition of registration, as they were not registered at all.

The regulation of the medical profession was one of the high priorities among the policies relating to health and medicine of the Nationalist government, but if the 830,000 TCM practitioners were not subject to regulation, one can see why the practitioners of Western medicine were reluctant to accept or implement any regulation. This aspect of the new health policies of the Nationalist government could hardly be considered successful.

Notes

[1] Ka-che Yip has researched this matter in some detail in *Health and National Reconstruction in Nationalist China — the Development of a Modern Health Service, 1928–1937,* University of Michigan, Association for Asian Studies, 1995.

[2] Chen Bangxian 陳邦賢, "Xiandai yiyao weisheng faling cunmu 現代醫藥衛生存目 Xiandai weisheng xingzheng shiliao zhi yi 現代衛生行政史料之一" [Cumulative index of contemporary regulations on modern medicine, pharmaceuticals and hygiene — Materials on the history of contemporary hygiene administration, Part 1] in *Yiyu jikan* 醫育季刊 [Medical Education Quarterly], Jiaoyubu yixue jiaoyu weiyuanhui 教育部醫學教育委員會, Vol. 4, No. 1, 1940, pp. 48–54.

[3] Paul U. Unschuld, *Medical Ethics in Imperial China — A Study in Historical Anthropology,* University of California Press, 1979, pp. 17–18.

4 Li Tao 李濤 "Zhongguo xiju zhong de yisheng" 中國戲劇中的醫生 [Doctors in Chinese opera], in *Yishi zazhi* 醫史雜誌, Vol. I, No. 3, Shanghai, 1948, pp. 1–16.

5 Zhao Hongjun 趙洪鈞 *Jindai Zhong-Xi-yi lunzheng* 近代中西醫論爭史 [The debate between Chinese and Western Medicine in Modern China], Anhui Kexue chubanshe, 1989, pp. 140, 261–62.

6 *Ibid.*, p. 141.

7 K. Chi Min Wong and Wu Lien-teh, *History of Chinese Medicine,* Shanghai: National Quarantine Service, 1936, p. 159; Ralph C. Croizier, *Traditional Medicine in Modern China — Science, Nationalism and the Tensions of Cultural Change,* Harvard University Press, 1968, p. 69.

8 Yip, *op. cit.*, p. 30.

9 Zhao, *op. cit.*, pp. 111–16; Wong and Wu, *op. cit.*, pp. 161–62.

10 Zhao, *op. cit.*, p. 149.

11 *Ibid.*, p. 115.

12 *Yijie chunqiu* 醫界春秋 No. 34, Shanghai, April 1929, p. 8.

13 "Guomin zhengfu wenguan chu fu Yijie chunqiu she han" 國民政府文官處復醫界春秋社函 [Reply from the Civil Service Office of the National Government to the *Yijie chunqiu*], *ibid.*, p. 20.

14 Kerrie L. Macpherson, *A Wilderness of Marshes: The Origins of Public Health in Shanghai, 1843–1893,* Oxford University Press, 1987, p. 149.

15 "Quanguo yishi lianhehui wei qing xiugai zanxing tiaoli shang dangju wen" 全國醫師聯合會為請修改暫行條例事上當局文 [Text of the submission of the National Association of Medical Practitioners to the authorities requesting revision of the provisional regulations] in *Yiyao pinglun* 醫藥評論, Shanghai, 1929, Vol. 1, No. 22, pp. 36–37.

16 Wang Qizhang 汪企張, "Du Kaoshiyuan gongbu zhi gaodeng kaoshi Xiyi yishi kaoshi tiaoli hou zhi wo jian" 讀考試院公布之高等考試西醫醫師考試條例後之我見 [My views after having read the regulations on advanced level examinations for Western medicine announced by the Examination Yuan] in *Yiyao pinglun,* Vol. 3, No. 51, Shanghai, 1931, pp. 1–2.

17 Zhuang Weizhong 莊畏仲, "Wu guo Xiyi yishi kaoshi wenti zhi shangque" 吾國西醫醫師考試問題之商榷 [Discussion on the question of examinations for practitioners of Western medicine in China], *ibid.*, pp. 3–5.

18 Song Ze 宋澤 (trans.), "Zhonghua Minguo weisheng shiye zhi xianzhuang — Guoji Lianmeng baojian buzhang baogao" 中華民國衛生事業之現狀 — 國際聯盟保健部長報告 [The current situation on the health industry in the Republic of China — A report by the Director of the Health Section of the League of Nations], in *Yiyao pinglun,* Vol. 3, No. 53, pp. 15–24, on pp. 20–21.

19 Unschuld, *op. cit.*, p. 22.

20 Wang Yugang 汪于岡 "Neizhengbu biantong yishi gei zheng banfa zhi xiaoguo" 內政部變通醫師給辯證法之效果 [Consequences of the *Modifications to the Certification of Medical Practitioners* issued by the Ministry of the Interior] in *Yiyao pinglun,* Vol. 4, No. 87, 1932, Shanghai, pp. 3–4.

[21] "Shanghai shi weishengju bugao Xiyi kaiye zhengming zhengmingshu banfa" 上海衛生局布告西醫開業證明證明書辦法 [Regulations on the Certification of Practitioners of Western Medicine issued by the Shanghai Municipal Health Bureau], in *Yishi huikan* 醫事匯刊, Vol. 5, No. 15, 1933, p. 56.

[22] Xie Junshou 謝筠壽, "Duiyu yishi dengji wenti zhi guanjian" 對于醫師登記問題之管見 [Some views on the issue of the registration of medical practitioners], in *Yiyao pinglun*, Vol. 4, No. 87, Shanghai, 1932, pp. 4–5.

[23] "Yishi zhenbie banfa" 醫師甄別辦法 [Means of examining and distinguishing medical practitioners], in *Yishi huikan*, Vol. 10, No. 29, Shanghai, 1936, pp. 461–63.

[24] *Yishifa* 醫師法 [Regulations on medical practitioners], in *Yixue daobao* 醫學導報, No. 5/6, Chongqing, 1946, pp. 43–53.

[25] Ju 菊 (pseud.), "Waiji Xiyi — shehui bai-wai xinli zhi yi" 外籍西醫 — 社會拜外心理之一 [Foreign western doctors — an example of the psychology of worshipping foreign things in society], in *Yiyao pinglun*, Vol. 4, No. 91/92, 1932, pp. 1–2; Chunren 蠢人 (pseud.), "Waiji jianghu-yi zhi xiezhen san pian — diguozhuyi yapo xia nü hushi zisha de huigu" 外籍江湖醫之寫真三篇 — 帝國主義壓迫下女護士自殺的回顧 [Three examples of the true face of foreign quack doctors — reflections on the suicide of a female nurse under imperialist oppression], in *Yiyao pinglun*, Vol. 6, 1934, pp. 5–12.

[26] Wang Qizhang 汪企張, "Waiji yishi gailun" 外籍醫師概論 [Overview on foreign western medical practitioners], in *Yishi gonglun* 醫事公論 No. 17, Nanjing, 1934, pp. 17–20.

[27] *Yixue yu yaoxue* 醫學與藥學, Vol. 3, No. 9, 1934, p. 91.

[28] "Quanguo yishi lianhe dahui xiaoxi zhixiang" 全國醫師聯合大會消息志詳 [Detailed report on news from the National Association of Medical Practitioners], in *Yiyao pinglun*, Vol. 1, No. 22, Shanghai, 1929, pp. 25–29, on p. 25.

[29] Song Guobin 宋國賓, "Qing quanguo yishi lianhehui zuzhi 'Yiye baozhang hui' yi baozhang quanguo yijie quanli bing chuli ge-di yishi jiufen an" 請全國醫師聯合會組織醫業保障全國醫界權利并處理各地醫事糾紛案 [A request to the National Association of Medical Practitioners to organize a medical insurance company in order to protect the interests of the medical practitioners of the whole country and to deal with disputes involving medical matters in various localities], in *Yiyao pinglun*, Vol. 6, No. 2, Shanghai, 1934, pp. 9–10.

[30] Yu Yunxiu 余雲岫, "Dajia tuanjie qilai" 大家團結起來 [Everybody should unite], in *Yixun* 醫迅 Vol. 1, No. 3, Shanghai, 1947, pp. 1–3.

[31] *Ibid.*, pp. 3–6.

[32] Yun 雲, "Yao fuxing minzu bixu tigao yishi zai shehui shang de diwei" 要復興民族必須提高醫師在社會上的地位 [In order to regenerate the nation we must raise the social status of medical practitioners], in *Yishi gonglun*, Vol. 2, 1933, pp. 14–15.

[33] Fan Shouyuan 範守淵, "Tuizhan tongdao de hu'ai jingshen" 推展同道互愛精神 [Develop the spirit of mutual respect amongst fellow medical practitioners], in *Yishi tongxun* 醫事通論, Vol. 1, No. 2, Shanghai, 1947, pp. 1–2; Lu Shuda 盧叔達,

"Yishi jiufen yuanyin de tuice' 醫事糾紛原因的推測 [A conjecture on the causes of medical disputes], *ibid.*, pp. 2–3.

[34] Song Ze, *op. cit.*, Vol. 3, No. 54, pp. 21–32, on p. 23.

[35] Shouzhi-zi-Wuji 壽芝自蕪寄, "Duiyu Shanghai tebie-shi weishengju gonggao zhi shangque" 對于上海特別市衛生局公告之商榷 [Discussion on the announcement of the Shanghai Municipal Bureau of Health], in *Yijie chunqiu*, No. 19, January 1928, Shanghai, pp. 1–2.

[36] "Beiping-shi tiaoji Zhong-Xi yishu zhi tongling" 北平市調劑中西藝術之通令 [General instructions on the regulation of Chinese and Western medicine in Beiping Municipality], in *Yijie chunqiu*, No. 29, Shanghai, November 1928, pp. 19–20.

[37] "Benshe shang weishengbu Xue buzhang dian wen" 本市上衛生部薛部長電 [Telegram from this journal to Minister Xue of the Ministry of Health], in *Yijie chunqiu*, No. 29, Shanghai, November 1928, p. 19.

[38] "Xue buzhang fu benshe dian wen" 薛部長復本社電文 [Telegram in reply from Minister Xue to this journal], *ibid.*

[39] Song Ze, *op. cit.*, p. 28.

[40] Jiang Deming 姜德明, "Zhou Zuoren yu Tang Erhe" 周作人與湯爾和 [Zhou Zuoren and Tang Erhe], in *Shutan mengxun* 書攤夢尋, Beijing, Yanshan chubanshe, 1996, pp. 102–105.

[41] Song Ze, *op. cit.*, pp. 12–13.

[42] *Ibid.*, p. 26.

[43] Wong and Wu, *op. cit.*, p. 745.

[44] Chen Bangxian, *op. cit.*, pp. 48–54.

[45] Song Ze, *op. cit.*, p. 24, Chen Bangxian, p. 50.

[46] Yip, *op. cit.*, pp. 132–34.

[47] Zhengxie Beijing wenshi ziliao yanjiuhui 政協北京文史資料研究會, ed., *Huashuo lao Xiehe* 話說老諧和 [Talks about the old Peking Union Medical College], Beijing, Zhongguo wenshi chubanshe, 1987, pp. 20–21. See also John Z. Bowers, *Western Medicine in a Chinese Palace — Peking Union Medical College, 1917–1951*, The Josiah Macy Jr. Foundation, 1972, pp. 89–91.

[48] Aitang 愛業, "Henan-sheng li Baiquan-xiang xiangcun yiyuan gaikuang" 河南省立白泉鄉鄉村醫院概況 [Overview of the village hospital in Baiquan county, Henan Province], in *Yiyao pinglun*, Vol. 7, No. 4, 1935, pp. 39–49.

[49] Ji Jilin 計濟霖, "You funü jiefang yundong shuo dao jinri zhuchanshi de chulu wenti" 從婦女解放運動說到今日助產士的出路問題 [A solution to the problem of midwives from the women's liberation movement], in *Yiyao pinglun*, Vol. 7, No. 4, 1935, pp. 5–6.

[50] Jiang Huiming 蔣晦鳴, "Bu-teng bu-yang de Zhong-Yi yundong" 不痛不癢的中醫運動 [The perfunctory Chinese medicine movement], in *Yixun* 醫迅, Vol. I. No. 9, 1948, pp. 1–2; Wang Juerong 王爵榮 "Xin yi-yao jie bixu da tuanjie" 新醫藥界必須大團結 [Practitioners of new [=Western] medicine and pharmaceuticals must unite], in *Yixun*, Vol. 1, No. 5, 1947, pp. 1–2.

The Transmission of Traditional Chinese Medicine (TCM) to England (Outline)

MA BO-YING
Shanghai University of TCM

ALICIA GRANT
London College of Traditional Acupuncture and Oriental Medicine

From the 16th Century to 1970

Many articles have described the history of the transference of acupuncture, moxa, herbs, and some theory of Chinese medicine to Europe and England. Dr. Joseph Needham's and Dr. Lu Gwei-Djen's book "Celestial Lancets, A History and Rational of Acupuncture and Moxa" is probably the most detailed in a Western language. Prof. Ma Kan-Wen has published a series of articles in Chinese which present much more information; unfortunately many scholars cannot read Chinese so his papers have not received their due attention.

TCM was a new kind of knowledge for Western people. Europeans, particularly the intelligentsia, were as always very receptive to new knowledge. They had open minds which were curious about different cultures; they read the available literature in Latin, French and Dutch as well as English translations. We would like to give an overview of their main publications and also some reports by English doctors through the centuries, some in the form of letters to medical journals, others from books they published which contain a passing comment on TCM modalities. In this way we will consider their attitude to the new form of treatment. Interest in TCM waxed and waned at different times over the centuries until we come to the latest period of major and sustained interest.

To the Jesuit missionaries in China we owe much early information. The great Jesuit missionary Matteo Ricci (1552–1610) gave the earliest organized medical work, his book pointing out that Chinese medicine is very different from Western but very effective. A little later the Swiss Jesuit Father Johannes Terrenz (1576 Constanz–1630) wrote an unfinished and unpublished work in 2 volumes, "Plinus Indicus", introducing 80 kinds of Chinese herbs with illustrations. Then the Jesuit priest Michel Boym (1612 Poland–1659), son of the physician to the king of Poland, wrote "Flora Sinensis", a short description of Chinese plants, published in Vienna in 1656. He finished writing "Clavis Medica ad Chinarum Doctrinam de Pulsibus" in 1658: this comprised four books on the pulse, a treatise on the tongue in different diseases and a description of simple drugs prepared according to Chinese prescriptions. This was published 24 years later in 1682 in Frankfurt by Andreas Cleyer, a surgeon in Batavia, under the title "Specimen Medicinae Sinicae" with the author's name suppressed — possibly due to an altercation between the Dutch East India Company and the Jesuits. In 1686 an edition was published with the author's name and the original title. As well as the pulse and tongue, which were illustrated, herbs and twelve regular acu-tracts were mentioned.

In 1658, the year Boym had actually finished writing the afore-mentioned books, Jacob de Bondt's (1598–1631) "Historia Naturalis et Medica Indiae Orientalis", the first book referring to acupuncture, was published. He wrote that the results with acupuncture "surpassed even miracles".

In 1671 a book appeared entitled "Les Secrets de la Medicine des Chinois, consistant en la parfaite Connoissance du Pouls", published in Grenoble and attributed to a French gentlemen of great merit. An interesting controversy surrounds the authorship, however it was translated into Italian and English, giving it a wide sphere of influence with its mention of 24 pulses.

In 1674 Hermann Buschof , a Dutch minister in Batavia, published "Het Podagra ...", the first book in the West on burning moxa (Artemisia Vulgaris), having benefited from the treatment himself for his gout from which he suffered for 14 years. He wrote, "This form of burning goes beyond all Remedies of Europe, against the Gout". The book is very clearly written, in question and answer format, and was translated into English and published in London under the title "Of the Gout" two years later.

It is interesting to note in passing that the first Western printed books on acupuncture, moxa, herbs and the pulse were all preceded by letters mentioning each of the topics some years before. The first book on acupuncture — de Bondt in 1658 is, according to Wolfgang Michel, pre-dated by a letter written 60 years before from Father Lourenco Mexica in Macao in 1584 relating that all illness are treated with silver needles and a burning herb. The first book in the west to mention the pulses, "Les Secrets de la Medicine Chinoise" printed in 1671 was, according to Grmek, pre-dated by a printed letter from the Spanish Father Francisco de Herrera Maldonado from 1621. In the same year Father Johannes Schreck in China wrote a letter to an apothacary friend in Rome about the important role of the pulse. An even earlier one from Father de Almeida in 1566 describes how he takes the pulse himself in the classical style. These letters also predate the book on the pulse Boym finished writing in 1658.

As the letter from Father Mexica from 1584 also mentioned moxa, it predates by 90 years the first Western printed book on this subject by Buschof in 1674.

In 1676 R. W. Geilfuss published a book in Marburg on moxa.

Then in 1682 J. A. Gehema published "Eroberte Gicht" in Hamburg, remarking that it was fast and safe.

In 1683 Willem ten Rhyne, a Dutch physician, published his book "Dissertatio de Arthritide, Mantissa Schematica de Acupunctura"; published in London in Latin, also in the Hague and Leipzig. A detailed treatise on acupuncture, it contained illustrations of acu-tracts with green points for applying needles and red points for moxa.

At the end of the 17th century we have a report in 1693 of Sir William Temple (1628–99), an Edinburgh surgeon, using acupuncture for treatment of an aortic aneurysm; also of the very famous clinician Thomas Sydenham (1624–89) who wrote the same year in his book "Miscellanea" with enthusiasm of "the Cure of the Gout by Moxa", following the cure of his own gout by moxibustion while attending an international conference in Holland in 1677. Ten years earlier he had been against the use of moxa. This raises a point that has been valid through the centuries, including the 20th century, that the effectiveness of TCM is the strongest reason for its adoption as a modality of treatment. Sydenham personally was against using acupuncture for dropsy, which nonetheless shows that it was then a current practice.

At the beginning of the 18th century Sir John Floyer published his "Physician's Pulse Watch", the first of 2 volumes in 1707; an essay to explain the old art of feeling the pulse. He had read the 1682 book edited by Cleyer with great interest, not knowing the true authorship, and was inspired to invent a stop-watch which ran for 60 seconds.

In 1712 a clear report of moxa and acupuncture was gained from Englebert Kaempfer's (1651–1716) dissertation in Latin at Leiden, "Amoenitatum Exoticorum". He named the acupuncture points in the treatments reported and included an illustration of 60 commonly used points.

The 18th-century surgeon Lorenz Heister wrote about the clinical uses of acupuncture in his book "Chirugie", published in Nurnberg in 1718, which was translated into six languages, making it a vehicle for dissemination of TCM . At the same time he regarded it as going out of fashion. Theoretical interest in the subject continued, however, as exemplified by Gvan Swieten writing in 1755 on the desirability of investigating the physiological communication involved in alleviation of pain using acupuncture and moxa.

Lady Mary Montague, wife of the English diplomat in Constantinople, returned to England and had her daughter inoculated for smallpox in 1721.

This had been the Chinese method of variolation used for centuries and led to Edward Jenner's (1749–1803) method of vaccination later in the century (1796).

In 1775 a Chinese businessman, Wan A-Tong, brought an acupuncture model made of copper to London. It had acu-tracts and acu-points marked on it; unfortunately its location is now unknown.

In 1779 we read an account by C. P. Thunberg, a naturalist well-travelled in East Asia, about printed tables showing the location of points for treatment and mentioning the technique of twirling the hair-thin needles.

A large work entitled "Memoires Concernant l'Histoire, les Sciences, les Arts, les Moeurs, les Usages, etc des Chinois" (1780–97?) had a great influence in Europe. Darwin (1809–1882) had read it and quoted some material to prove his theory, writing it into his book "The Origin of the Species" (1859). This knowledge is to be found in Boym's book "Flora Sinesis" and originated from the "Ben Cao Gong Mu", published 1596, the famous doctor Li Shi-Zhen's work which Darwin called the "ancient Chinese encyclopaedia".

The 18th century begins with an account by Dr. Coley, a country practitioner who treated a baby for abdominal flatulence with acupuncture in 1802. Often the surgeon and obstetrician Edward Jukes is credited with being the first person to use acupuncture in England, in 1821, on a patient with thigh pain.

In 1820 in Macao, Dr. Livingstone, working for the East India Company, founded a clinic; he put a Chinese, Dr. Li in charge of the dispensary with a complete assortment of Chinese medicines and gathered 800 volumes of Chinese medical books for the library.

Between 1821 and 1831 three general practitioners from England and Scotland wrote papers, published in the Lancet and the Edinburgh Medical & Surgical Journal, reporting more than 100 successful cases treated with acupuncture with no side effects, the cases including sciatica, rheumatism, tic douloureux (facial neuralgia) and dropsy.

During this same period J. M. Churchill wrote two books (in 1821 and 1828) with many case histories.

In Dublin William Wallace in 1827 employed the use of burning moxa on the end of the needles, a classic Chinese method.

In the 1840s the Leeds Infirmary became a renowned centre for acupuncture treatment of osteoarthritis and the medical museum there contains a section on acupuncture.

In the mid-19th century some surgeons used acupuncture to treat hernias, corneal opacity, varicose veins and aortic aneurysms.

In 1892 Dr. Lockhart wrote an article advocating the use of acupuncture for the skin condition of psoriasis. The same year the famous American surgeon William Osler wrote in his "Principles and Practice of Medicine" that acupuncture was the most efficient treatment for lumbago, using 3–4 inch needles retained for 5–10 minutes, and also reported his success with rheumatoid arthritis and intercostal neuralgia.

Looking back over the first half of the 19th century, acupuncture and moxa were popular in Germany, England and in the hospital clinics in France. There was some waning of interest in the second half of the century, but an increase in interest at the century's end. It was noted that moxa was difficult to obtain; there was no import of Chinese herbs, so this branch of TCM only developed later in the 20th century.

The first half of the 20th century finds both positive and negative remarks from doctors and medical missionaries; James Cantlie, who saved the life of Sun Yat Sen during his sojourn in England, was Dean of the Hong Kong Medical school and wrote in the Chinese Medical Journal in 1916 his observations on "needling", as he termed acupuncture. He was very impressed and adopted the technique for many ailments.

1929 saw the publication of the second edition of a book entitled "The Diseases of China" written by James Maxwell, a medical missionary, referring to "the deadly acupuncture needle, which is the favourite Chinese instrument of professional torture".

Far more positive was a book by Soulie de Morant (France 1878–1955), who had been the consul in Shanghai for 20 years, where he learned to practise acupuncture, returning to France to promote it. His book was published between 1939 and 1941, entitled "Precis de la Vrai Acupuncture Chinoise — A Summary of the True Chinese Acupuncture". He is reputed to be the first person to introduce acupuncture systematically to the West and was nominated for the Nobel Prize in 1950. His book was translated into English in 1994.

1951 saw the emergence of Dr. Paul Nogier's ear acupuncture, widely used in England. News of the first operation in China using acupuncture anaesthesia, in Shanghai at the First People's Hospital, for removal of tonsils, came in 1958.

Dr. Felix Mann, well-known London doctor, published a book in 1962 entitled "Acupuncture — Ancient Chinese Treatment Technique". Aldous Huxley's preface reflects the open attitude of a large section of intelligent observers — although modern physiology has no explanation, one can accept that strange effect which cannot be explained at this moment.

In April 1972 a patient wrote a letter to the "Sunday Times", a London newspaper, having had acupuncture anaesthesia for dental treatment. He noted it was not only good for people under Mao Tse Tung but also for a "pure Englishman".

The 1973 Royal Society Report included an article by W. B. Perlow which introduced acupuncture theory and reported 1000 case histories. We then have the book "Celestial Lancets" mentioned at the beginning and written in the 1970s.

According to Professor Ma Kan-Wen's count there were, before 1700, about 10 books published on Chinese medicine; about 60 between 1700–1840, and about 120 in 1840–1949.

TCM encounters the problem of scientific proof, verifiable in the laboratory. But for the many people who rejected it from this standpoint there were others who, in their writings, enthusiastically advocated an empirical treatment with effective results.

From the 1980s to Date

Due to the previous acupuncture tradition in the U.K., and in particular because of the influence of President Nixon whose visit to China was accompanied by news reports about acupuncture anaesthesia, many people went to China to learn acupuncture and then came back to England and started to practise. Acupuncture thus became quite popular there. When Dr. Joseph Needham's and Dr. Lu Gwei-Djen's book "Celestial Lancets, a History & Rationale of Acupuncture & Moxa" was published in 1980 by Cambridge University Press, the three pre-conditions for Chinese medicine to develop in the U.K. were in place: namely the media, practising acupuncturists, and a rational understanding of TCM.

This meant that not only acupuncture but also herbal treatment would soon be developed. However, CHANCE is still important. At that time there was only one Chinese herbal clinic in China Town in London, called Bao Shui Tan. Dr. Li Tian-Bao came from Hong Kong and his service was mainly for Chinese patients. During the Christmas holidays of 1981 a female Chinese doctor, Luo Ding-Hui, arrived in London. She graduated from Guangzhou TCM College in 1970 and had practised in Guangzhou First People's Hospital as a paediatrician. She started her clinic in 1982 in a small room behind her husband's travel agency on the first floor, and she saw mainly Chinese patients as well. She and her husband also supplied some patent medicines and needles for acupuncturists. At the beginning of 1983 her clinic moved to the present premises: 15 Little Newport Street, and was named "Hong Ning Tan" (Kan Nin Tan). She started to see some non-Chinese patients. The positive and interesting point here is that most patients came because the U.K. hospital experts had told them they had no further methods of treatment to help them. So the patients wanted to try Chinese medicine. In particular many patients were suffering from dermatological conditions such as eczema and psoriasis which are difficult cases. Fortunately they were cured or their suffering greatly alleviated. About 1985 Dr. Luo's reputation had started through "transmission" among patients. Several returned to the hospital to tell the experts how effective Chinese medicine was. Some dermatologists came to Dr. Luo to observe her treatment. One of them is Dr. David Atherton, a famous consultant in Great Ormond Street Hospital who graduated from Cambridge University. He has done special research since 1986 and published his article in 1990 in the *Lancet*, then in the *British Journal of Dermatology*. He reported that 80–90% of the patients who could not be cured in his hospital were cured by Dr. Luo. A letter from the general secretary of the Dermatology Association mentioned they received 480 telephone calls a day and 5,000 letters from people who wanted to be treated by Dr. Luo. Newspapers, radio and TV reports about Dr. Luo's clinic followed. In front of her clinic there was always a long queue. Usually Dr. Luo saw 50 patients a day but at that time she saw 70 a day.

More and more people know about Chinese medicine. More and more patients want to see Chinese doctors. Some clever people found this to be a good way to make money. Due to the Tiananmen Square events some young Chinese people came to U.K., and others already in U.K. were detained by the circumstances. They needed to make a living. According to English law, they can do business and employ Chinese doctors from China.

A Jewish businessman also opened a Chinese clinic as a profit-making venture. His doctor came from China through the Chinese embassy. More and more Chinese clinics opened subsequently, so that now there are more than 3,000 clinics in the U.K. Their herbs are imported from China — about 160 tons in 1991, but 9,900 tons in 1995. The herbal import companies have grown from 3 to 31. Many schools and colleges of TCM have appeared. Middlesex University and Westminster University started special courses for TCM and acupuncture training. In our college, the London College of Traditional Acupuncture and Oriental Medicine, there are 200 students and about 30 graduates every year.

Registered membership of the British Acupuncture Council stands at about 2,500, the Register of Chinese Medicine at about 350, and the Association of Traditional Chinese Medicine (herbs and acupuncture) at approximately 70. More Chinese herbalists did not join the above associations.

Since Dr. Luo arrived in England 16 years have passed. The position of TCM in the U.K. is totally different. There has been enormous change. There are 12 elements which have guided the changes, influenced or helped TCM to be successful in England.

(1) Effective treatment using TCM.
(2) Western medicine is not able to treat every disease.
(3) Antibiotics, hormones and chemical products (drugs) have encountered some problems.
(4) People want a return to Nature. Science had been re-defined.
(5) Human rights: "My body, my choice".
(6) Mixed cultures and respect for different cultures.
(7) The British tradition of alternative treatment.
(8) Government policy.
(9) Royal family's interest.
(10) Financial benefit from selling herbs.
(11) Veterinary acupuncture effective.
(12) WHO's status and the TCM research in China, Japan and U.S.A.

However, there are also 10 problems which do not augur well for the further development of TCM:

(1) Some practitioners do not possess good qualifications.
(2) Many clinics owners only pay attention to profits without listening to the opinion of the practitioners.
(3) A few accidents have happened.
(4) Some people do not know British law or disregard the law.
(5) Media reports are sometimes not correct.
(6) Some people such as Prof. Ezard Ernst are especially fighting TCM (from his position as director of Dept. of Complementary Medicine, Exeter University)!
(7) Foundations for endangered species of animals and plants are against the use of some herbs and some shops have really been selling tiger bone.
(8) Drug companies are not happy with TCM because it makes their market share smaller.
(9) Some doctors do not or do not want to understand TCM.
(10) E.U. wants to control the market for herbs and patent medicines with a high price for permits.

The practitioners are aware of the above problems and they are trying to correct or avoid them. In general, TCM has a brilliant future in the U.K. as well as elsewhere in the world, if it keeps to proper practices.

The Presentation of Traditional Chinese Medicine (TCM) Knowledge in Hong Kong

HOKARI HIROYUKI

Kawamura Gakuen Woman's University, Tokyo, Japan

The Chinese Medicine Ordinance, which is the first law to set a framework for control of Chinese medicine in Hong Kong, was passed on 14 July 1999. The Chinese Medicine Council will be formed to register practitioners of Chinese medicine, administer standardized examinations, and regulate the use of medicinal herbs. This registration system will be enforced by the year 2000. Those who are not qualified cannot practise Chinese medicine, with offenders facing a maximum of five years in jail.[1]

Recent sociological studies of health systems have pointed out that the dual utilization of traditional and modern medical systems is popular in Asian countries (including Hong Kong).[2] But the framework of analysis "tradition vs. modernity" cannot be applied to the case of Hong Kong after the 1990s, because the political and social situation there have largely changed. The Hong Kong Government adopted modern Western medicine as a legitimate medical system and excluded Chinese medicine throughout their colonial rule. The practitioners of Chinese medicine organized occupational associations independently to claim recognition of their legitimate status. Chinese medicine is now widely considered one of the most valuable parts of Chinese heritage. In a sense, the regulation of Chinese medicine in Hong Kong is the solution to a problem which colonialism left. But the social status of practitioners of Chinese medicine has not been improved by this regulation. On the contrary, practitioners have been compelled to go through new ordeals, that is, licensing examinations, registration assessments and competition with new rivals from the People's Republic of China (PRC).

Following the legitimization of Chinese medicine in the PRC, traditional Chinese medicine (TCM) professionalized in the late 1950s and early 1960s,[3] resulting in the presence of Chinese medicine practitioners in Hong Kong. At the same time, only in the 1990s has the regulation of Chinese medicine been started in Hong Kong. This study will focus on the process of regulating Chinese medicine in Hong Kong and consider its historical meaning.

Brief History of Chinese Medicine in Hong Kong[4]

Hong Kong Island was ceded by China to Great Britain by the *Nanjing treaty* in 1842, and the colonial government started to establish a series of institutions. But traditional customs of the Chinese survived. The Chinese secured free practice of Chinese medicine without any registration. Practitioners of Chinese medicine were only required to register at the tax office before they could practise. They could call themselves *zhongyi* (中醫), *zhongyishi* (中醫師), *guoyi* (國醫), *tangyi* (唐醫) etc. But they could not put up signboards saying "doctor" (醫師). They existed like merchants in a business city.

From the opening of the port of Hong Kong to the 1940s, the Chinese in Hong Kong enjoyed the benefits of Chinese medicine, but not of Western medicine. During this period the *Dong hua* Hospital (東華醫院; established in 1870) treated Chinese patients using Chinese medicine.[5] Fearing Western medicine, most of Chinese preferred to consult the *Dong hua* Hospital and not the modern Western hospitals. The *Dong hua* Hospital had influence on overseas Chinese communities as well as in Hong Kong. In Singapore, the *Tong ji* Hospital (同濟醫院) has provided Chinese medical services from the end of 19th century until the present. When they carried out employment examinations to recruit practitioners of Chinese medicine, they sent the answer sheets to Hong Kong and asked the *Dong hua* Hospital to examine them.[6] This practice suggests that the level of Chinese medical knowledge in Hong Kong was highly appraised in the Chinese communities in Southeast Asia. I would like to focus attention on the way the knowledge of Chinese medicine was evaluated. In the case of the *Dong hua* Hospital, the authorization of the knowledge level of Chinese medicine was achieved not in qualifying examinations by the government, but rather through reputation in the communities and networks of practitioners.

Because of the plague in Hong Kong in 1894, the *Dong hua* Hospital was compelled to employ a doctor of Western medicine. It was after that that Western medicine was gradually accepted by Chinese communities in Hong Kong. After the Second World War, the development of antibiotics gave Western medicine an advantage over Chinese medicine in the area of internal medicine. Thereafter, the *Dong hua* Hospital also adopted Western medicine entirely.

It is important to consider the relationship with mainland China in understanding the development of Chinese medicine in Hong Kong. Mainland China played an especially important role in providing talented practitioners of Chinese medicine. Because of the intrusion of Japanese troops in China from 1930s to mid-1940s, many practitioners escaped to Hong Kong. During the Cold War, China closed its door to the West. Thereafter, communications between Hong Kong and the mainland were restricted. But the stream of people escaping from political disorder in mainland China never stopped. Famous practitioners in North China moved to Hong Kong in the 1950s. And after the open door policy was adopted in mainland China in 1980, a lot of new types of TCM practitioners moved to Hong Kong. In the 1990s, the Hong Kong government started to regulate Chinese medicine. And in 1997, Hong Kong was handed over to China.

Recognition of Chinese Medicine by the Government[7]

In 1989, a citizen of Hong Kong took a Chinese medicine called *Guizhou longdancao* (貴州龍膽草) and fell into a coma. This case caused a great sensation in Hong Kong. After two month, the relatives of the victim requested the government to strengthen the control of Chinese herbal medicine. And the legislators also criticized the government for leaving the problems alone.[8]

But the control of Chinese herbal medicine was a difficult work, because the scope of problems extends to the examination and classification of Chinese herbal medicine, and to the qualifications of specialists. It was impossible to control these in Hong Kong, where there was no system of registration or education of Chinese medical practitioners.

Against this background, the government decided to examine issues relating to the usage and practise of Chinese medicine in Hong Kong. In August 1989, a Working Party on Chinese Medicine was appointed.[9] It consisted of experts from the academic sector and representatives of relevant government departments. But they considered it desirable to seek the views of practitioners and traders of Chinese medicine. In May 1990 they decided to form a Professional Consultative Committee. Their task was to collect as much information and data on the use, practice and trading of Chinese medicine as possible. In this respect, four major areas were identified:

(1) Information on the practice of Chinese medicine in Hong Kong.
(2) Information on the use of and trade in Chinese medicine in Hong Kong.
(3) Public attitudes towards the use of Chinese medicine and practitioners in Hong Kong.
(4) Relevant experience overseas.

In October 1994, the Working Party submitted a report and suggested the formation of a Preparatory Committee. The Preparatory Committee on Chinese Medicine ("the Committee") was established on 1 April 1995 to advise the government on how to promote, develop and regulate Chinese medicine. After two years, the Committee submitted a report to the government in late March 1997.[10]

The Committee's recommendations fell into two main areas:[11]

(1) *Regulatory framework*
- For better protection of public health, a statutory body should be set up to regulate the practice, use and trading of Chinese medicine.
- A system of accreditation and regulation, involving examination, registration and discipline, should be established for practitioners of Chinese medicine, with transitional arrangements for existing practitioners.
- A control mechanism, through registration, licensing and labelling, be set up to regulate the manufacture, distribution, retail and import/export of Chinese medicine.

(2) *Future development*
- A full-time education program on Chinese medicine should be developed and made available in Hong Kong.
- Scientific research and development in Chinese medicine should be encouraged and supported.
- Chinese medicine should be included, on a gradual basis, into Hong Kong's health care system.

Among the above recommendations, the Committee considered that priority should be given to the establishment of a proper control mechanism through legislation.

On 1 July 1997, Hong Kong was handed over to China, but administrative work on the regulation of Chinese medicine continued to progress. The Hong Kong Special Administrative Region (HKSAR)'s Chief Executive, Mr. Tung Chee-hwa, highlighted the importance of Chinese medicine and its practitioners in his maiden speech on 8 October 1997. He said:

> I strongly believe that Hong Kong has the potential to develop over time into *an international centre for the manufacture and trading of Chinese medicine*, for research, information and training in its use of Chinese medicine, and for the promotion of this approach to medicinal care ...[12]

It should be noted that the Committee's recommendations had no idea of "an international centre for manufacture and trading of Chinese medicine". But the Chief Executive stressed this point for future development. We can see that after the hand over, the SAR government has started to use

Chinese medicine as one of the important resources for the development of Hong Kong.

After consideration of the Committee's recommendations, the government proposed to accept the Committee's recommendation on the establishment of a regulatory framework and invited interested parties to express views and comments on it.

There were about 50 submissions from various organizations and individual members of the public from November to December 1997.[13] Most respondents supported:

The proposed establishment of a statutory framework for the regulation of Chinese medicine.

- The future regulatory body should consist of members from inside as well as outside the Chinese medicine sector.
- Establishment of a licensing system for Chinese medicine traders and a registration system for Chinese medicine.
- Development of formal Chinese medicine education program.
- Promotion of Chinese medicine research and development.

But there were divided views on the proposed transitional arrangements for the registration of the currently practising Chinese medicine practitioners. Some organizations pointed out that:

- The government should work out some feasible arrangements to ensure that the Chinese medicine practitioners can continue to provide service to the community.

On the other hand, some pointed out that:

- The Government should be extremely careful in designing the transitional arrangements to ensure that only practitioners of acceptable standard would be allowed for registration and to continue to practice.

The qualification problem is a matter of life and death to the practising Chinese medicine practitioners in Hong Kong. And there was debate regarding whether future registered Chinese practitioners should be allowed to issue valid sick leave certificates and other medical documents. This problem concerned the status of the Chinese medicine practitioners in the future medical system of Hong Kong. And this also showed that it would be difficult to coordinate with the existing system formed by Western medicine.

In 1998, the Government started to prepare the draft Chinese Medicine Bill. The Bill was passed at the Legislative Council on 14 July 1999.

Qualification of Chinese Medicine Practitioners

Modernism

On April 1992, nine Chinese medicine practitioners associations made a joint statement that they opposed the regulation of Chinese medicine by the Government. They pointed out as historical background that the proclamation on 1841 declared that the inhabitants of Hong Kong would be governed according to the laws, customs, and usages of the Chinese. The practice of Chinese medicine had been regarded as part of those "customs and usages". Therefore, Chinese medicine were not subject to certain controls by law. The opponents were afraid that current Chinese medicine practitioners would be put under restrictions by the registration, and tried to take Chinese medicine into sanctuary.

But the Government considered that because the social environment had changed and many residents were interested in their own health and safety as patients, it was therefore necessary to evaluate the level of Chinese medicine practices.

The Basic Law of Hong Kong SAR states that "The Government of the Hong Kong Special Administrative Region shall, on its own, formulate policies to develop Western and traditional Chinese medicine and to improve medical and health services" (Article 138). It is likely that this sentence provided a good ground for registration of Chinese medicine. And we can also see the consideration of balance with the medical system in mainland China in this sentence.

Concerning the qualification of Chinese medicine practitioners, different views were expressed from the beginning of consultation. In the proposed registration system for practitioners, it was decided that the licensing examination would be introduced, and only those who had passed the examination would be allowed to register as practitioners of Chinese medicine. But the difficult issue was how to assess the knowledge of practising practitioners in Hong Kong. The proposed Bill provided a transitional arrangement whereby practising practitioners could register with the Council without the need to sit for the licensing examination. The text follows:

Part IX Transitional Arrangement for Chinese Medicine Practitioners

93. Exemptions from Licensing Examination

A listed Chinese medicine practitioner who satisfies the Practitioners Board that he has fulfilled either one of the following criteria, namely —

(a) immediately before 3 January 2000, he has been practising Chinese medicine in Hong Kong for a continuous period of not less than 15 years; or

(b) (i) immediately before 3 January 2000, he has been practising Chinese medicine in Hong Kong for a continuous period of not less than 10 years; and

     (ii) he has obtained a qualification in Chinese medicine practice acceptable to the Practitioners Board,

shall be exempted from the Licensing Examination.

94. Registration assessment

(1) A listed Chinese medicine practitioner who satisfies the Practitioners Board that he has fulfilled either one of the following criteria, namely —

(a) immediately before 3 January 2000, he has been practising Chinese medicine in Hong Kong for a continuous period of not less than 10 years; or

(b) (i) immediately before 3 January 2000, he has been practising Chinese medicine in Hong Kong for a continuous period of less than 10 years; and

     (ii) he has obtained a qualification in Chinese medicine practice acceptable to the Practitioners Board,

shall be exempted from the Licensing Examination but shall be required to pass a registration assessment conducted by the Practitioners Board before he is qualified to apply to be registered as a Chinese medicine practitioner under section 69.

95. Requirement to undertake Licensing Examination

(1) A listed Chinese medicine practitioner who —

(a) satisfies the Practitioners Board that immediately before 3 January 2000, he has been practising Chinese medicine in Hong Kong for a continuous period of less than 10 years; or

(b) has failed the registration assessment,

shall be required to undertake the Licensing Examination, the passing of which shall qualify him to apply to be registered as a registered Chinese medicine practitioner under section 69.

One of the associations of Chinese medicine, the Kowloon Society of Practitioners of Chinese Medicine claimed that the proposed examination would be too tough for those who lacked any theoretical training. The chairman of the Society said:[14]

> I don't believe those who were trained under an apprenticeship are worse than those who received academic training as they possess sufficient experience. ... It will also be difficult for those herbalists who just attached themselves to a herbal shop and do not have any business registration of their own to prove their years of practice.

As the chairman said, it is very common that herbalists are working in a corner of a herbal shop in Hong Kong. It is likely that there are many practitioners who cannot prove their years of practice. And these types of practitioners are very familiar to ordinary people. If they lose their jobs, a lot of patients will suffer a loss.

The Government questioned about level of the knowledge and technique of practitioners of Chinese medicine who learned as apprentices. We can see the figure of modernism which requires academic qualification in the process of registration.

Regionalism

Concerning the qualification of practitioners of Chinese medicine, there is another problem in Hong Kong. That is the issue of the self-direction of Hong Kong in the political relationship with mainland China. That is the problem of the residency right of TCM practitioners from the PRC.

The Hong Kong Government has a plan to transform Hong Kong into an international centre for Chinese medicine. It is said that importing outstanding mainland professionals in TCM will further this plan.[15] But importing professionals from the mainland has been restricted. The pilot scheme of the entry of mainland professionals began in March 1994 and ended at the end of 1997. The scheme had a quota of 1,000 but only 602 came. The scheme was criticized on the grounds that the procedure of examination was too complex and required too long a time.

On January 1999, the newspaper reported that professionals from the mainland will be allowed to work in Hong Kong and become permanent residents after seven years under a scheme being drawn up by the government.[16]

After the handover to China, because of the financial crisis in Southeast Asia, the Hong Kong economy was not good. Some put the blame for the worsening unemployment situation on the importing of labor from mainland China. Others said it is possible that the importing of labor may put pressure on population growth and erode their quality of life.

On 4 November 1998, over 800 practitioners of Western medicine who were certificated in the mainland China held a demonstration parade to claim exemptions from licensing examinations.[17]

The chairman of the association said that those who were trained on the mainland are not worse than those who graduated from university in Hong Kong. One member of the association who has practised many years without certification in Hong Kong said that the examination has been affected by politics and is unfair, so he has refused to apply. He suspected that the Government has restricted the number of successful candidate from the mainland. One of the placards that they carried said, "Universal brotherhood, equal registration" (一視同仁, 平等注冊). It is clear that their demonstration was encouraged by the hand over to China. But the Hong Kong Government refused to recognize their claims. From this example, we can see the regionalism of Hong Kong mainly mainland China.

Training and Practice of Chinese Medicine Practitioners in Hong Kong

Traditional Training

Practitioners of Chinese medicine in early Hong Kong were mainly from mainland China. Before the Second World War, there were some training schools and night schools in Hong Kong. Because of the destruction of Guangdong city by the Japanese troops, many practitioners of Chinese medicine moved to Hong Kong. Then, private institutions (*Zhongguo guoyi xueyuan* 中國國醫學院 1947; *Qinghua xueyuan* 清華學院 1953 etc.) and part-time courses at associations of Chinese medicine were established.[18]

Recently, there are about 16 institutions which set up training courses in Hong Kong.[19] These institutions have played an important role in training practitioners of Chinese medicine in Hong Kong. But their facilities are not ideal. Lecturers are usually graduates of TCM colleges or universities in China with over 10 years experience. Most lecturers are practising practitioners and teach only on a part-time basis. The curriculum and years of each course vary. Some institutions have a certificate course, but they are not licensed by the Hong Kong Government.

Profile of Practitioners of Chinese Medicine

There are about 7,000 Chinese medicine practitioners in Hong Kong, some of whom are well trained while others received no more than verbal instructions. So it is difficult to evaluate their careers and specialized knowledge in practice.

The majority of practitioners in Hong Kong are men aged 45, born in China, most of whom had practised ten years or more in Hong Kong (about 70%). The principal forms of Chinese medicine practice in Hong Kong are:

(a) Chinese medical practitioners, commonly known as "Herbalists" (中醫師) 60%
(b) Bone-setters (跌打師) 20%
(c) Acupuncturists (針灸師) 10%
(d) Other Forms of TCM Practice 10%

Most of the practitioners in Hong Kong have gone through some form of apprenticeship (65%) or self-learning (31%), but many of them have later also attended part-time course (21%) in a college of Chinese medicine as their highest educational level. The forms of training are not mutually exclusive: 31% had apprenticeship as the only form of training, 22% a full-time course in addition to apprenticeship, 13% part-time study with an apprenticeship and 6% full-time training in both Western and Chinese medicine.[20] Both apprenticeship and institutional study are used by the practitioners in Hong Kong.

After 1980s, many practitioners who graduated from the TCM colleges or universities in China moved to Hong Kong. They used their officially qualified TCM knowledge in mainland China to their advantage in Hong Kong and started to teach TCM theory in private institutions and universities in Hong Kong. They are now indispensable people in the training of TCM practitioners in Hong Kong.

Besides the influx of TCM practitioners from the PRC, interchange between mainland China and Hong Kong has been flourishing. Some universities which have TCM departments on the mainland have established TCM courses in Hong Kong in cooperation with local institutions of Chinese medicine. And some universities of Chinese medicine have started advertising in Hong Kong to recruit students.

Organizations of Practitioners of Chinese Medicine

At present, there are about 13 associations of practitioners of Chinese medicine in Hong Kong. They aim at promoting the welfare of practitioners, improving the standard of practices and establishing contacts with other associations, both locally and overseas. It is common that practitioners join more than one association, although a substantial proportion of practitioners (42%) have not joined any association.[21]

Recently, some practitioners associations have come together as a federation. In October 1990, the Association of Practitioners of Chinese Medicine General Chamber Hong Kong Limited (全港中醫師公會聯合會), comprising five practitioners associations, was established with the aim of promoting the status of Chinese medicine in Hong Kong. Meanwhile, in December 1990, the Hong Kong Association of Traditional Chinese Medicine (香港中醫學會), comprising mainly TCM practitioners who graduated from TCM colleges or universities in the PRC, was formed to promote and develop TCM.

The Hong Kong Chinese Herbalist Association Limited (香港中醫師公會) has a long history in Hong Kong. Its forerunner association *Zhongguo guoyi xuehui* (中國國醫學會) was established in 1929, when the Chinese Government planned to abolish Chinese medicine. At present, they have about 3,000 members in Hong Kong and 6,000 overseas (over 3,000 in Taiwan). They have been closely related to Taiwan historically. But they started to have connections with mainland China around 1997. Their own school which trained TCM practitioners opened a TCM certificate course in cooperation with the College of Chinese medicine in Fujian (福建). When I visited their office in Hong Kong, a small poster celebrating the hand over in 1997 was up on the wall.[22]

In the area of training of TCM practitioners in Hong Kong, the TCM knowledge from PRC has gradually enlarged its influence.

Training in the Universities

The training of Chinese medicine practitioners in part-time courses at the universities started before 1997. The University of Hong Kong's School of Professional and Continuing Education set up a one year course in 1991. In 1997, they established a four year diploma course. They also opened a TCM course for doctors of Western medicine.

In 1995, the Chinese University of Hong Kong also established a TCM course in cooperation with the University of Chinese medicine in Chengdu (成都) in China. Later, the Baptist University's School of Continuing Education opened a diploma course in TCM in cooperation with the Royal Melbourne Institute of Technology in Australia.

The first full-time degree course in TCM was established in the Hong Kong Baptist University in September 1998. The President of the University was the chairman of the Preparatory Committee on Chinese Medicine in 1995. It is likely that the training system of TCM practitioners in universities will be made in accord with the policy of the Government.

The outline of the attempt at the Baptist University is as follows.[23] The course structure is a combination of basic biomedical sciences, basic Chinese medical sciences and basic Western medical sciences. The five-year program, which has been designed on a solid scientific foundation, will emphasize both basic and clinical research. The first four years of study will comprise lectures and laboratory work. Final-year students, in addition to writing their dissertation, will engage in clinical practices and undergo internship at the Beijing University of Chinese Medicine and Pharmacology.

The texts used in the course are almost all imported from mainland China (e.g. *Zhongyi jichu lilun* 中醫基礎理論, *Zhongyi zhenduanxue* 中醫診斷學, *Zhongyaoxue* 詞藥學).The five-year program of study is patterned after those in the PRC. The graduates will be able to further their study in mainland China as the course is designed to meet the requirements of the centralized examination there.

In the practice and training of practitioners of Chinese medicine in Hong Kong, the TCM knowledge from the PRC has expanded its influence.

Conclusion

It is important to note that regulation of Chinese medicine in Hong Kong is not the outcome of the historical development of Chinese medicine in Hong Kong. Rather, it is the result of a strong policy of the Government. It mainly depends on TCM theory from the PRC, which is strange for most practitioners of Chinese medicine in Hong Kong. We can see the counter relationship between Hong Kong and mainland China as well as between tradition and modernity in the world of Chinese medicine.

The counter relationship between Hong Kong and mainland China has appeared vividly in the area of politics and business. The Government is taking the initiative to promote Hong Kong as a worldwide centre for Chinese medicine. But it is impossible to realize this plan without the mainland as a market and a storehouse of talented people.

The new regulations would improve the public health and boost confidence in TCM in Hong Kong. But the ultimate aim is to provide a foundation for private enterprises to develop Chinese medicine-based industries. The establishment of a full-time degree course in TCM can be seen as a part of development strategy of the Government and the business community of Hong Kong. And the TCM theory from the PRC has been adopted in these courses.

TCM knowledge coming from mainland China is certified by the PRC Government. And they adopt Western scientific methods to carry out researches. On the other hand, the practitioners of Chinese medicine in Hong Kong did not have legislative status and scientific bases before, though they had their own network of Chinese medicine which guaranteed a certain level of standing and allowed the exchange of information with overseas Chinese communities. This kind of situation could be seen as a case of "professionalization from below" as Murray Last termed it. Last has pointed out that the major difference between the two perspectives on professionalization lay in the ways professional knowledge is perceived (whether bureaucratically from above or clinically from below).[24] It is likely that the development of Chinese medicine in Hong Kong showed that there may be another way of recognizing Chinese medical knowledge which is independent from the state.

Notes

[1] *South China Morning Post*, 15 July 1999.

[2] Stella R. Quah, ed., *The Triumph of Practicality* (Singapore: Institute of Southeast Asian Studies, 1989).

[3] Elisabeth Hsu, *The Transmission of Chinese Medicine* (Cambridge: Cambridge University Press, 1999).

[4] There are a few studies about the history of TCM in Hong Kong. The following is useful to this field: Xie Yongguang 謝永光, *Xianggang Zhongyiao Shihua* 香港中醫藥史話 (Hong Kong: Joint Publishing, 1998).

[5] Besides medical care, it managed many kinds of charitable activities. For example, accommodation and repatriation of refugees, help in funeral arrangements, sending coffins or bones back home, and fund raising campaigns for disaster relief, etc.; Elizabeth Sinn, *Power and Charity: The Early History of the Tung Wah Hospital, Hong Kong* (Oxford: Oxford University Press, 1989); Hiroyuki Hokari, "Hong Kong *Tung Wah* Hospital and the Cantonese Network" (in Japanese) *Toyoshi-kenkyu* 55.1 (1996).

[6] *Tongjiyiyuan Yibaiershizhounian Lishizhuanji* 同濟醫院一百二十週年歷史專集 (Singapore, 1989).

[7] Jiang Runziang 江潤詳 ed., *Zhongyi Luntan* 中醫論談 (Hong Kong: Shangwu Yingshu Guan, 1998).

[8] *Mingbao* 明報, 19 April 1989.

[9] *Working Party on Chinese Medicine Interim Report* (October 1991); *Report of the Working Party on Chinese Medicine* (October 1994).

[10] *Report of the Preparatory Committee on Chinese Medicine* (March 1997).

[11] *Consultation Document on the Development of Traditional Chinese Medicine in the Hong Kong Special Administrative Region*, Health and Welfare Bureau (November 1997).

[12] *Chief Executive's Policy Address* (8 October 1997).

13 *Report on the Legislation on Chinese Medicine*, Provisional Legislative Council, Panel on Health Services Meeting (9 March 1998).

14 *South China Morning Post*, 22 January 1999.

15 *Hong Kong Standard*, 7 January 1999.

16 *Ibid.*

17 *Xingdao Ribao* 星島日報, 5 November 1998.

18 Xie Yongguang, *Xianggang Zhongyiao Shihua*, pp. 114–16.

19 *Report of the Preparatory Committee on Chinese Medicine* (March 1997).

20 *Working Party on Chinese Medicine Interim Report* (October 1991), p. 14.

21 *Ibid.*, p. 16.

22 Visit to the Hong Kong Chinese Herbalist Association Limited (27 August 1998).

23 Hong Kong Baptist University, *New Horizons* (1996–99); *South China Morning Post*, 22 August 1998.

24 Murray Last, "Professionalization of Indigenous Healers", in T. Johnson & C. Sargent, eds., *Medical Anthropology* (New York: Greenwood Press, 1990), pp. 361–64.

Traditional Chinese Medicine (TCM) in Denmark: Conditions in a New Host Culture

ERLING HØG
Institute of Anthropology
University of Copenhagen

This paper addresses the cultural transmission of Chinese medicine, known as TCM, to the Scandinavian state of Denmark. Chinese medicine is undergoing a globalization process in which TCM becomes recontextualized in local knowledge traditions. Four central conditions for transcultural knowledge transmission will be discussed, concerning "knowledge in process", "self-regulation", "no tradition", and "slow transmission of knowledge". I argue that these four factors hinder the maturation of TCM as a local knowledge tradition in Denmark. This anthropological study will be aided by the analytical and theoretical ideas of knowledge traditions and narrative structures, as developed by Fredrik Barth (1989, 1993) and Edward Bruner (1986). Based on these discussions, I argue that it will take at least a generation to stabilize TCM in Denmark as the tradition is known in China.

Variations of Acupuncture in Denmark

During fieldwork in 1996/97 I visited six acupuncture schools and interviewed 26 acupuncture practitioners, discovering six major Danish variations: Classical Acupuncture, Five Element Acupuncture, SOMA Acupuncture, Physiological Acupuncture, Sri Lanka Acupuncture and TCM Acupuncture. There are of course always fuzzy boundaries between such practices, as I will discuss later using Barth's ideas of cultural streams and knowledge traditions, but first let me introduce some general working definitions.

Classical Acupuncture, which stems from the medical legacy of the Yellow Emperor, implies the use of ancient Chinese medical theories, philosophies and diagnostic methods, viewed as original and authoritative. Classical medical texts like *Neijing*[1] and *Nanjing* are employed generously to ensure the survival and practical application of age-old medical reasoning, the ideas of yinyang, five elements, and meridians being central. Classical acupuncture is employed for a wide range of illnesses and diseases for which it is viewed as superior to other variations.

Five Element Acupuncture in Denmark is a variation that stems mostly from the work of physical therapist Jack Worsley in England since the 1960s. As indicated by its name, the core teachings emphasize the interplay of the five elements and phases. While the principles of the five elements are included in comprehensive curricula, there are yet no distinctive Danish five element schools. Five element acupuncturists consider the mind, psyche, social relations, and the environment, in an inclusive search for etiological factors stemming from relations between micro- and macrocosms.

SOMA Acupuncture (*Systematized Oriental Medicine and Relaxation*) is a Danish innovation strongly inspired by five element teachings. Psychological etiology is emphasized in a way that resembles Western psychological disease causation, yet without declaring it an outgrowth of Freudian principles. Another important SOMA feature is the use of manual acupuncture, because needling practitioners lacking medical qualifications were outlawed in 1978. SOMA methods emphasize inner healing powers through the sensation of subtle and dynamic energies of the body. Meditative body means improve otherwise diffuse conditions.

Physiological acupuncture reached Europe during the 1970s, mostly among practitioners using biomedical explanatory models. This increasing interest in acupuncture was caused by US president Nixon's visit to China in 1971 (Steiner 1983: 60, Johannessen 1995: 43, Birch & Tsutani 1996: 172).[2] Physiological acupuncture is mostly employed for its anesthetic power and its effectiveness against simple pains in the motor functional system. While many physicians use other variations, physiological acupuncture remains their preferential method in both theory and practice.

Sri Lanka Acupuncture as it is known today in Denmark stems from the acupuncture variation developed at the University Hospital in Colombo, Sri Lanka. One Danish school of acupuncture is an obvious extension of the Sri Lanka philosophy, promoting folk acupuncture in the hands of the largest number of students and practitioners. Sri Lanka acupuncture emphasizes short training and practicability against common illnesses.

Traditional Chinese Medicine (TCM) was developed in communist China after 1949 as a revival of acupuncture and moxibustion in the new China (Cheng 1987: 8). Traditional Chinese Medicine was from then on written with capital letters, marking a new medical era different from traditional Chinese medicine. TCM was promoted in communist China as *The New Acupuncture,* attempting theoretical integration of diverse schools of thought (Shima 1992: 33).[3]

TCM gained ideological legitimacy during the late 1950s as a new medical construction within an emerging sociopolitical context that supported the revival of traditional medicine. This historical development has been documented extensively by many social scientists (e.g., Croizier 1968, Unschuld 1985, Farquhar 1994), while its transmission to Europe remains vaguely understood, or at least understudied compared to research on Chinese medicine conducted in China. TCM is most often known as TCM acupuncture, since acupuncture has been practised in Denmark on a larger scale for about 15 years more than herbal medicine. The combination of needles and herbs serves as a complementary method against many conditions, especially energetic imbalances.

Knowledge in Process

The knowledge in process argument states that the maturation of a foreign medical tradition in a recipient host culture needs a certain span. Rome simply was not built in a day. The transmission of a 'pure' TCM is complicated by overlapping variations and the ubiquitous disagreement about the proper content and form of Chinese medical training. The Danish field of acupuncture, both in terms of training and the foundation of schools, is highly based on individual agency, which complicates the visibility of TCM as a unique variation.

While seminars and informal sessions of *tai chi* and *qi gong* take place, and several training programs include TCM, acupuncture remains the attraction in Chinese medicine. TCM is included in training programs with great casualty, while its content is chosen freely, as I mentioned earlier the varied TCM semantics. In my earlier research I tried to delineate variations included in different training programs. Clear indications of a TCM program resembling the ones established in China during the late 1950s is found at two schools: the School of Traditional Chinese Medicine and Nordic College of Chinese Acupuncture (see Table 1).

Acupuncture is the most popular treatment modality in Denmark, witnessed by the fact that all variations, schools, and associations include the term 'acupuncture' in their names. Also official statistics on the use of alternative treatments include acupuncture as an independent therapy, while statistics on the use of Chinese medicine is absent (see Table 2).

The history of TCM in Denmark is a complex one, with many paths and avenues leading to the foundation of private schools resembling those developed in the PRC. Its Danish development has been heavily influenced by pioneering students and practitioners who travelled primarily to China and England in search of Chinese medical knowledge and experience. These people were from all walks of life, some pursuing their second or third career,

Table 1   Acupuncture Education in Denmark

School	Variation(s)	Hours*	Year
1. Danish Medical Society for Acupuncture, Hvalsø	TCM, Physiolog.	120	1974
2. Scandinavian College of Acupuncture, Søborg	Sri Lanka	200	1983
3. The Acupuncture School, Copenhagen	?	?	1983–198?
4. The Scandinavian Acupuncture School, Copenhagen	Classical	?	1983–1985
5. NIHAW Acupuncture School, Copenhagen	Classical	?	1984
6. The Acupuncture School in Denmark/Nordic Institute of Classical Acupuncture, Holstebro	? Classical, TCM	1350	198?–1996 1996
7. The Acupuncture School Alternative,Struer	?	?	198?
8. Danish Acupuncture Institute	?	?	1987?
9. The Acupuncture School in Århus	Classical, TCM	100	1988
10. SOMA School, Give	SOMA	240	1991–1996
11. School of Traditional Chinese Medicine, Alken	TCM	500	1992
12. The Acupuncture School, Odense	?	?	1993
13. Nordic College of Chinese Acupuncture, Valby	TCM	460	1997

Note: * Class instruction hours, basic training to open a clinic.

some alternative practitioners, some physicians and dentists, and many more. Only the most ambitious ones stayed more than two or three months, while some eager ones first visited acupuncture schools in England and then went on to China. Not forgetting influences from medical and therapeutic training in Holland and India, and Reichian psychology, the most influential schools have been the International College of Oriental Medicine in London and the Academy of Traditional Chinese Medicine in Beijing.

Table 2   Use of Alternative Treatments in Denmark, 1987 and 1994

Alternative Treatment	% 1987	% 1994
Herbal medicines	6	8
Reflexology	9	15
Relaxation therapy	3	4
Dietary therapy	2	3
Needle acupuncture	3	7
Laying on of hands	1	1
Massage/Manipulation	5	9
Magnetic stroking	1	2
Healing	1	3
Hypnosis	1	1
Other alt. treatment	3	3

Source: DIKE 1997: 125.

Many acupuncturists are unaware that TCM is a particular construction of traditional Chinese medicine, founded during the 1950s under Mao's leadership. TCM is synonymous with any variation and the semantic indifference is furthered by the liberal use of acronyms in English and Danish (e.g., tcm, TCM, TKM, tkm). Practitioners explain this indifference as commitment to the everyday practice of Chinese medicine in which medical labeling becomes secondary. On the other hand, practitioners who have attended schools or universities in China are very much aware of the TCM distinction. For example, the only Danish TCM doctor that I interviewed during my fieldwork, finally after many years of training in Denmark, Taiwan, and England, pursued her degree at the TCM College in Nan-chang, Jiangxi province in South Eastern China. Others will deliberately refer to the TCM universities in Shanghai, Nanjing, Beijing, and Chengdu.

On the other hand, TCM with a clear indication of PRC origin has since the late 1980s become visible as a distinct variation, stemming from collaboration between practitioners and teachers in Denmark, England, and China. For example, the Danish School of Traditional Chinese Medicine was started in 1992 by three Danish women, a physician, a *tai chi* teacher, and a medical herbalist, who since the late 1970s had learned various acupuncture variations in Denmark, Holland, England, and China.

Chinese medical textbooks in English have also been important for the emerging consciousness about TCM as a particular Chinese medical construction. Kaptchuk's *The Web That Has No Weaver* was an important publication for increasing interest in TCM. Another strong impetus for the emergence of TCM was acupuncturist and medical herbalist Giovanni Maciocia, who through prolonged collaboration with Chinese doctors published comprehensive Chinese medical textbooks in the early 1990s. Maciocia's *The Foundations of Chinese Medicine* and *The Practice of Chinese Medicine* are the central textbooks at one Danish TCM school.

While TCM strongly emphasizes theoretical insight, students wish for an early application of skills in professional clinical practice, tolerated within the loosely structured Danish context. Stepan (1985) and Last (1996) discuss 'tolerance' as a broad category for the regulation of health care, while Unschuld (1976) employs the term 'unstructured coexistence' of different medical systems. On this basis I argue that the Danish liberal professional environment is caused by a tolerant and unstructured system of health care that allows for an anarchy of knowledge that TCM struggles against to ensure its superiority and develop the institutional legitimacy it has known in China. From this perspective, knowledge in process inevitably concerns how Chinese medical knowledge is taught, received by students, and viewed by medical authorities.

Self-Regulation

Danish TCM taught at introductory levels serves to let only the most serious students advance to higher training at foreign schools, and from then on proceed to professional practice. Students take TCM courses out of curiosity, drop out due to changes in the family or in the primary job situation, or simply skip it, because TCM is too theoretical and too difficult to apply in practice. TCM attracts very few students, compared to the Sri Lanka variation, which is the most popular one in Denmark. For example, there were only six students at the Danish School of Traditional Chinese Medicine at my visit in 1996, while there were 25 students in one class attending the Sri Lanka school. It is also noteworthy that the Sri Lanka variation has three schools in different towns around Denmark.

If perhaps only one out of ten students in the Chinese medical market will graduate and pursue a professional career, then why not leave it unregulated? Who has got the work force to regulate a field for which they have no or very little insight other than that there are very few official complaints against malpractice? 'Malpractice' becomes a complicated matter considering that patients, according to Danish medical law, consult their preferential alternative practitioner at their own individual responsibility. Who would complain about bad treatment, when all one can claim is their own embarrassment and a stack of bills with no refund? Patient costs can swiftly climb sky-high, considering that the cost for one Chinese medical treatment in Denmark varies on average between 40 to 70 American dollars. Patient complaints center around everyday dissatisfactions that permeate patient and practitioner discourse but never reach the bureaucratic chambers in print.

The counter argument is usually that Chinese medical practitioners survive everyday clinical life by their own experience and practice, supported by the widespread belief in recommendation by *word of mouth*. However, this patient-to-patient recommendation implies a secure and indisputable way of dividing competent and incompetent practitioners. The positive recommendation may be a misleading justification which hides the fact that many patients leave with no or little alleviation, or even with worsened conditions. Patients experience the conditions of unregulated health care in a liberal market, where they have no consumer rights, individual responsibility, embarrassment, loss of money for useless health care, etc.

Why spend state budget resources on a field of health care with low state priority and little prestige? The student majority seeps into professional oblivion, thus implying that self-regulation is better than regulation out of place.[4] Moreover, Chinese medicine is foreign in epistemology and has yet to prove itself against the demands of biomedical scientific standards. It thus remains a subject with no demand for official educational regulation.

No Tradition

I argue that within the Danish tradition of biomedical domination other medical traditions have had a hard time growing to maturity, as compared to other host cultures. Biomedicine has for centuries been the supreme tradition officially recognized by Western nation states. Other traditions have been deemed deviant, alternative, or even quack medicine (see for example Leslie 1980).[5]

Chiropractice and reflexology provide two examples of how alternative medical knowledge has fared in the Danish health tradition. These forms of health care became popular in the 70s and 80s, but they have quite distinct histories. Chiropractice was learned at private schools or universities in England and the USA during these early years, and gradually gained a footing at Odense University Hospital. Basic training in chiropractice was ensured later as a platform toward accreditation and state recognition. This academic pursuit was accomplished by September 1997 when the Danish Minister of Health signed the official premises for a Masters degree in chiropractice. The condition for this advancement was unequivocally a common curriculum and not least its scientific biomedical reasoning. Chiropractice was consequently renamed 'clinical biomechanics'.

While chiropractice managed to compromise its internal pluralism of training and medical reasoning, resembling the biomedical organizational profile, the diversity within the field of reflexology remains intact. Reflexology is now divided into three major schools of thought, while the state authorities consistently demand one common training program for all coming practitioners. Regulated training may seem possible from an outsider's perspective, but it remains a problematic issue within a field of many opposing views, aims, and professional intentions.

Reflexologists have long viewed chiropractice as a textbook example, but recently the tide has turned toward realizing that official state-defined recognition may not be absolutely the only avenue. Official recognition means stricter regulation of training programs, making them conform to biomedical professional organization, and thus promising a form of bureaucratization that arrests the freedom held under the unregulated alternative health care market.

The field of Chinese medicine is in a similar situation that requires the choice between these two equally unpleasant choices of state affiliation to biomedical organization or, consequently, unauthorized, unregulated, alternative, and marginalized health care, restricted according to national medical law.

Slow Transmission of Knowledge

I will argue that lack of teaching capacities, textbook translations, economic support from the state, and rejection from the biomedical authorities, are the main factors delaying TCM transmission. First, lack of teaching capacities will inevitably occur in any new host culture in which a foreign medical tradition is introduced. Medical textbooks arrive long before the people capable of delivering their contents. However, from a different perspective, the tutorial shortage may also be caused by severe competition between private schools of Chinese medicine. Why not bring the lot together to strengthen the educational level? Private schools are run by one principal teacher who also owns the school and makes a profit. From this perspective, business counts more than proper training. Master teaching may not be a necessity for the distribution of useful Chinese medical ideas, but students may end with an array of inapplicable ideas and medical techniques. From this perspective TCM remains incomplete under the current educational circumstances.

Second, Chinese medical concepts are very difficult to convey to a Western audience. Translating Chinese textbooks into European languages is exceedingly problematic, since words may have many meanings. Medical semantics become obscured even more when textbooks are translated from a secondary to a tertiary European language, for example from German into English. Different variations and medical interpretations may even become mixed to an extent that they become unintelligible. One major problem of course is that most neophyte students are unaware of the different interpretations, and will not develop their Chinese medical consciousness until much later. The negligent translator may even skip the most difficult passages in Chinese textbooks to make them more accessible, but these omissions may be fatal for medical completeness and applicability. When TCM is taught in bits and pieces, it runs the risk of being discredited. Failing applicability suits any skeptic who sees Chinese medicine as a charlatan practice or quackery, who fail to include the possibility that practitioners missed important lessons.

Third, as Unschuld points out, there will always be a close relation between ideological legitimacy and political support (Unschuld 1985: 249). Chinese medicine in China serves as an example of survival through substantial state subsidies which laid the foundation for extensive university training after a long period of official rejection. Training programs in acupuncture, herbal medicine, qi gong, and tai chi are available in Denmark, surviving in the liberal medical market without any support from the state. Private enterprises remain niche industries in which small businesses compete for students.

From an outsider's point of view it seems possible that schools cooperate to ensure a complete training program, but history has shown that until now this has not been the case. Practitioners place the responsibility for ensuring complete and competent training on the state, which is seen as letting them down, while professional ambition hardly reaches beyond mere lip service. What practitioners say is not what they do.

Lack of economic support from the state for Chinese medical education is closely linked to the fact that Chinese medicine fails to meet the scientific criteria standard for biomedicine. Anyone may practise some form of medicine in Denmark, whatever educational background, as long as they do not claim to be biomedical doctors and follow other restrictions laid down in Danish medical law. The issues of TCM's lack of scientific legitimacy and the tolerant policy ensuring the right to practise may explain the missing educational regulation. TCM with its emphasis on both institutionalized educational form and content could gain political legitimacy, but the present social fabric of Chinese medicine in Denmark makes this scenario unlikely.

TCM as Local Knowledge Traditions

Fredrik Barth's ideas of 'cultural streams' and 'knowledge traditions' explain the coexistence of distinct yet overlapping acupuncture variations (see Høg 1998). Cultural streams define an open field of *social* processes (Barth 1989: 132), in which they show a certain number of empirical clusters in which local knowledge traditions develop (Barth 1989: 116–17). Barth invokes a thesis about the distribution of culture, identifying several cultural streams and knowledge traditions which become ideas and representations for life, specifying a discursive universe that shapes human consciousness (Barth 1993: 177). The very idea of cultural streams entails the coexistence of distinct knowledge traditions in a specific region, open to individual acquisition of facts in one or more knowledge traditions (Barth 1989: 131, Barth 1993: 305). Barth suggests an analysis of culture applicable in complex societies, implying constant migration and transmission of knowledge.

Barth's ideas may explain the development of TCM in Denmark. TCM in China created an ideological demarcation of the many strands of Chinese medicine. However, the TCM knowledge tradition transmits itself as cultural streams without the complete traditional features of the Chinese mother culture. TCM transmits in bits and pieces, while it also transgresses other local variations in education and practice.

The very idea of knowledge transmission implies not only a spatial but more significantly a sequential process through past, present, and future. Practitioners speak about their own careers and Chinese medicine within this period, and it will thus be equally useful to think with narrative structures employed by Edward Bruner (1986) to perceive the many paths of TCM.

TCM in the Past, Present, and Future

Barth's ideas of cultural streams and knowledge traditions may beneficially complement Edward Bruner's *dominant narrative structure*, characterized by a syntagmatic sequence through past, present, and future (Bruner 1986b: 140–41). While Barth adds a refreshing approach to the analysis of culture in complex societies, Bruner's idea of narrative structures may explain the TCM sequential process. I argue that TCM embraces a part-narrative structure which of course needs further explanatory elaboration.

Danish practitioners who claim TCM identity are exceedingly conscious about the distinctiveness of their chosen variation. TCM is viewed as superior medical knowledge, implying an advanced level in both theory and practice. Other variations are less tolerated, and often belittled as inferior knowledge with little theoretical, philosophical, or practical value.

Practitioners telling their stories about Chinese medicine imply a dominant narrative about its past, present, and future. The distant past is seen as glorious, referring to its great masters and their seminal accomplishments. The near past marked the 1970s introduction of acupuncture, beginning a new medical era in the West. Acupuncture was in these early days received in Denmark as little less than humbug and mumbo-jumbo, but its reputation gradually changed from 'illegal', then 'tolerable', and now perhaps 'semi-accepted'. These ambiguous terms should be noted with care, but they are useful indications of how the status of acupuncture has changed since its introduction into Danish health care in the early 1970s. The present concerns the problematic adjustment in medical law, education, training, and occupational organization. For example, a 1978 Supreme Court decision deemed acupuncture as surgery (Ugeskrift for Retsvæsen 1978: 926), thus applying legal restrictions against non-qualified needling practitioners. However, medical law hardly corresponds to medical practice, since acupuncture has spread rapidly among non-medically qualified practitioners. Competing interest groups complicate the scenario, also exacerbated by the ongoing debates over educational guidelines and proper training. Finally, the future will bring complete acceptance and recognition of Chinese medicine and acupuncture as an effective means of treating common and severe illness.

Part-narratives become inevitable comparing narrative structures and cultural streams. Part-narratives develop as competing stories that fit the dominant meta-narrative about acupuncture and its passage through time. Such competing narratives are obviously heaped with professional intentions, depending on the context in which they are told and thus appropriated (cf. Todorov 1984: 56–57).

I argue that any dominant narrative becomes an appropriated part-narrative following the actor's (individuals and groups) professional objectives (see also Høg 1998). In my view, TCM becomes accommodated as a part-narrative structure by referring to its PRC origin and institutional legitimacy. TCM practitioners know about its academic status in China, and are astonished and offended when witnessing negligent TCM training in Denmark. TCM is seen as the correct variation that will prevail, mature, and stand out as the most comprehensible in terms of form and content, however problematic this prediction will be, as witnessed by the advanced development of Chinese traditional medicine in North America and England. Here TCM has become sine qua non for the licensing of practitioners, while other kinds of training do not provide state recognized access to the medical market. TCM survives through the advantage of institutional legitimacy, i.e., its educational demarcation gives it a modern quality that fits Western ideas of academic medical training. However, how can a curriculum of five hundred teaching hours stand against five years of fulltime training at Chinese TCM universities, including theory, clinical practice, history, and Western medicine?[6] How will the professional ambition be accomplished, taking the conditions of overlapping knowledge traditions and competing narratives into consideration? Will complete TCM training be possible, considering the many acupuncture variations and struggles for recognition?

Remarks on TCM Conditions in Denmark

I have addressed late 20th-century TCM development in Denmark with a focus on its conditions in a new host culture. TCM is one knowledge construction among many variations, of which particularly Classical Acupuncture, Five Element Acupuncture, Physiological Acupuncture, Sri Lanka Acupuncture and SOMA Acupuncture are visible in the Danish context. Given these variations, the identity of TCM as a particular construction of Chinese medicine rooted in the history of the Peoples Republic of China is a difficult matter. Medical semantics become blurred in a sociocultural context in which TCM may mean anything from any independent therapy to all of Chinese medicine.

I discussed Danish TCM conditions by including four perspectives: the process of knowledge transmission, whether TCM should be part of a controlled and regulated market or remain self-regulating, whether TCM has entered a tradition that rejects unconventional medical knowledge, and finally the factors that decide the slow transmission of knowledge into a new society and culture. My sympathetic perspective agrees that the development of TCM in Denmark is in progress, and that its impact, not being very great, makes self-regulation a safe path. While the transmission of TCM into Denmark is based largely on individual agency, its educational content and form will likely mature over the years. Early textbook translations from the English and German language were perhaps of a poor quality, but newer translations of original Chinese TCM textbooks are constantly raising the quality of training programs for the benefit of students, practitioners, and most importantly, the patients. Teachers become more competent in this process and they are doing their best under the given circumstances.

Yet still writing from a sympathetic perspective, the issue of self-regulation pushes the argument toward critical scrutiny. I have presented several reasons why Chinese medicine is left unregulated in Danish health care. First, TCM is taught at introductory levels and perhaps only one in ten students becomes an advanced practitioner and pursues a professional career in Chinese medicine. Second, compared to other variations TCM remains intellectually most demanding and it therefore attracts few students. Third, since there are few complaints against malpractice, there would be no reason to interfere with a self-reliant and self-regulating health care niche. Fourth, Chinese medicine retains low priority and prestige on the state agenda for regulating health care in Denmark.

One critical factor on TCM conditions is whether Denmark as the new host culture has had a tradition of incorporating various medicines into officially recognized medical pluralism. Medical pluralism on the ground is sustained through a tolerant and unstructured medical system that simultaneously decides the prevalence of antagonistic systems and competing educational content and form. Chiropractice served as an example of how academic status in Denmark is invariably gained by reaching educational consensus, highly resembling that of biomedicine. Again, the complex nature of TCM training in Denmark complicates its advancement toward official recognition.

While the development of TCM in Denmark is perhaps in progress, its transmission is also slow. From a critical point of view, teachers are incompetent and do very little to improve their teaching skills. Introductory levels are offered year after year that are designed more to secure the teachers' livelihood than create competent practitioners. Slow transmission is also due to poor translations of Chinese medical textbooks that blur medical semantics. Finally, the lasting lack of political and economic support from the state serves only to let Chinese medicine and TCM in particular survive within a liberal and open medical market.

I have placed these four major arguments implying the issues of cultural transmission and time within the anthropological frame of cultural streams and knowledge traditions (Barth 1989, 1993) and narrative structures (Bruner 1986). The many variations of Chinese medicine found in Denmark may be understood as overlapping and antagonistic cultural streams in which TCM as a particular construction of Chinese medicine presents a weak profile.

By including the idea of narrative structures employed by Edward Bruner, I have highlighted central aspects of TCM past, present, and future in Denmark. While a general narrative structure of Chinese medicine depicts its glorious and mythical past, its problematic introduction and transmission to a new host culture in the present, and its coming maturation in the future, then the part-narrative structure of TCM has only a small chance of coming true. The four major factors discussed in this paper impede the maturation of TCM as a local knowledge tradition in Denmark. Based on my sympathetic and critical perspectives on TCM in Denmark I will predict that it will take at least a generation before TCM as it is known in China will emerge in this

Scandinavian country. I hope to have conveyed stimulating thoughts for both compromisers and hardliners concerning TCM transmission to new host cultures, while there should be no doubt that I find maturation of TCM in Denmark in the near future not merely unrealistic but simply utopian.

Notes

[1] However, *Neijing* is a compilation of schools of thought rather than a homogenous medical ideology. Diverging theories have never been a problem for Chinese scientific development. As Allchin points out (1996: S111) incomparable interpretations of Chinese medicine have been embraced in a largely pragmatic Chinese culture in which practical efficiency counts more than mere explanation.

[2] While acupuncture was in use among European doctors in earlier decades, it was quite unknown among Danish doctors until the mid-1970s.

[3] According to Croizier (1968: 180) the PRC in 1956 decided to build four university hospitals for Chinese medicine, stepping away from apprentice medicine surviving through family traditions. The first TCM hospitals were placed in Beijing, Shanghai, Guangzhou, and Chengdu, each providing five year training programs.

[4] No reason for state pollution, thanks to Mary Douglas, see Douglas 1996.

[5] However, the reception of foreign medicines is not uniform in Western countries. For example, the UK has had longer contact with Asian medical cultures (former colonial power, long tradition of homeopathy) and the UK thus serves as fertile ground for the growth of TCM.

[6] I thank Dr Elisabeth Hsu for information and comments on TCM training in China. Also thanks to Dr Hsu for comments on earlier versions of this paper.

References

Allchin, Douglas (1996), Points East and West: Acupuncture and Comparative Philosophy of Science. *Philosophy of Science* 63 Suppl: S107–S115.

Barth, Fredrik (1989), The Analysis of Culture in Complex Societies. *Ethnos* 54 (3–4): 120–42.

————— (1993), Balinese Worlds. Chicago: The University of Chicago Press.

Birch, S. and K. Tsutani (1996), A bibliometric study of English-language materials on acupuncture. *Complementary Therapies in Medicine* 4: 172–77.

Bruner, Edward M. (1986), Ethnography as Narrative. In The Anthropology of Experience. V. W. Turner and E. M. Bruner, eds., pp. 139–55. Urbana and Chicago: University of Illinois Press.

Cheng, Xinnong (1987), Chinese Acupuncture & Moxibustion. Beijing: Foreign Languages Press.

Croizier, Ralph C. (1968), Traditional Medicine in Modern China. Science, Nationalism, and the Tensions of Cultural Change. Cambridge, Massachusetts: Harvard University Press.

Dike (1997), Danskernes sundhed mod år 2000: DIKE Dansk Institut for Klinisk Epidemiologi (Danish Institute of Clinical Epidemiology).

Douglas, Mary (1996), Purity and Danger. An Analysis of the Concepts of Pollution and Taboo. London: Routledge.

Farquhar, Judith (1994), Knowing Practice: The Clinical Encounter of Chinese Medicine. Oxford: Westview Press.

Høg, Erling (1998), Omstridte Grænser: En Antropologisk Analyse af Kompetence & Legitimitet i Dansk Akupunktur (Contested Boundaries: An Anthropological Analysis of Competence & Legitimacy in Danish Acupuncture). MA thesis. University of Copenhagen.

Johannessen, Helle (1995), Alternativ Behandling i Europa: Udbredelse, Brug og Effekt — et litteraturstudie: Sundhedsstyrelsens Råd vedrørende alternativ behandling.

Kaptchuk, Ted J. (1983), Chinese Medicine: The Web that has no Weaver. London: Rider Books.

Last, Murray (1996), The Professionalisation of Indigenous Healers. In Medical Anthropology. Contemporary Theory and Method. Revised Edition. C. F. Sargent and T. M. Johnson, eds., pp. 374–95. Westport, Connecticut: Praeger Publishers.

Leslie, Charles (1980), Medical Pluralism In World Perspective. *Social Science & Medicine* 14B: 191–95.

Maciocia, Giovanni (1994), The Practice of Chinese Medicine: The Treatment of Diseases with Acupuncture and Chinese Herbs. 2 vols. Edinburgh, London, Madrid, Melbourne, New York and Tokyo: Churchill Livingstone.

———— (1995), The Foundations of Chinese Medicine: A Comprehensive Text for Acupuncturists and Herbalists. 2 vols. Edinburgh, London, Madrid, Melbourne, New York and Tokyo: Churchill Livingstone.

Shima, Miki (1992), Getting Acupuncture Education Back on Track: What Our Training Has Been Missing and How We Can Benefit from the Japanese Empirical Schools. *American Journal of Acupuncture* 20(1): 33–42.

Steiner, R. Prasaad (1983), Acupuncture — cultural perspectives. 1. The Western view. *Postgraduate Medicine* 74(3): 60–67.

Stepan, Jan (1985), Traditional and Alternative Systems of Medicine: A Comparative View of Legislation. *International Digest of Health Legislation* 36(2): 283–341.

Todorov, Tzvetan (1984), Mikhail Bakhtin: The Dialogical Principle. Godzich, Wlad, transl. 13 vols. London: University of Minnesota Press.

Ugeskrift for Retsvæsen (1978), Akupunkturvirksomhed i behandlingsøjemed sidestillet med operativt indgreb i lægelovens forstand. Ikke strafbortfald. Ugeskrift for Retsvæsen 926 ("Administration of Justice Weekly").

Unschuld, Paul (1976), Western Medicine and Traditional Healing Systems: Competition, Cooperation or Integration? *Ethics in Science and Medicine* 3(1): 1–20.

———— (1985), Medicine in China: A History of Ideas. Berkeley: University of California Press.

'Improving' Chinese Medicine: The Role of Traditional Medicine in Newly Communist China, 1949–53

KIM TAYLOR

Needham Research Institute, Cambridge

This article will describe an early and uncertain stage in the recent history of Chinese medicine, before its eventual institutionalization and standardization in CCP China.[1] Many have regarded the Chinese Communist Party's commitment to Chinese medicine as a matter of course, yet its support was not immediate, and it was in fact a consequence of many unlikely factors. The period 1949–53 marks a transformation from a general policy of promoting Chinese medicine to one of wholesale support.[2] This transformation was to a great extent linked with Mao Zedong's attitude toward the Ministry of Health, and his determination to subordinate specialist intellectual skills to the might of the CCP machine. Prior to 1954 there were no explicit directives issued from the Central Committee regarding the future path of the development of Chinese medicine. Only the most general statements from Mao made it clear that Chinese medicine was to be maintained and fostered through the use of the slogan 'unification of Chinese and Western medicines' (*zhongxiyi tuanjie* 中西医团结). This article will attempt to show how Mao's instructions were interpreted by physicians of Western medicine within the Ministry of Health, who were responsible for the implementing of health care policy across the nation. They were soon to be condemned due to suspicion of their Party loyalties, yet their vision of a new, world medicine, the creation of which Chinese medicine would play a major part, came to be endorsed by Mao Zedong himself, and was to continue to characterize the next stage in the development of Chinese medicine in PRC China.

The main protagonist in this history is He Cheng 贺诚 (1901–), originally from Sichuan province. He Cheng was a physician trained in Western medicine who had played an important role in health care planning in the National Revolutionary Army as director of the General Medical Department of the Red Army since 1945. His status within the army ranks was demonstrated when on August 1, 1949, He Cheng was made director and

Party committee secretary to the Ministry of Health of the Central Committee's Military Commission.[3] His influence then enabled him to be elected to the position of deputy director of the newly established Central Party's Ministry of Health on November 1, 1949. Croizier describes He Cheng as "probably the most important figure in the Ministry".[4] This was despite being second in ranking to Li Dequan 李德全 (1896–1972), the director of the Ministry of Health. Li Dequan was a keen supporter of women and children's welfare but had no background in medicine. Lampton describes her as "the least important major personality" in the Ministry of Health.[5] Rather, it was He Cheng who wielded the most power within the Ministry of Health, and it was he who devised and implemented government policy concerning Chinese medicine during the period 1949–53.

Up until 1949, government policy towards Chinese medicine had been one of general acceptance. Mao had called in 1944 for the 'co-operation of Chinese and Western medicines' (*zhongxiyi hezuo* 中西医合作), which Party workers had interpreted as 'the scientification of Chinese medicine and the popularisation of Western medicine' (*zhongyi kexuehua, xiyi dazhonghua* 中医科学化, 西医大众化).[6] After 1949, however, the slogan changed to the 'unification of Chinese and Western medicines'. This slogan was initially put forth by Mao at the First National Meeting on the Administration of Public Health in October 1949. In this speech, Mao called for people to "unite the new, old, Chinese and Western, every part of the medical and hygiene workers, in order to form a consolidated united front, and in this way develop the great health of the people and [continue the] struggle".[7] The Communist Party's ideal of 'unification' therefore implied not only the 'unification' of the scattered Chinese medical profession as a whole, but also the 'unification' of this body of workers with the rest of the new revolutionary society, and in particular, with the profession of Western medicine.[8]

It was up to the government ministers within the Ministry of Health to understand and implement such an order of unification. In the Communist Party machine, the Party leaders did not deal with the technicalities of policy at all. Schurmann has described the organization as a three-tier affair — with the policy makers in the top tier, the policy translators and implementers in the middle tier, and the workers in the bottom tier.[9] The CCP at this time functioned through what has been called 'practical' ideology. Schurmann defines 'practical' ideology as "an idea, for example a policy, which they [the leaders] expect their followers to implement".[10] However, such policies could not be implemented without working out the "correct thought", i.e. the ideological jargon of the higher powers had to be "interpreted". Once interpreted, it then had to be implemented by various Party organs. Diversity in the implementation of Party directives was a necessary consequence of this organizational structure. And we shall shortly see just how open to interpretation policy on Chinese medicine could turn out to be.

He Cheng, with his Western medical training, most likely understood Mao's requests in the context of the current health care situation in China. Like most Westerners, what he probably saw was a country with only the rudiments of an urban-based Western hospital system and the very beginning of public health technology. Estimates have it that with a population of around 540 million in 1949 there was a death rate of 30/1000, a birth rate of 40/1000 and an infant mortality rate of 200/1000.[11] Malnutrition and infectious diseases were responsible for the great majority of these deaths, preventive medicine was practically unheard of, and modern Western medicine remained only in the big cities.[12] Lampton states that the CCP's budget for the Ministry of Health only came to 2.6% of the total budget between 1950–56.[13] The introduction of the Labour Insurance Act of March 1951 entitled 3.3 million workers to free health care services (*gongfei yiliao* 公费医疗),[14] further stretching to breaking point the Ministry of Health's already very limited resources. The Ministry of Health tackled this problem by placing the emphasis on prevention, prevention being cheaper than cure, and by their emphasis on local talent. From the viewpoint of the primary health care services, the idea was thus "to use all available talent with little concern for ideological purity".[15] Chinese medicine was thus to be used at the least "medical" level of Western medicine — public health.

Therefore, on the most basic level, practitioners of Chinese medicine were needed to supplement the numbers. With such a problem of health care distribution in rural China, every man with therapeutic skills counted. Mao himself noted that "you only have ten to twenty thousand doctors of Western medicine. This strength is frail. You must properly unite the Chinese medical doctors".[16] This was in comparison to a possible 500,000 Chinese medical doctors.[17] There was also a very real concern about social stability (*shehui anding* 社会安定). If the Chinese medical practitioners were ignored, not forcibly integrated into the new Communist society, or their medicine not encouraged, it would mean hundreds of thousands of people would be without a livelihood, and therefore without any means of support. It is possible that in the mind of the Party, the importance of keeping people usefully employed within society outweighed the dangers of supporting a not-fully understood medicine. Doubts concerning the medical integrity of Chinese medicine, however, were only short-term. For the long-term, the Party had much more ambitious plans for the future of Chinese medicine.

He Cheng's plan for Chinese medicine was explicitly articulated at the First National Health Convention, held in Beijing in August 1950. At this meeting he emphasized once more how Chinese medicine had to learn scientific theory. Yet Western medicine too, he felt, needed to learn from Chinese medicine to better approach the masses. If the two medicines could co-operate and work together, they would best be able to solve the health problems of the Chinese people. The result of this would be that "the distance between Chinese and Western medicine would day by day grow smaller, and the differences between Chinese and Western medicines would day by day disappear", until the two "open-heartedly and inseparably unite".[18] In other words, He Cheng was leaving no space for either a pure Western or Chinese medicine in the future. Both were to fuse together as the party slogan of 'the unification of Chinese and Western medicines' implied, and form one new medicine designed to serve the needs of the Chinese people. This He Cheng defined as the pathway for the new medicine of the new China.

Such a 'unification' of the medicines was to take place in the setting of the 'Chinese medical improvement schools' (*Zhongyi jinxiu xuexiao* 中医进修学校). He Cheng had called for the establishment of such schools a few months earlier on May 30, 1950 at the opening ceremony of the new Beijing Chinese Medical Institute (*Beijing zhongyi xuehui* 北京中医学会). The establishment of the Beijing Chinese Medical Institute gave Chinese medicine a new status in the medical world. Practitioners of Chinese medicine had previously only belonged to small localized organizations, and had had no national representation at all. At the opening ceremony He Cheng discussed the need for Chinese medicine once again in manpower terms, and then stressed the fact that Chinese medicine needed to improve on its scientific knowledge. In this meeting he called for Chinese medical improvement schools to be set up around the country. He Cheng defined the role of these schools as providing courses in contagious diseases, and other science-based classes, in order that this scientific knowledge be absorbed into Chinese medicine.[19]

The purpose of the Chinese medical institutions was to "strengthen the unification of Chinese and Western medicine and to promote the scientification of Chinese medicine".[20] The Chinese medical institutions were to be the 'bridge' (*qiaoliang* 桥梁) for the unification of Chinese and Western medicines. The Ministry of Health was very involved in the teaching programs at these schools and on December 27, 1951, it issued guidelines on the curricula to be taught in the Chinese medical improvement schools and classes.[21] The guidelines specified that courses at the Chinese medical improvement schools were to last 12 months. In the topics of preventive medicine, acupuncture research or first-aid techniques, Chinese medical improvement classes were to last only six months (159 hours), and could in fact be given as specialized classes lasting three months (87 hours). In addition, all Chinese medical improvement schools were to provide the use of laboratory equipment, practical demonstration teaching rooms etc. in their courses. This equipment, however, was generally supplied by uniting the local, relatively large hospitals or medical schools and sharing their facilities.

By early 1952, there were 17 improvement schools set up around China, and 101 improvement classes. However, Croizier writes that "scattered provincial figures suggest that well under 25% of the traditional physicians had taken such courses by late 1954".[22] Superficial though these training programs might have been, they still represented the government approved version of Chinese medicine. It is therefore significant that the scientification of Chinese medicine, which was meant to raise Chinese medicine to another level, was to occur only in the subdivisions of acupuncture and moxibustion (*zhenjiu* 针灸), bonesetting (*zhenggu* 正骨) and massage (*anmo* 按摩).[23] These were the only representations of Chinese medicine to be taught at these Chinese medical 'improvement' schools.[24]

The Beijing Chinese Medical Improvement School was the first such Chinese medical improvement school to be set up, and it opened on May 1950 on the site of the old North China National Medical Academy (*Huabei guoyi xueyuan* 华北国医医院), under the auspices of the Beijing Chinese Medical Institute. The North China National Medical Academy was originally set up in 1932 as an offshoot of the Beiping National Medical College (*Beiping guoyi xueyuan* 北平国医学院), which had been established in 1930 by three of the four 'Beijing Four Great Medical Masters (*Beijing sida mingyi* 北京四大名医)' — namely, Xiao Longwen 萧龙文 (1870–1960), Kong Bohua 孔伯华 (1884–1955) and Shi Jinmo 施今墨 (1884–1955). These three men were a strong representative force of the Chinese medical talent present in Beijing in the Republican era, and the choice of such an historic location for the siting of the new school did emphasized the considerable amount of Chinese medical talent available in Beijing at the time of Liberation.

The Beijing Chinese Medical Improvement School gives us a representative example of the type of knowledge Chinese medical practitioners were expected to learn, and the way in which they were to learn it.[25] The curriculum, however, contained little in the way of Chinese medicine. Teaching materials for the Beijing Chinese Medical Improvement School were the 'General Outline of Physiology', 'Modern Pharmaceutics', 'General Outline of Pathology', 'Practical Bacteriology', 'Infectious Diseases etc. The majority of these books were published by the Western medical publisher

'Society of [Western] medical matters and livelihood' (*yiwu shenghuoshe* 医务生活社). We can see that the books tended to be 'practical' (*shiyong* 实用) or 'general outlines' (*dagang* 大纲), or 'rudimentary' (*chuji* 初级). The subject of public hygiene, however, was based on the Soviet system of medicine and health care. There was little Chinese medical content, with the exception of the history of medicine, where it would appear that the students were at least taught the history of Chinese medicine,[26] and courses in acupuncture, bone-setting and massage.

For the school in Beijing, each course lasted 26 weeks, divided into two terms. Each term would be 12 weeks long with a week long holiday in between and an extra week at the end for a general summing up of the program. Because clinic hours for Chinese medical doctors tended to be in the mornings, these classes would be held in the afternoon, every day from 3 p.m. to 6 p.m. except on Sundays.[27,28]

The first term put an emphasis on physiology and the second term put an emphasis on pathology. Added to this were the courses of public health, infectious disease and simple clinical techniques. The whole curriculum carried out the principle of 'prevention first', and both terms included classes on the social sciences. The first term consisted of classes on anatomy, physiology, pharmacology, medical history and the social sciences, which included the history of the development of society and current political reports. The second term consisted of classes on pathology, bacteriology, intestinal parasites, clinical diagnosis (simple treatment techniques), public health and the social sciences, which included the theory of the New Democracy and current political reports.[29]

Therefore we can see that the aim of the Chinese medical improvement schools was not so much to propagate Chinese medicine, but to control it. These were not schools designed to teach Chinese medicine to young students, but rather to have Chinese medical practitioners as students. It was stated that, 'The main aim of the improvement of Chinese medical practitioners is to improve their awareness of politics and to improve their level of professional work."[30] The lecturers, meanwhile, would be doctors of Western medicine.[31] At the Beijing Chinese medicine improvement school, for example, the majority of the teaching staff were all Western medical staff brought in from outside universities such as the Beijing University Medical School (*Beida yixueyuan* 北大医学院), the Union Medical College (*Xiehe yixueyuan* 协和医学院) and the Ministry of Health research centre (*Zhongyang weisheng yanjiuyuan* 中央卫生研究院).[32] There were only two full-time teaching staff at the school. These were Ma Jixing 马继兴[33] as assistant teacher for anatomy and Zhao Fangzhou 赵芳洲 as assistant teacher for parasitology and lab technician.[34] The life span of the Chinese medical improvement schools continued until around 1958, although they did linger on longer in more out-of-the-way areas of China until Academies of Traditional Chinese Medicine were built to replace them.[35]

Chinese medicine outside of such institutions, however, was to be strictly controlled. In October 1952 the Ministry of Health introduced a new medical licensing exam which all those in the health profession, from doctors (traditional or otherwise) to dentists and pharmacists, were obliged to sit.[36] Statistics from these exams were not positive. Out of 165 counties and 92 large and middle-sized cities which registered for the exam in 1953, only 14,000 participants passed. The large majority of Chinese medical practitioners were banned. In Shanxi province, in Yuncheng prefecture, out of 18 counties, not one Chinese medical practitioner qualified.[37] This was allegedly because a large part of the content of the exam was in Western medicine, but in big cities where the Chinese medical doctors had slightly more exposure to such education, the results were hardly better. In Huabei region, out of 68 counties, over 90% of the doctors who took the exam failed.[38] In Tianjin, out of 530 doctors who took the exam, only 55 passed. This meant that the majority of Chinese medical practitioners were unable to legally practise Chinese medicine under the new government.

Other restrictions which handicapped the Chinese medical profession were the fact that the newly instituted health insurance system did not include treatment by Chinese doctors,[39] and one source has even said that factories which produced the drugs and herbs used by Chinese herbalists were shut down, so that there was no supply of medicaments and Chinese medical doctors could not practise.[40] He Cheng's policies thus resulted in greatly restricting the Chinese medical profession. Those who survived the veritable mincing process of examinations, licensing, improvement schools etc. were not accorded equal status with doctors of Western medicine. 'Qualified' Chinese medical practitioners were not allowed to work in Western medicine dominated hospitals. Some were allocated to teach in turn in the Chinese medical improvement schools, others had to seek work as assistants (*yizhu* 医助) to doctors of Western medicine. Those who preferred to carry on practising their medicine found newly imposed restrictions squeezing their livelihood which resulted in many traditional practitioners having to change profession (*gaihang* 改行).[41]

Such miserable conditions for the practising Chinese medical physician coincided with a peak in He Cheng's grand national scheme concerning Chinese medicine. This was announced at the Second National Health Administrative Convention in December 1952. Here He Cheng reaffirmed his views on the future of Chinese and Western medicines in China. He said that while some were worried that Chinese medicine would soon die out, others were also worried that both medicines would continue to exist in China. He Cheng declared that at the moment there was no other choice but for both medicines to exist together in China. But it would not be right, if in the long-term, the two medicines were to diverge, and continue along their separate ways. He proclaimed that "in the future, there will only be one medicine in the world ... the problem from now on is how we can make a greater contribution to the progressive medical treasures of the world, in order to carry out more work for the health of the people. This is the path we need to take."[42] He Cheng thus advocated that all medicines of the world had to work together, and learn from each other, and in this way China would make her own particular mark on the future medicine of the world.

The Central Committee, however, was about to intervene radically. During early 1953, a three-month long investigation into the work of the Ministry of Health was ordered. The lead-up to the attack on the Ministry of Health has not hitherto been clearly explained, but it now appears that the investigation was triggered by what began as a complaint against lack of proper leadership among the Departments of Health across the nation. A new political affairs post had been set up in early 1953 in the Military Commission's Ministry of Health. This post was taken up by a man called Bai Xueguang 白学光, former editor of the *August 1 Journal* (*Bayi zazhi* 八一杂志). Bai Xueguang subsequently made an inspection of the health situation in various provinces and, in his report to the central party, declared that "as for the hygiene work of the entire army, and the vocational work of the subordinate work units, these are lacking in leadership, to the point where there is simply no leadership".[43] The upshot of the investigation was that in October 1953 Mao published a 'Criticism of the Ministry of Public Health', a monologue apparently aimed at He Cheng, describing the Ministry of Health as a "total mess".[44]

The attack on the Ministry of Health was soon translated into whole-scale condemnation of the current state of Chinese medicine in China. While this was the fat in the fire, the rancour caused by the ongoing feud of Red vs. Expert should not be underestimated.[45] We will however, probably never know exactly why and for whom He Cheng was being targeted, but the issue of Chinese medicine was to be the metaphorical sword by which he was to die. By the end of 1953 He Cheng was being "blamed for the Ministry's mistakes regarding traditional medicine".[46] Chinese critics accused him of having "too stringent standards for the improvement and reform of Chinese medicine. He was not operating from the [standpoint of] maintaining the traditional theory of Chinese medicine and its special treatment properties, but rather was making demands on Chinese medicine from a Western medical viewpoint".[47] He Cheng was also accused of trying to substitute Chinese medicine eventually with Western medicine, such that the future of Chinese medicine would be "from the city into the countryside, and from the countryside it will naturally die out."[48] Such criticism was to gain momentum in the hostile political climate of the mid-1950s, forcing He Cheng to write a self-criticism in November 1955 and to step down from office.

The irony though, is that He Cheng's intentions for the future of Chinese medicine were to be adopted as future medical policy. It was his means, not his aims, which were found to be at fault. While the investigation into the Ministry of Health had been triggered by claims that it was seeking autonomy from the Central Committee, it was on the grounds of his handling of Chinese medical manpower that He Cheng was mainly accused. Yet regarding the actual ideological pattern into which he was trying to fit Chinese medicine, He Cheng was not criticized, except in so far as he had not yet managed to bring about this unification.[49] His ideological outlook was successful to the point that Mao was to endorse his view on Chinese medicine and prolong it as a major criterion in the further development of Chinese medicine. For Mao himself was to state in 1953, "with extreme foresight" (*gaozhanyuanzhu* 高瞻远瞩) according to Ma Boying, that "in the future there will only be one medicine; that is to say a [single] medicine guided by the laws of dialectical materialism, and not two [separate] medicines".[50] The aim of creating a single, world medicine through the 'unification of Chinese and Western medicines' was to continue.

Notes

[1] This article is an excerpt taken from the second chapter of my PhD thesis, "Medicine of Revolution: Chinese Medicine in Early Communist China, 1945–63", University of Cambridge, 2000, which I was privileged to present as part of a panel at the 9th ICHSEA Conference in Singapore. My thanks go to Andrew Cunningham, Hans van de Ven and Elisabeth Hsu who were kind enough to read over earlier versions of the chapter and give me their comments.

[2] The period 1949–53 has been regarded as the heyday of Mao's authority over the Party and country. This period marked the propagation of a 'Mao cult', when Mao was "accorded imperial and even superhuman qualities". It was a time when Mao overrode Central Party hesitations in decisions over land reform, and the Korean War. The ensuing successes of both served to establish his authority within the Party leadership. For a more detailed discussion, see, Sullivan, Lawrence R., "Leadership and Authority in the Chinese Communist Party: Perspectives from the 1950s" in *Pacific Affairs* 59(4) (1986–87): 605–633.

3   Zhu Kewen 朱克文, Gao Enxian 高恩显 and Gong Chun 龚春 (eds.), *A History of China's Military Medicine* 中国军事医学史 (Beijing: Renmin junyi chubanshe, 1996), p. 316.

4   Croizier, Ralph C., *Traditional Medicine in Modern China* (Cambridge, Mass.: Harvard University Press, 1968), p. 171.

5   Lampton, David M., *The Politics of Medicine in China: The Policy Process 1949–1977* (Folkestone, Kent: WM Dawson and Sons, 1977), p. 23.

6   For a description of the Communist Party's initial dealings with Chinese medicine during the years of guerrilla warfare, and how the medicine came to be promoted by, in fact, doctors of Western medicine within the Party, see Taylor, Kim, "Civil War in China and the New Acumoxa" in Hsu, Elisabeth (ed.), *Innovation, Convention and Controversy in Chinese Medicine* (Cambridge: Cambridge University Press, 2001).

7   "团结新老中西各部分医药卫生工作人员, 组成巩固的统一战线, 为开展伟大的人民卫生工作而奋斗", Zhu Chao 朱潮 and Zhang Weifeng 张慰丰 (eds.), *A History of Medical Education in the New China* 新中国医学教育史 (Beijing: Beijing yike daxue and Zhongguo xiehe yike daxue, 1990), p. 40.

8   The ideal of 'unification' carried on the criteria of the 'New Democracy' which were to characterize these first years of the new regime. Goldstein writes that "the CCP's initial commitment to the moderate tasks of 'New Democracy' [i.e. 'unity', 'new' and 'scientific'] helped nurture the popular appeal of the regime" in Goldstein, Avery, *From Bandwagon to Balance-of-Power Politics: Structural Constraints and Politics in China 1949–1978* (Stanford: Stanford University Press, 1991), p. 60.

9   Schurmann, Franz, *Ideology and Organization in Communist China* (Berkeley: University of California Press, 1968), pp. 68–73.

10 Schurmann (1968), p. 21.

11 Chabot, H. T. J., "The Chinese System of Health Care", *Tropical and Geographical Medicine* 28 (2) (1976): S99. This was as compared to UK 1991 figures of a death rate of 11.3/1000, birth rate of 13.8/1000 and infant mortality rate of 18/1000.

12 Sidel, Ruth and Sidel, Victor, *The Health of China: Current Conflicts in Medical and Human Services for One Billion People* (London: Zed Press, 1982), p. 26.

13 Lampton (1977), p. 13.

14 Hillier, S. M. and Jewell, J. A., *Health Care and Traditional Medicine in China 1800–1982* (London: Routledge and Kegan Paul, 1983), pp. 75–77.

15 Lampton, David M., "The Succession of Health Policy-Making Systems and Chinese Political Development", *Michigan Papers in Chinese Studies* (18) (1974), p. 94.

16 "你们的西医只有一两万, 力量薄弱, 你们必须很好地团结中医", Zhu Chao (1990), p. 40.

17 Croizier (1968), p. 158.

18 "中西医之间的距离就会日渐缩短, 中西医之间的差别就会日渐消除 ... 开诚布公地紧紧团结", He Cheng, "Chinese Medical Section of the Summarising Report of the First National Health Convention 第一届全国卫生大会总结报告中的中

医部分" (August 1950) in Department of Chinese Medicine of the People's Republic of China Ministry of Health (*Zhonghua renmin gongheguo weishengbu zhongyisi* 中华人民共和国卫生部中医司) (ed.), *Collection of Documents on Chinese Medical Work (restricted distribution)* 中医工作文件汇编 (内部发行) 1949–83 (Beijing: Zhonghua renmin gongheguo weishengbu zhongyisi, July 1985), pp. 4–5. (Hereafter this document is referred to as *CCMW*.)

[19] He Cheng, "The Issues of the Unification of Chinese and Western Medicines and the Improvement of Chinese Medicine 中西医团结与中医进修问题", *People's Daily* (June 13, 1950), p. 5.

[20] "加强中西医的团结, 与推进中医科学化" in "Directives of the Ministry of Health regarding the organisation of the Chinese medical institutions 卫生部关于组织中医学会的指示" (December 27, 1951), *CCMW* (1985), p. 18.

[21] "Circular of the Ministry of Health's "Regulations regarding the organisation of Chinese medical improvement schools and improvement classes" 卫生部＜关于组织中医进修学校及进修班的规定＞的通知" (December 27, 1951), *CCMW* (1985), p. 23.

[22] Croizier (1968), p. 266, fn. 30.

[23] "Progress Report of the Beijing Chinese Medical Academy during the Year 1950 北京中医学会一九五零年工作报告" in Central People's Government's Medical Policy Unit of the Ministry of Health (*Zhongyang renmin zhengfu weishengbu yizhengchu* 中央人民政府卫生部医政处) (ed.), *Chinese Medical Work* 中医工作 (Beijing: Zhongyang renmin zhengfu weishengbu yizhengchu, 1951), p. 30.

[24] The formerly predominant field of Chinese pharmaceutics (*zhongyao* 中药) was not included. This does not mean that this field was being ignored, for a great deal of scientific research work was being carried out on the herbs at this time in the hope that it would provide clues to the eventual goal of the 'scientification' of Chinese medicine.

[25] See Fig. 1.

[26] This course was taught by Li Tao 李涛 (1901–1959). Originally trained in Western medicine, Li Tao came from Beijing and during the 1930s worked at the medical history section of the Beijing University Medical College (北京大学医学院医史学科) and worked as curator of the Chinese medical books section of the Beiping Union Medical College Library (北平协和医学院图书馆的中医藏书局). His area of expertise was in the history of China's medicine, and he was not a trained practitioner of the medicine.

[27] In other words, the doctors were expected to be able to support themselves throughout the duration of the course. Otherwise the Chinese medical improvement schools were supported by government subsidies. The financial costs of the Chinese medical institutes were to be met by monthly membership fees and revenue from publications.

[28] Chinese improvement classes held in Shanghai were run from 6 p.m. to 10 p.m. and were held in rooms at the local medical colleges and hospitals, *People's Daily* (May 27, 1952), p. 3.

[29] Taken from *Zhongyang renmin zhengfu weishengbu yizhengchu* (1951), pp. 6–16.

30 "中医进修的主要目的在于提高政治觉悟和业务水平" in "Resolutions of the Third National Health Administrative Convention 第三届全国卫生行政会议觉" (produced 25 February 1954). Unpublished manuscript. In author's possession. Also available at the Needham Research Institute, Cambridge.

31 The Shanghai Chinese medical improvement classes were also only staffed with doctors of Western medicines, *People's Daily* (May 27, 1952), p. 3.

32 This was to change after 1953 and the attack on Ministry of Health policies, when the Chinese medical content in the majority of the improvement schools rose to an estimated 40–60% of the overall curriculum, Zhu Chao (1990), p. 43.

33 Ma Jixing had already carried out significant scientific research into acupuncture while working from 1946–50 at the physiology department of the Beijing University Hospital. It would thus appear that his background was in Western biomedicine and he was not by trade a Chinese medical practitioner. Today, Ma Jixing is a very prominent figure in the history of Chinese medicine, based at the Beijing Research Academy of Traditional Chinese Medicine.

34 *Zhongyang renmin zhengfu weishengbu yizhengchu* (1951), p. 17.

35 I have records of publications from various Chinese medical improvement schools from as late as 1959. For example, the *Popularised Handbook of Acupuncture Therapeutics* 针灸疗法普及手册, published by the Hunan Province Chinese Medical Improvement School (湖南省中医进修学校) in 1959.

36 The Ministry of Health issues "Temporary Examination Procedures for Doctors, Chinese Medical Doctors, Dentists and Pharmacists" 卫生部发布置: 医师, 中医师, 牙医师, 药师考试暂行办法" 2 (October 4, 1952), *CCMW* (1985). pp. 26–27.

37 Zhu Chao (1990), p. 40.

38 Zhu Chao (1990), p. 105.

39 Zhu Chao (1990), p. 41.

40 Interview, Lu Bingkui 吕炳奎, former director of the Department of Traditional Chinese Medicine, Ministry of Health, in Beijing, April 10, 1998. Lampton describes how traditional herbs were allocated less and less acreage by government agencies who were emphasizing industrial and food crops. As a result, Chinese herbal medicaments became very expensive, in Lampton (1977), p. 51.

41 Interview, Yin Huihe 印会河, senior clinician of the Sino-Japanese Friendship Hospital and honorary professor, Beijing University of Traditional Chinese Medicine, in Beijing, December 12, 1997.

42 "将来世界的医学只有一个。今后的问题, 是如何使我们对世界进步的医学宝库, 多贡献力量, 为人民的健康事业作更多的事情, 这是我们应走的道路", He Cheng, "Chinese medical section of the Summarising Report of the Second National Health Administrative Convention 第二届全国卫生行政会议上的总结报告中的中医部分" (December 1952), *CCMW* (1985), pp. 28–29.

43 "对全军的卫生业务以及直属单位的业务工作缺乏指导, 甚至根本没有指导", Zhu Chao (1990), p. 106.

44 Mao Zedong, "Criticism of the Ministry of Public Health", in Kau, Michael Y. M. (ed.), *The Writings of Mao Zedong*, Vol. 1 (Sep. 1949 – Dec. 1955) (London: M. E. Sharpe, 1986), pp. 425, 441.

[45] Lampton describes it as the 'theory of leadership error', where rather than blaming internal structures such as lack of resources, or internal authority conflicts, the party leadership saw "the Ministry's major defect, ... as being divorced from Party leadership" in Lampton (1977), p. 46.

[46] Croizier (1968), p. 171.

[47] "对中医的提高和改造要求过急过高, 不是从保持中医传统的理论和医疗特色出发, 而是以西医的观点来要求中医", Zhu Chao (1990), p. 105.

[48] "由城到乡, 由乡走向自然淘汰", Zhu Chao (1990), p. 105.

[49] "[They] had not yet put into effect the Central Party's directive concerning the unification of Chinese and Western medicine, the genuine unification of Chinese and Western medicines had also not yet been realised and this was wrong 对中央关于团结中西医的指示未贯彻, 中西医的真正团结也还未解决, 这是错误", Ma Boying *et al.* (1993), p. 578.

[50] "将来只有一个医学, 即唯物辩证法作指导的一个医, 而不是两个医", Ma Boying *et al.* (1993), p. 577.

An Introduction to the History of Traditional Medicine and Pharmaceutics in Vietnam

LAN TUYET CHU

The Institute of Sino-Nom Studies, Hanoi

Vietnam is a country with a long history of traditional medicine and pharmaceutics. Traditional medicine and pharmaceutics in Vietnam were developed on the basis of Eastern theoretical principles and protected the health of the people throughout the development of the nation. This paper seeks to give a brief introduction to the history of Vietnamese traditional medicine and pharmaceutics, present extant sources for their study, and discuss recent research on sources.

Features of the History of Traditional Medicine and Pharmaceutics in Viet Nam

Based on Chinese records of the herbs and plants in Giao Chi (ancient Viet Nam), since the third century B.C. over 100 types of herbs used for medicinal purposes have been discovered and categorized in Viet Nam. Under the reign of Thuc An Duong Vuong (257–179 B.C.), the Vietnamese people already knew how to distil alcohol for drinking and medicinal purposes. They had also mastered techniques for casting bronze and making poison arrows. This data shows that before the second century B.C., the Vietnamese people already had some knowledge of using and processing various substances for medical purposes.

From 179 B.C. until 938, Vietnam was under Chinese domination. There are no extant documentary records about medicinal practices from that time. However, according to Chinese books, for example 中越兩國人民的友好關係和文化交流/陳修和. 北京: 中國清年出版社 (*Zhong Yue liang guo renmin de youhao guanxi he wenhua jiao liu*), 1957 and 中国药学大词典 (*zhongguo yaoxue dacidian*) … , many medicinal herbs were imported from Vietnam and recorded in Chinese medicinal manuscripts along with their place of production and experience of application in each locale. During this period, a number of Chinese traditional medical doctors arrived to cure the diseases of the colonial mandarins and aristocratic families of Vietnam. This phenomenon accompanied the dissemination of culture from the North, so we can guess that traditional Vietnamese medicine was shaped since that time by the theoretical principles common in the countries of East Asia, although there are no extant medical works.

Under the Ly dynasty (1010–1224), there were many professional medicinal doctors, and the court established a Royal Physician Service to care for the health of the king. At that time, however, treating diseases by incantations was still widespread among the people (see appendix 1), so medicine was unduly influenced by superstitious beliefs and hence remained underdeveloped.

During the Tran dynasty (1225–1399), the ideology of Confucianism pushed back Buddhism, so medicine had a better climate in which to progress. The Tran dynasty established an Institute of Royal Physicians which oversaw medicine throughout the country, recruited physicians with examination degrees in 1261, and established guidelines for the population during plagues (from 1362). From the 13th century, the Tran dynasty established the cultivation and gathering of regional medicinal herbs (Southern medicine) to be used in the Institute of Royal Physicians and supplied to the army physicians. The head of the Institute of Royal Physicians was Pham Cong Ban who used his own medicine to treat poor people and displayed medical ethics by treating patients without regard to whether they were rich or poor. During the 14th century, the monk Tue Tinh opened clinics for treating people who lived in temples throughout Giao Thuy and Da Cam (present-day Cam Giang), his residence and native land. The regional medicinal herbs (Southern medicine) used for treatment were cultivated in temples, pagodas, and families' gardens and then harvested in accordance with the motto *Vietnamese medicine cures Vietnamese people* (see appendix 2).

Tue Tinh taught medicine to all the Buddhist clergy and collected methods of local medicinal treatments from the people. His lessons were edited and compiled as *Nam duoc chi nam* 南藥指南 (*Nan yao zhi nan*) and passed down after his death. The work of Tue Tinh is no longer extant. What remains today is the collection *Nam duoc than hieu* 南藥神效 (*Nan yao shen xiao*) edited by Buddhist Monk Ban Lai with original documents from Tue Tinh and published in 1761. The collection includes 499 manuscripts about local medicinal herbs written in rhyme as well as 10 branches of treatment using 3932 prescriptions to cure 184 types of diseases (see appendix 3). In addition, there is the collection *Hong Nghia giac tu y thu* 洪義覺斯醫書 (*Hongyi juesi yishu*) by the Institute of Royal Physicians of the Le dynasty republished in 1717, which includes lessons on applying local medicines, and Tue Tinh's

Thap tam phuong gia giam 十三方加減 (*Shu san fang jia jian*), which was translated into the Vietnamese language.

Nam duoc than hieu 南藥神效 (*Nan yao shen xiao*) together with all the prescriptions in local herbal medicines portray several local herbal medicines discovered not later than the 14th century.

The science of acupuncture was already developed by the 15th century. During the Ho dynasty (1400–1406), the *Quang Te thu* 廣濟署 (*Guang Ji Shu*) was founded to expand medical treatments, particularly acupuncture, for the army and the people residing in all the major regions. Nguyen Dai Nang, a resident of Kinh Mon (present-day Hai Hung) and a leading acupuncturist in this organization, left the book *Cham Cuu tiep hieu dien ca* 針灸捷校演歌 (*Zhenjiu Jiexiao Yange*) for the treatment of 130 types of diseases with 140 acupuncture points. Among those points, there are 15 points discovered by Vietnamese doctors.

Under the reign of the Le dynasty (1428–1788), which restored independence following the expulsion of the Ming invaders, national medicine was reorganized and developed. During this reign, there was an Institute of Royal Medicine (太醫) for treating the illnesses of the king and his court, and in the provinces there were clinics (濟生堂) for treating low-ranking mandarins and prisoners as well as for providing disaster relief and plague prevention for the people. In the countryside there was a system for caring for handicapped people and orphans who didn't have anyone else to look after them. The army medical corps had its own units among the troops as well as inside the citadel. The reign of Le followed a hygenic code such as forbidding the sale of spoiled food, severely punishing the disposal of poisons and toxins, outlawing the use of tobacco, abortion, marriage of under-age youth, and popularizing hygenic standards and physical exercise for the people. Moreover, the Le dynasty advocated developing pharmaceuticals and the study of medicine. Education officials were posted in the provinces and districts to supervise medicinal studies and there were medical examinations combined with doctoral laureate examinations of literature in 1747.

Extant Ancient Books and Manuscripts on the Traditional Medicine and Pharmaceutics, and the Activities of the Branch

The achievements of national medicine up to the 18th century have been displayed in the great collection *Hai Thuong Lan Ong y tong tam linh* 海上懶翁醫宗心領 (*Haishang Lanweng yizong xinling*) by Hai Thuong Lan Ong (Le Huu Trac) (see appendix 4) who was raised in Van Xa (present-day My Van, Hai Hung) and resided in Huong Son (present-day Nghe Tinh). The contents include the system of fundamental theoretical principles of eastern medicine, the study of over 800 specimens of both local and northern herbal medicines, and the use of internal medicine, external medicine, gynecology, pediatrics, traditional and original prescriptions including popular medicinal wisdom. Lan Ong, in particular, summed up the methods of physical exercise and preventive care relevant to the everyday living conditions of the Vietnamese people in his work *Ve sinh yeu quyet* 衛生要訣 *(Wei sheng yao jue)* (Vital secrets of hygiene) and the method of nutritional food preparation in the book named *Nu cong thang lam* 女工勝覽 *(Nu gong sheng lan)*. Lan Ong invented many remedies prescriptions for external ailments according to the health conditions in Vietnam at that time.

In the south of Vietnam, a few doctors wrote several medical treatises of which the extant ones are as follows:

1. Hoang Nguyen Cat's treatise *Quy vien gia hoc* about case histories;
2. Hoang Danh Suong's treatise *Lac sinh tam dac* 樂生心得 *(Lesheng xinde)* about the study of veins;
3. Nguyen Du's ancestors wrote *Nam duong tap yeu* 南陽集要 *(Nanyang jiyao)* about the new method of categorizing diseases according to their symptoms.

Under the Tay Son reign (1789–1802) the Royal Institute of Medicine established the Office of Local Medicine in order to conduct research on local medicines. Nguyen Hoanh headed this organization, and today the treatises *Nam duoc 500 vi (Nanyao wu bai wei)*, *Gia truyen bi thu* 家傳祕書 *(Jia chuan mi shu)* and *Kinh nghiem luong phuong* 經驗良方 *(Jing yan liang fang)* are still extant.

Nguyen Gia Phan compiled *Ly am thong luc* 理音通綠 *(Li yin tong lu)* on gynecology, *Ho nhi phuong phap tong luc* 護兒方法總錄 *(Hu er fang fa zong lu)* on pediatrics, and *Lieu dich phuong phap toan tap* 療疫方法全集 *(Liao yi fang fa quan ji)* on contagious diseases.

Nguyen Quang Tuan, also known as La Khe, compiled *Thap tam thien quoc am ca* 十三篇國音歌 *(Shi san pian guo yin ge)* about traditional treatments and *Kim Ngoc quyen* 金玉卷 *(Jin yu juan)* including family prescriptions which had been passed down from generation to generation.

Under the reign of the Nguyen dynasty (1802–1883) the study of national medicine maintained its previous organization. The Royal Institute of Medicine *Thai Y Vien* (太醫院) *(Tai Yi Yuan)* had a teaching division, and in the provinces there were *Ty Luong Y* (司良醫) *(Si Liang Yi)* and *Duong Te Su* (養濟事) *(Yang Ji Shi)*. But with the arrival of the French a colonial system of medicine was instituted, and from 1905 all the institutes of herbal medicines were dismantled.

During this time, the remaining documents concerning medicine are the following:

1. *Nam duoc quoc am tap nghiem* 南藥國音集驗 (*Nan yao guo yonji yan*) by Nguyen Quang Luong in Yen Ninh, Hanoi.
2. *Trach vien mon truyen tap yeu y thu* 宅園門傳集要醫書 (*Zhai yuan men chuan ji yao yi shu*) by Nguyen Huan in Dong Anh, Hanoi.

Under the French, the colonial medical service cared mainly for the ruling class and only a few people in the urban areas. The majority of people in the countryside and the mountainous regions continued to treat their own diseases with the national medicines as before. However, though the practice of eastern medicine languished under the colonial system, Vietnamese medicines were still developed and used for export.

An Association of the Study of Medicine in the central region as well as similar organizations in the northern and southern regions, were established in order to sustain and restore national medicine.

The works concerned with national medicine are as follows:

1. Ve sinh yeu chi (*Wei sheng yao zhi*) by Bui Van Trung in Giao Thuy, Nam Dinh.
2. Bi truyen tap yeu (*Mi chuan ji yao*) by Le Tu Thuy in Duy Tien, Ha Nam.
3. Trung Viet duoc tinh hop bien (*Zhong Yue yao xing he bian*) by Dinh Nho Chan in Ha Tinh.
4. Ngoai khoa bi yeu y ly phuong dong (*Wai ke mi yao yi li*) and Nam duoc bo (*Nan yao bu*) by Nguyen An Cu in the South.
5. Viet Nam duoc hoc (*Yue Nan yao xue*) (Vietnam Pharmaceutics) by Pho Duc Thanh, the Medical Association of Vinh.
6. Y hoc tung thu (Medical Collection) (*Yi xue cong shu*) by Nguyen An Nhan, Hanoi.

Supervising organizations were established for national medicine in July, 1957: the Department of Eastern Medicine, the Institute of Eastern Medicine (now the Department of National Medicine, the Institute of National Medicine) supervise the field as part of national health care. The Association of Eastern Medicine in Vietnam (now the Association of Traditional National Vietnamese Medicine), established in 1957, has gained by now around 230,000 members by means of a system extending from the metropole to the provinces, cities, and communes, assembling and supervising traditional doctors in 3879 national medicinal quarters.

Throughout the cities and towns, there are clinics of national medicine, and in all the general hospitals there are departments of national medicine offering treatment alongside modern medicine. The Institute of Pharmaceuticals was founded in 1961 in order to research and develop guidelines for the pharmaceutical profession throughout the country. The Institute of National Medicine in Ho Chi Minh City founded in 1976, and the Institute of Acupuncture, founded in 1982, have attested to the development of the profession of national medicine as part of popular health care. In the army hospitals and army clinics there are Departments of Eastern Medicine to provide treatment by national medicine. And in the metropole there is also the National Medicine Hospital to conduct research and provide treatment for serious diseases with national medicine. The study of national medicine is taught as part of the curriculum in medical schools, pharmaceutical schools, and in the civilian as well as the army medical services. Many written works on the heritage of national medicine have contributed to the development of the profession and the well-being of the people, some of which are mentioned below:

1. *The Life and Work in Medicine* of Hai Thuong Lan Ong. Hanoi: Medical and Sport Publishing House, 1970. 398 p., 19 cm
2. *Tue Tinh and the Foundation of Vietnamese Traditional Medicine.* Hanoi: Medical Publishing House, 1975. 403 p., 19 cm
3. *Chu Van An and Essential Interpretations of Medicine.* Hai Hung: The Hai Hung Medicine Association, 1976. 144 p., 19 cm
4. *The Acupuncture Tradition of Nguyen Dai Nang.* Hai Hung: The Hai Hung Medicine Association, 1979. 215 p., 19 cm
5. *Nguyen Dinh Chieu with the Exchanges between the Fisherman and the Firewood Cutter.* Hanoi: Medical Publishing House, 1983. 163 p., 19 cm
6. *Hai Thuong y tong tam linh* 海上懶翁醫宗心領 (Hai Shang lan weng Yi zong xin ling) by Lan Ong (12 Tomes). Hanoi: Medical Publishing House, 1993. 578 p., 27 cm
7. *Ve Sinh Yeu quyet.* 衛生要訣 (Wei sheng yao jue) and *Nu cong thang lam by Lan Ong.* Hanoi: Medical Publishing House, 1996. 184 p., 19 cm
8. *Nam duoc than hieu* 南藥神效 (*Nan yao shen xiao*) by Tue Tinh. Hanoi: Medical Publishing House, 1993. 421 p., 27 cm
9. *Hong nghia giac tu y thu* 洪義覺斯醫書 (*Hong Yi jue si yi shu*) by Tue Tinh. Hanoi: Medical Publishing House, 1978. 319 p., 24 cm
10. *Tue Tinh toan tap* (*Tue Tinh collected writings*). Hanoi: Health Ministry, 1998. 506 p., 27 cm

11. *Cham cuu liep hieu dien ca by Nguyen Dai Nang*. Hanoi: Medical Publishing House, 1981. 203 p., 19 cm

12. *Hoat nhan toat yeu tang bo* 活人撮要增補 (Huo ren cuo yao zeng bu) by Hoang Don Hoa. Hanoi: Medical Publishing House, 1980. 75 p., 19 cm

13. *Case studies of Prescriptions* by Tran Ngo Thien. Hanoi: Medical Publishing House, 1986. 106 p., 20 cm

14. *Dictionary of Medicinal Cases in Vietnam* by the Ministry of Health. Hanoi: Medical Publishing House, 1977. 946 p., 27 cm

15. *Methods of Sanitoriums* by Nguyen Van Huong. Hanoi: Medical Publishing House, 1982. 223 p., 19 cm

16. *450 medicinal herbs* by Pho Duc Thanh

17. *Eastern pharmaceutical making* by the Institute of Eastern Medicine.

18. *Acupuncture the Southern Way* by the Institute of Eastern Medicine. Ha Bac: The Institute of Ha Bac Medicine, 1977. 95 p., 27 cm

19. *The Study of Acupuncture* by the Institute of National Medicine in Hanoi (2 Tomes). Hanoi: Medical Publishing House, 1984. 501 p., 27 cm

20. *Handbook of Herbs* by Do Huy Bich, Bui Xuan Chuong. Hanoi: Medical Publishing House, 1980. 563 p., 19 cm

21. *Herbal Medicines and Vietnamese Remedies* by Do Tuat Loi. Hanoi: Science & Technique Publishing House, 1977. 1184 p., 24 cm

22. *Planting, Cultivating, and Using Herbal Medicines* by Le Tran Duc (3 Tomes). Tome I: Hanoi: Agriculture Publishing House, 1983. 276 p., 19 cm; Tome II: Hanoi: Agriculture Publishing House, 1986. 466 p., 19 cm; Tome III: Hanoi: Agriculture Publishing House, 1987. 309 p., 19 cm

Ancient Books and Manuscripts on Traditional Medicine and Pharmaceutics and Introduction of Research on Vietnamese Traditional Medications

In the treasury of ancient books and manuscripts on science and technology in Vietnam, the materials concerning traditional medicine and pharmaceutics are the most numerous and valuable. To date we have counted 373 books on traditional medicine and pharmaceutics along with a great number of copies that are presently stored in libraries both in the country and abroad.

The major works include *Nam duoc than hieu* 南藥神效 (*Nan yao shen xiao*), *Hong Nghia giac tu y thu* 洪義覺斯醫書 (*Hong yi jue si yi shu*) by

Tue Tinh and *Hai Thuong Y tong tam linh* 海上懶翁醫宗心領 (*Hai Shang Lan Weng yi zong xin ling*) by Hai Thuong Lan Ong Le Huu Trac.

The documents concerning ancient Vietnamese medicinal studies mentioned above are a precious collection on theories of medicine and pharmaceutics which have been summed up on the basis of the country's practical circumstances, with rich experience alongside influences and exchanges with Chinese medicine.

It is a pity that with the exception of a few major works that have been introduced, edited, researched and applied, the bulk of the extant works on traditional medicine and pharmaceutics still remain unexplored in libraries and archives. The Institute of Han-Nom Studies has around 400 books on traditional Vietnamese medicine and pharmaceutics, which makes it the largest collection in Vietnam, but there are no specialists in this area to make proper use of this collection. We think that with the several hundred books and several tens of thousands of manuscript pages, immersion in the work on traditional medicine and pharmaceutics in Vietnam is a worthy investment of time, resources, and expertise.

At this conference today, we enthusiastically invite all the scientists and researchers in the East Asian region to strengthen our mutual concerns in the work of improving research on medicinal studies. We believe that all of us here today will have much to say to each other because we all live in one geographical, climatic, cultural region, and more importantly, we all have common traditions of medicinal practice.

References

An Outline of Traditional Medicine History of Vietnam/Le Tran Duc. *Review of History Studies*, No. 3, 1990.

The Study of Sciences and Technology in Vietnamese History. Hanoi: Social Sciences Publishing House, 1979.

Vietnamese Culture Identity. Phan Ngoc, Hanoi: Information & Culture Publishing House, 1998.

Tran Nghia and Francois Gros. *Di san Han-Nom thu muc de yeu — Catalogue des livres en Han-Nom*. Hanoi: Social Sciences Publishing House, 1993 (3 vols.).

Two Medicals Writings in the 15th Century/Lam Giang. *Han-Nom Review*, No. 2 (27), 1996.

Books on Accupuncture in the Collections of the Vietnamese Traditional Medicine/ Lam Giang. *Han-Nom Review*, No. 2 (31), 1997.

Medical Works by Nguyen Gia Phan/Lam Giang. *Han-Nom Review*, No. 1 (34), 1998.

APPENDIX 1

(Extracted from *Knowledge of Vietnam in the past through 700 pictures*)

359- Mother curing her child's hicough by sticking the end of a betel leaf on his forehead - H. Oger - Dán lá trầu chūa nấc cho con.

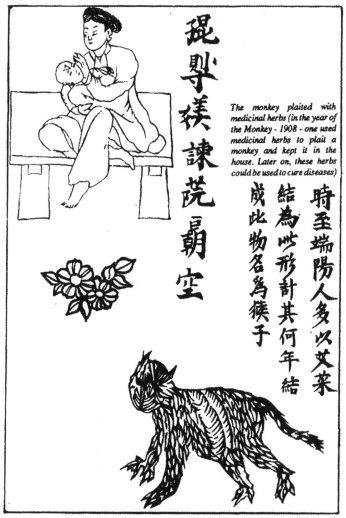

The monkey plaited with medicinal herbs (in the year of the Monkey - 1908 - one used medicinal herbs to plait a monkey and kept it in the house. Later on, these herbs could be used to cure diseases)

360- Búa lá cây làm thuốc - H. Oger Origin Ng-M-Hung : 359 - 360.

APPENDIX 2

(Extracted from *Knowledge of Vietnam in the past through 700 pictures*)

353- Physician prepares Chinese medicinal herbs - Dumoutier
Ông lang làm thuốc

354- Sino-Vietnamese Pharmacy - Dumoutier, Hàng thuốc bắc.

APPENDIX 3
(Extracted from *Knowledge of Vietnam in the past through 700 pictures*)

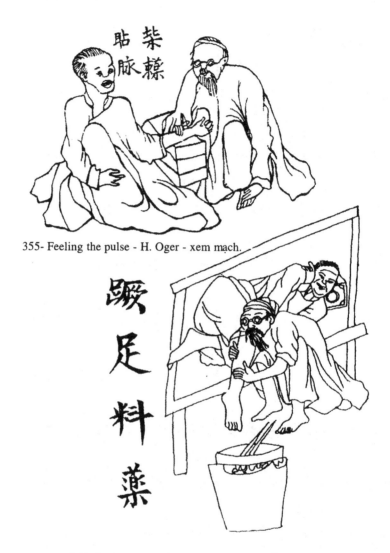

355- Feeling the pulse - H. Oger - xem mạch.

356- Setting a bone in place - H. Oger - nắn xuong sai khóp

APPENDIX 3   *(cont'd)*

357- chaffing one's body with saffron
after giving birth. - H. Oger - 1905 -
Xoa nghệ sau khi dê

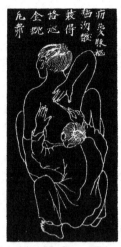

358- Keratitis - H. Oger - Tri mông mắt

APPENDIX 4
(Extracted from *Anecdote of Vietnamese Confucians*, p. 203)

LÊ HỮU TRÁC
(Hải Thượng Lãn Ông)
(1720 ? - 1791)

PART IV

Western Scientific Influence

Westernization from Different Angles: Review of the Historiography of Science from the Viewpoint of Colonial Science

TSUKAHARA TOGO

Kobe University, Japan

In this paper, I would like to examine the formation of Western science from different angles, and hope to provide some methodological remarks on the historiography of post-colonial science.

In conventional historiography, Western science in non-Western cultures is described as a process of introduction, namely, a process of Westernization. However, as a result of more recent scholarship, those conceptions which have entered history through modernization theory, such as the linear development of the sciences and the simple accumulative nature of scientific knowledge have been fundamentally questioned. We come to realize that modernity was not, is not, and probably never will be what everyone sought, seeks, and will seek. Those who live in modern societies have already become used to technological and scientific progress, but at the same time, have been forced to appreciate that such progress is not a simple business, and that techno-scientific innovation brings its problems.

Review and criticism of scientific developments have also affected our understandings of the history of Western sciences in non-Western cultures. To understand the introduction of science as a part of a modernization process is too one-sided; we become more aware of the different aspects of rejection and reception in the process of adaptation. Much attention is now being paid to recipient strategies of acceptance of the Western sciences in non-Western cultures, as well as the negotiation of knowledge-trading between the two or more parties involved in knowledge transfer.

When technology is involved, these argument is easier to exemplify. When we look at military artifacts brought by Western imperialists in the modern period, the issue of modernization is clearly not simple. Here is a history of war and dominance, power and enforcement. Such objects as steam boats and guns are easily classed as "Tools of Empires" (Headrick), and the introduction of military technology is never seen as without serious consequences for the recipient. There always have been, and still are, many war-mongers, so-called "merchants of death", and the transfer of military technology and militarization were sometimes forced on recipients against their wills.

But when "science" is involved in this story, such "dirty" aspects of history are suddenly changed into a "bright" story of modernization and enlightenment. There is still a prevailed view of science as universal and objective truth, and it goes beyond the boundaries of nations, race, ethnicity, and cultures. Post-Kuhnian historiography of science is opposed to such an understanding. Science is culture-bound, and the social background of science is not only a factor "behind" it, but a crucial and necessary condition. Recent sociological discourse about science is more radically revealing the socially embedded nature of science, and claims have been made for science's social construction.

Needless to say, when science is transferred from one culture to another, socio-cultural factors, including economic and political conditions, are more than important. To name a few, the prohibition of exchange or foreign trade by the Edo government in 16th- to 18th-century Japan — the so-called Sakoku (seclusion) policy — gave the Japanese development of science and technology its particular shape. The exceptional Dutch traders played a very important role that they had originally not anticipated. Even the small amount of exchange through Japanese and Dutch colonial traders resulted in the introduction of Western knowledge into feudal Japan, and its consequence was to give the Japanese a full preparation for the later acceptance of Western technology in the Meiji era. The warrior Samurai class culture of Japan was also regarded to be optimal for the acceptance of then-advanced military technology, and at the same time, some of the relatively enlightened Japanese Confucian schools and practical-minded medical doctors were a cultural force ready to accept Western curiosities. The entangled background of the Japanese introduction of science could be the most suitable case to exemplify the socio-cultural nature of the transfer of scientific knowledge.

Regarding such cultural problems of scientific transfer, I would also like to remark on methodological issues.

This problem, as approached in the framework of the history of science, and the history of transfer of knowledge in general, can be described generally as and A → B problem (Figure 1). This is the most simplified model of our problematique. This A to B might be growth within a collective itself, or the transformation or change of an original A to a different B. When we regard this as a model of time, it can represent historical development, while if this as a model of geography, it is seen as scientific transfer. Our interests are to inquire how and why A becomes B, and to decide where it occurred and when

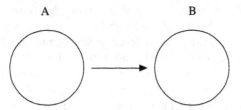

Time scale: Historical Development
Geographic scale: Scientific Transfer

Fig. 1.

it is realized. We also seek out the mechanism of change: who did it and the functions and effects of their work.

Thomas Kuhn's paradigm can be also understood using this model (Figure 2). Suppose A is representing Newtonian mechanics and B is Quantum mechanics. Incommensurability is Kuhn's concept that each component of A and B, such as time, length, and weight have different meanings and they

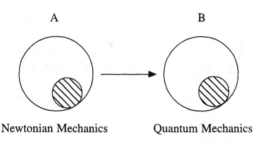

Newtonian Mechanics   Quantum Mechanics

Fig. 2.

are not common to each other, because different meanings and locations are given to each form of knowledge.

If we convert this model to geography, it may possibly brings us other insights. Take as an example the transfer of knowledge from the West to non-Western cultures, and assume A as Western culture and B as a culture that is non-Western. What would the place and meaning of science be? I would like to discuss this problem from a methodological viewpoint, with the key conception being not only the Western and non-Western but colonial science. My argument is that we can distinguish four different patterns in the historiography of the transfer of Western science into non-Western cultures.

Given such a scheme, did scientific knowledge flow from A to B or B to A, and should one pay attention to the A side of affair or the B side of the endeavor? Is it also possible to describe the interaction of A and B? This problem is a matter of the positioning of the historian.

Four patterns of how the transfer of scientific knowledge is commonly viewed are presented using an A-B model in Figures 4, 5, 6 and 7.

Figure 4 shows the first pattern in historiography. This first pattern pays most attention to the Western side (A) of the affair of the flow of knowledge. Such authors as L. Pyenson, D. Headrick and W. van der Schoor take this position. What they discuss are Western scientists' activities in non-Western settings. This position has its weak point in that science transfer is seen from the dominant side; and non-Western recipients of science are more or less ignored. When interests are included in the aspect of the colonialists'scientific activities, either they are exploitative or enlightenment ideals, such as the "civilizing mission". Some of these studies are intended to be critical of colonialism, and describe the colonialists' limitations and hypocrisy.

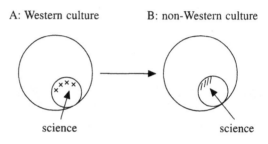

Fig. 3.

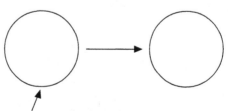

Discussion focus: colonial science by the dominant, discussed by Pyenson, Headrick and Schoor

Fig. 4.

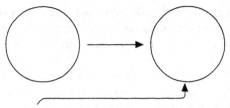

Discussion focus: colonial science by the subordinates, discussed by Kumar, Krishna, Sangwan *et al.*

Fig. 5.

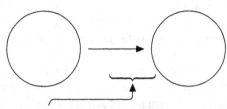

Discussion focus: Independent initiative to acquire scientific curiosities, discussed by the cases of Turkey by Ihsanoglu, Japan by Nakayama, Yoshida *et al.*

Fig. 6.

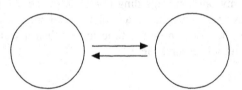

Discussion focus: Interrelation, interaction and encounter.

Fig. 7.

Research from the opposite view point is characterized by the excellent work of our Indian colleagues, D. Kumar, V. V. Krishna and S. Sangwan. Their history of science in India is firmly based on indigenous Indian scientific activities. Colonial science is described from the subordinated side, which had been often neglected and underestimated. The merits of this approach are numerous, the most important one being that such a view point revealed that underestimation of the indigenous scientific capacity was not due to ignorance or simple neglect, but to a systematically manipulated political discourse and ideological bias inherent in colonialism. Science was the intellectual symbol of colonial dominance, and at the same time, it was the colonialists' most valuable tool for dominance.

A slightly different pattern is described in Figure 3. It is often practised by historians in places, different from India, that were not colonialized, and mostly focuses upon independent initiative in inquiring and adopting Western science to their respective cultures. Examples are Turkey's (the Ottoman Empire's) acquisition of Western science as described by E. Ihsanoglu and the Japanese Rangaku case as discussed by S. Nakayama, T. Yoshida and other researchers. Unlike the case of Indian science, however, relative political freedom allowed the indigenous intellectuals to pursue then somewhat exotic Western scientific knowledge. Science in these cases is viewed not as a politically manipulated tool of dominance, but primarily as an object of curiosity in an early stage of interaction. Later, as the direct confrontation with Imperialism began, the meaning of science would change according to the particular historical situation.

Finally, the fourth pattern shown in Figure 7 represents scientific exchange as a mutual resonance and interaction between colonial science and indigenous reaction. In the history of science, some such cases of cultural encounter are fully discussed in depth, such as the Jesuit missionaries in China, and mutual stimuli given by Arabic mathematics and Medieval Indian mathematics. In my opinion, regarding our topic of the colonial encounter with modern Western science in the age of Imperialism, such a perspective has yet to be reached. It is now the time to establish a wider and deeper historical vision in our research field.

The College of São Paulo in Macao: A Background (16th and 17th Centuries)

LUIS SARAIVA
Center of Mathematics and Fundamental Applications (CMAF)
University of Lisbon
HENRIQUE LEITÃO
Condensed Matter Physics Center (CFMC)
University of Lisbon

The Educational Reform of King D. João III

In 1521 D. João III was crowned King of Portugal. From very early in his reign he aimed to reform the Portuguese education system.[1] One of his first moves was to establish an elite of scholars who would help him carry out the actual reform. The University of Paris had always been the preferred university for Portuguese students studying abroad,[2] and with D. João III the number of such students increased dramatically. The King talked to Diogo de Gouveia, a respected Portuguese scholar in charge of the Sainte-Barbe College in Paris since 1520, and a plan was drawn up to give scholarships to Portuguese students to study grammar and arts in Gouveia's College. Their studies there were to be followed by theology studies at the Sorbonne.

More than 50 students were in this way sent to Paris, with very substantial scholarships. Among them were André de Gouveia, nephew of Diogo de Gouveia, João da Costa and Diogo Teive. These three students entered the Sainte-Barbe College in 1527.

The King also had the idea of moving the University from Lisbon to Coimbra. The motives for this are still unclear. Although it has been said that it happened because the University teachers in Lisbon did not openly support the King (they had not been represented at his coronation and had not appointed him as Protector of Studies, as was then the custom), this seems unlikely, as the choice was given to the Lisbon Professors of either moving to Coimbra or being appointed to alternative well-paid jobs, or even retiring on a generous salary.

The establishment of the University in Coimbra was the best-known decision in a whole program of educational reforms aimed at updating and improving the medieval type of learning that then characterized Portuguese institutions. Other major points of this reform were the explicit separation of elementary and advanced studies, making the former much more complete and self-sufficient; and the change from the medieval trivium/quadrivium curriculum to a curriculum of humanities, arts and higher studies.

The elementary (or preliminary) level consisted of four years of humanities, to be followed by three years of arts. The humanities curriculum consisted of languages (Latin, Greek, and Hebrew) and literature (poetics and rhetoric). This corresponded to the trivium, but was more comprehensive, as in the latter only Latin was studied, and at a more elementary level than was intended in the King's reforms. Here the student not only learned its grammar and could read and write, but was also supposed to go through the literary analysis of a vast selection of Latin texts. In arts the student had two years of Aristotelian logic, plus one year of philosophy, which included metaphysics, natural philosophy (with physics and some astronomy) and moral philosophy (ethics and politics). There were no significant changes in the advanced level, where the courses of theology, law, and medicine were maintained.

As with any educational reform, its final outcome is difficult to assess, but it was certainly not as successful as its promoters had anticipated. The King was mainly counting on the collaboration of the people he sent to study in Paris, but their response was very disappointing: only two of the scholars who graduated accepted the King's request to teach in Portugal: the others were either not interested or thought that the salary offered was insufficient. Therefore the teaching staff suffered from serious shortcomings. Bad planning also hindered the efficiency of the reform. The buildings where the classes were to be held were insufficient for the number of students, last minute changes being needed in order to accommodate all of them. There was also bad planning of the courses themselves, as some subjects started with considerable delays in their originally scheduled times. On the positive side was the student response to the reform: they turned up in great numbers. By the beginning of the 1540/41 academic year there were more than 600 students attending the courses.

The King then tried a different approach to have his reform successfully established: as he saw that he could not count on the collaboration of his former bursars in Paris, he appointed one scholar whom he trusted to choose a team of foreign scholars to teach at a school in Coimbra organized in a similar way to the College de Sainte-Barbe, which would centralize preliminary education in Portugal. In 1542 he contacted André de Gouveia,[3] who had been Principal of the Guyènne College in Bordeaux since 1534. Five years later André de Gouveia and a team of ten teachers chosen by him arrived in Coimbra to teach at Colégio das Artes. They were to teach two main courses: "Latinity" (*Latinidade*) was the most important one, with ten classes, from the most elementary level of reading and writing in the tenth class, to rhetoric through the writings of Latin writers.[4] Greek was taught to students in the last five classes, and mathematics — Euclid's Elements and the Treatise on the Sphere — was taught in the last two. Hebrew was supposed to be taught also, but at a very elementary level (grammar only) and it is not certain in which class. This course could take ten years or less, depending on the ability of the students. The other main course was philosophy, a four-year course. Logic was taught in the first two years (including some mathematics) and physics, ethics and metaphysics in the final two.

As the Regulations of the Colégio das Artes established that the teaching staff had to be sixteen, another six were appointed. They were known as *os parisienses*, the Parisians, as opposed to those from Bordeaux, *os bordalenses*. The latter group ruled the College, deciding on the content of the subjects and on who would teach what. There was a tension between the two groups, which was made more intense by religious differences: while the *bordalenses* had a much broader view of religion (indeed some of them shared views with the Lutherans), the *parisienses* were a much more orthodox Catholic group. André de Gouveia, using his enormous prestige in intellectual circles, was to be a moderator between these two groups, but this was shortlived, as he died in June of 1548. The conflict between the two groups increased, and only ended when the *bordaleses* were excluded from the Colégio das Artes.[5] In 1552 the new Regulations of the College were published, containing instructions to the teachers on which books should be read. All were based on Aristotelian doctrine. So there ended, again short of expectations, another stage in the King's efforts to establish a successful reform of the education system.

After the successive failures of the attempted collaboration with Portuguese scholars trained in France, and with the team of foreign scholars chosen by André de Gouveia, the way was open for the Society of Jesus to enter the scene. Ignacio de Loyola, the founder of the order, had studied at the Sainte-Barbe College in the period 1528–33, and in 1538 Diogo de Gouveia had written to the King saying how useful the Society could be to the Portuguese in order to spread Catholicism in the East. In 1539 the King wrote to the Portuguese ambassador at the Vatican, asking him to gather information on the Society, and to contact it in order for the Society to contact him. In 1540 Loyola sent two missionaries to the mission in India. One of them, Simão Rodrigues, stayed in Portugal, while the other, Francisco Xavier, went to India. In 1541 the monastery of Santo Antão in Lisbon was given to the Jesuits, the first house in the world to be owned by them.[6] In 1542 a new Jesuit College was established: the Colégio de Jesus in Coimbra,[7] which at first was only a residential house, but from 1547 onwards it was also a private Jesuit College.

Another initiative by King D. João III in order to increase knowledge and improve teaching in Portugal can be seen in his protection of Pedro Nunes (1502–1578) and the creation of a nautical class, the Aula de Náutica. The King recognized the exceptional value of Pedro Nunes as a mathematician, so from very early on he protected him, giving him the best possible working conditions.[8] In 1547 Nunes was appointed Cosmographer-in-Chief of the Kingdom, a position he held until his death. His duties were to examine the pilots' knowledge and check the accuracy of nautical instruments and maps. He also taught on the Aula de Náutica, an essentially practical course for pilots and ship crews in which they were taught how to use nautical instruments and maps for navigation purposes. This course was taught at the *Armazéns da Guiné e da Índia* (Guinea and India Warehouses), which was the body in charge of the ships for Royal service, including maintenance (gunnery, food, etc.) and crews. They also stored the sea maps and had cartographers working for them. Unfortunately little is known about the activities there because their archives were destroyed in the 1755 Lisbon earthquake. There is no surviving copy of the 1559 Cosmographer-in-Chief Regulations,[9] but according to A. Teixeira da Mota[10] this was not significantly different from the 1592 Regulations.[11] The 1592 Regulations described the contents of the Mathematics course as follows:

1. Elementary astronomy needed for navigation;
2. Construction and use of sea maps;
3. Rules for the knowledge of moon phases and tides;
4. Use of the nautical astrolabe for sun measurements; computation of latitudes;
5. Use of the cross-staff and the quadrant for nocturnal measurement of the Pole Star; and
6. Computation of needle variations, use of the sundial.

However, from the records of the pilots' examinations, we can appreciate that their training was always very basic. Also, if there were above average students, the cosmographer could teach a sphere course and present more advanced instruments. It is also explicitly stated that the cosmographer should have in the class all the instruments required to make his lessons clear.

The Beginning of the Jesuit Education Network

As is well known, the Society of Jesus had enormous success with the opening of public schools: in Gandia, in Spain, in 1546; in Messina, Italy, in 1547; and in Rome, at the Collegio Romano, in 1551. The Society soon adapted to this new situation, and from then onwards teaching became an essential component of its work.

In 1551 Loyola wrote to Simão Rodrigues, advising him to start a public College in Portugal.[12] Two years later, Jerome Nadal, founder and rector of the Messina College, came to Portugal to assist the Portuguese Jesuits in starting their own public schools. In the same year two Jesuit Colleges were set up: the Colégio de Santo Antão in Lisbon, and the Colégio do Espírito Santo in Évora. They were unable to establish a Public College in Coimbra, however, because secondary teaching in the country's second city was the monopoly of the Colégio das Artes. The Jesuits used their influence to have the King give them control of the College. They managed this in 1555, and Colégio das Artes became a Jesuit College. At the beginning things were difficult for the Society, as they did not have enough competent teachers to maintain the high standards previously set by André de Gouveia. Therefore initially there was a worrying drop in teaching standards in the Colégio das Artes.

Against the will of most of the Lisbon Council, who wanted the Santo Antão College to admit only the sons of the nobility, the Society established that the College was open to everyone, with no restrictions concerning the social status of its students. In six months, the number of students rose to over 330. The success of this school persuaded parents to take their children out of other schools and enrol them in Santo Antão.

In 1559, after obtaining authorization from the Pope, a University was established in Évora. It had four faculties: humanities, arts (i.e., philosophy), scholastic theology, and moral theology. From its foundation the University depended solely on the Society of Jesus.

The spread of Jesuit schools continued: Colégio de S. Paulo in Braga and Colégio de S. Lourenço in Oporto, both in 1560, Colégio de Jesus in Bragança in 1561, Colégio de S. Manços in Évora in 1563. Recognizing that something should be done in order to maintain a minimum standard of teaching in their Colleges,[13] the Second General Congregation, held in 1565, decided to establish training colleges for teachers.

The study plans in the Jesuit Colleges had a sequence of three courses, to be held in such a way that each provided the knowledge needed for the next one. The first corresponded to "Latinity" in André de Gouveia's plan for the Colégio das Artes, and was called *curso de letras*. Although not explicitly stated, it most probably lasted ten years, and included grammar[14] classes at three levels (lower, medium, and higher), where simultaneously Greek was taught; humanities classes, which used as textbooks anthologies of Latin authors; and rhetoric.[15] All texts used in the courses were previously carefully censored and edited by the Society. The second course was arts, and lasted three years and seven months. It included dialectics, logic, physics, and metaphysics (mostly Aristotle's *Works*). In the second year some elementary mathematics was taught: arithmetic, geometry, and perspective. In the third year students would learn Sacrobosco's *Sphere*. Usually the teacher would dictate the lesson. The Society thought that this method could lead to incorrect interpretation of the written works studied, so it recommended that textbooks be written. This is the origin of the celebrated *Commentarii Conimbricensis Societatis Jesu*, published between 1592 and 1598.[16] Finally there was a four-year Theology course, which included scholastic theology, scriptures and Hebrew. For those students who were considered able to become theology teachers, there were a further two years of studies.

The Jesuit public Colleges were a success in terms of numbers of students: in Colégio de Santo Antão there were 500 students in 1553, 1,300 in 1575, 2,000 in 1588, and 1,800 in 1598; in the Colégio das Artes there were 1,000 students in 1558, 1,500 in 1578, and 1,200 in 1598; in Évora (Colégio do Espírito Santo and University of Évora) there were 600 in 1559, 1,000 in 1571 and 1,600 in 1592; in the Colégio de S. Paulo in Braga there were 200 students in 1560, 1,000 in 1585, and 1,200 in 1591; that is, by the end of the 16th century, these five schools had about 6,000 students.[17]

The Society was well aware of the social importance of education, and so it tried to provide a minimum standard of quality in its Colleges. However, the high number of students is not only due to the quality of the teaching. It is also a social fact connected to the excitement of the era of the discoveries; the notion that many important things were being discovered sparked the need to know more. With the Jesuit Public Schools it was a two-way relation: the general public saw in them the possibility of getting an education which until then was not at all easily accessible; the Society of Jesus saw in them a means to reach an enormous audience.

To complete this general overview of the Portuguese education system, it must be said that the relations between the Society of Jesus and the University of Coimbra were always tense. The University felt threatened not only by the constant growth of the Society's education network, and by what was seen by the academics as an aggressive policy followed by the Jesuits, but also by the constant increase in its material resources (money, lands, buildings), which consequently meant an increase in its political power. The founding of the Jesuit University in Évora meant direct competition with Coimbra, and the royal decision to grant the Jesuits control of the Colégio das Artes was seen as a sign of a royal bias towards the Society of Jesus, which could only have worried the University even more. In addition to this the University could not accept that secondary level teachers were better known in Europe than the University teachers, a fact that was helped by the then well-established Jesuit knowledge network.

However, the Jesuits were never in charge of the University, and only in theology and mathematics were there isolated Jesuit teachers. In the case of the latter this was a direct consequence of the state of the mathematics course, which did not exist until the reform of the University in 1772. At the end of the 17th century, a time when relations between Portugal and Spain were tense, the Portuguese King felt that it was necessary to have a course in military engineering at the University. As there was no competent teacher of this subject in the University, King D. Pedro II, ignoring the conflict between the University and the Society of Jesus, asked the latter to provide a specialist to teach at the University. The Society appointed the Swiss Johann König, who taught in the University from 1682 to 1685.[18]

The Society of Jesus and Its Policy Towards the East: The Portuguese Assistancy

By the end of the 16th century, the Society of Jesus had five Assistancies: France, Germany, Italy, Portugal, and Spain.[19] Considered together with their overseas territories, Portugal and Spain were the largest. In the case of Portugal, this posed a problem, because, of all the Assistancies, the Portuguese had the smallest number of members, so it was impossible for the Mission to supply the Portuguese missions with Portuguese Jesuits only. So in Portuguese missions we have a mixture of Portuguese and non-Portuguese Jesuits.

To understand the background and standpoint of the Jesuits who went to Asia we must bear in mind the Society's policy towards the East, and in particular China. For the Jesuits, China would become the most important scientific region outside Europe. This was initially because Chinese scholars and rulers were interested in western science; that the Jesuits thought of the Chinese scientific tradition as one which would welcome scientific traditions different from its own; and also because they saw China as a very hierarchical and centralized society, so action at the top that would impress the elites could have beneficial short-term consequences for the Society. Jesuit scientific activities in China developed from initial personal relations, via massive translation projects, and finally to an institutional presence at the Court of the Emperor. Mastery of mathematics, and later of astronomy, was seen as the most effective way of impressing local elites. But one should bear in mind that notwithstanding the importance of scientific training, this was not the fundamental requirement for missionary work. Jesuits were chosen for the East and Far East missions not only according to their scientific competence, but also according to their political and practical skills, decision-making ability, and activity. This explains why good Jesuit mathematicians who had applied to leave Europe for the East were refused (Gregoire de Saint-Vincent, Riccioli, Cysat), and why other good Jesuit mathematicians were not appointed to the more important missions (J. Galli, C. Borri).

Portugal and Spain were the last Assistancies to organize the teaching of mathematics in their schools. At least from 1586 there were private mathematics lessons in the Colégio de Jesus, by Richard Gibbons, a pupil of Clavius; and from 1590 onwards in Colégio de Santo Antão, in the Aula da Esfera. Here the first teacher, João Delgado, was also a pupil of Clavius. Until the reform of the Portuguese University in 1772, this was in fact the only mathematics course taught in Portugal for very long spells, seeing that the University of Coimbra rarely had any mathematics in this period.

The influence of Clavius' Academy on Jesuit teaching of mathematics in the Santo Antão school was paramount.[20] The teachers were either his students or his students' students; studies were planned according to Clavius' Collegio Romano; and Clavius' commentaries and handbooks were the basic texts used, even contents of Portuguese origin being known from Clavius' texts, like the part on Optics from Pedro Nunes' *De Crepusculis*, or even his *Algebra*.

By comparing the programs of the courses taught at the Aula de Esfera in Santo Antão during the 17th century we can point out some general characteristics.[21] The courses always included a section on nautical science, although from a more theoretical perspective than the one we assume was taught at the Aula de Náutica (no notes from courses taught there are known today); geography and hydrography (the latter meaning here the description of the oceans and the islands and coasts at their boundaries), which were part of the early courses, disappeared later on; cosmography and construction and the use of the globe was always taught, and underwent no significant changes over time; astrology, from a dominant place on courses in the first half of the 17th century, became a minor subject later on, although it did not disappear from the list of subjects taught; by the 1640s, as there was an urgent need for army engineers, due to the wars with Spain, subjects needed for military architecture were introduced at the King's request.

We can thus see that the contents of the subjects taught did not depend only on the Jesuits; on the contrary, frequently (almost as a rule) they resulted directly from the King's demands and the needs of the State. As a general observation, one can say that Jesuit mathematical courses in Portugal were always highly influenced by factors external to the Society: up to 1640 there was a decided concentration on nautical matters, and in the second half of the 17th century there was a shift to military engineering and architecture. The lack of a skilled class of Portuguese mathematicians is certainly a factor in this lack of autonomy of the Jesuit mathematicians: faced with the almost total absence of mathematical studies in the university, they did not have to establish their intellectual credibility in the mathematical sciences, therefore they could teach whatever the state rulers wanted in this domain, and could duly take advantage of this situation.

The most important characteristic in the Mathematics course at Santo Antão, which did not exist in any other Jesuit College, was the inclusion of nautical science — essential for Portuguese navigation, and probably included in the course at the King's request. This is an important divergence from what is prescribed in the *Ratio Studiorum,* and meant that some subjects usually taught in Jesuit schools had to be excluded in Santo Antão. These included both pure higher mathematics and the mathematics needed for the new physical-mathematics sciences. This meant above all mathematical astronomy and theoretical mechanics. Consequently it is no wonder that when the Asia missions started to ask Rome for missionaries who were competent astronomers, they had to be non-Portuguese. Also, this request, and the need for good Portuguese astronomy experts, entailed the replacement of the Jesuit lecturers at Santo Antão, until 1615 predominantly Portuguese, by non-Portuguese teachers. This is why, if we follow Ugo Baldini (see note 19) and compute the number of Portuguese and non-Portuguese lecturers of the Aula da Esfera, we have a very different picture for the situation up to 1615 compared to after 1615. In the period 1590–1615, and considering only the years for which there are records of teaching staff, we have Portuguese teachers for 18 years: João Delgado (1590–97, 1598–99, and 1605–1608), Francisco da Costa (1591–92, 1597–99, both as Delgado's assistant, and 1602–1604), António Leitão (1597–98), Francisco Machado (1604–1605), Sebastião Dias (1610–1614), and only one non-Portuguese, who taught for three years: Christoph Grienberger, between 1599 and 1602.[22] After 1615 there are mainly non-Portuguese teachers on the Aula da Esfera. Between 1615 and 1640, the year Portugal regained its independence from Spain, for 23 of those 25 academic years there are non-Portuguese lecturers: Giovanni Paolo Lembo (1615–17), Johan Chrysostomus Gall (1620–27), Cristoforo Borri

(1627–28), Ignace Stafford (1630–36), and Simon Fallon (1638–40). There is only one Portuguese lecturer in this period: Dionísio Lopes, for two academic years (1617–18 and 1618–19), and this was because health problems forced Lembo to return to Italy in 1617. After 1640 this trend continues, with teachers like Heinrich Uwens, John Roston, and Valentim Stancel.[23]

To reiterate what has been written above, we can also note that most of these lecturers had specialized astronomical knowledge:[24] Lembo was mainly concerned with astronomical instruments; he built the Collegio Romano's first telescope in 1610. He was one of the four signatories of the Roman Jesuit mathematicians' answer to Cardinal Bellarmine's questions about Galileo's telescopic observations (the other signatories were Clavius, Grienberger, and Maelcote). Gall, while studying at Ingolstad, had very competent astronomy teachers: J. Lanz, C. Scheiner, J. B. Cysat, whom he assisted in astronomical observations. He was aware of the shortcomings of the Ptolemaic system, and of the impact of the new research and theories. His lessons show he accepted Galileo's observations and adopted Tycho Brahe's system. He was convinced that the celestial spheres did not exist, and taught this publicly. Borri taught mathematics in Mondovi (ca. 1607–1610), and then Philosophy, and perhaps mathematics in Milan (ca. 1611–14). In 1614 he was suspended from teaching for voicing a public attack on the solid spheres. General Acquaviva decided to send him to Goa. He returned to Europe in 1624 and was in Coimbra from the summer of 1626 to November 1627. In March 1628 he was already in Lisbon. He wrote *A Arte de Navegar*, the first part of his 1627–28 course, considered to be one of the best writings on nautical science produced in the Santo Antão College. Stafford studied from 1620 to 1625 in the English College in Valladolid. He published *Elementos matematicos*, a mathematics handbook written in Spanish, on Euclidean Geometry, and in 1633 wrote *Tratado da Fábrica e Usos do Globo Geográfico* (*On the construction of the geographic globe and on the problems that can be solved with its use*), presumably a part of his Santo Antão course. Fallon was an influential figure in the Portuguese technical and scientific community. It is thought that he studied at Colégio de Jesus in Coimbra in 1628, and from 1630 to 1633. When Portugal regained its independence in 1640, King D. João IV appointed him *engenheiro-mor do Reino*, that is, he became the general supervisor of all military constructions in Portugal.

Conclusions

Having briefly examined the main aspects of the establishment of the Jesuits in Portugal and the constitution of their learning centres, it is important to stress a few points.

Firstly, the fundamental aspect that differentiates the Jesuit scientific tradition in Portugal from the other regions in Europe is the almost complete absence of lay mathematical teaching. That is, in the final decades of the 16th century, and throughout the 17th, the Society of Jesus was the sole institution maintaining regular teaching of mathematical matters in Portugal.

Secondly, the political and geographical circumstances created by its enormous empire placed a tremendous demand on Portuguese society at all levels, and obviously also on education. It is not surprising, therefore, to observe that most of the technical and scientific teaching in Portugal in this period was devoted to nautical matters. This specific character of Portuguese technical education was bound to affect the Society of Jesus. Any Jesuit trained in a Portuguese College would therefore have a rather unusual training when compared to other Assistancies of the Society: he would certainly have a reasonable command of applied knowledge related to nautical matters; a thorough grounding in natural philosophy and cosmological matters of the type found in the sphere literature; and some practical experience of observational astronomy. But he most certainly would have only a very rudimentary knowledge of the more theoretical aspects of mathematics or astronomy. This training was sufficient for most of the needs in all the Missions of the Portuguese Assistancy, with the obvious exception of China. For the missionaries with scientific responsibilities in China a much more profound training in mathematics was required.

Thirdly, we can see that King D. João III supported the Jesuits not only because they had similar points of view with regard to the education reform the King wanted to initiate, with a very comprehensive preliminary level with a strong emphasis on Humanities, but also because of his experience of previous failures. The Society's efficiency, with its proverbially stern discipline, was thought of as a guarantee of success precisely where the Paris bursars and André de Gouveia's group had failed. This relation between the State and the Society of Jesus was a constant in this period, a two-way relation, clearly shown in the support the King gave the Jesuits in all their conflicts with Coimbra University, the inclusion in Jesuit courses of subjects needed for the country's policy (nautical science, military engineering) and the appointment of Santo Antão teachers to important state positions (Fallon as *engenheiro-mor do Reino* in 1640, etc.).

Fourthly, we see how the subject contents at the Aula da Esfera at Santo Antão, the most important mathematics course in Portugal during the 17th century (and the paramount influence of Clavius), determined the first mathematics brought to the East, with a strong emphasis on Euclidean geometry. It also determined, when competent astronomers were needed for the Asia missions, that they could not be chosen from among the students of the College, and in fact this need led to a significant change in the nationality of the teachers, who after 1615 were mainly non-Portuguese.

In a future paper we intend to analyze the significance and importance of the S. Paulo College in Macao in the 16th and 17th centuries, relating it to the background which is the main theme of this paper.

Notes

Luis Saraiva would like to thank Fundação Oriente, for sponsoring his participation in the 9th ICHSEA Conference. The research for this paper was supported by FCT, FEDER, PRAXIS XXI, and Project PRAXIS/2/2.1/MAT/125/94. Henrique Leitão would like to thank Fundação Oriente for financial support.

[1] For more information on King D. João III and his education policies, see José Sebastião da Silva Dias, *A Política Cultural da Época de D. João III*, Universidade de Coimbra, 1969.

[2] On Portuguese bursars in Paris, see Luís de Matos, *Les Portugais à l'Université de Paris, 1500–1550*, Universidade de Coimbra, 1951.

[3] Gouveia enjoyed an enormous reputation. Montaigne calls him "le plus grand et le plus noble principal de France" (in *Essais, De l'institution des enfants*).

[4] The books to be used in each class were mentioned by such authors as Cato, Cicero, Terence, Ovid, and Lucan. The grammar by Van Pauteren is also quoted.

[5] On this subject see Mário Brandão, *A Inquisição e os Professores do Colégio das Artes*, Universidade de Coimbra, 1948.

[6] In Rómulo de Carvalho, *História do Ensino em Portugal*, Fundação Calouste Gulbenkian, 1986, p. 287.

[7] There were 12 founders: Diogo Mirão, from Valencia; Francisco de Vilanueva and Francisco Rojas from Castille; Ponce Cogordan and François Onfroy from France; Angelo Paradisi, Isidoro Bellini, Martim Pezzano and Jacob Romano from Italy; Manuel Coutinho, Manuel Fernandes and António Cardoso, from Portugal.

[8] This is why Pedro Nunes taught very little at Coimbra University, although he was part of the teaching staff. Most of his classes were taught by his assistants, as he spent most of the time in Lisbon, working on nautical or cartographic problems. Hence his influence on the Portuguese University was non-existent.

[9] Pedro Nunes is presumed to be either the sole or the main writer of this Regulation, as he was living in Lisbon from 1557 to 1560.

[10] A. Teixeira da Mota — *Os regimentos do cosmógrafo-mor de 1559 e 1592 e as origens do ensino náutico em Portugal*, Memórias da Academia das Ciências de Lisboa (Classe de Ciências),13 (1969), pp. 227–91.

[11] Idem, pp. 272–73.

[12] On the Portuguese Assistancy of the Society of Jesus see Francisco Rodrigues, S. J., *História da Companhia de Jesus na Assistência de Portugal*, 7 volumes,

Apostolado da Imprensa, Porto, 1931–50, a comprehensive, but dated, work; see also Dauril Alden, *The Making of an Enterprise: the Society of Jesus in Portugal, its Empire, and Beyond: 1540–1750*, Stanford University Press, 1996.

[13] Diogo de Mirão, who was the first director of the Colégio Jesus in Coimbra, was one of the first to be aware of this situation: he had written a letter to Loyola in 1544 about the lack of competence of some of the teachers at his College, and said that in his opinion it would be useful to have in the College teachers from Coimbra University.

[14] Until 1572 the adopted textbook was a grammar by Van Pauteren, then from 1572 onwards, Manuel Álvares' *De Institutione Grammatica Libri Tres*.

[15] From 1562 onwards, the adopted textbook was Cipriano Suarez's *De Arte Rhetorica Libri Tres*.

[16] Its contributors included Manuel de Góis, Cosme de Magalhães, Baltazar Álvares, and Sebastião do Couto.

[17] Figures quoted from Francisco Rodrigues, S. J., *História da Companhia de Jesus na Assistência a Portugal*, II, vol. 2, Livraria Apostolado da Imprensa, Porto, 1938, p. 15, and Rómulo de Carvalho, *História do Ensino em Portugal*, Fundação Calouste Gulbenkian, Lisboa, 1986, p. 325. When the Jesuits were expelled from Portugal in 1759 their Colleges had, in a reasonable estimation, above 20,000 students. It also must be noted that, although the Society had control of the majority of the schools, it did not have a national monopoly; this only stood in some cities: Coimbra, Évora, Bragança, Angra.

[18] He then left the University to work for the King on the production of a map of Portugal, and was replaced in the University post by another Jesuit: Manuel de Amaral, who taught in Coimbra until 1690. He was followed by another Portuguese Jesuit, Francisco Barbosa, and then by a Bohemian Jesuit, Albert Buckowski.

[19] The basic references throughout this section are Ugo Baldini, *As assistências ibéricas da Companhia de Jesus e a actividade científica nas Missões Asiáticas (1578–1640). Alguns aspectos culturais e institucionais*, Revista Portuguesa de Filosofia, 54 (1998), pp. 195–245, and Ugo Baldini, *The Portuguese Assistancy of the Society of Jesus and Scientific Activities in its Asian Missions until 1640*, Proceedings of the Conference: "Portugal and the East", Fundação Oriente, Lisboa, 2000.

[20] This point, generally overlooked, has recently been stressed by Ugo Baldini's works on the scientific teaching in the Portuguese Assistancy. See Note 19.

[21] On this see Luis de Albuquerque, *A "Aula da Esfera" do Colégio de Santo Antão no século XVII*, Anais da Academia Portuguesa de História, 2ª série, vol. 21 (1972), pp. 337–91. Also in *Estudos de História*, vol. 2 (1974), pp. 127–200.

[22] Grienberger was the most important Jesuit mathematician who taught at Aula da Esfera. He had been a mathematics teacher at Vienna College before coming to the Collegio Romano, where he was one of the most important collaborators of Clavius. After Clavius' death, he became its director. He was the Society's main scientific expert in the period 1615–33. Compelling reasons must have existed for Grienberger to come to Lisbon, but unfortunately the ARSI correspondence for the relevant years is incomplete, so we do not know the reasons either for the Society sending him to Lisbon, or for his return to Rome.

[23] By the end of the 17th century the Society was finding it difficult to provide quality teachers to Santo Antão, as is shown by the case of J. Gelarte, who taught there for two spells, in the 1680s and 1690s. The fact that, despite gaining a reputation as a bad teacher during his first spell at Santo Antão, this did not prevent his recall in the 1690s, speaks for itself.

[24] Detailed biographical information can be found in Baldini's works referred to in Note 19.

A Jesuit Scientist and Ancient Chinese Heavenly Songs: Antoine Gaubil's Research and Translation of 步天歌 (Bu Tian Ge)

ISAIA IANNACCONE 儀賽雅
Istituto Universitario Orientale, Naples

Bu Tian Ge 步天歌

Bu Tian Ge 步天歌 (Songs of Pacing the Heavens) of ca. 590 is considered to be the work of Wang Ximing 王希明 (Tang 唐 Era). However, since Wang used the name Dan Yuanzi 丹元子, the book can sometimes be found cited as *Dan Yuanzi Bu Tian Ge* 丹元子步天歌. According to a different interpretation, the author actually is Dan Yuanzi, a Taoist hermit who lived in the Sui 隋 Era, whereas Wang Ximing is the person who widely distributed the text during the Tang Dynasty.[1] Whoever the author is, *Bu Tian Ge* can be dated to between the end of the 6th century and the first decades of the 7th.

Bu Tian Ge is important in the history of Chinese astronomy because it definitively codifies the subdivisions in the celestial skies into 3 *yuan* 垣 (enclosures or barriers[2]) and 28 essential *xiu* 宿 thus delineating the original theoretical organization of ancient Chinese astronomy. One must note that during these last years of the Sui Dynasty, there were several Indian astronomers who were active.[3] From the *Sui Shu* 隋書 (Book of the Sui Dynasty), which was completed in 636 by Wei Zheng 魏徵, one learns that Brahmin astronomical texts were also circulating, such as *Po Luo Men Tian Wen Jing* 婆羅門天文經 (Canon of Brahmin Astronomy), and *Po Luo Men Yin Yang Suan Li* 婆羅門陰陽算曆 (The Calculation of the Calendar According to Brahmin and *Yin-Yang* Methods). Although these texts have since been lost, but it could be these very texts which influenced or suggested how to reorganize and formalize the 28 *xiu*.[4]

From the literary point of view, *Bu Tian Ge* uses the poetic form to present the stars in the sky. The verses are composed of 7 syllables, and are in rhyme. The descriptions are simple and realistic, easy to memorize; these characteristics lead one to deduce that this text was destined for broad

299

distribution instead of being a true scientific document. One could quite plausibly conclude that between the Sui and the Tang Dynasties, the study of the sky was not only destined to an exclusive circle of experts, but that astronomy enjoyed wide popularity.[5] The illustrations that accompany the rhymes and describe the stars and constellations in detail further substantiate this conclusion.

The contents of the *Bu Tian Ge* is presented in the table of contents (*mu lu* 目錄) in the following order:[6]

(a) title: *Tian Wen Bu Tian Ge* 天文步天歌;

(b) the three Enclosures (*san yuan* 三垣): The Secret Purple Enclosure (Zi Wei Yuan 紫微垣)[7], the Enclosure of the Great Secrets (Tai Wei Yuan 太微垣), and the Enclosure of the Celestial Market (Tian Shi Yuan 天市垣);

(c) the seven *xiu* of the Eastern Region (Dong Fang *qi xiu* 東方七宿): Jiao *xiu* 角宿, Kang *xiu* 亢宿, Di *xiu* 氐宿, Fang *xiu* 房宿, Xin *xiu* 心宿, Wei *xiu* 尾宿, Ji *xiu* 箕宿;

(d) the seven *xiu* of the Northern Region (Bei Fang *qi xiu* 北方七宿): Dou *xiu* 斗宿, Niu *xiu* 牛宿, Nü *xiu* 女宿, Xu *xiu* 虛宿, Wei *xiu* 危宿, Shi *xiu* 室宿, Bi *xiu* 壁宿;

(e) the seven *xiu* of the Western Region (Xi Fang *qi xiu* 西方七宿): Kui *xiu* 奎宿, Lou *xiu* 婁宿, Wei *xiu* 胃宿, Mao *xiu* 昴宿, Bi *xiu* 畢宿, Shen *xiu* 參宿, Zui *xiu* 觜宿[8];

(f) the seven *xiu* of the Southern Region (Nan Fang *qi xiu* 南方七宿): Jing *xiu* 井宿, Gui *xiu* 鬼宿, Liu *xiu* 柳宿, Xing *xiu* 星宿, Zhang *xiu* 張宿, Yi *xiu* 翼宿, Zhen *xiu* 軫宿.

This is followed by the celestial map of the polar stars (Figure 1), and the description of the *yuan* and *xiu* according to the following sequence:

Zi Wei Yuan (from p. 2r to p. 3r);
Tai Wei Yuan (from 3v to 4v);
Tian Shi Yuan (from 5r to 5v);
Dong Fang *qi xiu* (from 5v to 8v);
Bei Fang *qi xiu qi* (from 8v to 12v);
Xi Fang *qi xiu* (from 13r to 15v);
Nan Fang *qi xiu* (from 16r to 18v).

The illustrations are distributed on the following pages (where not specified, one per page): 1v (cited above), 3v, 4v, 5v, 6r, 6v, 7r, 7v, 8r (2 illustr.), 8v, 9r, 10r, 10v, 11r, 11v, 12v (2 illustr.), 13r, 13v, 14r, 14v, 15r, 15v (2 illustr.) 16v, 17r (2 illustr.), 17v, 18r, 18v.

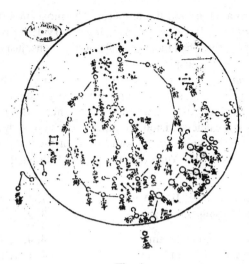

Fig. 1.

Aside from the literary aspect in *Bu Tian Ge*, there are also many interesting scientific aspects. Since the work is a catalogue of the stars, it allows us to understand the astronomical observations of the period. Furthermore, the richness of the illustrations allows one to verify the extent to which the representation of the sky was formalized. There is a third aspect, however, which is even more interesting: *Bu Tian Ge* is a very rich source for learning about Chinese *organicist* thought, and is testimony to how old its codification is. As is well known, the tradition of Chinese philosophic and scientific thought has its foundation rooted in what was defined as *organicism*.[9] From the organicist vision of the universe follows the conviction that the world of nature and that of human society are directly linked. Specifically, events on Earth are a reflection of those in the skies, and each disturbance in one affects the other and vice-versa. Thus, the main goal of Chinese science was to determine the links between natural phenomena and events on Earth; the study of the former was therefore always very accurate as opposed to the preconceived and ahistoric analysis of the latter. Knowledge of the dynamics and rhythms in nature was the greatest concern of the Chinese governing class. The Son of the Heaven was supposed to govern in harmony with the universe, otherwise he would no longer hold *the celestial mandate*. The abundance of astronomical phenomena made it necessary to choose astronomy as the science of the state, and the emperor was directly associated with the North Star: just as the celestial sphere rotated around it, the nation and the people rotated around the emperor. The most important political offices, military installations, rituals, legendary figures, provincial administration offices etc., had the names of constellations, just as the naming of imperial locations was inspired by the stars.

The *Bu Tian Ge* gives one the chance to verify relationships and facts: macrocosm and microcosm, sky and Earth, astronomy and politics, stars and humans are all tied together in the text highlighting the indispensable relationships in the organicist vision. Looking at the start of the first section, that is, the first nine columns (upper and lower) of the text (see Figure 2), the one dedicated to Zi Wei Yuan (the translation follows the quotation in *pinyin* 拼音 [10]):

(I)	Zhong Gong Bei Ji Zi Wei Gong	1, 1, 3, 2, 3, 2, 1
	Bei Ji wu xing zai qi zhong	3, 2, 3, 1, 4, 2, 1
(II)	Da Di zhi zuo di er Zhu	4, 4, 1, 4, 4, 4, 1
	di san zhi xing Shu Zi ju	4, 1, 1, 1, 4, 3, 1
(III)	di yi hao ri wei Tai Zi	4, 1, 4, 1, 2, 4, 3
	si wei Hou Gong wu Tian Shu	4, 2, 4, 1, 3, 1, 1
(IV)	zuo you si xing shi Si Fu	3, 4, 4, 1, 4, 4, 3
	Tian Yi Tai Yi dang Men lu	1, 1, 4, 1, 3, 2, 4
(V)	Zuo Shu You Shu jia Nan Men	3, 1, 4, 1, 2, 2, 2
	liang mian Ying Wei yi shi wu	3, 4, 2, 4, 1, 2, 3
(VI)	Shang Zai Shao Wei liang xiang dui	4, 3, 4, 4, 3, 1, 4
	Shao Zai Shang Fu ci Shao Fu	4, 3, 4, 3, 4, 4, 3
(VII)	Shang Wei Shao Wei ci Shang Cheng	4, 4, 4, 4, 4, 4, 2
	hou Men dong bian Da Zan fu	4, 2, 1, 1, 4, 4, 3
(VIII)	Men dong huan zuo yi Shao Cheng	2, 1, 4, 4, 1, 4, 2
	yi ci que xiang qian Men shu	3, 4, 4, 4, 2, 2, 3
(IX)	Yin De Men li liang huang ju	1, 2, 2, 3, 3, 2, 4
	Shang Shu yi ci qi wei wu	4, 1, 3, 4, 2, 4, 3

(I)   The Central House (where the) North Pole (is), is (surrounded by) the Secret Purple Enclosure.
The five stars of the North Pole are in the middle.

(II)   The Great Emperor's throne is the second pearl.
The third star contains the seat of the Office of the Crown Prince.[11]

(III) The first (star) is called The Crown Prince.
The fourth is The House of the Empress, and the fifth is The Heavenly Pivot.

(IV) To the left and the right there are four stars: The Four Advisors. The Heavenly Unity and The Great One block the entrance to the (Southern) Door.

(V) The Left Pivot and The Right Pivot stand at the Southern Door. The encampments of the Guards on both sides add up to fifteen.

(VI) The Prime Steward and the Vice-Commander are facing one other. The Vice-Steward and the First Advisor follow the Vice-Advisor.

(VII) The First Guard and the Additional Guard follow the Grand Coadjutor.
East of the Rear Door is the Office of the Grand Secretary of the Commission for the Administration of the Provinces.

(VIII) The Vice-Coadiutore was called east of the [Rear] Door. After that one has to go toward the front Door.

(IX) Inside the Door of the Virtue of Yin two yellow stars unite. The Ministers come in order in their five positions.

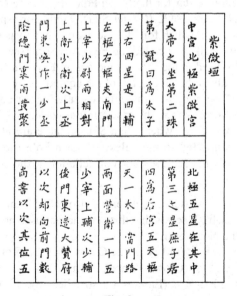

Fig. 2.

The protagonists in these lines of text are titles and places: emperor, prince, civil servants, soldiers, and locations in the Forbidden City. Each one fulfills a specific role that cannot and should not be changed. But these names and titles also belong to the stars and constellations as has been mentioned above; if the sovereign is associated with the North Star, the Forbidden City is the Central House, that is, the section of sky where the polar stars are lodged. The representatives of the imperial family and its court also have names of stars or constellations. In analogy, just as the Imperial City is protected by a long purple wall, the Central House is surrounded by the Secret Purple Barrier, a certain number of stars that crown the perimeter.

Proceeding with the identification of the stars associated with the official positions and with the imperial locations, an alphabetical list is shown below in relation to the passage cited in *Bu Tian Ge*. A note must be made that some of the stars are quite small and close to one another thus rendering their recognition uncertain.[12]

Bei Ji 北極:

North Pole. To identify Bei Ji, Figure 1 from *Bu Tian Ge* is used. At the center is the star Tian Shu (Heavenly Pivot) with the following caption: *Bei Ji yi ming Tian Shu* 北極亦名天樞: "Bei Ji is also called Tian Shu." Thus Tian Shu is the pointing star for the North Pole, that is, the North Star. Tian Shu, also called Niu Xing 紐星 (Center Star), would correspond to: 4339 or 32^2 H, or Σ 1694 Cam, which for Chinese astronomers was surrounded on three sides by four small stars called the Four Advisors (see Si Fu). This was the polar star also used in the Han 漢 and Jin 晉 Eras. Yu Xi 虞喜 (active from 307 to 338) valued the precession of the equinoxes by 1° every 71.75 years. Thus in the Tang and Sui Eras, after calculating for corrections, Tian Shu was still the closest star to the true celestial north.

In the area of sky where Tian Shu is, there are also four other stars, generally assimilated to the celestial North Pole (note that the Chinese astronomers did not see the same constellation as the one designated in the West as Umi):

(a) Da Di 大帝 (Grand Emperor), also called Tian Di 天帝: β Umi (Kochab). It seems that this was the star used around 1,000 B.C. as the North Star;

(b) Shu Zi 庶子 (Office of the Crown Prince): 5 or a 3233 Umi;

(a) Tai Zi 太子 (Crown Prince): γ Umi;

(b) Zheng Fei 正妃 (Empress), also called Hou 后 or Hou Gong 后宮 (House of the Empress): A A or b 3162 Umi.

Da Zan Fu 大贊府:

Office of the Grand Secretary for the Commission of the Administration of the Provinces: this star (or constellation) is not shown in the Figure 1.

Hou Gong 后宫:

See Bei Ji 北極.

Hou Men 後門:

Rear Door: it is not a star or constellation; in Figure 1 it is the area at the upper entrance of Dong Fan and Xi Fan (see Zi Wei Yuan) 紫微垣.

Nan Men 南門:

Southern Door: it is not a star or constellation; in Figure 1 it is the area at the lower entrance of Dong Fan and Xi Fan (see Zi Wei Yuan) 紫微垣.

Shang Cheng 上丞:

See Zi Wei Yuan 紫微垣.

Shang Fu 上輔 (two distinct stars):

See Zi Wei Yuan 紫微垣.

Shang Shu 尚書:

Minister (in Han time: master of writing, secretary): five stars southeast of Zi Wei; the stars are all in Dra.

Shang Wei 上衞 (two distinct stars):

See Zi Wei Yuan 紫微垣.

Shang Zai 上宰

See Zi Wei Yuan 紫微垣.

Shao Cheng 少丞:

See Zi Wei Yuan 紫微垣.

Shao Fu 少輔 (two distinct stars):

See Zi Wei Yuan 紫微垣.

Shao Wei 少衛 (two distinct stars):

See Zi Wei Yuan 紫微垣.

Shao Wei 少尉 (two distinct stars):

See Zi Wei Yuan 紫微垣.

Shao Zai 少宰:

See Zi Wei Yuan 紫微垣.

Shu Zi 庶子:

See Bei Ji 北極.

Si Fu 四輔:

The Four Advisors. In the Zhou tradition, it was the four advisors (or ministers) who served the emperor; they each had their own names and during audience they deployed themselves in a specific way in relation to the sovereign, and they each had precise functions: Ni 擬 in front of the emperor who was responsible for general questions; Zheng 丞 was behind and was responsible for administrative questions; Fu 輔 was to the left and was responsible for questions regarding civil matters; Bi 弼 was to the right of the emperor and was responsible for military questions. During the intermediary reign of Wang Mang 王莽 with his short-lived Xin 新 Dynasty (9–23 A.D.), four functions were officially instituted which were inspired from the Si Fu of the Zhou, but with different names: Tai Shi 太師 (Superior Teacher), Tai Fu 太傅 (Superior Master), Guo Shi 國師 (State Teacher), Guo Jiang 國將 (State General).[13] From the astronomical point of view, it has not yet been possible to identify the Four Advisors with four stars to a point of certainty; according to the most recent hypotheses, they could all belong to the Uma constellation, but it is difficult to identify them. Other interpretations hypothesize that these stars were: 32 H, 207 B, 223 B, and the Piazzi XIIIh 133 Cam; or 29 H and 30 H Cam, and 1 H Dra.

Tai Yi 太一:

The Great One, also called Tai Yi 太乙: 8 Dra; according to one legend, when the mythical Huang Di 黃帝 (Yellow Emperor) died after having reigned for 100 years, his spirit went to heaven: one part of him was transformed into the divinity Tai Yi, and the other part changed into some of the stars called Xuan Yuan 軒轅 (another name for Huang Di, which would indicate his birthplace in Henan 河南), and 17 stars in LMi and LMa which make up the shape of a dragon.

Tian Shu 天樞:

See Bei Ji 北極.

Tian Yi 天一:

Heavenly Unity, also called Tian Yi 天乙: 7 Dra.

Yin De 陰德:

Virtue of the Yin: two stars (not identified) perhaps from Uma, west of Shang Shu (see). It is the invisible power exerted by the *qi* 氣 vital spirit of Tai Yi 太一 (see) and Tian Yi 天一 (see). According to *Sui Zhi* 隋志 (Annals of the Sui), Yin De is formed of two stars: Yin De 陰德 and Yang De 忖德 (Virtue of the Yang); among these there is Yin De Men 陰德門 (Door of the Virtue of the Yin).

Yin De Men 陰德門:

See Yin De 陰德.

Ying Wei 營衛:

Encampments of the guards: it is another name that is given to the 15 stars that make up Zi Wei Yuan 紫微垣 (see); they represent the officials and the troops responsible for the frontier safety. Ideally, it is as if all the characters (or stars) that make up Zi Wei Yuan were protected by a group of soldiers.

You Shu 右樞:

See Zi Wei Yuan 紫微垣.

Zi Wei 紫微:

Purple House: the portion of the sky that is associated with the imperial palace; it is surrounded by the Secret Purple Enclosure (see Zi Wei Yuan 紫微垣).

Zi Wei Yuan 紫微垣:

The Secret Purple Enclosure. It encloses the Purple House (Zi Wei 紫微). It is composed of 15 stars with eight to the east (Dong Fan 東藩, Eastern Frontier) and seven to the west (Xi Fan 西藩, Western Frontier). In order, the ones that constitute Dong Fan are:

(1) Zuo Shu 左樞 (Left Pivot): ι Dra;
(2) Shang Zai 上宰 (First Steward): θ Dra;
(3) Shao Zai 少宰 (Vice Steward): η Dra;
(4) Shang Fu 上輔 (First Advisor): ξ Dra;
(5) Shao Fu 少輔 (Vice Advisor): φ Dra;
(6) Shang Wei 上衛 (First Guard): χ Cep;
(7) Shao Wei 少衛 (Adjunct Guard): γ Cep;
(8) Shang Cheng 上丞 (Grand Co-adjunct): 21 Cas.

In order, the stars on the Western Frontier (Xi Fan) are:

(1) You Shu 右樞 (Right Pivot): α Dra;
(2) Shao Wei 少尉 (Vice Commander): χ Dra;
(3) Shang Fu 上輔 (First Advisor): λ Dra;
(4) Shao Fu 少輔 (Vice Advisor): *d* 2106 Uma;
(5) Shao Wei 少衛 (Adjunct Guard): 43 Cam;
(6) Shang Wei 上衛 (First Guard): 9 Cam;
(7) Shao Cheng 少丞 (Vice Co-adjunct): 1 H^1 Cam.

Other names used to indicate Zi Wei Yuan are: Chang Yuan 長垣 (Long Walls), Qi Xing 旗星 (Celestial Flags), Tian Ying 天營 (Celestial Encampments), Ying Wei 營衛.

Zuo Shu 左樞:

See Zi Wei Yuan 紫微垣.

The analysis of this first passage of *Bu Tian Ge* shows at least two things: first, the detailed knowledge of the skies that the Chinese had, and the high level of development of organicist thought. Only a scientific reading designed to research all the microcosm-macrocosm correlations that make up the work could bring back the depth of its complexity. In light of what has been said, it becomes strikingly evident that the Chinese astronomical tradition cannot be reduced to a mishmash of *absurd things*.[14]

One of the first to realize this, was the Jesuit astronomer, Antoine Gaubil, a missionary in China in the first half of the 18th century. Gaubil was the first Westerner to translate *Bu Tian Ge*. The careful analysis that Gaubil made of *Bu Tian Ge* will be demonstrated below. The translation of the entire text with a commentary from Gaubil will be presented in a volume to be published in the near future.

Antoine Gaubil and *Bu Tian Ge*

Antoine Gaubil was one of the important Jesuit scientists got to China as a missionary between the 17th and 18th centuries. Certain aspects distinguish him and make him a character to be put front and center in the history of the Europe-China cultural exchanges. First and foremost, his scientific training was excellent. He remained in China for an extremely long time (from 1723 to 1759) allowing him to research and bring his knowledge to bear in a profound way. Aware that rapports with civilizations different from one's own must be transmitted through language, he studied Chinese and Manchurian thoroughly, becoming the greatest linguist among Europeans who trod Chinese soil in the 18th century. Finally, being a man who loved comparison and exchange, he always maintained contact with scholars in Europe, a condition greatly favouring the divulgation of Chinese science in the Old Continent, and European science in China (since August 22th, 1750 he was official correspondent of the French Academie Royale des Sciences). His abundant diaries,[15] the very many scientific facts he witnessed, and the translations of classic Chinese texts to which he gave life, are the best evidence for his importance.

On July 25th, 1734, Gaubil writes to Joseph-Nicolas Delisle from Beijing:

> Dear Sir,
> I am sending you two packages. The first is a roll of waxed canvas. Within, you will find what I have gathered on Chinese stars, the solstices, and the meridian shadows. I have also added a few reflections and answers regarding what you did me the honor of asking. There are also two letters for the Reverend Father Souciet. The second package is covered in a blue canvas. Within, you will find the celestial maps in Chinese by Father Grimaldi. There, you will see what our Fathers have added to the Chinese

stars and what they have corrected. There is a copy for you and for Father Souciet. Aside from this, there is also a small Chinese book which dates back to the Souy Dynasty.[16] This contains the characters of the Chinese constellations of which the description is done in verse ...[17]

This was the first time that Europe gained knowledge of *Bu Tian Ge*. Thus, from Gaubil the book arrived in the hands of Delisle. In the hand of the latter comes the following note on a piece of paper, summarizing the work:

Chinese book titled Pou tien Ko.[18] It is a description in verse of all the Chinese constellations, including maps, which was written under the Souy Dynasty (toward the year 500 A.D.). Father Gaubil, who sent me this book along with the letters of the 13th and the 25th of July 1734, did the translation and gave the explanation which he also sent me. He would then like me to give it to Father Souciet once I have read it. I took note of it, just as I did of the characters of these same constellations in larger size which Father Gaubil says to have copied from a sample of this book which can be found in a large collection of various works .... Father Gaubil ... says that this catalogue was done at the beginning of the 7th Century of Our Lord, or at the end of the 6th, and that the author was named Tan Yue Tse.[19]

Along with the copy of *Bu Tian Ge* (now in the Library of Observatoire Astronomique in Paris) sent by Gaubil there are also:

(a) the handwritten note by Delisle cited above;
(b) a brief handwritten note by Gaubil saying: "The things that I am sending you are taken from a copy which can be found in a great collection of various works. The author of the small work lived at the same time as the Souy Dynasty";
(c) another handwritten note by Delisle which generally summarizes what has already been quoted above; 31 numbered tablets done by the hand of Gaubil which represent the Chinese constellations just as they are illustrated in *Bu Tian Ge*; the voluminous handwritten copy of Gaubil's manuscript, made up of 95 pages (with some errors of numbering), with the translation and technical-scientific comment of *Bu Tian Ge*; handwritten paper of 34 pages about the method of reduce Chinese days to Europeans days.

The translation with commentary that Gaubil gave in the French language is full of technical explications. That is, it is configured by the annotations even to refer to the Chinese commentaries from the classical books. As was mentioned, we will soon present Gaubil's work in the form of a book. In this work we will like to focus briefly on his scientific attitude and the impartiality with which he faced this work on Chinese traditional astronomic knowledge.

In three pages which acted as a preface, Gaubil firstly presents the designs. This hasty excursus is named "About Chinese illustrations of stars" ; he starts with the admission that the illustrations from the edition are "without doubt... changed" compared to the originals. At the same time he gives detailed information to help Western astronomers identify Chinese stars and constellations; his conclusion is that only "some Chinese stars correspond to European stars". He adds also, later on, that he will produce all necessary indications, which (is) why a catalogue of latitudes and longitudes is "unnecessary".

On astronomical references therefore, Gaubil does not preserve any particularities. The Chinese sky is very different from the Western sky; but thanks to Gaubil's commentaries it takes on geometrical shape and boundaries and becomes more and more clear. All the Central Empire's celestial bodies were named for their relationship to the imperial family, or bureaucratic offices, or military ranks and so on. In this way they assumed well-known names and aspects. Consequently, all stars represented in *Bu Tian Ge* find the right position in the Western astronomy of the 18th century. Thus the Jesuits, in spite of his preceding declaration, compiled an almost twenty page table (par. 43–57) with notes in which each star has a Chinese name and its distance from Pole and Equator are recorded. With the same regard he describes other peculiarities of Chinese Astronomy, for example the 28 *xiu* system (par. 36), other catalogues of Chinese stars (parr. 62 sgg), and the procedures applied by the Chinese astronomers to measure the shadows.

Another important aspect that Gaubil studies in *Bu Tian Ge* is its Chinese cultural background, since the stars and constellations have names related to human life. In explaining these denominations, he does not even reserve linguistic details which could digress from Chinese culture, history and tradition. For example, he clarifies that *zi* "means a color composed by black and red, it means imperial" or he writes and translates the character "*Tay Yi*" (Ta Yi) and *Tian*, and he adds that Confucius alluded *Ta* as "*coelum est unum et magnum*" (par. 1). Constantly, to clarify some aspects of Chinese language, he brings up-to-date translations and interpretations given by other missionaries. One case (par. 1) is the constellation "*Keou Tchin*" [20] translated by Noel as "*hannus extensus*". For Gaubil these characters are changed to "Ladies' palace and apartment" and they "represent the empress and ordinary residence of the great emperor". Then, "It is possible to see 4 stars named Ladies of the Palace ..." and they are the "81 emperor's wives".

For appreciation of the rich details that Gaubil gives in the commentary of the translation, we could cite many of other passages. One in particular is about the star Zao Fu[21] in the "middle of the Milky Way"; Gaubil writes: "Tsao Fu is one name of the ancestor from the Tsin emperors A. C. The Tsao Fu looked after the Emperor Mourang's coach from the dynasty of the Tcheou.[22] He had a great knowledge of horses, he yielded great services to emperor ... The 17th year of the Mourang Empire is the 989 year B.C. It is in this year that the history speaks about Tsao Fu.[23] Tsao fu is one of the ancestors of the princes Tsin[24]; they were China emperors before the Han dynasty".

About Chinese customs, we can read (note to par. 7): "In the early Chinese monarchy there were colleges and academies where it was possible to learn correct customs, sciences, arts, and so on. Emperors and Rulers came to the colleges; they made ceremonies to honour to ancient sage and erudites; in this places they served food to old men, famous for their rectitude and for their knowledge; these ceremonies was made in the presence of Princes and Greats, in the presence of students, to teach the whole Empire the esteem and respect for the science of virtue and wisdom, and for erudites, above all the teachers. Since the age of these ancient ceremonies at colleges were gone. As a result the custom was to have in every city of the Empire a college or an academy which honours Confucius as a sage and erudite man; but this article about Chinese colleges and academies merits to be treated at length; it means that I will do it in another occasion".

Concerning Chinese astrology (par. 38): "... there are many things about this astrology which are in this text, [and] which I did not translate ... the editors of this catalogue have the particular conviction that the stars and the planets give influence to delightful events, and there is a mutual correspondence between stars, planets, comets and humans, and the celestials' phenomena create or indicate something governing humans. In view of this, during the Tcheou Dynasty, some provinces, cities and countries were associated with certain stars, that is why in some countries they paid more attention than in others to planet's movement through the zodiac and the stars. The origin of this astrological aspect could be derived from some antique practice of researching the sky for correct terrestrial customs, practice of the parallax's knowledge to foresee the Sun, the stars' and the planets' eclipses caused by the Moon, ... This catalogue [Bu Tian Ge] supposed that the spirit or the soul of certain great people resided in certain stars; it is not strange to see, the honor accorded to Being and the belief that there is the power to procure faster well-being or illness. The Chinese literati have had constantly faraway from this idea." This kind of interventions enrich the scientific data, and ease to present them as completed in a involved, and attractive culture.

Conclusions

As mentioned earlier, the translation and the comment to *Bu Tian Ge* are one of the Gaubil's astronomical studies in China; he informed his European contemporaries about his researches. His observations, calculations, translations, and comparisons arrived in Europe constitute an indispensable reference in the history of meeting between European and Chinese science. The Gaubil's work on *Bu Tian Ge* can be considered one of the ancestor of this kind of studies that allowed the understanding of the ancient and complex Chinese astronomy and scientific culture. For his peculiarities, Gaubil was different from all his predecessors (Jesuit scientists and missionaries), except for Johann Schreck, Terrentius, Giacomo Rho, Johan Adam Schall von Bell, and Ferdinand Verbiest. The founder of the Chinese mission, Matteo Ricci, has been excluded from this list because of his lack of scientific curiosity and his prejudices toward the Chinese and their culture.[25]

Notes

[1] Cfr. *Jian Ming Tian Wen Xue Ci Dian* 簡明天文學詞典 (Simple and Clear Dictionary of Astronomy), Shanghai 上海 1986, p. 316.

[2] In the sense of surrounding wall.

[3] Cfr. Yabuuchi Kiyoshi, *Zui To Rekihoshi no Kenkyu* 隋唐曆法史の研究 (Research on the history of the calendar in the Sui and Tang Eras), Tokyo 1944.

[4] The question regarding the origin of the *xiu* is notoriously still open since other civilizations used a similar system for subdividing the celestial sphere as well: the 28 Arab *al-manazil* and the 28 Indian *nakshatra*. Since the Arab system is relatively more recent (even if it pre-dates the *Koran*), the main diatribe still ongoing is to define whether the Chinese or the Indian system is more ancient. A less fiscal approach and with broader view believes that the *xiu* and *nakshatra* each had a separate development and did not come in contact with one another until later (in the period when the *Bu Tian Ge* was written) which would have determined a reciprocal contamination. There is also a hypothesis which states that they had a common derivation from the lunar stations from the remote Babylonian Astronomy [this question is summarized in Iannaccone I., *Misurare il cielo: l' antica astronomia cinese (Measuring the Sky: Ancient Chinese Astronomy)*, Napoli 1991, pp. 78–79; for the problemes of identification of the Chinese *xiu*, cfr. Sun Xiaochun & Kistemaker, *The Chinese Sky during the Han*, Leiden 1997, pp. 60–67.

[5] Cfr. Chen Zungui 陳遵嬀, *Zhong Guo Tian Wen Xue Shi* 中國天文學史 (History of Chinese Astronomy), vol. I, Taipei 1988, p. 219.

[6] Cfr. *Bu Tian Ge*, p. 1r.

[7] Since the color purple is the dominant color in the buildings in the Forbidden City, the character *zi* 紫 could also be translated as " imperial". Thus Zi Wei Yuan becomes " Secret Imperial Enclosure".

[8] Note that Zui *xiu*, instead of preceding Shen *xiu* here follows it.

[9] The point of departure (and arrival) of each philosophical speculation was the conviction that the universe is one immense container in which all things, all living beings, all organizations and all phenomena are all related: an organism that pulses, was moved by dynamics and equilibriums in which opposites alternate, thus determining double and multiple influences which are reciprocally interdependent. The philosophical thought was founded on this organicist vision and research into natural phenomena was done with the aim of discovering why they happened, but much more so with an eye toward a much wider meaning which illuminated the effects that these could provoke on other phenomena, on the chain reactions that these could set off, or in which these very ones were already established (Cfr. Iannaccone I., *Misurare il cielo ...*, *op. cit.*, p. 1).

[10] A partial translation of *Bu Tian Ge* ("... *only a Polar division* ..." was furnished by Soothill in 1951, but it used the copy of the work presented in the *Tu Shu Ji Cheng* 圖書集成 (Imperial Encyclopedia, 1726), which was quite different from the original (cfr. Soothill W. E., *The Hall of Light. A Study of Early Chinese Kingship*, London 1951, pp. 244–51 and 273–77).

[11] One must remember that Shu Zi 庶子 generally indicates the child of one of the emperor's wives of second rank, but in the Sui Era the characters were used to indicate the Office of the Crown Prince (cfr. Hucker C. O., *A Dictionary of Official Titles in Imperial China*, Stanford 1985, p. 437).

[12] To understand the official titles (when it is possible with the Sui Era use), Hucker C. O., *A Dictionary of Official Titles ...*, *op. cit.* was used. For the astronomical identifications, we referred to:

    (1)   Allen R. H., *Star Names, Lore and Meaning*, New York 1899 (re-published, New York 1963);

    (2)   Boss, B. *General Catalogue of 33342 Stars for the epoch 1950*, 5 vols. Washington 1936–37;

    (3)   Chen Zungui 陳遵嬀, *Heng Xing Tu Biao* 恆星圖表 (Fixed Stars Atlas), Shanghai 1937;

    (4)   Feng Shi 馮旅, "Zhongguo zaoqi xingxiangtu yanjiu" 中國早期星象圖研究 (Research on Early start charts in China), in *Ziran kexuexi yanjiu* 自然科學史研究 (Researches in History of Natural Science), 9, 1990, N.2;

    (5)   Ho Peng Yoke, *The Astronomical Chapters of the Chin Shu*, Paris 1966;

    (6)   Schlegel G., *Uranographie Chinoise*, Leyde 1875 (re-published So-Wen, Milano 1977);

(7) Sun Xiaochun & Kistemaker J., *A Study of the Chinese Sk ...*, *op. cit.*;

(8) Wu Qichang 吳其昌. "Han yiqian hengxing faxian cidi kao" 漢以前恒星發現次第考 (A Study on the identification of the fixed stars known before the Han), in *Zhenli Zazhi* 真理雜志 (Magazine of the Truth), 1, 1943, N. 3;

(9) Xi Zezong 席澤宗, "Dunhuang xingtu" 敦煌星圖 (AStar Map from Dunhuang), in *Wenwu* 文物 (Cultural Objects), 1966, N.3;

(10) Yi Shitong 伊世同, *Zhong Xi hengxing duizhao tu biao* 中西恒星對照圖表 (Fixed Stars in China and West Correlative Catalogue and Atlas), Beijing 北京, 1981.

[13] Sun Xiaochun & Kistemaker J, *The Chinese Sky ...*, *op. cit.*, p.164.

[14] Cfr. D' Arelli F., " Matteo Ricci S. J.: le 'cose absurde' dell astronomia cinese: genesi, eredità ed influsso di un convincimento, tra i secoli XVII e XVIII" (Matteo Ricci S. J.: the 'absurd things' regarding Chinese astronomy. Genesis, heredity, and influx of a conviction between the 16th and 17th Centuries), in Iannaccone I. & Tamburello A. (ed.), *Dall' Europa alla Cina: contributi per una storia dell' astronomia* (From Europe to China: Contributions for a History of Astronomy), Napoli 1990, pp. 85–127.

[15] Gaubil's voluminous diaries were collected and published by R. Simon, with a preface by P. Demieville and the appendix by J. Dehergne (cfr. Gaubil A., *Correspondance de Pekiin. 1722–1759*), Geneve 1970.

[16] Sui Dynasty (561–618 A.D.).

[17] Cfr. manuscript in Library of Observatoire. Astronomique in Paris, Boite d'archive A.B. 1.6, Tome XI, 164; 2a *via*, T. XI, 211c.

[18] Clearly, this is *Bu Tian Ge*.

[19] Dan Yuanzi.

[20] Gou Chen 鈎陳.

[21] Zao Fu 造父, five stars, one of them is d Cep.

[22] Mu Wang 穆王 (1001–946 B.C.), King of the Western Zhou 周 dynasty; his legendary charioteer was named Zao Fu.

[23] In accordance to the legend, in 989 B. C. Zao Fu drove as fast as possible Mu Wang to suppress a revolt in the Zao 造 province; the grateful King gave his charioteer the govern of this province and in this occasion the man was named Zao Fu.

[24] Qin 秦.

[25] One must remember, however, that Matteo Ricci's missionary experience in China was the first conscious encounter between European Renaissance science and Chinese science, even if his declared primary aim was not scientific exchange, but the religious colonization of China, or the evangelization. He introduced algebra, Euclidean Geometry with its applications to planetary motion and topography, the concept of the Earth as a sphere divided into meridians and parallels, the reform of the calendar, and techniques for constructing spheres, clocks, globes, sextants,

and quadrants. But, he lived culturally isolated from the Chinese experience because the differences did not inspire his curiosity, and was thus always enclosed within the pride of his faith. Ricci's isolation and his propensity toward prejudice did not affect his intuition for *realpolitik*: if one wanted to establish significant contact with the Chinese, one had to take advantage of the fallacy of the calendar and correct it. It was on the basis of this reflection that the Jesuit era was launched in China, and the encounter was brought to fruition not in religious ideology, but in science and technology. It must be underlined that if it had not been for Matteo Ricci and his political intuition, perhaps there may never have been a significant encounter between European science and Chinese science in the following centuries (cfr. Iannaccone I., *Matteo Ricci and the Eclipses in the Chinese Scientific Tradition: Prejudices and Transmissions*, conference given at the Beijing Institute for Foreign Languages in September 1996, in publication).

Two Spanish Cosmographers in the Philippines: Andrés de Urdaneta and Martín de Rada

JOSÉ ANTONIO CERVARA

University of Zaragoza, Spain

Introduction

It is very well known that Catholic missionaries, especially the Jesuits during the 17th and 18th centuries, played an important role in the introduction of European science to China. However, Jesuits were not the only missionaries to go to China. Neither were they the only ones to carry their scientific knowledge to East Asia.

One of the most important missionaries was the Spanish Dominican Juan Cobo (?–1592). He visited the Philippines and wrote the book *Shi Lu*, published in Manila in 1593, the first book written in Chinese that introduces some elements of European science. In this book Cobo shows that the Earth is round, for example, and presents arguments to prove it. He also provides some notions of astronomy, physics, geography, geology, botany and zoology.[1]

Several other missionaries to the Philippines wrote important scientific works. Dominicans were on the Islands in 1587, but they were not the first ones. In 1565, a Spanish expedition led by López de Legazpi established the first Spanish settlement in the Philippines, and with it some Augustinian friars arrived. In this paper, two important Spanish Augustinian *cosmographers*,[2] will be introduced: Andrés de Urdaneta and Martín de Rada.

Andrés de Urdaneta (1508–1568)

Urdaneta was born in Villafranca de Oria (Guipuzcoa, Spain) in 1508. As a young man he was a soldier, but later studied mathematics, astronomy and navigation. He went to East Asia for the first time in 1525, with the expedition of Frey Jofre de Loaysa. He arrived to the Moluccas in 1526, and stayed there for eight years. After having returned to Spain in 1538, he travelled to Mexico, where he stayed for 16 years and in 1553 entered the Order of St. Augustin. Finally, in November 1564, he left for the Philippines as a member of the expedition of López de Legazpi. It was then when he made his most celebrated discovery: the *tornaviaje* or return way from the Philippines to Mexico.

Until that moment, the Spanish had made several attempts at reaching Mexico from East Asia.[3] But the return to Mexico had proved impossible. The Spanish had been forced to look for the way back through India and Africa, both controlled by the Portuguese, and for this reason the establishment of a permanent Spanish colony in Asia had not been possible. The expedition of López de Legazpi in 1564–65 was a new attempt. The Spaniards arrived in the Philippines on February 13th, 1565, and built their first permanent settlement in Cebu on April 27th. Soon thereafter, an expedition led by Urdaneta left Cebu (on June 1st, 1565). It travelled first to the north, near Japan, and afterwards to the east reaching America. On October 1st, 1565, they arrived at the Mexican harbor of La Navidad, and one week later to Acapulco. The *tornaviaje* from the Philippines to Mexico had been found, and thanks to this discovery, the Spaniards stayed in the Philippines for more than three centuries.

Urdaneta went back to Spain and later again to Mexico in 1567. He died in Mexico city on June 3rd, 1568.

Scientific Contributions of Andrés de Urdaneta

Among the scientific works of Andrés de Urdaneta, his most important contributions are related to navigation. We have different letters or reports about his travels and about the geography of the lands discovered in the Pacific Ocean. Some authors give him the honor of having produced the first study from a scientific point of view of hurricanes. It was also on his advice that the name of the main Spanish port in the West Coast of Mexico was changed from La Navidad to Acapulco. He is therefore responsible for the development of the city of Acapulco and the route between Manila and Acapulco, one of the most famous maritime routes during the next centuries.

One of Urdaneta's most interesting scientific works is the *Parecer sobre la demarcación de Filipinas* (Report about the demarcation of the Philippines), in which he uses geography and astronomy in order to prove that the Philippine Islands were inside the Spanish zone of the Earth according to the Treaty of Tordesillas (1494), and not in the Portuguese area. In this manuscript of 1566, he uses the astronomical observations made by Martín de Rada along with his own knowledge to show that the Philippines are to the east of the Tordesillas line in the Pacific. He was in fact wrong (the Philippines were in the Portuguese zone), but the important point about this *Parecer* is the use by Urdaneta of several astronomical books of his time, including that of Copernicus, published only 23 years before. At the time, Copernicus was not very well known, even in Europe. The fact that his book could be found on the newly discovered lands of East Asia, and was being used by some Spanish Augustinian friars, gives us an idea about the great importance of science to these two men, Andrés de Urdaneta and Martín de Rada.

Martín de Rada (1533–78)

Martín de Rada is one of the most outstanding friars in the history of early scientific relations between Asia and Europe. He was born in Pamplona (Navarra, Spain) in 1533. He went to the University of Paris, where he studied both Greek and sciences, but he soon showed an exceptional ability in mathematics, geography and astronomy. He continued his studies in Salamanca, where he joined the order of St. Augustin. He decided to go to Mexico and from there to the Far East. He travelled in the same ship as Legazpi and Urdaneta, in 1565. He stayed in Cebu until 1572, when he went to Manila.[4]

One of his most important contributions took place between 1567 and 1568, while he was in Cebu. A Portuguese captain arrived there with the task of asking the Spaniards to leave the Philippines, which were in the Portuguese zone. But according to the calculations by Rada, as has been said, the Philippines belonged to the Spanish demarcation. The scientific prestige of Rada was so large that the Spaniards did not leave the Islands. Perhaps if Rada had not been there, the whole history of South East Asia, especially of the Philippine Islands, would have been very different.

The most important incident in Rada's life was probably his journey to China, in 1575.[5] At the time, as we know, it was nearly impossible for foreigners to enter China, but friars made several attempts, using every means they found. The opportunity arrived in 1574 when a Chinese pirate attacked Manila. A Chinese captain arrived later to fight him with the help of the Spaniards. Afterwards, he proposed that they send some Spanish people to the continent. Four Spaniards (two soldiers and two friars, one of them being Martín de Rada) were chosen for this purpose. They left Manila on June 12th, 1575, and arrived to Amoy[6] on July 5th. They stayed at several cities in Fujian province until October 11th, 1575, when they were forced to return to the Philippines.

This voyage is very important for several reasons. On the one hand, Rada wrote an accurate description of the parts of China he visited.[7] On the other hand, we have evidence that Rada brought a large number of Chinese books to Manila, and these books could have represented a large contribution to the knowledge of China and Chinese culture in the West.

Rada tried to go back to China again next year but did not succeed. In 1578, he went to Borneo with the expedition led by the Governor of the Philippines, Francisco de Sande, and he died on the way back, in June 1578.

Scientific Contributions of Martín de Rada

We know that Martín de Rada was the author of many works, but, unfortunately, many of them have not reached us. We have a large number of letters written by him. He seems to have written some works in the various languages of the Philippine Islands. We have some references to a Chinese Grammar and Vocabulary which he could have composed, but it has not been found. According to some Augustinian authors, Rada had some knowledge of the Chinese language, probably learned in the Philippines from some Chinese traders, but we do not have any solid evidence for this.

In any case, the most important work by Rada, as mentioned above, is his *Relación* (Report) of his travel to China in 1575. This manuscript, which has survived, can be divided into two very different parts. In the first, Rada describes the voyage to China undertaken by him and his colleagues; in the second, he describes China, its geography, history, traditions, policy, rites, etc. In order to compose this description, Rada could have used some of the Chinese books he was supposed to have taken to Manila from China.[8]

Rada's work has been reproduced several times. Probably the most important example is the work of the Spanish Augustinian Juan González de Mendoza (1545–1618). In 1585 he published his *Historia del Gran Reino de la China* (History of the Great Kingdom of China). In this book, Mendoza makes a general description of Chinese history, geography, philosophy, customs, religions and policy, and he also describes Rada's voyage to China, as well as two other voyages made by Spanish Franciscans some years later. In his book, Mendoza uses information from Rada and some other early missionary travellers. The *Historia del Gran Reino de la China* was a best-seller in his time. By 1600, 38 editions had appeared and it had been translated from Spanish into English, Italian, French, Latin, German and Dutch. We can consider this book the most important work for knowledge about China in Europe at that time.

We know that Martín de Rada wrote some scientific works on geography, geometry, astronomy, astrology and watch-making, but they were all lost during his lifetime, as he mentions in one of his letters. In the same letter he asks for some scientific books from Europe, since he writes, he *only* has the works by Euclid, Archimedes, Ptolemy, Copernicus, and some others. We have to remember that Rada was travelling thousands of kilometers and months away from Europe. The fact that he had so many scientific books, some of them quite new, such as the already quoted *De Revolutionibus* by Copernicus, gives us an idea of the scientific importance of this author. We know, according to other sources, that Rada also wrote some books on navigation, on the latitude and longitude of different places, and composed some astronomical tables, which have also been lost. We have references about the construction by Rada of some astronomical instruments. He was also the first European to give notice about the Strait of Anian or Bering.

Finally, we must mention one of the most important contributions made by Martín de Rada: the identification of Cathay with China. Usually, the discovery that the old mythical kingdom of Cathay, described by Marco Polo, was the same country as the modern China where the Portuguese had arrived has been attributed to Matteo Ricci and his Jesuit colleagues. But, in fact, some decades before Ricci, Rada had made the same assertion in his *Relación* about his travel to China. This is, as far as we know, the first European text to clearly state that Cathay and China are the same country.

Notes

[1] The figure of Juan Cobo, with his book *Shi Lu*, is very important for the History of the relations between European and Chinese science. His work was introduced during the *20th International Congress of History of Science* (Liège, July 1997) and the *8th International Conference on the History of Science in China* (Berlin, August 1998).

[2] Today, we would say *geographers.*

[3] The first complete circumnavigation of the world by Magallanes and Elcano in 1519–22, or the expedition of Jofre de Loaysa in 1525–26 are only two examples.

[4] Rada was the first European who spoke about the small town of Manila, in a letter of 1569, two years before it was chosen by the Spaniards to transfer the center of their power in the Philippines from Cebu.

[5] During the first years of Spanish presence in East Asia, their main objective from an economic and religious point of view was not the Philippines, but China. The Philippines were considered only an intermediate step to the continent, and only later, after they were convinced that a permanent settlement in China was unattainable, did they decide to remain on the Islands.

[6] Today, Xiamen, in the coast of Fujian Province.

[7] Some modern authors, like the French Jesuit Henri Bernard, consider this trip the first scientific exploration of China by Westerners.

[8] It is not clear if Rada was able to read Chinese, or if some Chinese people living in the Philippines translated some books into Spanish. There is some discussion about this point between the scholars studying the early history of missionaries in the Philippines, but the truth is that there are no clear proofs for either of the two hypotheses.

References

Alonso, C. (1989), *Primer viaje misional alrededor del mundo (1542–1549). Una gesta agustiniana*. Valladolid, Estudio Agustiniano.

Bernard, H., S. J. (1933), *Aux portes de la Chine. Les Missionnaires du Sixième Siècle. 1514–1588*. Tientsin, Hautes Etudes.

———— (1936), *Les Iles Philippines du Grand Archipel de la Chine. Un essai de conquete spirituelle de l'Extreme Orient. 1571–1641*. Tientsin, Hautes Etudes.

Boxer, C. R. (1953), *South China in the Sixteenth Century*. London, The Hakluyt Society.

De Castro, A. M., O. S. A. (1954), *Misioneros Agustinos en el Extremo Oriente (1562–1780)*. Collection "Biblioteca Missionalia Hispanica", Series B, VI. Madrid, Consejo Superior de Investigaciones Científicas, Instituto de Santo Toribio de Mogrovejo, Ediciones Jura. Edition of the original, written by the author 1780.

Galende, P. G., O. S. A. (1980), *Martín de Rada O. S. A. (1533–1578) Abad frustrado, misionero y embajador real*. Manila, Arnoldus Press.

Gonzalez De Mendoza, J., O. S. A. (1990), *Historia del Gran Reino de la China*. Collection "Biblioteca de Viajeros Hispánicos", 6. Madrid, Miraguano Ediciones y Ediciones Polifemo. Edition of the original published in Rome in 1585.

Martinez, B., O. S. A. (1918), *Historia de las misiones agustinianas en China*. Madrid, Imprenta del asilo de huérfanos del S. C. de Jesús.

Perez, E. J., O. S. A. (1901), *Catálogo bio-bibliográfico de los Religiosos Agustinos de la Provincia del Santísimo Nombre de Jesús de las Islas Filipinas, desde su fundación hasta nuestros días*. Manila, College of Sto. Tomas.

Picatose, F. (1891), *Apuntes para una Biblioteca Científica Española del Siglo XVI*. Madrid, Imprenta y Fundición de Manuel Tello.

Rodriguez, I. (1978), *Historia de la Provincia Agustiniana del Smo. Nombre de Jesús de Filipinas*. Manila, Arnoldus Press.

Rodriguez, I. & Alvarez, J. (1992), *Diccionario Biográfico Agustiniano. Provincia de Filipinas*. 2 volumes. Valladolid, Estudio Agustiniano.

Santiago Vela, G. de, O. S. A. (1913–31), *Ensayo de una Biblioteca Ibero-Americana de la Orden de San Agustín*. 8 volumes. Madrid. Imprenta del Asilo de Huérfanos del Sagrado Corazón de Jesús (6th volume, 1922); El Escorial, Imprenta del Monasterio (8th volume, 1931).

Sanz, C. (1958), *Primitivas relaciones de España con Asia y Oceanía*. Madrid, Librería General Victoriano Suárez.

Sebes, J., S. J. (1988), "The precursors of Ricci". Included in: Ronan, C. E., S. J. & Bonnie B. C. Oh. (ed.), *East meets West. The Jesuits in China, 1582–1773*. Chicago, Loyola University Press.

PART V

Mathematics

The Development of Interpolation Methods in Ancient China: From Liu Zhuo to Hua Hengfang

JI ZHIGANG 纪志刚

Mathematics Dept., Xuzhou Normal University

A Brief Introduction

Around 600 A.D., in his *Huangji Calendar* (皇极历, 604A.D.), Liu Zhuo used a new algorithm — now called interpolation — which improved accuracy and coherency. Subsequently, this methodology was widely adopted in compiling and further improving calendars. Indeed, Liu Zhuo's contribution to calendrical science ushered in a new era in the history of mathematical astronomy in ancient China. This paper will expound the development of this interpolation from calendrical science to mathematics.

In the science of calendars, it was Li Chunfeng who first inherited Liu Zhuo's work and further developed it. Using the new application, Li Chunfeng calculated the sun's shadow at noon in the *Lider Calendar* (麟德历, 664A.D.). In the *Dayan Calendar* (大衍历, 728A.D.), the algorithm was given the form of an unequal interval by Yi Xing, and another form in Xu Ang's *Xuanming Calendar* (宣明历, 822A.D.). When the *Chongxuan Calendar* (崇玄历, 893A.D.) came into existence, Bian Gang applied this algorithm to the equal interval form again and made it simpler. Meanwhile, in the *Chongxuan Calendar*, the new algorithm "*xiang jian xiang cheng*" (a kind of parabolic interpolation) was skilfully applied by Bian Gang. By means of this new method, the interpolation was given an extra interpretation. Finally, the cubic interpolation by Wang Xun and Guo Shoujing in the *Shoushi Calendar* (授时历, 1280A.D.) pushed Liu Zhuo's theory even further.

Interpolation was not even mentioned in the world of mathematics before Qin Jiushao's *Shu Shu Jiu Zhang* (数书九章, 1247A.D.). But Qin's discussion of "Calculating the position of the planets with *Zhuishu* (缀术)" still belonged to an earlier calendrical science. Zhu Shijie was the first mathematician to touch upon the mathematical problems of interpolation in his *Si Yuan Yu Jian* (四元玉鉴, 1303A.D.). In the *Qing* Dynasty, the traditional subject of interpolation did not receive much attention until about 1850. Li Shanlan was the one who made a study of it. Another mathematician after him, Hun

Hengfang, wrote an important monograph on interpolation known as *Ji Jiao Shu* (积较术, 1882A.D.). He elaborated the limited difference, established its various principles, and set up a related systematic theory and a series of formulas.

Tracking Down the Development of the Interpolation

Liu Zhuo's Formula and its Development in the Tang Dynasty

The first point to be noted is that the interpolation methods were given in the form of algorithmic language. So it is necessary to translate them into modern mathematical symbols. In the *Huangji* calendar, Liu Zhuo wanted to calculate the data of the Sun's apparent motion each day, which was called the 'slow' or 'fast' data (迟速数). This data on the days of 24 *Qi* (24 solar terms) was listed by observation.

Now, let $f(x)$ be the function to be determined. l is one length of the 24 solar terms, when near the winter solstice (from Autumn to Spring Equinox), $l = 14.55$ days, in this interval, and the sun's motion becomes f(x) [fast]. When near the summer solstice (from Spring to Autumn Equinox), $l = 15.45$ days; in this interval, the sun's motion becomes f(x) [slow]; $t \in [0, l]$, n = 0, 1, 2, ... 11. Let:

$$\Delta_i = f(\overline{i + 1l}) - f(il)$$

Supposing $\Delta_1 > \Delta_2$, Liu Zhuo's algorithms are as follows:

(1) Calculate the end ratio of the *Qi* (气末率): $a = (\Delta_1 + \Delta_2)/2l$;
(2) Calculate the total difference (总差): $b = (\Delta_1 - \Delta_2)/l$
(3) Calculate each day's difference (别差): $\alpha = (\Delta_1 - \Delta_2)/l^2$
(4) Calculate the first day's ratio of the *Qi* (气初率): $\beta = a + b$
(5) Calculate the rising or dropping ratio of the first day (初日陟降数): $\gamma = \beta - \alpha/2$;
(6) Calculate the rising ratio or dropping of each day in the *Qi* (每日陟降数): $\delta_i = \beta - i\alpha$
(7) Calculate the summary (随算其数): $S(t) = \delta_1 + \delta_2 + ... + \delta_t$
(8) The slow or fast data of the *t* day (迟速数): $f(nl + t) = f(nl) + S(t)$

Now, we get Liu Zhuo's equal interval formula:

$$f(nl + t) = f(nl) + \frac{t}{2l}(\Delta_1 + \Delta_2) + \frac{t}{l}(\Delta_1 - \Delta_2) - \frac{t^2}{2l^2}(\Delta_1 - \Delta_2)$$

By the same methods we get:

Yi Xing's formula, unequal interval

$$f(U + t) = f(U) + t\frac{\Delta_1 + \Delta_2}{l_1 + l_2} + t\left(\frac{\Delta_1}{l_1} - \frac{\Delta_2}{l_2}\right) - \frac{t^2}{l_1 + l_2}\left(\frac{\Delta_1}{l_1} - \frac{\Delta_2}{l_2}\right)$$

Xu Ang's formula, unequal interval

$$f(U + t) = f(U) + t\frac{\Delta_1}{l_1} + t\frac{l_1}{l_1 + l_2}\left(\frac{\Delta_1}{l_1} - \frac{\Delta_2}{l_2}\right) - \frac{t^2}{l_1 + l_2}\left(\frac{\Delta_1}{l_1} - \frac{\Delta_2}{l_2}\right)$$

Bian Gang's formula, equal interval

$$f(nl + t) = f(nl) + \frac{t}{l}\Delta_1 + \frac{t}{2l}(\Delta_1 - \Delta_2) - \frac{t^2}{2l^2}(\Delta_1 - \Delta_2)$$

Of these four formulas, the latter are simpler than the former in form; it is just this change that shows us the development which took place during the *Tang* Dynasty.

The Algorithm of "xiang jian xiang cheng"

This method first appeared in Cao Shiwei's *Futian* Calendar (符天历, 780),

$$f(x) = \frac{x(182 - x)}{3300}; 0 \leq x \leq 182$$

And then Bian Gang developed it in his *Chongxuan* Calendar. For example, in order to calculate the polar-latitude, he used the combined functions as follows:

$$f(x) = \begin{cases} \dfrac{(91 - x)x}{5600}; 0 \leq x \leq 30 \\[2mm] \dfrac{(91 - x)^2}{11500}; 30 \leq x \leq 152 \\[2mm] \dfrac{(182 - x)(x - 91)}{5600}; 152 \leq x \leq 182 \end{cases}$$

Wang Xun and Guo Shoujing — A Cubic Interpolation

$$f(x) = x[\alpha - x(\beta + x\gamma)]$$

α: *dingcha* (定差), β: *pincha* (平差), γ: *licha* (立差)

Zhu Shijie's Interpolation Formula

$$f(x) = x\Delta + \binom{x+1}{2}\Delta^2 + \binom{x+2}{3}\Delta^3 + \binom{x+3}{4}\Delta^4$$

It is important to point out that Zhu Shijie first set up the general form of the coefficient of the interpolation.

Hua Hengfang and *Ji Jiao Shu* (The Method of Limited Difference)

Much attention should be paid to Hua Hengfang's *Ji Jiao Shu*, because this book was an important monograph on interpolation.

Basic Difference Table

In *Ji Jiao Shu*, Hua Hengfang made a rectangle difference table, with the middle line chosen as basic, and called "0-side difference".

Table 1

y_n	$\cdots$	y_3	y_2	y_1	y_0	y_{-1}	y_{-2}	y_{-3}	$\cdots$	y_{-n}
Δy_n	$\cdots$	Δy_3	Δy_2	Δy_1	Δy_0	Δy_{-1}	Δy_{-2}	Δy_{-3}	$\cdots$	Δy_{-n}
$\Delta^2 y_n$	$\cdots$	$\Delta^2 y_3$	$\Delta^2 y_2$	$\Delta^2 y_1$	$\Delta^2 y_0$	$\Delta^2 y_{-1}$	$\Delta^2 y_{-2}$	$\Delta^2 y_{-3}$	$\cdots$	$\Delta^2 y_{-n}$
$\Delta^3 y_n$	$\cdots$	$\Delta^3 y_3$	$\Delta^3 y_2$	$\Delta^3 y_1$	$\Delta^3 y_0$	$\Delta^3 y_{-1}$	$\Delta^3 y_{-2}$	$\Delta^3 y_{-3}$	$\cdots$	$\Delta^3 y_{-n}$
$\vdots$	$\cdots$	$\vdots$	$\vdots$	$\vdots$	$\vdots$	$\vdots$	$\vdots$	$\vdots$	$\cdots$	$\vdots$
n	$\cdots$	3	2	1	0	-1	-2	-3	$\cdots$	$-n$

Elementary Properties

Hua made some elaborations on the elementary properties of limited difference. Such as:

(1) $\Delta^n(x^n) = n!$
(2) $\Delta^k(ax^n) = a\Delta^k(x^n)$
(3) If $y_n = f(x) = \Sigma a_i x^{n-i}$, then: $\Delta^k y_n = \Sigma a_i \Delta^k x^{n-i}$
(4) $\Delta^n y_n = $ con.; $\Delta^{n+1} y_n \equiv 0$

Fundamental Formulas

By reckoning, Hua got a series of formulas, such as:

$$y_n = y_0 + n\Delta y_0 + \frac{n(n+1)}{2}\Delta^2 y_0 + \frac{n(n+1)(n+2)}{3!}\Delta^3 y_0$$

$$\Delta y_n = \Delta y_0 + n\Delta^2 y_0 + \frac{n(n+1)}{2}\Delta^3 y_0$$

$$\Delta^2 y_n = \Delta^2 y_0 + n\Delta^3 y_0$$

$$\Delta^3 y_n = \Delta^3 y_0$$

$$y_{-n} = y_0 - n\Delta y_0 + \frac{n(n-1)}{2}\Delta^2 y_0 - \frac{n(n-1)(n-2)}{3!}\Delta^3 y_0$$

$$\Delta y_{-n} = \Delta y_0 - n\Delta^2 y_0 + \frac{n(n-1)}{2}\Delta^3 y_0$$

$$\Delta^2 y_{-n} = \Delta^2 y_0 + n\Delta^3 y_0$$

$$\Delta^3 y_{-n} = \Delta^3 y_0$$

Their general forms are as follows:

$$y_n = \sum_{i=0}^{n}\binom{n+i-1}{i}\Delta^i y_0 \qquad y_{-n} = \sum_{i=0}^{n}(-1)^i\binom{n}{i}\Delta^i y_0$$

They are now called "Hua Hengfang's interpolation formulas", and are different from Newton's.

How Interpolation was used — Examples from Original Works

The uses of interpolation in calendars were wide-ranging. For example: calculating the sun's apparent motion; the moon's motion (by use of those two algorithms, the true moon can be calculated); the apparent solar declination; the lunar polar-latitude; the motion of the planets; and the length of the sun's shadows at noon. Another example comes from Hua Hengfang's *Ji Jiao Shu*: how to solve the numerical coefficient higher degree equation by using the limited difference.

In *Ji Jiao Shu*, for the numerical coefficient higher degree equation:

$$f(x) = a_0x^n + a_1x^{n-1} + \ldots\ldots + a_{n-1}x + a_n = 0$$

the algorithm for solving roots is as follows:

First, calculating the "0-side difference" of $f(x)$: $f(0)$, $\Delta f(0)$, $\Delta^2 f(0)$, ..., $\Delta^n f(0)$; and then constructing the difference table of $f(x)$, for $x= 0, 1, 2, \ldots\ldots$, k; if $f(i) = 0$, when i is the root of $f(x)$; if $f(i) \neq 0$, and $f(0), f(1), \ldots, f(i)$ have the same sign, but $f(i + 1)$ has the different sign, this means that there is a root in the interval $(i, i + 1)$. Now regarding $f(i), \Delta f(i), \Delta^2 f(i), \ldots\ldots, \Delta^n f(i)$ as "0-side difference", reckon its function $f_1(x)$, then repeat the operation all over again, and so on till the required result is obtained.

Example: Solve

$$f(x) = x^3 - 190x^2 - 2627x - 2436 = 0 \tag{1}$$

(see *Ji Jiao Shu*, Vol. 2, p. 21)

Solution: Because the constant's absolute value is larger, so the transformation $x = 100y$ is necessary:

$$f(y) = 1000000y^3 - 1900000y^2 - 262700y - 2436 = 0 \tag{2}$$

Calculating the "0-side difference" for $f(y)$ (see note 1):

$$f(0) = -2460, \ \Delta f(0) = 2637300, \ \Delta^2 f(0) = -9800000, \ \Delta^3 f(0) = 6000000$$

Constructing its difference table (the first line is 0, and the pace is 100):

9109464	−127836	−1165136	−2436
9237300	1037300	−1162700	2637300
8200000	2200000	−3800000	−9800000
6000000	6000000	6000000	6000000
			pace
300	200	100	0

Obviously there is a root in (200, 300); now regarding 200 as "0-side difference", reckoning its function $f_1(y)$:

$$f_1(y) = y^3 - 410y^2 - 31373y - 127836 = 0 \tag{3}$$

In fact, we can prove that from $f(x)$ to $f_1(y)$, the transformation $y = x - 200$ is made.

In order to get the second digit of the root, the operation repeats from the beginning again as follows:

Calculating the "0-side difference". For $f_1(y)$:

$$f_1(0) = -127836, \ \Delta f_1(0) = 40946, \ \Delta^2 f_1(0) = -814, \ \Delta^3 f_1(0) = 6$$

Constructing the difference table (now the pace is 1, and the first line is 200):

0	–43442	–86052	–127836
43442	42610	41784	40946
832	826	820	814
6	6	6	6
			pace
203	202	201	200

So, 203 is one root of the $f(x)$.

Notes

In *Ji Jiao Shu*, there are many tables. The two most important are as follows:

Table 2 h_k^n: table of $\Delta^i x^n$

1					
0	1				
0	–1	2			
0	1	–6	6		
0	–1	14	–36	24	
0	1	–30	150	–240	120

Table 3 H_k^n: inverse table of h_k^n

1					
0	$\frac{1}{1}$				
0	$\frac{1}{2}$	$\frac{1}{2}$			
0	$\frac{2}{6}$	$\frac{3}{6}$	$\frac{1}{6}$		
0	$\frac{6}{24}$	$\frac{11}{24}$	$\frac{6}{24}$	$\frac{1}{24}$	
0	$\frac{24}{120}$	$\frac{50}{120}$	$\frac{35}{120}$	$\frac{10}{120}$	$\frac{10}{120}$

The relation of tables (h_k^n) and (H_k^n) is $(h_k^n) \cdot (H_k^n) = I$. In other words, (h_k^n) and (H_k^n) are a couple of mutual inverse tables.

The function of the table (h_k^n) is to calculate the "0-side difference" of $f(x)$, and the table of (H_k^n) is to calculate the coefficients of $f(x)$ from the $f(0)$, $\Delta f(0)$, $\Delta^2 f(0)$, ..., $\Delta^k f(0)$;

For note 1:

$$f(y) = 1000000y^3 - 1900000y^2 - 262700y - 2436 = 0$$

In order to get its "0-side difference", Hua Hengfang designed an algorithm which is called "multiplying-table-and-adding". In fact this method proved to be the same as the matrix multiplication. Here is an example: taking the first six lines of the table (h_k^n)

$\otimes\rightarrow$ $\oplus\downarrow$

−2436	1					−2436			
−262700	0	1				0	−262700		
−1900000	0	−1	2			0	1900000	−3800000	
1000000	0	1	−6	6		0	1000000	−6000000	6000000
						−2436	2637300	−9800000	6000000

Thus, $f(0) = -2436$, $\Delta f(0) = 2637300$, $\Delta^2 f(0) = -9800000$, $\Delta^3 f(0) = 6000000$.

For note 2:

$f(200) = -127836$, $\Delta f(200) = 1037300$, $\Delta^2 f(200) = 2200000$, $\Delta^3 f(200) = 6000000$.

Taking the first six lines of the table (H_k^n)

$\otimes\rightarrow$ $\oplus\downarrow$

−127836	1				−127836			
1037300	0	$\frac{1}{1}$			0	1037300		
2200000	0	$\frac{1}{1}$	$\frac{1}{2}$		0	1100000	1100000	
6000000	0	$\frac{2}{2}$	$\frac{2}{3}$	$\frac{1}{6}$	0	2000000	3000000	1000000
		$\frac{2}{6}$	$\frac{3}{6}$	$\frac{1}{6}$	−127836	4137300	4100000	1000000

So we get

$$f(z) = 1000000z^3 - 4100000\,z - 4137300z - 127836 = 0$$

Let: $z = y/100$, thus:

$$f_1(y) = y^3 - 410y - 41373y - 127836 = 0$$

References

Collected Monographs on Astronomy and Calendrics from the Dynastic Histories, Beijing: Zhonghua Press, 1976.

Qin Jiushao, *Mathematics Treatise in Nine Sections*, 1247.

Zhu Shijie, *Precious Mirror of the Four Elements*, 1248.

Hua Hengfang, *The Method of Limited Difference*, 1882.

Chen Meidong, *A New Exploration to Chinese Classical Calendars*, Liaoning Education Press, 1995.

Chen Zungui, *A History of Chinese Astronomy*, Shanghai People's Press, 1984.

Cuulen C., *Astronomy and Mathematics in Ancient China: the Zhou Bi Suan Jing*, Cambridge University Press, 1996.

Ho Pengyoke, *Li,Qi and Shu: An Introduction to Science and Civilization in China*, Hong Kong University Press, 1985.

Li Jimin, *A Study on the Oriental Classic Nine Chapters of Arithmetic and Their Annotations by Liu Hui*, Shanxi People's Education Press, 1990.

Li Yan, *A Study on the Interpolation in Ancient China*, Science Press, 1957.

Libbrecht U., *Chinese Mathematics in the Thirteenth Century*, London, 1973.

Needham J., *Science and Civilization in China*, Vol. 3, Cambridge,1959.

Qian Baocong, *A History of Chinese Mathematics*, Science Press, 1964.

_____ , *Selected Works on the History of Science by Qian Baocong*, Science Press, 1983.

Qu Anjing, Ji Zhigang and Wang Rongbin, *An Examination of Mathematical Astronomy in Ancient China*, Northwest University Press, 1994.

Yabuuchi Kiyoshi, *Researches on Sui and Tang Calendrical Science*, Tokyo, 1944.

Why Interpolation?

QU ANJING 曲安京
Northwest University, Xian, China
Visiting Scholar, Department of the History of Science
Harvard University

A New Solar Theory in Medieval China

In about 560 AD, and after a 30-year observation, Zhang Zixin (张子京) found that both the solar and the planetary movements were irregular. He pointed out that there are deviations in the mean speeds of the apparent solar and the five visible planetary motions.[1] Since the correction of irregular solar motion has a big influence on the prediction of solar eclipses, Zhang's discovery was recognized as so important that quite a few astronomers threw themselves on this issue for further investigation. In a previous study I have noted that, at least in 576 AD, the correction of the solar equation of centre had been used by Liu Xiaosun (刘孝孙)in his *Wuping li* (武平历).[2]

Zhang's new solar theory and its development were recorded in *The Astronomical Chapter of the Sui History* (《隋书天文志》). It is said that, in 576 AD, the North Qi Dynasty (北齐) received three new calendars, two of them compiled by the students of Zhang Zixin. Coincidentally, a solar eclipse was expected to take place in the 6th month of that year. The emperor organized a competition among the new and old calendars. He asked astronomers to forecast the time of this solar eclipse. The results showed that Liu Xiaosun's prediction was the most accurate.

Unfortunately, the North Qi Dynasty was destroyed by the North Zhou Dynasty (北周) the very next year, so the new calendars missed their chance to be officially employed. All of them have been lost.[3]

It is well known that irregular lunar motion was discovered in 2nd-century China. In 206 AD, Liu Hong (刘洪) made use of a kind of linear interpolation method for calculating the lunar equation of centre in his *Qianxiang li* (乾象历). The method he developed was to change the lunar speed from day to day while the moon kept in a mean motion within each day.[4] Liu Hong's algorithm changed the curve of lunar speed from a straight line to a zigzag pattern. This algorithm was adopted by the calendar-makers for the lunar motion aftermath.

In the calendars of Liu Xiaosun (576 AD) and Zhang Zhouxuan (张胄玄, 597 AD), Liu Hong's idea was employed to deal with the irregular solar motion. They let the solar speed change from qi (气) to qi while the sun remained in a mean motion during the period between two consecutive qi.[5] This algorithm changed the curve of the solar speed from a straight line to a zigzag. This is a kind of piecewise linear interpolation.

Although the algorithm of the solar motion in Liu Xiaosun and Zhang Zhouxuan's calendars had made great progress from the old theory, there were still obvious shortcomings and opportunities for improvement. The core idea which Liu Zhou promoted was to change the solar speed curve from a zigzag pattern to an oblique line. More precisely, he was going to change the solar speed curve from the straight line to an oblique one on each interval between two consecutive qi. This was the change from a linear to a parabolic interpolation.

Let us take the first two qi after the Winter Solstice as an example to show what Liu Zhou did (see Figure 2). In constructing such an oblique line of solar speed, first of all, Liu let the solar speed change as an arithmetic sequence from day to day in the interval OM. He then summed up this arithmetic progression. The summation was a parabolic function with a variable of days after the initial time of this interval. This parabolic function was the so-called quadratic interpolation. It calculates the value of the solar equation of centre.

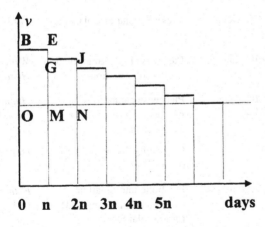

Fig. 1. Linear Interpolation

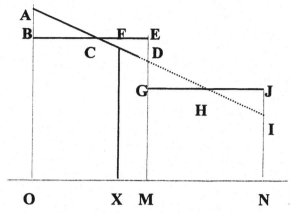

Fig. 2. Liu Zhou's Quadratic Interpolation

Table 1   A Chronological Table for the Evolution of Interpolation in China

Time	Astronomer	Events	Source
c. 2nd century	LI Fan 李梵 SU Tong 苏统	Irregular lunar motion.	*Later Han History* 《后汉书》
+206	LIU Hong 刘洪	Piecewise linear interpolation for the Lunar motion.	*Qianxiang li* in *Jin History* 《晋书》
c. +560	ZHANG Zixin 张子信	Irregular solar and planetary	*Sui History* 《隋书》
+576	LIU Xiaosun 刘孝孙	Piecewise linear interpolation for the solar motion.	*Wuping li* in *Sui History* 《隋书》
+600	LIU Zhuo 刘焯	Piecewise parabolic interpolation for the Sun and planets.	*Huangji li* in *Sui History* 《隋书》
+724	Yi-xing 一行	Correcting the solar theory of Liu Zhou	*Dayan li* in *Tang History* 《唐书》
+780	CAO Shiwei 曹士为	Parabolic interpolation (minus and times method) for the solar motion	*Futian li* 符天历
+1280	GUO *Shoujinf* 郭守敬	Cubic interpolation	*Shoushi li* in *Yuan History* 《元史》

In fact, Liu Zhou used the 24 *qi* to break a tropical into 24 small intervals, and based on the observing data of the 24 *qi*, he constructed a parabolic interpolation for each interval. We call the 24 parabolic interpolation functions of a tropical year together a piecewise parabolic interpolation.

This method was adopted widely to deal with the irregular movements of the sun, the moon and the planets in later calendar-making systems in China. It actually formed the mainstream of numerical method in medieval Chinese mathematical astronomy.[6]

Evolution of the New Solar Theory

Where did the new solar theory come from? Around the time when Zhang Zixin declared his discovery, there was a substantial exchange of trade and culture between China and her neighboring countries. Of course this also fostered an exchange of scientific knowledge. It is well known that both Indian and Islamic astronomers knew of the irregular movement of the sun and planets much earlier than did those in China. Concerning the new solar theory after Zhang's discovery, one might ask whether this was influenced by Indian or Islamic astronomical theory. Such a question might be expressed in two parts: was Zhang's discovery learnt from Buddhist or Moslem astronomers?; and, did Liu's quadratic interpolation come from another civilization?

To answer these questions, a discussion of the evolution of the new solar theory after the 6th century is needed. Since Liu Xiaosun's *Wuping li* (576 AD) has been lost, Zhang Zhouxuan's *Daye li* (大业历, 597 AD) is considered the first Chinese calendar which shows us the calculation the solar equation of centre in the 6th century. Zhang's method is a kind of piecewise linear interpolation. For constructing his algorithm, the 24 *qi* were used to break a tropical year into 24 small intervals. He let the length of each interval have the same value *n*. According to the observation, he listed the data of the solar equation of centre of the 24 *qi* in a constructed solar table. By the use of this data, he derived the mean speed of the solar motion for each interval.

The curve of the solar speed in the *Daye li* is a zigzag line. A conspicuous feature of this curve's pattern is that it is not only unsymmetrical but also irrational. For instance, the speed of the sun makes a big jump around the points of the spring or the autumn equinox. The interesting point is that Liu Zhou adopted Zhang's data with a pure mathematical adjustment. In Liu's solar theory, its speed was supposed in an absolutely symmetric pattern. The most irrational thing what Liu inherited from Zhang is that the solar speed reached its maximum on the day before the spring equinox, and fell to its minimum one day later. It fell to its minimum again on the day before the autumn, and reached its maximum one day later.[7]

The Buddhist monk Yi-xing criticized Liu's solar theory in a long article which is recorded in the *Calendar Chapter of the Tang History* (《唐书历志》). He said:

> In the Northern Qi Dynasty, from his experience of observing eclipses, Zhang Zixin found that there is a deviation from the mean speed of the solar motion. But he did not obtain the correct details of this deviation. Up to Liu Zhou, a method for the solar equation of centre was established ...

> When the sun moves to its southernmost direction (the point of the winter solstice), the speed is at its maximum, and then it gets gradually slower. It will reach its mean speed at the spring equinox, and then move slower than the mean speed. When the sun moves to its northernmost direction (the point of the summer solstice), the speed is at its minimum, and then it gets gradually faster. It will reach its mean speed at the autumn equinox, and then move faster than the mean speed. ...

> In Liu Zhou's solar theory, the solar motion reaches its fastest speed on the day before the spring equinox, and its slowest one day later. It is at its slowest speed on the day before the autumn equinox, and at its fastest one day later. The fastest and slowest values (around the equinoxes) are the same with the values of the winter and the summer solstices, and on the day of the equinox the sun is at its mean motion. His solar theory is not correct. ...[8]

From these words of Yi-xing, we know that he had understood a rational pattern for solar motion. Given that the patterns of the solar motion in Zhang Zhouxuan and Liu Zhou's theory are so strange, I would like to postulate that the new solar theory after Zhang Zixin was developed by Chinese calendar-makers independently.

The Importance of Liu's Interpolation

In ancient and medieval times, mathematical astronomy was one of the most important branches of learning. It was a well-developed subject both in the medieval west and China. In comparing Western and Chinese sciences, mathematical astronomy may be a subject worthy of consideration in detail.

We know that there are quite different academic traditions behind Western and Chinese science. Concerning the precision of predictions of celestial bodies' movements in medieval time, it is hard to say which calendar-making system is better. The scope of the two systems is also rather too large for a single paper to make a detailed study. A possible way to make a comparison between the two would be to choose a few cardinal points of medieval mathematical astronomy.

Generally, elements which determined the precision of a calendar-making systems included the following:

1. Dynamic system, which in pre-modern times was simply the construction of a geometric model.
2. Astronomical observation and instruments.
3. Numerical method.

In the Western tradition, the most important thing for an astronomer was to build up a cosmic model for his calendar-making system. Chinese calendar-makers, by contrast, never declared what kind of cosmic model they used to help them establish their algorithm system, even if they did occasionally make use of a few geometric models of celestial bodies. Actually, the impression we have of Chinese astronomy is that no Chinese calendar-makers mentioned that they had intentionally constructed a cosmic model for the use of their calendar-making system.

In the Imperial Observatory, Chinese calendar-makers consisted of two groups of experts: those who were called astronomers dealt with the astronomical instruments and observations; those who were called mathematicians took charge of constructing algorithms. There were no positions for those who were interesting in cosmic model building. Chinese people set out to invent a numerical method such as the interpolation, not a celestial model such as Ptolemy's epicycle-deferent system. In the aspect of astronomical observation and instruments, there was no substantial difference between the accuracy of the two astronomies. Both of them reached a very high standard in their time.

In Ptolemy's system, there is a well-built geometric model for the movements of celestial bodies. By the use of this model, one could obtain a few relations among the astronomical items it illustrated. Since spherical trigonometry was not yet built up at that time, a highly precise *Table of Chord* created by Ptolemy was used for exchanging the length of arc on a celestial sphere to its corresponding chord. This was one of the main numerical algorithms in the medieval Western calendar-making system.

Creating mathematical tables for calculating some astronomical items was also a method used in medieval Chinese calendar-making systems. For instance, in early 8th-century China, the Buddhist monk Yi-xing created a mathematical table for calculating the length of the solar shadow at noon. In this table, the length of a shadow of an 8 *chi* high gnomon was given in each integer *du* (Chinese degree) of the solar zenith distance from 0 to 81. Since this data of shadows was 8 times the value of a tangent function, it is regarded as a tangents table by modern historians of science. This table was used by Yi-xing to calculate the length of the meridian line.[9]

Although Yi-xing constructed a tangents table, it seems to us that it did not have a substantial influence on Chinese calendar-making systems. The most important methods that people used were some numerical functions. For instance, in the *Qintian li* (钦天历, 956 AD), the length of Yi-xiang's meridian line was used by Wang Pu (王朴) in his new algorithm of the solar shadow.[10] If Yi-xing's shadow table could be regarded as a tangents table, Wang's algorithm gave us a kind of tangent function.

Even if they calculated the time of eclipses from the first (or second) contact to the middle eclipse, Chinese astronomers use the numerical method to construct a quadratic function instead of a geometric model. This method was kept without change until the *Shoushi li* (授时历, 1280 AD). The mainstream of algorithms in the Chinese calendar-making system was Liu's piecewise parabolic interpolation. This method was used so extensively that it appeared everywhere it could be used.

The piecewise parabolic interpolation is such an important numerical algorithm that it is still used widely in modern science. It is well-known that an nth order interpolation function is normally constructed by n+1 interpolation points. In other words, in case one chooses n+1 nodes as interpolation points, there is one and only one determined interpolation function whose order is no more than the nth.

One may think from his intuition that the higher the order of an interpolation function is, the more precise it is. Unfortunately, this is far from the truth. In computational mathematics, there is a famous example which is called the Runge Phenomenon. Runge told us that the higher the order of an interpolation function, the bigger the amplitude of vibration around the interpolation node (see Figure 3). This is the reason why a textbook of numerical analysis recommends that engineers use piecewise lower order interpolation rather than a higher order one.

Therefore, the most important idea that Liu Zhou told us does not only include a higher-order interpolation than the linear, or the piecewise. In fact, from the technical point of view, as a numerical algorithm, the piecewise interpolation has been good enough for dealing with the calculations necessary to medieval calendar-making. The reason why Chinese calendar-makers did not reach the high standard we might expect with this method is simply due to their rather too rough data of astronomical observation.

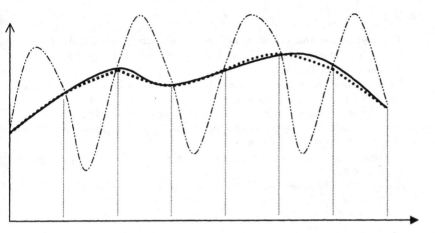

Fig. 3. The Range Phenomenon of Higher Order Interpolation

A Conjecture

Based on a series of calculations of the solar eclipse and the planetary movements with the methods of medieval Chinese and Ptolemy-Islamic mathematical astronomy, we found that prediction in the Chinese calendar-making system is as good as the results in the Ptolemaic-Islamic system. What was the real reason behind such a balance? I propose here a conjecture shown in Table 2.

Table 2   A Comparison between Chinese and Western Calendar-making

	Subject	China	West
1	Dynamic system or Geometric model	+	+ + +
2	Observation & Instruments	+ +	+ +
3	Numerical method	+ + +	+

Notes

[1] The original text which recorded of Zhang's discovery is found in *ACH*, vol. 2, p. 599. A study of this historical material is given in Qu 1994, pp. 226–28.

[2] See Qu 1994, pp. 228–30.

[3] Some fundamental astronomical constants in these calendars, such as the tropical year and synodic month, are re-discovered in Qu 1994, pp. 157–63.

[4] An entire day is defined here between two consecutive mid-nights.

[5] *Qi* is a special term in the traditional Chinese astronomy. Roughly speaking, when people break the ecliptic circle into 24 equal length arcs, each node is named a *qi*. For instance, the winter and summer solstice, and spring and autumn equinox points are four of the 24 *qi*.

[6] See Li and Qu 1994, pp. 181–91.

[7] See Qu 1994, pp. 226–35.

[8] See *ACH,* vol. 7, p. 2203.

[9] See Qu 1998.

[10] See *ACH*, vol. 7, pp. 2415–16.

References

ACH. 1976. *Lida Tianwen Luli Dengzhi Huibian* (历代天文律历等志汇编 Collected Monographs on Astronomy and Calendrics from the Dynastic Histories), vols. 2, 5–10. Beijing: Zhonghua shuju.

Li, Yan. 1957. *Zhongsuanjia de Neichafa Yanjiu* (中算家的内插法研究 Studies on the Interpolations in Old China), Beijing: Science Press.

Qu A., Ji Z. and Wang R. 1994. *Zhongguo Gudai Shuli Tianwenxue Tanxi* (中国古代数理天文学探析 An Examination of Mathematical Astronomy in Ancient China), Xian: Xibeidaxue Press, pp. 157–235.

Qu Anjing. 1998. "Applications of the Tangent Table in Yi-xing's Meridian Survey of the 8th Century", *Chinese Studies* (汉学研究) 16(1): 91–109.

An Investigation of the *Yíng-Bùzú* 盈不足 Method

CHEN CHENG-YIH 程貞一
Department of Physics, University of California, La Jolla, California
LI ZHAO-HUA 李兆華*
The Wei-Kung 為公 Institute, La Jolla, California

Introduction

The method called *yíng-bùzú* 盈不足 has probably multiple origins. Its earliest extant description is found in the 'Yíng-Bùzú' 〈盈不足〉 chapter of the *Jiǔ-Zhāng Suàn-Shù* 《九章算術》 (*Mathematics in Nine Chapters*). The discovery in 1985 of a *zhújiǎn* 竹簡 (bamboo-slip) mathematical book entitled *Suàn-Shù Shū* 《算數書》 in a Western Hàn 漢 tomb at Zhāngjiāshān 張家山[1] offers another early source of the method.[2] Mathematically, the *yíng-bùzú* method deals basically with systems of two linear equations with two unknowns, y and z,

$$x_1 y + z = c_1 \qquad (1a)$$

$$x_2 y + z = c_2 \qquad (1b)$$

In this sense, it is related to the *fāng-chéng* 方程. Unlike the *fāng-chéng* method, the linear equations are not solved numerically by an algorithm of elimination,[3] instead the *yíng-bùzú* method makes use of the formal solution to express the unknowns, y and z, in terms of the constants, x_i and c_i.

The most important mathematical innovation in the *yíng-bùzú* method is to visualize a functional relationship between x_i and c_i by introducing the expression[4]

$$c_i = f(x_i) \qquad (2)$$

This leads to the functional relationship:

$$ax + b = f(x) \qquad (3)$$

345

where the unknowns, $y \equiv a$ and $z \equiv b$, are visualized as constants as the function f(x) changes with respect to its variable x. In terms of the functional relationship given by Eq. 3, Equations 1 can be viewed as Equation 3 at two distinct values, x_1 and x_2, of the variable x:

$$ax_1 + b = f(x_1) \tag{4a}$$

$$ax_2 + b = f(x_2) \tag{4b}$$

The formal solutions for Equations 4 then take the form:

$$a = \frac{f(x_2) - f(x_1)}{x_2 - x_1} \tag{5a}$$

$$b = \frac{x_1 f(x_2) - x_2 f(x_1)}{x_1 - x_2} \tag{5b}$$

By introducing the functional relationship given by Eq. 2, the early Chinese mathematicians were able to change the nature of the mathematical problem. Instead of examining the system of linear equations with two unknowns (Eq. 1), they would deal with a linear function with one variable (Eq. 3).

The significance of this formalism in transforming a system of linear equations with two unknowns into a linear function with one variable lies in the new mathematical interpretations it facilitates. By analyzing the behavior of the linear function f(x) with respect to its variable x, the early Chinese mathematicians were able to make the following interpretations:

$$
\begin{array}{lll}
f(x) > 0 & \text{'yíng 盈'} & \text{(excess)} \\
f(x) = 0 & \text{'zú 足'} & \text{(sufficiency)} \\
f(x) < 0 & \text{'Bùzú 不足'} & \text{(deficiency)}
\end{array} \tag{6}
$$

This is probably the earliest known mathematical analysis of a functional behavior. At the point $x = x_0$ where f(x_0) = 0, the linear function has a solution given by the equation $ax_0 + b = 0$. Hence, one has, with the help of Eqs. 5,

$$x_0 = -\frac{b}{a} = \frac{x_1 f(x_2) - x_2 f(x_1)}{f(x_2) - f(x_1)}. \tag{7}$$

This is the well-known *yíng-bùzú* formula given in the 'Yíng-Bùzú' 〈盈不足〉 chapter of the *Jiǔ-Zhāng Suàn-Shù* 《九章算術》 (*Mathematics in Nine Chapters*), known later in Europe as the Rule of False Position.[5,6]

The *Yíng-Bùzú* 盈不足 Method[7]

We now examine how the *yíng-bùzú* method is presented in the *Jiǔ-Zhāng Suàn-Shù* 《九章算術》 (*Mathematics in Nine Chapters*). In order to account for the rules of signs, the author of the 'Yíng-Bùzú' 〈盈不足〉 chapter divided the method into two cases based on the signs of the function $f(x)$ (see Eq. 6) and presented them separately. In the first case, called the 'one-*yíng*-and-one-*bùzú*' method (盈，不足術), the two functional values $f(x_1)$ and $f(x_2)$ are of opposite signs (i.e. one positive and one negative). In the second case, called the 'two-*yíng*-or-two-*bùzú*' method (兩盈，兩不足術), the two functional values are of the same sign (i.e. both positive or both negative). Accordingly, we will examine these two cases separately.

The Opposite-Sign Case

For the case in which the functional values $f(x_1)$ and $f(x_2)$ are opposite in signs (i.e. one positive and one negative), the rules for the *wéi-chéng* 維乘 operations is stated as follows:

置所出率，盈、不足各居其下。
Set out the selected values x_1 and x_2 for the variable x, and below them place their corresponding excess and deficiency values [i.e. functional values $f(x_1)$ and $f(x_2)$], yielding the configuration

$$\begin{array}{cc} x_1 & x_1 \\ f(x_1) & f(x_2) \end{array} \tag{8}$$

令維乘所出率，并以為實。并盈，不足為法。實如法而一。
From this configuration, one obtains $\pm\{x_1|f(x_2)| + x_2|f(x_1)|\}$ by adding (*bìng* 并) the products obtained in cross multiplication. One then takes this quantity $\pm\{x_1|f(x_2)| + x_2|f(x_1)|\}$ as the numerator (*shí* 實). Similarly, one obtains $\pm\{|f(x_2)| + |f(x_1)|\}$ for the denominator (*fǎ* 法) by adding their corresponding *yíng* and *bùzú* values. The solution is then given by uniting the numerator and denominator into one, i.e.

$$x_0 = \frac{x_1|f(x_2)| + x_2|f(x_1)|}{|f(x_2)| + |f(x_1)|}. \tag{9}$$

This is the *yíng-bùzú* 盈不足 formula for the case in which the functional values, $f(x_2)$ and $f(x_1)$, are in the opposite signs.

The Same-Sign Case

For the case in which the functional values $f(x_1)$ and $f(x_2)$ are of the same sign (i.e. both positive or both negative), the rules for the *wéi-chéng* 維乘 operations is stated as follows:

置所出率，盈、不足各居其下。

Set out the selected values x_1 and x_2 for the variable x, and below them place their corresponding functional values [i.e. functional values $f(x_1)$ and $f(x_2)$], yielding the configuration

$$x_1 \qquad x_2$$
$$f(x_1) \quad f(x_2).$$

令維乘所出率，以少減多，餘為實。兩盈、兩不足以少減多餘為法。實如法而一。

From this configuration, one obtains $\pm\{x_1|f(x_2)| - x_2|f(x_1)|\}$ by subtracting (*jiǎn* 減) the smaller from the larger of the products obtained in cross multiplication. One then takes this quantity $\pm\{x_1|f(x_2)| - x_2|f(x_1)|\}$ as the numerator (*shí* 實). Similarly, one obtains $\pm\{|f(x_2)| - |f(x_1)|\}$ for the denominator (*fǎ* 法) by subtracting the smaller from the larger of the two functional values. The solution is then given by uniting the numerator and denominator into one, i.e.

$$x_0 = \frac{x_1|f(x_2)| - x_2|f(x_1)|}{|f(x_2)| - |f(x_1)|}. \tag{10}$$

This is the *yíng-bùzú* 盈不足 formula for the case in which the functional values, $f(x_2)$ and $f(x_1)$, are in the same sign.

In the above two cases, the over-all signs $\pm$ for the numerator and for the denominator are not explicitly mentioned in the texts. This is because these over-all signs are cancelled out in the formulas. The *yíng-bùzú* 盈不足 formulas for the two cases can be summarized as below:

$$x_0 = \frac{x_1|f(x_2)| + x_2|f(x_1)|}{|f(x_2)| + |f(x_1)|} \quad \text{for} \quad \begin{array}{l} f(x_1) = +|f(x_1)| \text{ and } f(x_2) = -|f(x_2)| \\ f(x_1) = -|f(x_1)| \text{ and } f(x_2) = +|f(x_2)| \end{array},$$

$$x_0 = \frac{x_1|f(x_2)| - x_2|f(x_1)|}{|f(x_2)| - |f(x_1)|} \quad \text{for} \quad \begin{array}{l} f(x_1) = +|f(x_1)| \text{ and } f(x_2) = +|f(x_2)| \\ f(x_1) = -|f(x_1)| \text{ and } f(x_2) = -|f(x_2)| \end{array}.$$

Obviously, the formulas given in the 'Yíng-Bùzú' 〈盈不足〉 chapter of the *Jiŭ-Zhāng Suàn-Shù* can be combined into the general formula

$$x_0 = \frac{x_1 f(x_2) - x_2 f(x_1)}{f(x_2) - f(x_1)},$$
(11)

It is also obvious that the reason for separating the presentation of the formula into these two cases is to specify the 'rules of signs' in performing the *wéi-chéng* 維乘 operations.

Positional Notation: Matrix and Determinant

In the positional notation of the *fāng-chéng* method as given in the 'Fāng-Chéng' 〈方程〉 chapter of the *Jiŭ-Zhāng Suàn-Shù*, the system of two linear equations with two unknowns, *a* and *b*,

$$ax_1 + b = f(x_1)$$
(4a)

$$ax_2 + b = f(x_2),$$
(4b)

is represented by the following configuration:

$$
\begin{array}{cc}
x_1 & x_2 \\
1 & 1 \\
f(x_1) & f(x_2)
\end{array}
$$
(12)

The configuration given in Equation 12 originated from the positional notation created by the ancient Chinese mathematicians to represent mathematical expressions in terms of the layouts of the counting rods in their mechanical computational scheme called *chóu-suàn* 籌算.

The derivation of the *yíng-bùzú* 盈不足 formulas given in the *Jiŭ-Zhāng Suàn-Shù* 《九章算術》 as discussed above calls for the setting of the following configurations for values x_1 and x_2 and their corresponding functional values $f(x_1)$ and $f(x_2)$:

$$
\begin{array}{cccc}
x_1 & x_2 & 1 & 1 \\
f(x_1) & f(x_2) & f(x_1) & f(x_2)
\end{array}
$$
(13)

It is of interest to note that these configurations came directly from the configuration given by Eq. 12 for the two linear equations. By the *wéi-chéng* 維乘 operations with appropriate consideration of the rules of signs, one obtains from the left configuration

$$\pm\{x_1|f(x_2)| + x_2|f(x_1)|\} \qquad \text{for the opposite-sign case}$$

$$\pm\{x_1|f(x_2)| - x_2|f(x_1)|\} \qquad \text{for the same-sign case} \qquad (14)$$

Similarly, from the right configuration

$$\pm\{|f(x_2)| + |f(x_1)|\} \qquad \text{for the opposite-sign case}$$

$$\pm\{|f(x_2)| - |f(x_1)|\} \qquad \text{for the same-sign case} \qquad (15)$$

This lead to the expressions $x_1 f(x_2) - x_2 f(x_1)$ and $f(x_2) - f(x_1)$ from the two configurations of Eq. 12.[8]

It is well-known that if M is a 2×2 matrix, in current notation,

$$\begin{pmatrix} m_{11} & m_{12} \\ m_{21} & m_{22} \end{pmatrix},$$

then its determinant is defined as $\det M = m_{11}m_{22} - m_{12}m_{21}$, often written as

$$\begin{vmatrix} m_{11} & m_{12} \\ m_{21} & m_{22} \end{vmatrix}. \qquad (16)$$

This implies that the left configuration given by Eq. 13 can be considered as a 2×2 matrix with elements $m_{11} = x_1$, $m_{12} = x_2$, $m_{21} = f(x_1)$, and $m_{22} = f(x_2)$, since $x_1 f(x_2) - x_2 f(x_1)$ is exactly the determinant of such a matrix. Similarly, the right configuration with its determinant $f(x_2) - f(x_1)$ is also a 2×2 matrix. Thus, the configurations given by Equation 13, although missing brackets, are in effect matrix configurations, since they have the function of a matrix.

The observation of the first appearance of a matrix and its determinant in the *yíng-bùzú* 盈不足 method of the *Jiŭ-Zhāng Suàn-Shù* 《九章算術》 (*Mathematics in Nine Chapters*) is of great interest since here one finds the precursors of the Seki Kowa's determinants. Earlier attempts to demonstrate the possible appearance of a determinant in the *yíng-bùzú* method failed to take into consideration the proper procedures for the *wéi-chéng* 維乘 operations with respect to the rules of signs.[9] In performing the *wéi-chéng* multiplication, one must consider the relative signs of the elements of the configuration (i.e. the matrix elements) in combining the products of the multiplication. Before the development of a convenient symbol for the sign, the early Chinese mathematicians turned to the division of the opposite-sign and same-sign cases to implement the rules of sign.[10]

Cramer's Rules

In the positional notation of the *fāng-chéng* method as discussed in the last section, the system of two linear equations with two unknowns, *a* and *b* given by Eqs. 4 is represented by the following matrix configuration:

$$
\begin{array}{cc}
x_1 & x_2 \\
1 & 1 \\
f(x_1) & f(x_2)
\end{array}
\tag{12}
$$

Other than the missing brackets, the matrix configuration given by Eq. 12 differs from current notation only in its orientation.

In the 'Yíng-Bùzú' 〈盈不足〉 chapter of the *Jiǔ-Zhāng Suàn-Shù*, we have the instructions for the determination of the unknowns *a* and *b*. In the instructions, the unknowns are identified with those that appeared in the selected examples. The identifications that *a* = *wùjià* 物價 and *b* = *rénshù* 人數 are adopted in the following instructions:

副置所出率，以少減多，餘以約法、實。實為物價，法為人數。

By setting out the selected values x_1 and x_2 for the variable x, and by subtracting (*jiǎn* 減) the smaller from the larger, one obtains the *yú* 餘, (i.e. $\pm|x_2 - x_1|$). By dividing $\pm|x_2 - x_1|$ into the *shí* 實 (i.e. $\pm\{x_1|f(x_2)| + x_2|f(x_1)|\}$ or $\pm\{x_1|f(x_2)| - x_2|f(x_1)|\}$), one obtains for the unknown *b* the formula:

$$
b = \frac{x_1 f(x_2) - x_2 f(x_1)}{x_1 - x_2} = \frac{\begin{vmatrix} x_1 & x_2 \\ f(x_1) & f(x_2) \end{vmatrix}}{\begin{vmatrix} x_1 & x_2 \\ 1 & 1 \end{vmatrix}}
\tag{17}
$$

By dividing $\pm|x_2 - x_1|$ into the *fǎ* 法, (i.e. $\pm\{|f(x_2)| + |f(x_1)|\}$ or $\pm\{|f(x_2)| - |f(x_1)|\}$), one obtains for the unknown parameter *a* the formula:

$$
a = \frac{f(x_2) - f(x_1)}{x_2 - x_1} = \frac{\begin{vmatrix} f(x_1) & f(x_2) \\ 1 & 1 \end{vmatrix}}{\begin{vmatrix} x_1 & x_2 \\ 1 & 1 \end{vmatrix}}
\tag{18}
$$

From these expressions for the unknowns a and b, one obtains the formal *yíng-bùzú* 盈不足 formula:

$$x_0 = -\frac{b}{a} = -\frac{\begin{vmatrix} x_1 & x_2 \\ f(x_1) & f(x_2) \end{vmatrix}}{\begin{vmatrix} f(x_1) & f(x_2) \\ 1 & 1 \end{vmatrix}} = \frac{\begin{vmatrix} x_1 & x_2 \\ f(x_1) & f(x_2) \end{vmatrix}}{\begin{vmatrix} f(x_2) & f(x_1) \\ 1 & 1 \end{vmatrix}} \tag{19}$$

It is now well-known that the following are determinant identities

$$\begin{vmatrix} x_1 & x_2 \\ f(x_1) & f(x_2) \end{vmatrix} = \begin{vmatrix} x_1 & f(x_1) \\ x_2 & f(x_2) \end{vmatrix},$$

$$\begin{vmatrix} f(x_1) & f(x_2) \\ 1 & 1 \end{vmatrix} = \begin{vmatrix} f(x_1) & 1 \\ f(x_2) & 1 \end{vmatrix}, \begin{vmatrix} x_1 & x_2 \\ 1 & 1 \end{vmatrix} = \begin{vmatrix} x_1 & 1 \\ x_2 & 1 \end{vmatrix}$$

This implies that the formulas given in the *yíng-bùzú* 盈不足 method can be transformed into the Cramer's form using these identities. One then has for a, b, and x_0 the identical formulas based on Cramer's rules.

$$a = \frac{\begin{vmatrix} f(x_1) & 1 \\ f(x_2) & 1 \end{vmatrix}}{\begin{vmatrix} x_1 & 1 \\ x_2 & 1 \end{vmatrix}}, b = \frac{\begin{vmatrix} x_1 & f(x_1) \\ x_2 & f(x_2) \end{vmatrix}}{\begin{vmatrix} x_1 & 1 \\ x_2 & 1 \end{vmatrix}},$$

$$x_0 = -\frac{b}{a} = -\frac{\begin{vmatrix} x_1 & f(x_1) \\ x_2 & f(x_2) \end{vmatrix}}{\begin{vmatrix} f(x_1) & 1 \\ f(x_2) & 1 \end{vmatrix}} = \frac{\begin{vmatrix} x_1 & f(x_1) \\ x_2 & f(x_2) \end{vmatrix}}{\begin{vmatrix} f(x_2) & 1 \\ f(x_1) & 1 \end{vmatrix}} \tag{20}$$

This observation is of interest since here one finds the precursors of Cramer's rules in the *yíng-bùzú* 盈不足 method of the *Jiǔ-Zhāng Suàn-Shù* 《九章算術》 (*Mathematics in Nine Chapters*).[11] However, this formal approach was not extended beyond the 2×2 matrix for systems of two linear equations. For systems of more than two linear equations, progress was in algorithm of elimination along the *fāng-chéng* 方程 method of the *Jiǔ-Zhāng Suàn-Shù*.

Yíng-Bùzú 盈不足 and Linear Interpolation

The beauty of the *yíng-bùzú* method is not only in its mathematical concepts and formalism but also in its potential development in the mathematics of approximation, a very important topic in applied mathematics.[12] It was in the applications of the method for the calculation of values in between two determinations (which are not linearly related) that led to the *yíng-bùzú* method for linear interpolation. Such applications are found in three problems (Problems 11, 12 and 19) in the 'Yíng-Bùzú' 〈盈不足〉 chapter of the *Jiǔ-Zhāng Suàn-Shù*. These applications are seminal developments in the mathematics of approximation. Although empirical estimations of values between observations probably were made earlier in astronomy, using procedures similar in nature with interpolation,[13] it was mathematicians with the proper formalism and appreciation of the mathematical concepts of functions and variables who facilitated the development of the *yíng-bùzú* method for mathematical interpolation.

In this section, we examine such a linear interpolation application of the *yíng-bùzú* method by analyzing Problem 19 of the 'Yíng-Bùzú' 〈盈不足〉 chapter. The problem is produced and translated below:

今有良馬與駑馬發長安至齊。齊去長安三千里。
良馬初日行一百九十三里，日增十三里·駑馬初日行九十七里，
日減半里。良馬先至齊，復還迎駑馬。
問幾何日相逢及各行幾何。

Given a good horse and an old horse to be dispatched from Chángān 長安 to Qí 齊, a distance of 3,000 *lǐ* 里. The good horse travels a distance of 193 *lǐ* 里 in the first day and an increase of 13 *lǐ* 里 each day; the old horse travels a distance of 97 *lǐ* 里 in the first day and a decrease of 0.5 *lǐ* 里 each day. The good horse reached Qí 齊 first and then returned to meet the old horse. At which day will the two horses meet and by then what is the distances travelled by each horse?

This is a very interesting problem not only in mathematics but also in the physics of motion.[14] Based on the information given in the problem, one can construct a function $g(t)$ with the variable t (i.e. time in days) as follows:

$$g(t) = s_g(t) + s_o(t) - 6000 \qquad (21)$$

where $s_g(t)$ and $s_o(t)$ are the distances travelled by the good horse and the old horse as a function of time, respectively, and the functions, $g(t)$, $s_g(t)$, and $s_o(t)$, as well as the constant 6000, are all measured in *lǐ* 里 . By calculating the distance travelled by each horse each day, one can easily tabulate $g(t)$ versus t as given in the first and second columns of Table 1.

Table 1   A Comparison of Function g(*t*) with its Linear Interpolation I(*t*)

t (day)	g(t) (*lĭ*)	I(t) (*lĭ*)	ΔI(t) (*lĭ*)
11	−2122.5	−2247.5	477.5
12	−1695.0	−1770.0	477.5
13	−1255.0	−1292.5	477.5
14	−802.5	−815.0	477.5
15	−337.5	−337.5	477.5
16	+140	+140.0	477.5
17	+630	+617.5	477.5
18	+1132.5	+1095.0	477.5
19	+1647.5	+1572.5	477.5
20	+2175	+2050.0	

It is seen from the tabulation that a sign change occurred between the 15th and the 16th day. On the 15th day, the total distance travelled by the two horses is less than 6000 *lĭ* and on the 16th day more than 6000 *lĭ*. One has t_1 = 15 days, $g(t_1)$ = −337.5 *lĭ* and t_2 = 16 days, $g(t_2)$ = +140 *lĭ*. Using the formulas given by Eqs. 14 and 15 (or Eqs. 5), the two unknowns can be calculated, we have a = 477.5 and b = 7500. Hence, we have the linear function I(*t*) to represent g(*t*),

$$I(t) = 477.5 \ t + 7500. \tag{22}$$

A comparison of I(*t*) with g(*t*) is also given in the tabulation. It is seen that I(*t*) approximates g(*t*) well at points near the sign change. Solving $I(t_0) = 0$ yields $t_0 = 15 \frac{135}{191}$ which is the solution given for Problem 19 in the *Jiŭ-Zhāng Suàn-Shù*.

It is not explained in the answer given for Problem 19 whether the distances travelled by the two horses as a function of time [i.e. $s_g(t)$ and $s_o(t)$] were determined numerically or using certain rules. Liú Huī 劉徽, in his commentary of +263, made used of the summation rules for arithmetic progression to calculate the values of $s_g(t)$ and $s_o(t)$:

$$s_a(t) = 193 \ t + \frac{13}{2} t \ (t - 1). \tag{23}$$

$$s_a(t) = 97 \ t - \frac{1}{4} t \ (t - 1). \tag{24}$$

This implies that no later than the time of Liú Huī the nonlinear nature of the function $g(t)$ for Problem 19 was noticed. We have, with the help of Eqs. 23 and 24, the following expression for $g(t)$:

$$g(t) = 6.25 \ t^2 + 283.75 \ t - 6000. \tag{25}$$

This implies that, no later than the time of Liú Huī, it was realized that the answer given for the problem is only approximate. The linear function $I(t)$ obtained according to the formulas given in the *Jiǔ-Zhāng Suàn-Shù* is not in agreement with the function $g(t)$ obtained by known summation rules at the time. Although Liú Huī noticed that the answer given for the problem is only approximate, he did not, however, solve Equation 25 for the exact answer.

In view of its early date, this problem constitutes an amazing example of mathematical interpolation. It reveals the attributes of the *yíng-bùzú* method for linear interpolation. The nonlinear nature of the functions, $s_g(t)$ and $s_o(t)$, obtained by Liú Huī are in themselves of great interest not only in mathematics but also in physics since the result reveals that motion with a constant acceleration or de-acceleration has a second-order time dependence. This probably is the earliest demonstration of such a time dependence, and is indeed a major accomplishment in the physics of motion.[15]

Acknowledgement

The authors are grateful to the Lee Foundation 李氏基金 for the support of the Panel on Interpolation in which this paper was presented at the 9th International Conference on the History of Science in East Asia held at Singapore, 23–27 August 1999.

Notes

* On leave from the Department of Mathematics, Tiānjīn 天津 Normal University, China.

1 Zhāngjiāshān Zhú-Jiǎn Zhěng-Lǐ Xiǎo-Zǔ 張家山竹閒整理小組 (1985).

2 The book contains problems in the topic of Yíng-Bùzú 盈不足, Péng Haò 彭浩, private communication with CYC, 1999, see also Cheng-Yih Chen (1987), p. 32. For a further discussion on the content of the book see Dù Shí-Rǎng 杜石然 (1988).

3 Described in the 'Fāng-Chéng' 〈方程〉 chapter of the *Jiǔ-Zhāng Suàn-Shù* and in the Liú Huī's 劉徽 Commentary.

4 Cheng-Yih (Joseph) Chen (1980).

5 See for example, David E. Smith (1953), pp. 347–441 .

6 For a discussion on the transmission of the *yíng-bùzú* 盈不足 method see Qián
   Bǎo-Cóng 錢寶琮 (1983), pp. 83–96.
7 Based on the Lecture Notes of Cheng-Yih (Joseph) Chen (1980).
8 It is important to emphasize that in the derivation of the *yíng-bùzú* 盈不足 formulas
   as discussed above, the Chinese mathematicians were not describing these two
   expressions by themselves but describing a formula which is given by the ratio of
   these two expressions. Since the over-all signs ± for the two expressions cancelled
   out exactly in the ratio, they were not explicitly mentioned. With the proper
   procedures for the *wéi-chéng* 維乘 operations, the 'rules of signs' is always
   correctly satisfied.
9 See for example, Ulrich Libbrecht (1973), pp. 163–71.
10 Cheng-Yih (Joseph) Chen (1980).
11 *Ibid.*
12 For a review of Chinese development in interpolation, see Lǐ Yǎn 李儼 (1957).
13 The students of astronomy was probably first to face the task to devise a
   way to find values between those for which measurements (or observations) are
   available in the tabulation. A possible earlier example of linear interpretation
   is found in the work of Chén Zǐ 陳子 in the 'Róng Fāng Wèn Chén Zǐ 〈榮方
   陳子〉 of the *Zhōu-Bì Suàn-Jīng* 《周髀算經》 (*Mathematics Classics of Zhōu
   Gnomons)'.
14 Cheng-Yih Chen (1996), pp. 9–10.
15 Cheng-Yih Chen (1996), pp. 6–10.

References

A. Books and Documents Before +1800

Jiǔ-Zhāng Suàn-Shù 《九章算術》 (*Mathematics in Nine Chapters*), Zhōu 周.
   Restored by Zhāng Cāng 張蒼 (fl. -210 – -162, d. -152) in c. -165 based on Zhōu
   周 fragments and edited by Gěng Shòu-Chāng 耿壽昌 (fl. -75 – -49). The earliest
   extant commentary is that by Liú Huī 劉徽 of the +3rd century.
Kai Fukudai no Hō (*Methods of Solving Problems by Determinants*), 1683, Seki Kowa
   (Seki Takakusu 關孝和).
Suàn-Shù Shū 《算數書》, Western Hàn *zhújiǎn* 竹簡 (bamboo-slip), a mathematical
   book unearthed in 1984 from a Xī Hàn 西漢 tomb at Zhāngjiāshān 張家山.
Xià Xiǎo Zhèng 《夏小正》 (*Lessor Annuary of the Xià Dynasty*), Xià 夏.
   Recompiled in the Zhōu 周 based on pre-Zhōu 周 materials. Author and
   compiler unknown.
Zhōu Bì Suàn-Jīng 《周髀算經》 (*Mathematics Classics of Zhōu Gnomons*). Compiled
   in the –3rd century. Incorporated in the book are two dialogues, 'Zhōu Gōng Wèn
   Shāng Gāo' 〈周公問商高〉 of the Xuī Zhōu 西周 and 'Róng Fāng Wèn Chén
   Zǐ' 〈榮方問陳子〉 of the –5th century or earlier.

B. Books and Articles Since +1800

Chen, Cheng-Yih (Joseph) 程貞一, *History of Mathematics in Chinese Civilization* (Lecture Notes for Chinese Studies 170, University of California, San Diego, 1980).

_____ , ed. *Science and Technology in Chinese Civilization* (World Scientific, Singapore, 1987).

_____ , *Early Chinese Work in Natural Science* (Hong Kong University Press, 1996).

Dù Shí-Răng 杜石然, 'Jiānglíng Zhāngjiāshān Zhú-Jiǎn *Suàn-Shù Shū* Chū-Tàn' 〈江陵張家山竹簡《算數書》初探〉('A Preliminary Investigation of the Bamboo-Slip Book *Suàn-Shù Shū* of Zhāngjiā-shān), 《自然科學史研究》 (Studies in the History of Natural Sciences) 7, No. 3 (1988), pp. 201–204.

Qián Băo-Cóng 錢寶琮, *Qián Băo-Cóng Kē-Xué-Shǐ Lùn-Wén Xuǎn-Jí* 《錢寶琮科學史論文選集》 (Science Press, Beijing, 1983).

Lǔ Yǎn 李儼, *Zhōng-Suàn-Jiā de Nèi-Chā-Fǎ Yán-Jiū* 《中算家的內插法研究》 (Science Press, Beijing, 1957).

Libbrecht, Ulrich, *Chinese Mathematics in the Thirteenth Century: The Shu-shu chiu-chang of Ch'in Chiu-shao* (The MIT Press, Cambridge, 1973).

Needham, Joseph. *Science and Civilisation in China* (The University Press, Cambridge, 1959), vol. 3.

Smith, David E. *History of Mathematics* (Ginn and Company, 1952), vol. II. *Special Topics of Elementary Mathematics*.

Zhāngjiāshān Zhú-Jiǎn Zhěng-Lǔ Xiǎo-Zǔ ,張家山竹簡整理小組, 'Jiānglíng Zhāngjiāshān Hàn-Jiǎn Gài-Shù' 《江陵張家山漢簡概述》 ('A Description of the Hàn Bamboo Slips Unearthed at Zhāngjiāshān in Jiānglíng') in *Wén-Wù* 《文物》 (*Cultural Relics*) No. 1, pp. 9–15 (1985).

The Influence of Yang Hui's Works on the Mathematical Mainstream in the Ming Dynasty

GUO SHIRONG 郭世荣

The Institute for the History of Science
Inner Mongolia Normal University, Huhhot, China

The research direction of traditional Chinese mathematics during the Ming dynasty (1368–1644) changed so greatly that mathematicians became interested in very different research fields. The character of mainstream mathematics in the Song and Yuan dynasties (960–1368) could be described as "theoretic", while the emphasis of mathematical research in the Ming dynasty was on the application and popularization of mathematics, or mathematics characterized as "practical". Generally speaking, Chinese mathematics experienced a stagnant and backward period in the Ming dynasty, especially from the middle of the 14th to the end of the 16th centuries, compared to the Song and the Yuan dynasties. Meanwhile, not only were most of the important mathematical achievements that flourished in the Song and the Yuan dynasties lost or forgotten, but almost no important new result or method was achieved.[1]

Why did mathematics during the Ming dynasty not developed as before? This is a difficult question which has puzzled historians of Chinese mathematics for a long time. Undoubtedly, we can study the question from different angles. Most important is to study the development and transformation process of knowledge internal to mathematics. First it should be made clear which parts of the mathematical achievements of the Song and Yuan dynasties were taken up by the Ming mathematicians and which parts were ignored. To do this, it is necessary to know the mathematical character of both periods. In other words, the mathematical mainstreams of both periods should be analyzed. Here, the mathematical mainstream of a period implies the main mathematical methods and approaches, main research fields and mathematical problems, the main points of view and thoughts on mathematics, and the main forms given mathematical results.

In the present paper, we focus on the relation between the mainstream of the Ming dynasty and that of the Song and Yuan period. The results will show that Yang Hui's works played a crucial role in the formation and development of the mathematical mainstream in the Ming dynasty, while the other mathematical works of the Song and Yuan period had little influence.

The Mathematical Mainstream and Tradition in the Song and Yuan Period

Chinese mathematics before the Song dynasty was centered in the tradition of the *Jiuzhang Suanshu* (《九章算术》 *Nine Chapters on Mathematical Arts*). From the *Jiuzhang Suanshu* to the Tang dynasty's mathematical text books *Suanjing Shishu* (《算经十书》 *Ten Mathematical Works*), both mathematical theory and its application had been emphasized, although various problems from economy, agriculture, astronomy, survey, social activity, and gaming were studied in mathematical works. The mathematical tradition had not changed until the late 11th century when the works of mathematicians Liu Yi (刘益), Jia Xian (贾宪) and Shen Kuo (沈括) ushered in a new era of mathematics in China. From the late 11th century onward, mathematical theory was emphasized more than application. Theory therefore became the mainstream of Chinese mathematics. Mathematical theory was of most concern to mathematicians such as Liu Yi, Jia Xian, Jiang Zhou (蒋周), Qin Jiushao (秦九韶), Yang Hui (杨辉), Li Ye (李冶), Zhu Shijie (朱世杰), and others who were active in northern China during the Jing and Yuan dynasties. They developed a series of results in algebra, geometry and number theory. The following are some examples:

(a) *zencheng kaifang fa* (增乘开方法 method of extraction by addition and multiplication) first appeared in Liu's and Jia's works and applied to extracting the root of a high degree equation. This method was later popularized by other mathematicians, and was developed by Qin Jiushao in 1247 until it became as effective as the so-called Horner's method, which appeared in Europe in 1819.

(b) The triangle of the coefficients in the binomial theorem, the so-called Pascal's triangle in the West, had been studied by Jia Xian and was used in the extraction of roots of high degree equations. Yang Hui and Zhu Shijie continued this study.

(c) Qin Jiushao studied the method of solving the simultaneous linear congruent equations or undetermined problems.

(d) Zhu Shijie developed a group of formulae for solving problems of *duoji* (垛积 packing problem) and *zhaocha* (招差 interpolation problem) in his works of 1299 and 1303. Before him, the method had been applied to astronomical calculations for a long time, but

Zhu's study was deeper than other's. Mathematically, this study was half influenced by the triangle in Jia's work and half by Shen Kua's *xiji shu* (隙积术 method of calculating "interstitial-volumes"). It was followed by Yang Hui.

(e) Li Ye mastered and effectively developed the algebraic method *tianyuan shu* (天元术 method of the heaven element) and published his *Ceyuan Haijing* (《测圆海镜》 *Sea Mirror of Circle Measurement*) which continued a research tradition in northern China. Zhu Shijie formulated *siyuan shu* (四元术 method of the four elements) in 1303. From Zu Yi's (祖颐) description of the developmental process from *tianyuan shu* to *siyuan shu* in his postscript to Zhu's work, we know that there were many mathematicians working in the field in north China.

The research method and approach to mathematics also changed. The engagement of the method of *yanduan shu* (演段术) and *tianyuan shu* made great progress in algebra. The method of number theory appeared in Li Ye's[2] and Qin Jiushao's works. Therefore, we can say that mathematical thought in the period changed and developed greatly.

On the other hand, the tradition of popularization of mathematics still remained. Mathematicians such as Qin Jiushao, Li Ye and Zhu Shijie all paid attention to the application of mathematics. Yang Hui devoted himself to promoting the popularization of mathematics. He not only proposed a general outline for mathematical studies, and prepared textbooks for beginners, but also wrote on how to find simplified calculation methods that were necessary for popularization.

Thus we can say that the mathematical mainstream in the Song and Yuan consisted of two component parts. The theoretical studies of Qin Jiushao, Li Ye, and Zhu Shijie, among others, stood for the first one, and the practical studies of Yang Hui stood for the second one. Because both the two components were developed in the same time-frame, mathematics of the Song and Yuan period have been regarded as a highlight in the history of mathematics in China.

Mathematical Works in the Ming Dynasty

The mathematical works from the middle of the 14th century to the appearance of the Chinese translation of Euclid's *Elements* in 1607 can be classified into four classes:

(a) Comprehensive mathematical works: The main part of the works in this class were arranged according to the format of *Jiuzhang Suanshu*;

in other word, mathematical knowledge was arranged under nine headings of *Fangtian* (方田 Plane Measurement), *Sumi* (粟米 Proportion), *Cuifen* (衰分 Fellowship), *Shaoguang* (少广 Evolution), *Shanggong* (商功 Solid Measurement), *Junshu* (均输 Alligation), *Yingbuzhu* (盈不足 Surplus and Deficiency), *Fangchen* (方程 Linear Equation), and *Gougu* (勾股 Right-angle Triangle). Before the main parts there usually were preliminary chapters introducing some kind of algorithms, rhymes for expressing some algorithms, methods for studying mathematics, and so on. Problems that could not be attached to any one of the nine headings were arranged at the end of those books. The books in this class are as follows:

Jiuzhang Tongming Suanfa (《九章通明算法》 *Understandable and Bright Arithmetical Methods in Nine Chapters*), by Liu Shilong (刘仕隆), 1424;

Jiuzhang Bilei Suanfa Daquan (《九章比类算法大全》 *Fully Comprehensive Collection of Arithmetical Methods in Nine Chapters with New Analogous Problems and Rules*), by Wu Jing (吴敬), 1450;

Gujin Suanxue Baojian (《古今算学宝鉴》 *Precious Mirrors on Arithmetic from Ancient to Today*), by Wang Wensu (王文素), preface in 1524;

Suanfa Tongzong (《算法统宗》 *Systematic Treatise on Arithmetic*), by Cheng Dawei 程大位, 1592.

(b) The popular works of applied mathematics: This class includes mainly those works about calculation of measurements, capacities, weights, areas of planes, large and small numbers, changes of weights, and so on, and applications of mathematics to various daily activities. No important progress in mathematics was made in these works. Examples of these works are:

Ding Ju's Suanfa (《丁巨算法》 *Jing Ju's Methods of Computation*), 1355;

Suanfa Quanneng Ji (《算法全能集》 *Collection of Universal Methods of Computation*) by Jia Heng (贾亨);

Toulian Xicao (《透廉细草》 *Mathematical Curtain Pulled Aside*), author unknown, 1372;

Tongyuan Suanfa (《通源算法》 *Original of Mathematics*), by Yan Gong (严恭);

Xiangming Suanfa (《详明算法》 *Explanations of Arithmetic*), by An Zhizhai (安止斋) and He Pingzhi (何平之);

Jinnang Qiyuan (《锦囊启源》 *Brocaded Bag (instructions) of Enlightening Origin*), author unknown.

All these works were writings of the 14th century. These books were designed to satisfy demands of popular readers and were welcomed by them.

(c) Works on the abacus: In the 16th century the studies on the abacus and its algorithms flourished. Several works on the abacus appeared:

Panzhu Suanfa (《盘珠算法》 *Arithmetical Method on Abacus*), Edited by Xu Xinlu (徐心鲁),1573;

Shuxue Tonggui (《数学通轨》 *Rules of Mathematics*), by Ke Shangqian (柯尚迁), 1578;

Yihong Suanfa (《一鸿算法》 *Yihong Arithmetical Method*), by Yu Kai (余楷), 1584;

Suanfa Zuanyao (《算法纂要》 *Summarized Methods of Arithmetic*), by Cheng Dawei, 1598;

Suanfa Zhinan (《算法指南》 *A Guide to Methods of Arithmetic*), by Huang Longyin (黄龙吟), 1604.

(d) Works discussing a special mathematical problem:

Gougu Suanshu (《勾股算术》 *Arithmetic of Right-angle Triangle*), by Gu Yingxiang (顾应祥), 1533;

Hushi Suanshu (《弧矢算术》 *Arithmetic of Arcs and Sagittae*) by Gu Yingxiang, 1552;

Ceyuan Haijing Fenlei Shishu (《测圆海镜分类释术》 *Classified Study on the Sea Mirror of Circle Measurement*), by Gu Yingxiang, 1550;

Ceyuan Suanshu (《测圆算术》 *Arithmetic of Circle Measurement*), by Gu Yingxiang, 1550;

Suanxue Xinshuo (《算学新说》 *A New Account of the Science of Calculation*), by Zhu Zaiyu (朱载), 1584;

Shendao Dabian Lizong Suan Hui (《神道大编历宗算会》 *Assembly of Computing Methods Connected with Calendar*), by Zhou Shuxue (周述学), 1558.

Comparatively, the works in the class (a) are more important than the others. They cover the contents of almost all the other three classes and contain almost all the research results achieved from the middle of the 14th to the 16th centuries. Therefore, for our purpose, it is sufficient to analyze the works in class (a).

The Mainstream of Mathematics in the Ming Dynasty

Based on the study of the mathematical works of the Ming dynasty, we can understand the mainstream of mathematics in the Ming dynasty more comprehensively.

First, the mathematicians of the Ming dynasty paid more attention to the practical problems that were connected with practical activities and closely met demands of non-mathematicians. In their works, problems from the measurement of fields, survey, abacus and its usage, changes of weights, and other daily activities, were concerned about more than theoretical research. The theoretical results of the early period were either lost or ignored. A typical example was *tianyuan shu*, which was not understood and even given up by Gu Yingxiang and Tang Shunzhi (唐顺之) who were celebrated mathematicians in the 16th century.

Second, pithy formulae, or mathematical rhymes and poems were popularly used in mathematical books. The use of the mathematical rhymes was of great importance for the computation of the abacus. During the Ming dynasty, it was in fashion to use rhymes and poems in mathematical books. Almost all arithmetical rules, formulae, and methods were matched by rhymes or poems. Moreover, problems were also composed in this style.

Third, the abacus became very popular in the 16th century.

Therefore, the direction of mathematical research during the period changed greatly. The development of mathematics tended toward the trends of popularization, application, and rhymes. The general level of mathematical research of the period could not be compared to that of the early one.

Summary of Yang Hui's Mathematical Works

Yang Hui was a mathematician in the 13th century. The following is a list of his mathematical works:[3]

(1) *Xiangjie Jiuzhang Suanfa* (《详解九章算法》 *A Detailed Analysis of Mathematical Methods in the "Nine Chapters"*) and *Jiuzhang Suanfa Zuanlei* (《九章算法纂类》 *Reclassification of Mathematical Methods in the "Nine Chapters"*) in 1261. These two books made up a total of 12 chapters. Eighty problems out of 246 problems in *Jiuzhang Suanshu* were selected as typical ones. These problems concerning methods and algorithms were explained in detail and new problems of the same type were added. Yang Hui also reclassified the 246 problems, methods and algorithms from his own point of view.

(2) *Xiangjie Suanfa* (《详解算法》 *A Detailed Analysis of the Mathematical Methods of Computation*) in 1261. Some rules of multiplication and division were dealt with in the book.

(3) *Riyong Suanfa* (《日用算法》 *Arithmetical Methods for Daily Use*) in 1262. It was written for beginners. The arithmetical subject was the method of multiplication and division, including "addition"-a method of multiplication through addition, and "subtraction"-a method of division through subtraction. Problems in the book were from the practice of measuring and weighing.

(4) *Cheng Chu Tongbian Suanbao* (《冻顺通变算宝》 *Precious Reckoner for Variations of Multiplication and Division*) in 1274. This book consisted of three chapters, the third of which was written in conjunction with Shi Zhongrong (史仲荣). Six methods of multiplication and two methods of division were discussed in the first chapter. In the other two, various methods of "addition", "subtraction", "*qiuyi*" (求一 doubling and halving) and a substitution for other methods of multiplication and division were introduced. In addition, a general outline of mathematical studies was included in the first chapter.

(5) *Tianmu Bilei Chengchu Jiefa* (《田亩比类乘除捷法》 *Practical Rules of Arithmetic for Surveying*) in 1275. In this book, Yang Hui devoted himself to a method of measuring areas in various shapes and a reversal problem-solving the quadratic equations.

(6) *Xu Gu Zhaiqi Suanfa* (《绪古摘奇算法》 *Continuation of Ancient Mathematical Methods for Elucidating the Strange [Properties of Numbers]*) in 1275. This is a collection of rare problems from various

mathematical texts and forgotten literature contained in old prints. Magic squares and indeterminate problems, among others, are included.

The existing volumes of book (1) are incomplete, and books (2) and (3) are not handed down to us. The last three works formed a collection titled *Yang Hui Suanfa* (《杨辉算法》 *Yang Hui's Methods of Computation*). Although other mathematicians' research work was cited in the above books and some theoretical results were reported, generally speaking, all these books emphasized the application and the popularization of mathematics.

Yang Hui's Influence

As all know, the trend of emphasizing application and population of mathematics appeared in the late Tang dynasty and dozens of mathematical works with this purpose were written. Yang Hui's work made the study of practical mathematics a component part of the mathematical tradition of his time. Mathematicians of the Ming dynasty took up the tradition through Yang Hui's works, which had played an important role in the transformation of the mainstream of mathematics in the Ming dynasty. Yang Hui's influence on the mainstream of mathematics in the Ming dynasty can be understood in several aspects.

Yang Hui's works were cited and referred to implicitly or explicitly again and again by the mathematicians of the Ming dynasty, in contrast to the works of mathematicians such as Qin Jiushao, and Zhu Shijie of the Song and Yuan period, who were little touched by them, and Li Ye's *Ceyuan Haijing* which was only given attention by Gu Yingxiang. Comparing Yang Hui's works with those of Wu Jing, Wang Wenshu, and Cheng Dawei, among others, it can be seen that almost all main subjects taken up by mathematicians of the Ming originated with Yang Hui's works. One of the most important subjects of the Ming mathematics was the measurement and determination of plane areas of peculiar shapes and their reverse problems. The problems of the subject appeared in various mathematical works, especially those of Wu, Wang, and Cheng. This subject was taken up in Yang Hui's *Tianmu Bilei Chengchu Jiefa*. Another concern in the Ming was to find simplified methods of computation. "Addition", "subtraction", "*qiuyi*"(doubling and halving) and other substitutions for multiplication and division (which were the subjects of Yang Hui's book [4]) were cited by many mathematical books in the 15th and 16th centuries. Most problems and methods of Yang Hui's book (6) can be found in the works of Wu, Wang, and Cheng, among others. In addition, many of Yang Hui's geometrical illustrations appeared in the works of the Ming dynasty.

As did Yang Hui, the mathematicians of the Ming paid much attention to how computation should be taught and studied. The characteristic engaging in pithy formulae, or mathematical rhymes and poems, was also influenced by Yang Hui. The first use of pithy formulae and poetry in mathematics was in the Tang dynasty. Yang Hui and Zhu Shijie popularized it. During the Ming dynasty it became fashionable.

Yang Hui's studies on methods of computation also promoted the development of the abacus.

The style and structure of the works of Wu, Wang, and Cheng were also influenced by Yang Hui's work. The *Jiuzhang suanshu* referred to by them was mainly after Yang Hui's book (1), instead of the old original text. As Yang did, Wu Jing and others also added new problems.

Conclusion

Undoubtedly, the stagnant and backward mathematics of the Ming dynasty was caused by many factors, but from the mathematical point of view, one of the most important was the transformation of the mathematical tradition. It was the transformation of the mathematical tradition from a theoretical to a practical direction which resulted in mathematicians in the Ming dynasty emphasizing popularization. They were concerned about whether their works were understood and accepted by readers who were not mathematicians. In this way, the use of the abacus and its algorithms, the rhythms of mathematics, and simplified calculation methods became more and more popular. Mathematicians became more and more concerned with applications of mathematics. This study has shown that the situation was a result of the influence of Yang Hui's works on the mathematicians of the Ming dynasty interested in the same research subjects, for both of them paid great attention to the application and popularity of mathematics. With this background, it is easier to understand why Yang Hui's works were praised highly and were referred to more than others of the Song and Yuan periods.

Notes

[1] Mei Rongzhao (梅荣照). *Ming Qing Shuxue Shi Gailun* (明清数学史概论 "An Outline of the History of Mathematics during the Ming and the Qing"), in *Ming Qing Shuxue Shi Lunwen Ji* (《明清数学史论文集》 *Selected Papers on the History of Mathematics during the Ming and the Qing Dynasties*, edited by Mei Rongzhao), Nanjing, 1991.

[2] K. Chemla. *Li Ye ceyuan haijing de jiegou jiqi dui shuxue zhishi de biao shu* (李冶《测圆海镜》的结构及其对数学知识的表述 "The Textural Construction of

the Sea Mirror of Circle Measurement and its Expression of Mathematical Knowledge"), in *Shuxue Shi Yanjiu Wenji* (《数学史研究文集》 *Collection of Research Papers on the History of Mathematics*, edited by Li Di), No. 5, Huhhot and Taibei, 1993.

3 About Yang Hui and his mathematical research work there have been many bibliographies. The most important ones are: Yan Dunjie (严敦杰). *Song Yang Hui Suanshu Kao* (宋杨辉算书考 *A Study on Yang Hui's Mathematical Works in the Song Dynasty*), in *Songyuan Shuxue Shi Lunwen Ji* (《宋元数学史论文集》, *Collection of Paper on the History of Mathematics in the Song and Yuan Dynasties*, edited by Qian Baocong (钱宝琮), Beijing, 1966); Ho Peng Yoke. Yang Hui, in *Dictionary of Scientific Biographies* [WTBZ], New York, 1970, Vol. XII; Lam Lay Yong. *A Critical Study of the Yang Hui Suan Fa*, Singapore, 1977; Guo Xihan (郭熙汉). *Yang Hui Suan Fa Dao Du* (杨辉算法导读), Wuhan, 1997.

On the Basic Rules for Reconstruction of the Calendar Used in the State of Lu During the Spring-Autumn Period

CHEN MEIDONG

Institute for History of Natural Science
Chinese Academy of Sciences

At least 20 scholars have at different times tried to reconstruct the calendar used in the State of Lu during the Spring and Autumn period — the 393 records of year, month, and date in the 60-cycle being the main basis for their work. Because of different opinions on the basic situation of the calendar, and different evaluations of the reliability of the recorded dates, different rules for the reconstruction work were adopted and different results consequently given. So the crucial problem is to find basic rules for the reconstruction. We set out the following 5 basic rules:

(1) The reconstructed calendar must conform to reliable records of the years, months, and cyclical dates of the new moon, and the intercalary months recorded in the *Chun-qiu* text. There are a total of 32 records of this kind (see Table 1).

(2) Through analysis of the 32 records, it can be seen that: from [11]–[12] and [19]–[20], there are small months that contain 29 days, while from [16]–[17], there are big months that contain 30 days. The period [6]–[7] contains 88 days, which conforms to the arrangement style of small month, big month, and small month. From [9] and [31] we know that there are leap months, which were arranged after the twelfth months. From the above we can reach the conclusion that the calendar used in the State of Lu was a solar-lunar calendar, taking the new moon as the first day of a month, and with a non-leap year containing 12 months, and a leap year containing 13 months. Leap months were arranged at the ends of years. A big month consisted of 30 days, and a small month consisted of 29 days. Big months and small months were arranged alternatively, and the mean length of a lunar month (H) was about 29.5 days.

It is known that the period between [4] and [5] has 3278 days (Julian day 1482415–479137). If 29.5 days per month is adopted, there are about 111 months (3278/29.5 = 111.12). If there are 110 months, then H = 29.800 days, if 112 months, H = 29.268 days, and if 111 months, H = 29.531 days. For another example, the period between [26] and [27] has 2776 days, so there are about 94 months. If there are 93 months, then H = 29.849 days, if 95 months, then H = 29.221 days, and if there are 94 months, H = 29.532 days. Similarly, take two records separated by more than 2000 days in sequence (if less than 2000 days, the error may be too great), and calculate in the same way, the result is listed in Table 2. Table 2 shows that in column A and column C, the discreteness is remarkable, and evidently cannot be the values of H used in the State of Lu. On the contrary, column B shows great consistency, so the H used in the Lu calendar must be between 29.527–29.536 days. The mean value of the above 14 terms is 29.531 days, which should be very close to the true H used in the Lu Calendar.

(3) Through analysis of the 32 records it can also be seen that in a given period, the Lu calendar arranged regularly months, leap months and continuous big months, etc. That information, hidden in the Lu calendar, cannot be recovered.

In Table 1, the "number of months" refers to the number of months between each year in question and the first month [1]. It can be obtained by dividing the difference between two Julian days by 29.531 and taking the integer part.

The "number of leap months" refers to a number that should exist between two neighboring recorded years or months. It can be obtained by calculating the difference between two "number of months", minus "W". "W" is calculated in this way: the number of years between two recorded neighboring years times 12, then plus or minus the difference of months in the two years (supposing that a year has 12 months). From Table 1 it can be seen that the 6th year of Duke Wen and the 5th year of Duke Ai should have leap months. The following conditions could also be deduced: from [2]–[3], that the 25th year of Duke Zhuang could not be a leap year; from [12]–[13], that the 16th year of Duke Cheng could not be a leap year; from [15]–[16], that the 20th year of Duke Xiang must be a leap year; from [17]–[18], that neither 21st nor 22nd year of Duke Xiang are leap years; from [18]–[19], that the 23rd year of Duke Xiang must be a leap year; and from [20]–[21], that neither 24th, nor 25th, nor 26th year of Duke Xiang are leap years. It can be seen that in the 7 years from the 20th to the 26th of Duke Xiang, earlier, an

Table 1 Some Basic Points and Conditions Which Must be Met by the Lu Calendar

No.	Year	Mth	Date	JD	Number of month	Number of leap month	D	Number of continuous big month	First month	Last month
1	3rd Huan	7	Ren-chen, nm 709BC, 7.17	1462659	141	4	7	8	S	S
2	25th Huan	6	Xin-wei, nm 669BC, 5.27	1477218	493	14	31	30	B	B
3	26th Huan	12	Gui-hai, nm 668BC, 11.10	1477750	511	0	2	2	S	B
4	30th Huan	9	Geng-wu, nm 664BC, 8.28	1479137	558	2	1	2	S	S
5	5th Xi	9	Wu-shen, nm 655BC, 8.19	1482415	669	3	7	7	B	S
6	15th Xi	9	Ji-mao, bnm (geng-chen, nm) 645BC, 9.27	1486107	794	4	9	8	B	B
7	16th Xi	1	Wu-shen, nm 645BC, 12.24	1486195	797	0	0	0	S	S
8	22nd Xi	11	Ji-si, nm 638BC, 10.10	1488676	881	2	6	5	B	B
9	6th Wen	12 leap								
10	15th Wen	6	Xin-chou, nm 612BC, 4.28	1498008	1197	9	20	20	S	B

Table 1 (Continued)

No.	Year	Mth	Date	JD	Number of month	Number of leap month	D	Number of continuous big month	First month	Last month
11	16th Cheng	6	Bing-yin, nm 575BC, 5.9	1511533	1655	14	28	28	S	B
12	16th Cheng	6	Jia-wu, bnm (yi-wei, nm,) 575BC, 6.7	1511562	1656	0	0	0	S	S
13	17th Cheng	12	Ding-si, nm 574BC, 10.22	1512064	1673	0	1	1	B	S
14	14th Xiang	2	Yi-wei, nm 559BC, 1.14	1517262	1849	6	12	11	B	S
15	20th Xiang	10	Bing-chen, nm 553BC, 8.31	1519683	1931	2	4	5	S	S
16	21st Xiang	9	Geng-xu, nm 552BC, 8.20	1520037	1943	1	0	0	B	S
17	21st Xiang	10	Geng-chen, nm 552BC, 9.19	1520067	1944	0	0	0	B	
18	23rd Xiang	2	Gui-you, nm 550BC, 1.5	1520540	1960	0	2	1	B	B
19	24th Xiang	7	Jia-zi, nm 549BC, 6.19	1521071	1978	1	0	0	S	B
20	24th Xiang	8	Gui-si, nm 549BC, 7.18	1521100	1979	0	0	0	S	
21	27th Xiang	12	Yi-hai, nm 546BC, 10.13	1522282	2019	0	4	3	B	B

(cont'd overleaf)

Table 1 (Continued)

No.	Year	Mth	Date	JD	Number of month	Number of leap month	D	Number of continuous big month	First month	Last month
22	7th Zhao	4	Jia-chen, nm 535BC, 3.18	1526091	2148	5	7	8	S	S
23	15th Zhao	6	Ding-si, nm 527BC, 4.18	1529044	2248	2	6	6	B	S
24	21st Zhao	7	Ren-wu, nm 521BC, 6.10	1531289	2324	3	6	5	B	B
25	22nd Zhao	12	Gui-you, nm 520BC, 11.23	1531820	2342	1	0	1	S	S
26	24th Zhao	5	Yi-wei, nm 518BC, 4.19	1532322	2359	0	1	1	B	S
27	31st Zhao	12	Xin-hai, nm 511BC, 11.14	1535098	2453	3	6	6	B	S
28	5th Ding	3	Xin-hai, nm 505BC, 2.16	1537018	2518	2	5	4	B	B
29	12th Ding	11	Bing-yin, nm 498BC, 9.22	1539793	2612	2	4	5	S	S
30	15th Ding	8	Geng-chen, nm 495BC, 7.22	1540827	2647	2	3	2	B	B
31	5th Ai	12 leap year								
32	14th Ai	5	Geng-shen, nm 481BC, 4.19	1545847	2817	5	10	10	S	B

Column 4: nm = new moon, bnm = the date before new moon.
Column 10 and Column 11: S = small month (29 days), B = big month (30 days)

Table 2

Area	A	B	C
1–4	29.583	29.530	29.478
4–5	29.800	29.532	29.268
5–6	29.774	29.536	29.302
6–8	29.872	29.529	29.193
8–10	29.625	29.532	29.483
10–11	29.595	29.531	29.466
11–14	29.684	29.531	29.379
14–21	29.704	29.529	29.537
21–22	29.758	29.527	29.300
22–23	29.828	29.530	29.238
23–26	29.800	29.532	29.268
26–27	29.849	29.532	29.221
27–29	29.715	29.528	29.344
29–32	29.676	29.532	29.388

intercalary month was arranged every 2 years, while later, it was arranged at least every 4 years. From [24]–[25] it can be seen that the 21st year of Duke Zhao must be a leap year, and from [25]–[26], that neither the 22nd nor 23rd year of Duke Zhao is a leap year. The above is all that can be known about the arrangement of leap months. In addition, from [22]–[23] it can be seen that in the 8 years from the 7th to the 14th of Duke Zhao, there were only 2 leap years; but from [29]–[30], in the 3 years from the 12th to the 14th of Duke Ding, there were 2 leap years. The former example shows that in the Lu calendar, two neighboring leap years can be separated by as long as 4 years, while the later example shows that they can be separated by as short as 1 year. These situations show that the Lu calendar arranged leap months irregularly.

Column D refers to the appropriate number of continuous big months between two neighboring recorded years and months. But for the true number of continuous big months, the following different situations should be taken into account:

Suppose there are S months (i.e., difference of "number of months") between two neighboring years, months, and dates, and there are P days (the difference in Julian Day). Their relationship with D can be expressed as follows:

$$P = 29.5 \, S + 0.5 \, D$$

So
$$D = 2(P - 29.5 \, S)$$

When the first month in the period is a big month and the last month is a small one, or when the first is a small month and the last is a big one, then the value D calculated in this way is the true number of continuous big months. When both the first and the last months of the period are big months, then (D − 1) is the true number of continuous big months. When both the first and the last months of this period are small months, then (D + 1) is the true number of continuous big months. If the value of S/D is around 16, it must be the first situation. If the value of S/D is smaller than 16, it must be the second situation. If the value of S/D is larger than 16, it must be the third situation. Dividing 2871, the "number of months" in the period [1]–[32], by 172, total of D, we get 16.4, which is the mean number of months for the Lu Calendar to arrange continuous big months.

Through analysis of calendar dates from the 21st to 24th years of Duke Xiang, the fact of continuous big months arranged first after 15 months, then after 21 months can be detected. If the calendar is calculated by a fixed length of lunar month (29.531 days), continuous big months must be arranged every 15 or 17 months. The result we got above is obviously not this. The conclusion can be drawn that since there exist continuous big months after 21 months, we cannot use a fixed length of lunar month to reconstruct the calendar used in the State of Lu, so we must try to find another way.

(4) The reconstruction must match most of other dating records in *chun-qiu*.

Some of the dates recorded in *chun-qiu* have been tested and found wrong, the reason for which probably lies in mistakes made in scribal transmission. If we do not admit this, we can only obtain a calendar without any regulars, which denies the 2nd and 3rd rules above.

(5) Though a large number of dates are recorded in *chun-qiu*, it is not enough to reconstruct the calendar for 244 years, especially as there is the problem that the dating records are not well-distributed. So some reasonable assumption must be adopted. For example:

Continuous big months should have been arranged fairly evenly according to the mean value, but different regulars should have been adopted in different periods. They must match the first and last months as well as the number of continuous big months shown in Table 1.

There existed only two months of continuous big months, neither three months of continuous big months nor continuous small months.

Wang Tao of the Qing dynasty pointed out that before Prince Xi and Prince Wen, the first month of a year was often the month after the winter solstice, while after that, the first month was often the month containing the winter solstice. This is an important discovery, which must be taken into account in arranging leap months. In addition, leap months must be considered to have been arranged evenly in general. Since both a lack and surplus of leap months do not occur very often, without sufficient testimony we choose to believe that the Lu calendar arranged leap months in proper regulars. For example, two leap months should not be arranged in continuous years.

We believe that only when based on these basic rules, a calendar like that used in the State of Lu can be reconstructed.

A Reversion Study on the Solar Shadow Algorithm (SSA) of the *Qintian* Calendar

WANG RONGBIN 王榮彬
Beijing Planetarium, Beijing

Foreword

Wang Pu 王朴 (A.D. 915–959), the author of the *Qintian* 钦天 calendar, wrote in the memorial presenting his work:[1]

> There are four parts in the *Qintian* calendar. They are 1 volume of *lijing* 历经; 11 volumes of *li* 历; 3 volumes of *cao* 草; and 1 volume of the calendar table of the third year of the *Xiande* 显德 period (954–960 A.D.). *Lijing* is also composed of four parts in order to calculate the motion of the sun, moon, the five-planets and the problem of the so-called *falian* 发敛.

Only that first volume of *lijing* has come down to us today; the other parts are all lost. Generally speaking, the *lijing* illustrates the general method or algorithm of the calendar. Unfortunately, the *lijing* of the *Qintian* calendar is too simple in content to fully explain any of these general methods. The methods or algorithms explain how to do the linear interpolation within the period of a day based on the *licheng-biao* (立成表, see ready-reference table). The large magnitude *licheng-biao* and their composing methods were in those now-lost 11 volumes of *li* and 3 volumes of *cao*. Therefore, the substances of the methods and algorithms in Wang Pu's *Qintian* calendar are now unknown. So is the Solar Shadow Algorithm (SSA).

We can only read the *shu* (术, the text of an algorithm) of SSA in the *Qintian* calendar as follows:[2]

> Multiply *wuzhong-ruli-fen* 午中入历分 with *sunyi* 损益 (decreasing or increasing) rate of the day. Divided the product by *tongfa* 通法 (the number of the fen in a day, and 1 day = 7200 fen), The quotient's unit is *fen*, 10 *fen* is 1 *cun* 寸. Add the quotient to or subtract the quotient from the *zhonggui* 中晷 constant of that day to obtain the *ding* 定 (lit. Exact) value of solar shadow of what one wants to calculate.
> 置午中入历分，以其日损益率乘之，如统法而一为分，分十为寸。用损益其下中晷数，为定数也。

Let N stands for the number of the date after the winter or summer solstice; n stands for the number of the *fen* in the day. Δ_N stands for the *sunyi* rate of the date N. Then the former *shu* results in the following formula:

$$f\left(N + \frac{n}{7200}\right) = f(N) + \frac{n}{7200} \cdot \Delta_N \qquad (1)$$

in which, $\Delta_N = f(N + 1) - f(N)$, and "$N + n/7200$" is right *wuzhong-ruli-ri-fen*, $n/7200$ is *wuzhong-ruli-fen*, $f(N)$ and Δ_N can be obtained in the *licheng-biao*. But this *licheng-biao* is lost.

Despite this, we still can get some general knowledge of the *Qintian* calendar's SSA by proceeding along the following lines.

Analysis of Historical Documents

Wang Pu also wrote in his memorial:

> When the Da Zhou Dynasty was founded, The City of Bian was selected as its capital. Then the observatory (namely Yuetai 岳台) was founded in the City of Bian, and the gnomon and *louke* 漏刻 were set to examine the solar shadow. And the data of the solar shadow obtained at Yuetai was accepted for the standard constant.
> 大周建国，定都于汴。树圭置箭，测岳台漏刻，以为中数。

This is to say that Wang Pu had the new solar shadow data at hand when he made his *Qintian* calendar.

In the period of *Huangyou* in the Northern Song Dynasty (A.D. 1049–1054), Zhou Cong 周琮 and his colleagues had improved the SSA and compared their new calculating results with the values that were obtained by Wang Pu's algorithm. Zhou Cong and his colleagues did their comparison by measuring the data of the solar shadow at Yuetai on every day of *Jieqi* (节气). They calculated the solar shadow on the same day according to the algorithm of Wang Pu and their own one respectively. They recorded this in a classic named *Huangyou-guibiao* 皇祐圭表,[3] where they gave out 44 group records of the solar shadow value. Each group had three sets of data. The first one was the observational data; the second one was the result of Wang Pu's SSA (called *Wang Pu-suan*); and the third one was the result of Zhou Cong and his colleagues' SSA. Here, the Wang Pu-suan is the measurement of whether the data of Zhou Cong and his colleagues' new algorithm is or is not better. This is not only because Wang Pu and Zhou Cong's SSA were both made on the basic data of solar shadows at Yuetai, but also indicates that Wang Pu's SSA was very important in the eyes of Zhou Cong and his colleagues.

In the *Qintian* calendar, *wuzhong-ruli-fen* is defined as follows.[4]

On the winter solstice day, if the point of the true winter solstice is before noon, the distance from that point to noon is called *wuqian-fen* 午前分; *wuqian-fen* is right to be *wuzhong-ruli-ri-fen* of *yingli* 盈历, if the point of the true winter solstice is after noon, then called the distance *wuhou-fen* 午后分, and *wuzhong-ruli-ri-fen*, which is in the *suli* 缩历 now, is the remainder of "182.62–*wuhou-fen*". Add 1 to the *wuzhong-ruli-ri-fen* of the winter solstice day continuously to obtain every day of the *wuzhong-ruli-ri-fen*. When the *wuzhong-ruli-ri-fen* is more than 182.62, subtract 182.62 from it, then *yingli* and *suoli* are exchanged.
置天正中气午前分，便为午中入盈历日分，其在午后者，以午后分减岁中，为午中入缩历日分。累加一入，满岁中既去之，盈缩互命，为每日午中入历也。

In Figure 1, the distance from points W to S is the so-called *yingli*, and from the points S to W is the distance of the so-called *suoli*. That is to say, the *licheng-biao* in the *Qintian* calendar must have 182 columns after the winter and summer solstices respectively. Then it is the same as the *zhousui-li* 周岁历, a table given by Zhou Cong in *Huangyou-guibiao*.

We know that the *zhousui-li* in *Huangyou-guibiao* is obtained by computing the result of the formula of SSA in the *Mingtian* 明天 calendar. Then we come to a problem: how did the *licheng-biao* of the *Qintian* calendar come into being?

Fig. 1(1). The case of *yingli*   Fig. 1(2). The case of *suoli*

Fig. 1. Diagram of *wuzhong-ruli-ri-fen*

In fact, for the SSA in table form, such as in the *Linde* 麟德 calendar and in the *Dayan* 大衍 calendar, the algorithm is in 24 segments, one for each *Jieqi*. But for the *Qintian* calendar, its SSA is in only two segments: *Yingli* and *Suoli*. It is not a coincidence that all the SSA in the formula form we can see today are designed in two segments.

On the other hand, the SSA of *zhousui-li* has almost the same text of the *shu* as what we see in the *Qintian* calendar today, although we cannot find a similar *licheng-biao* in the *Qintian* calendar now. The text of Zhou Cong's algorithm appeared as follows:

> Multiply the *wuzhong-zhiyu* 午中之余 with *sunyi-chafa* 损益差法 of the day. Divide the product by *fa* 法. Add the quotient to or subtract the quotient from the shadow constant of that day to obtain the ding value of the solar shadow of what one wants to calculate.
> 若用周岁历，直以其日晷景损益差分，乘其日午中之余，满法约之。
> 乃损益其下数，既其日午中定晷。

Here *wuzhong-zhiyu, sunyi-chafa, fa* are the same thing as *wuzhong-ruli-fen, sunyi* rate, and *tongfa* respectively as in the *Qintian* calendar.

Scholars have appropriately called the SSA in *Huangyou-guibiao* "the table algorithm based on the formula operation results".[5] Summarizing the information above, we can surmise that the SSA of the *Qintian* calendar must be the same one as that of *zhousui-li*.

Analysis of the Data on Wang Pu-san

From the first to the fourth years of *Huangyou* in the Northern Song Dynasty (A.D. 1049–1054), Zhou Cong and his colleagues obtained 44 records of the solar shadow value at Yuetai and also produced 44 pieces of data for the *Wang Pu-suan* (see Table 1).

By analyzing this *Wang Pu-suan* data of the solar shadow in detail, we get some interesting information.

First, we can see the values for the solar shadow of the winter and summer solstices in the *Qintian* calendar. They are

$$L_w = 12.86 \ chi \ (尺); \text{ and, } L_s = 1.51 \ chi.$$

Second, the 44 datapieces of the *Wang Pu-suan* have a visible contradiction. According to them, we come to:

$$f(N + n_1) = f(N + n_2) \tag{2}$$

But according to formula (1), formula (2) must be wrong. It seems to us that when they calculated the data of *Wang Pu-suan*, Zhou Cong and his colleagues only referred to the *licheng-biao*, and did not use the formula (1) to make the data of *Wang Pu-suan* perfect. But they used the formula (1) to make their own data perfect!

Since the 44 datapieces in Table 1 are calculated from SSA of the *Qintian* Calendar, we may probe into the SSA by further analyzing the data.

Table 1 44 Data of Wang Pu-suan in *Huangyou-guibiao*

Jieqi	The first year		The second year		The third year		The fourth year	
	Wuzhong ruli (day)	Yingchange (chi)	Wuzhong ruli (day)	Yingchang (chi)	Wuzhong ruli (day)	Yingchang (chi)	Wuzhong ruli (day)	Yingchang (chi)
Xiaoxue	153.114	11.39			152.619	11.47		
Daxue			167.870	12.45				
Dongzhi	0.492	12.86	0.241	12.86				
Xiaohan	15.492	12.48			15.000	12.48		
Dahan			31.241	11.44				
Lichun					45.000	10.15		
Yushui					61.241	8.50	60.000	8.61
Jingzhe			76.492	6.85	76.241	6.85		
Chunfen			9.492	5.27			91.000	5.27
Qingming			107.492	3.89			106.000	3.96
Guyu					122.241	2.96	121.000	3.10
Lixia			137.492	2.30	137.241	2.30	136.000	2.34
Xiaoman			152.492	1.86	152.241	1.86		
Mangzhong			167.492	1.60	168.241	1.59	167.000	1.60
Xiazhi			0.870	1.51			.0378	1.51
Xiaoshu					15.619	1.60		
Dashu			30.870	1.85	30.619	1.85	31.378	1.87
Liqiu			45.870	2.29	46.619	2.33		
Chushu					61.619	3.00		
Bailu								
Qiufen					91.619	5.21		
Hnalu					106.619	6.80	107.378	6.91
Shuangjiang			121.870	8.45			122.378	8.56
Lidong			137.870	10.10	137.619	10.10	137.378	10.10

The Reversion of the SSA Formula

Because the *Qintian* is between the *Chongxuan* and the *Yitian*, the formula of SSA in the *Chongxuan* and the *Yitian* Calendars are in the same form (3–4). When x is in the quadrants before or after the winter solstice,

$$y = L_w - (a - bx)x^2 \cdot 10^{-6}. \tag{3}$$

When x is in the quadrants before or after the summer solstice,

$$y = L_s - (c - dx)x^2 \cdot 10^{-7}. \tag{4}$$

So, suppose the *Qintian*'s SSA formula takes the form of formula (3), when x is in the quadrants before or after the winter solstice, in which $L_w = 12.86$, $x \in (0, 59)$. Substituting the relevant data of *Wang Pu-suan* in formula (3), we get the different values of a and b by the different x and y. Let $b = 12$, then, we get a series value of a_i by each pair of x and y (see Table 2). We can see the change range is very small. The average value of a_i is: $a = 1872$. We may approximately regard the formula of the *Qintian*'s SSA, when x is in the quadrants before or after winter solstice, as follows:

$$y = 12.86 - (1872 - 12x)x^2 \cdot 10^{-6}. \tag{5}$$

Let's still take the the *Qintian*'s SSA formula as the form of formula (4), when x is in the quadrants before or after summer solstice, in which $L_s = 1.51$, $x \in (0, 123.62)$. By substituting the relevant data of *Wang Pu-suan* in formula (4), we found that x is far from the summer solstice, c is increasing monotonically (see Table 3). That is to say, the value c in formula (4) cannot be the constant.

Table 2   Analysis of the Data of *Wang Pu-suan* Around the Winter Solstice

No.	x (day)	y (chi)	a_i
1	−45.62	10.10	1874
2	−30.62	11.47	1850
3	−29.62	11.49	1917
4	−15.62	12.45	1868
5	15	12.48	1869
6	31	11.44	1850
7	45	10.15	1878

Table 3   **Analysis of the Data of *Wang Pu-suan* Around the Summer Solstice**

No.	x (day)	y (chi)	c_i	r_i	k_i
1	−122.62	8.61	5213	12.67	18.38
2	−121.62	8.50	5212	12.76	17.99
3	−106.62	6.85	5124	13.73	13.79
4	−91.62	5.27	4846	12.94	13.20
5	−76.62	3.96	4480	10.70	16.30
6	−75.62	3.89	4464	10.63	16.33
7	−61.62	3.10	4434	12.56	7.84
8	−60.62	2.96	4188	8.71	22.37
9	−46.62	2.34	4005	7.40	23.30
10	−45.62	2.30	3978	6.97	47.03
11	−30.62	1.86	3855	6.37	82.76
12	−15.62	1.60	3751	5.83	91.88
13	−14.62	1.59	3801	9.64	4.02
14	15	1.60*	—	—	—
15	30	1.85	3898	7.93	15.54
16	31	1.87	3870	6.77	40.26
17	45	2.29	4032	8.27	19.82
18	46	2.33	4059	8.67	17.23
19	61	3.00	4248	9.64	16.76
20	91	5.21	4832	12.88	13.23
21	106	6.80	5132	13.87	13.47
22	107	6.91	5145	13.88	13.58
23	121	8.45	5224	12.93	17.46
24	122	8.56	5225	12.83	17.86

Then consider the form of:

$$y = 1.51 + (c - d'x + y')x^2 \cdot 10^{-7} \qquad (6)$$

In which c is still the constant, let y' be the function of x. Consider the orderliness of c_i in Table 3, there is no harm in taking $c_0 = 3660$, when $x = 0$. Let

$$r_i = \frac{c_i - c_0}{|x|} \qquad (7)$$

Substitute each c_i in formula (7), then we have the relevant r_i. See Table 3. Because r_i in Table 3 appears symmetrical, it proves that $c_0 = 3660$ is a good selection.

Comparing formula (4) and (6), we have

$$y' = c_i - c_0 = r_i x \qquad (8)$$

From Table 3, we know that y' may not be a linear function. That means r_i is not constant, either. Let

$$y' = \left(r_0 + \frac{x}{k} \right)x \tag{9}$$

then

$$k_i = \frac{|x|}{|r_i + r_0|} \tag{10}$$

Substitute each r_i in formula (10) to obtain k_i. And the average value of k_i is: $k = 17$, then $y' = (102 + x)x/17$. Thus, we get an approximate formula below for the *Qintian*'s SSA when x is in the quadrants before or after summer solstice.

$$y = 1.51 + \left[3660 - 4x + \frac{(102 + x)}{17} \right] \tag{11}$$

Because we have not enough materials to support our reversion work, it may not be an ideal result. We need further exploration in the future.

Acknowledgement

The author is grateful to the Lee Foundation 李氏基金 for the support of the Panel on Interpolation in which this paper was presented at the *9th International Conference on the History of Science in East Asia* held at Singapore, 23–27 August 1999.

Notes

[1] *Qintian calendar*, in *Lidai tianwen lü li deng zhi huibian* 历代天文律历等志, No. 7, edition of Zhonghua Shuju, 1976, p. 2409.

[2] *Qintian calendar*, in *Lidai tianwen lü li deng zhi huibian*, No. 7, edition of Zhonghua Shuju, 1976, p. 2414.

[3] *Songshu lizhi* (Vol. 9), in *Lidai tianwen lü li deng zhi huibian*, No. 8, edition of Zhonghua Shuju, 1976, pp. 2699–743.

[4] *Qintian* calendar, in *Lidai tianwen lü li deng zhi huibian*, No. 7, edition of Zhonghua Shuju, 1976, p. 2413.

[5] Chen Meidong. A Study on the Algorithm of Solar Shadow in *Huangyou* and *Congnin* Periods, *Studies on the History of Natural Science*, Vol. 8, No. 1, 1989. pp. 17–27.

PART VI

Astronomy

Fake Leap Months in the Chinese Calendar: From the Jesuits to 2033

HELMER ASLAKSEN

National University of Singapore

Introduction

It is not my purpose to give a detailed discussion of the Chinese calendar or the history of the contribution of the Jesuits. I will just focus on one consequence of the reform of the Chinese calendar carried out by the Jesuits in 1645, namely the "fake leap months".

For details about the Chinese calendar see my paper ([As1]), the books by Dershowitz and Reingold ([DR]) and Tang ([Ta]), and the articles by Doggett ([Do]) and Sivin ([Si]). For details about the history of the contribution of the Jesuits see the article by Chu ([Ch]) and the book by Spence ([Sp]). For the computations in this paper I used the Mathematica version of the code from the book by Dershowitz and Reingold ([DR]). Their astronomical functions are based on the book by Meeus ([Me1]). The conversion from Lisp to Mathematica was done by Robert C. McNally. Additional functions are in my Mathematica package Chinese Calendar ([As2]).

Fake Leap Months and the Jesuits

The calendar has always been very important in Chinese society. The Chinese emperor based his authority on being the "Son of Heaven". In that case, it was very embarrassing if the calendar was not in harmony with the heavens. Unfortunately, with a lunar or lunisolar calendar, errors are much more obvious than with a solar calendar. A solar calendar can be off by a couple of weeks without anybody noticing. The only reason why the Catholic church had to reform the Julian calendar was because the rules for computing Easter had frozen the March equinox on March 21. That meant that Easter was drifting noticeably towards summer. Otherwise, nobody would have noticed the drift of the March equinox. But with a lunar calendar an error of even a couple of days is a serious problem. Every peasant could see each month that the new moon was visible near the end of the previous month or that the old moon was visible in the next month. Why should they pay taxes and serve in the army if the emperor did not know the secrets of the heavens?

Since the 4th century B.C.E., Chinese astronomers knew that the motion of the moon was not uniform, and by the 6th century C.E. they knew that the sun's motion was also irregular ([CJJ]). Before the Tang dynasty, they used mean values for the motion of the sun, *ping qi* (平氣), and the moon, *ping shuo* (平朔). This means that they assumed that the time intervals between the 24 *jie qi*'s and the length of the synodic month were both constant. Leap months were determined using the *zhong qi* (中氣) rule, saying that a month without any *zhong qi* was a leap month. This meant that leap months would come at regular intervals, and that each month was equally likely to have a leap month.

Chinese astronomers started using the true moon, *ding shuo* (定朔), in the calendars in 619 at the beginning of the Tang dynasty. The length of the synodic months used to be fixed at 29.53, but now it could range between 29.27 and 29.84. It was not until 1645, however, at the beginning of the Qing dynasty that the Jesuits introduced the true sun, *ding qi* (定氣), into the Chinese calendar. In the old calendar, the time between two *zhong qi*'s was constant at 30.44 days, but under the new system, the time could range between 29.44 and 31.44 days, and the *zhong qi*'s were closer together during the winter.

The true sun caused serious changes in the calendar. Under the new system, leap months were more likely to occur during the summer. In exceptional cases, it was now possible to have two *zhong qi*'s in a month, and to have months without any *zhong qi*'s that were not the result of a "drift" of the *zhong qi*'s but just a compensation for a nearby month with two *zhong qi*'s.

Such cases require clear rules about which of the months without a *zhong qi* should be counted as a leap month. The modern rule is to consider the "astronomical year" from the new moon immediately before or on the same day as a December solstice. The solar year from one December solstice to the next is called the *sui* (歲) in Chinese astronomy, but for convenience, I will also call the astronomical year a *sui*. A *nian* (年), however, is the Chinese year from one Chinese New Year to the next.

Let me clarify some terminology. When I talk about the Chinese year 2033, I mean the *nian* from Chinese New Year 2033 to Chinese New Year 2034. The problem with this convention is that dates in the 11th or 12th months may fall in the following Gregorian year. For example, the 12th month of the Chinese year 2033 starts in January 2034. The *sui* 2033 is the *sui* from the new moon immediately before or on the same day as the December solstice in 2032.

If the *sui* contains 13 months, I will call it a leap *sui*, and the first month without a *zhong qi* in it is the leap month. Any month without a *zhong qi* that is not a leap month is called a *fake leap month*. This is the modern rule as described by Liu and Stephenson ([LS1]). It is possible that the Jesuits used a variation of this rule.

I think it is important to try to understand why it took so long for the Chinese to introduce the true moon and the true sun into the calendar. First of all, there is a big difference between knowing that the motion is not uniform and knowing how to predict the motion. The ancient Chinese calendars were based on arithmetical resonance patterns, and the Tang reform came about because of Indian influences. As explained above, the need for introducing the true moon was more pressing than the need for using the true sun, since using the mean moon created discrepancies that were noticeable to everybody, while the errors caused by using the mean sun were only noticeable to astronomers. Given the computational complexities associated with using the true sun, the Chinese astronomers choose to stay with the old method. The Jesuits, however, needed to demonstrate their superiority over the Chinese and Muslim astronomers. Changing the calendar by using the true sun was a way of asserting control over the calendar.

It is therefore somewhat ironic that this backfired. Many aspects of Chinese astrology were based on the mean sun, and months with three *jie qi*'s or fake leap months were very alien to Chinese scholars of the time. Part of the reason why the Jesuits were thrown into jail in 1664 was because Yang Guang Xian (楊光先) and Wu Ming Xuan (吳明炫) noticed that the 11th month in 1661 had three *jie qi*'s (including two *zhong qi*'s), something that was an obvious error in the old system. (Chu ([Ch]) writes that this was the 12th month, but this must be a typo. In fact, the last *zhong qi* should have been in the following month, because it occurred 39 minutes after midnight — the Jesuits made an error.) Moreover, both the 7th and the 12th month had no *zhong qi*. The 7th was a leap month, but the 12th was a fake leap month. Fake leap months did not exist under the old system, so again Yang claimed that the Jesuits were obviously wrong. These were two of the complaints that Yang raised, and which eventually led to the 1664 calendar case in the Imperial Court ([Ch]).

The 2033 Error in the Chinese Calendar

Up until the early 1990s, all Chinese calendars had a leap month after the 7th month in 2033, while in fact it should follow the 11th month. In this section I will discuss this error in detail. Let us start by looking at the times for the *zhong qi*'s and new moons during the end of 2033. This is given in Table 1. I denote the n'th month (or the n'th new moon) by "Mn", and I denote the new moon after Mn by "Mn+" and the new moon after that by "Mn++". The reason is that before I have compared the *zhong qi*'s, I cannot tell whether any of them are leap months or not. I denote a leap month after Mn by "Mn-leap". I will denote the n'th *zhong qi* by "Zn". (The December solstice is Z11.)

Table 1   *Zhong qi*'s and New Moons During the Winter of 2033/34

M7:	2033 7 26, 16h 11m	Z7: 2033 8 23, 3h 0m
M8:	2033 8 25, 5h 38m	Z8: 2033 9 23, 0h 50m
M9:	2033 9 23, 21h 38m	Z9: 2033 10 23, 10h 26m
M10:	2033 10 23, 15h 27m	Z10: 2033 11 22, 8h 14m
M11:	2033 11 22, 9h 38m	Z11: 2033 12 21, 21h 44m
M11+:	2033 12 22, 2h 45m	Z12: 2034 1 20, 8h 25m
M12:	2034 1 20, 18h 0m	

It is known that the *zhong qi*'s all occur between the 19th and 23rd of the month. The date of the new moon, however, more or less regresses through the Gregorian month. If you write out the Gregorian calendar with the months as columns, and mark the new moons and the *zhong qi*'s, you see that the *zhong qi*'s form a more or less horizontal line, while the date of the new moon climbs upwards until it reaches the top and jumps to the bottom and starts climbing again. Leap months occur when the new moon curve crosses the *zhong qi* curve. Most of the time you get a "clean" crossing, but sometimes the curves might get intertwined in complex ways. In 1998 (Table 2) the *zhong qi*'s fell before the new moon until June, in July they coincided, and from August on the *zhong qi*'s fell after the new moon. This clean crossing gave a "normal" leap year.

Table 2   Position of the *Zhong qi*'s and New Moons in 1998

	June	July	August
19			
20			
21	Z5		
22			M7
23		Z6/M6	Z7
24	M4-leap		

In 2033 (Table 3), however, the *zhong qi*'s fall before the new moon until August, and for seven months between September and March they either coincide, or the *zhong qi* fall earlier. Not until April do the *zhong qi*'s fall after the new moon.

Table 3 Position of the *Zhong qi*'s and New Moons in 2033/34

	Aug.	Sep.	Oct.	Nov.	Dec.	Jan.	Feb.	Mar.	Apr
18							Z1		
19							M1		M3
20						Z12/M12		Z2/M2	Z3
21					Z11				
22				Z10/M11	M11-leap				
23	Z7	Z8/M9	Z9/M10						
24									
25	M8								

The number of *zhong qi*'s for the winter months of 2033/34 is given in Table 4.

Table 4 Number of *Zhong qi*'s for the Winter Months of 2033/34

2033 M8:	0
2033 M9:	1
2033 M10:	1
2033 M11:	2
2033 M11-leap:	0
2033 M12:	2
2034 M1:	0

We see that the 9th month takes Z8, the 10th month takes Z9, and the 11th Month takes Z10. But the 11th month also holds on to the December solstice, Z11. The fact that the 8th month does not have a *zhong qi* is compensated for by the fact that the 11th month has two. Hence the *sui* 2033 has only 12 months, while the *sui* 2034 has 13 months. In other words, *sui* 2033 is not a leap *sui*, while *sui* 2034 is a leap *sui*. It follows that the month after the 7th month is not a leap month, because there is no room for a leap month in the *sui*. The 8th month is a "fake" leap month, in the sense that it does not contain any *zhong qi*, but is not a leap month.

Table 5 Fake Leap Months

Year	Leap year	Leap *sui*	Leap month	Month with 2 *zhong qi*'s	Fake leap month
1832	Yes	Yes	9-leap	11	
1833	No	No			1
1851	Yes	Yes	8-leap	12	
1852	No	No			2
1870	Yes	Yes	10-leap	11	12
1984	Yes	Yes	10-leap	11	
1985	No	No			1
2033	Yes	No	11-leap	11,12	8
2034	No	Yes			1

It is clear that fake leap months are closely related to months with two *zhong qi*'s. Table 5 shows all such months between 1800 and 2100. In 1832, 1851, 1870 and 1984, a month with two *zhong qi*'s caused a month with no *zhong qi* in the next *sui* (in 1870 in the same *nian*, in the other three cases in the next *nian*). In all these cases, the December solstice was early in the 11th month, making the year both a leap year and leap *sui*. But 2033 is unique in that the December solstice is the second *zhong qi* in the 11th month. Since there is a leap *sui* if M11+ falls within about 11 days of Z11, we see that this is the reason why 2034 is a leap *sui* while 1833, 1852, 1871 and 1985 are not. The fake leap month in 2034 is the first fake leap month in a leap *sui* since 1645. The next will occur in 2129.

It also follows that Chinese New Year is the third new moon after the December solstice in 2034. Notice also that 2033 contains two months with two *zhong qi*'s. It is interesting to observe that in the Indian calendar, a fake leap month is counted as a leap month, but when a month has two *zhong qi*'s they skip a month!

A month with two *zhong qi*'s will of course have three *jie qi*'s. Sometimes there are months with three *jie qi*'s where one is a *zhong qi* and two are odd *jie qi*'s. This happened in the 10th month in 1999. They are not so interesting since they do not affect the leap months.

A year is said to have "double spring", *shuang chun* (雙春), if it contains "beginning of spring", *li chun* (立春) at both its beginning and end. It is easy to see that this happens if and only if the year is a leap year. In the same way, a year is said to have "double spring, double rain", *shuang chun shuang yu* (雙春雙雨), if the *nian* contains both "beginning of spring", *li chun*, and "rain water", *yu shui* (雨水), at both its beginning and end. This is considered significant in Chinese astrology. Between 1645 and 2644, this happened only

15 times. It happened in 1699, 1832, 1851 and 1984, and will happen again in 2033 and 2053. We see that these years are almost the same as the exceptional years we have discussed earlier.

References

[As1] Helmer Aslaksen, ChineseLeapMonths.nb, Mathematica package available at http://www.math.nus.edu.sg/aslaksen/.

[As2] Helmer Aslaksen, ChineseLeapMonths.nb, Mathematica package available at http://www.math.nus.edu.sg/aslaksen/.

[CJJ] Chen Jiujin, Chinese Calendars, in *Ancient China's Technology and Science*, Foreign Language Press, Beijing, 1983, pp. 33–49.

[Ch] Chu Pingyi, "Scientific Dispute in the Imperial Court: The 1664 Calendar Case", *Chinese Science*, 14 (1997), pp. 7–34.

[DR] Nachum Dershowitz and Edward M. Reingold, *Calendrical Calculations*, Cambridge University Press, 1997.

[Do] L. E. Doggett, Calendars, in *Explanatory Supplement to the Astronomical Almanac*, P. Kenneth Seidelmann (ed.), University Science Books, 1992, pp. 575–s608.

[LS1] Liu Baolin and F. Richard Stephenson, *The Chinese calendar and its operational rules*, manuscript.

[Me1] Jean Meeus, *Astronomical Algorithms*, 2nd ed., Willmann-Bell, 1998.

[Ne] Joseph Needham, *Science and civilisation in China*, v. 3 *Mathematics and the Sciences of the Heavens and the Earth*, Cambridge University Press, 1959.

[Si] Nathan Sivin, "Cosmos and computation in early Chinese mathematical astronomy", *T'oung Pao* 55 (1969), pp. 1–73. Reprinted in: Nathan Sivin, *Science in Ancient China*, Variorum, 1995.

[Sp] Jonathan Spence, *To Change China, Western Advisors in China 1620–1960*, Little, Brown, 1969.

[Ta] Tang Hanliang (唐漢良), *Lishu bai wen bai da* (歷書百問百答), *Jiangsu kexue jishu chubanshe* (江蘇科學技術出版社), 1986.

Originality and Dependence of Traditional Astronomies in the East

ÔHASHI YUKIO 大橋由紀夫
Tokyo, Japan

Introduction

There are two old traditions of astronomy in the East, namely Indian and Chinese. Later, Islamic astronomy also plays a very important role. These three traditions are almost independent in their origin, although there were some interrelations.

There are several other small traditions of astronomy in the East, which were more or less influenced by these three large ones. For example, Korean and Japanese traditional astronomies were largely influenced by that of China. There are also Tibetan and South-East Asian astronomies which were influenced by India and China. Islamic astronomy was also introduced into South-East Asia. I would like to examine the originality and dependence of some of these traditional astronomies.

Similarity Does Not Mean Interrelationship — Chinese Gaitian and Indian Sumeru Mountain Models

The Chinese Gaitian (蓋天) model is an ancient cosmology which was formed in about the Former-Han (Western-Han) period. It is explained in the *Zhoubi-suanjing* (周髀算經).[1] According to this model, the heaven and the earth are flat (or convex) and parallel. This model later gave way to the Huntian (渾天), where the heaven is a sphere, in the late Former-Han and Later-Han (Eastern-Han) periods.

The Indian Sumeru (or Meru) Mountain model is a traditional cosmology in Hinduism, Jainism, and Buddhism. According to this model, the earth is flat, and the Sumeru Mountain is at its centre. In Hindu Classical astronomy, the Sumeru Mountain model gave way to that of the spherical earth.

The Chinese Gaitian model and Indian Sumeru Mountain models look similar at first sight. The earth is flat in both, and the heaven is above the earth. So one may suspect some kind of interrelationship between them. However, there are important differences, such as:

(1) In the Chinese Gaitian model, the heaven is a single plane. Its height was determined by the observation of the terrestrial latitudinal difference of the midday gnomon-shadow.

The Indian Sumeru Mountain model has a multi-layered heaven. The sun and moon revolve around the middle of the Sumeru Mountain. The *Abhidharma-kośa* (III.60 a) of Vasubandhu reads:[2] "*Ardhena meroś candrārkau*" (The moon and sun [revolve] around the middle of Meru).

Determination of the height of heaven by the observation of the gnomon-shadow, which is the foundation of Chinese Gaitian model, is foreign to the Indian Sumeru Mountain model. This difference is important.

(2) In the Chinese Gaitian model, the orbit of the sun is concentric circles whose diameter changes in accordance with seasons.

In the Indian Sumeru Mountain model, the orbit of the sun is eccentric circles whose diameter is constant. Seasonal change of the sun's orbit in the Buddhist Sumeru Mountain model is explained in detail in the *Lishi-apitan-lun* (立世阿毘曇論), chapter 19, "Riyuexing-pin" (日月行品),[3] a Chinese version of a Buddhist work translated by Zhendi (真諦) (= Paramārtha) in the middle of the 6th century A.D.

The explanation of the difference of seasons is therefore completely different between the Gaitian model and Sumeru Mountain model. This is also important.

(3) In the Chinese Gaitian model, the earth is basically a single square. In the Buddhist Sumeru Mountain model, there are nine concentric mountains and eight concentric oceans, and the plane we are supposed to inhabit is only an island.

From the above three differences, it is now clear that the Chinese Gaitian and Indian Sumeru Mountain models are different in their origins. Even if there are some similarities, there may not be an actual interrelationship. This is what we should keep in mind.

As regards the possibility of the influence of Indian astronomy upon Chinese astronomy in the Later-Han (Eastern-Han) period, I have discussed this at the *8th International Conference on the History of Science in East Asia* (26–31 August 1996, Seoul),[4] and have shown that there is a possibility that certain Indian calendrical theories were introduced into Later-Han China, although Indian theory did not enjoy popularity at that time.

The Areas Between India and China

There are some small astronomical traditions in the East, which were influenced by both Indian and Chinese astronomy. One of these is Tibetan astronomy. It was further transmitted to Mongolia and Bhutan. As regards the origin of Tibetan astronomy, I have discussed this at the *3rd International Conference on Oriental Astronomy* (27–30 October 1998, Fukuoka).[5] The Tibetan calendar is basically based on Kālacakra astronomy, which was produced by Esoteric Buddhists in India in the 11th century A.D. Its astronomical constants are based on the Ārdharātrika school of Hindu astronomy, which was popular in East India. Kālacakra astronomy uses the South Indian system of a 60-year cycle, but the North Indian system of a 60-year cycle seems to have been wrongly incorporated at the place called Hijra, and produced a 2 years' error.

Chinese influence upon Tibetan astronomy is the use of the 60-year cycle called "*ganzhi*" (干支) (which is different from the Indian 60-year cycle), and the use of serial numbers for the names of months. This nomenclature is called "*Hor-zla*" (Mongolian month) in Tibetan, probably because it was introduced at the time of the Yuan (元) dynasty, which was ruled by the Mongols.

Another tradition which was influenced by both India and China is South-East Asian astronomy. This tradition is very important, but little investigation has been done so far.

The Traditional Astronomy in Mainland South-East Asia

Introduction

The traditional calendars of Burma, Thailand, Laos, Cambodia, and also of Tai (= Dai [傣]) people in Sipsong-panna (= Xishuang-banna [西双版纳]) etc. in Yunnan (雲南) province of China are largely based on the Indian calendrical system, but there are some differences. There is also a certain Chinese influence, such as the use of the cycle of 60 numbers "*ganzhi*", the animal names of the 12-year cycle etc.

The main differences between the traditional Hindu and South-East Asian calendars are as follows.

(1) In the Hindu calendar, the name of a month is determined by the position of the sun in zodiacal signs at the time of the new moon. Therefore, intercalary months are automatically produced when two successive new moons occur, during which the sun remains in the same zodiacal sign.

In the South-East Asian calendar, a 19-year cycle is usually used for intercalation, and 7 intercalary months which consist of 30 days are distributed after the definite month of the year. As we shall see later, this 19-year cycle is not harmonious with the length of the year and month used in the South-East Indian calendar.

(2) In the Hindu calendar, one month is divided into 30 *tithis*, which are the periods during the longitudinal difference between the sun and moon change by 12 degrees. The number of the *tithi* at the beginning of the calendrical day (usually the sunrise) becomes the name of the calendrical day. If the beginning of the calendrical day is not included in a *tithi*, the *tithi* becomes an "omitted *tithi*". Therefore, the number of days in a month is automatically determined.

In the South-East Asian calendar, the 29-day and 30-day months are usually distributed one by one at definite times of the year, and 11 intercalary days are distributed in 57 years in definite months. Actually this method is slightly inexact, and is not exactly harmonious with the length of a month used in the South-East Asian calendar.

(3) One more difference may be mentioned here. In the early Burmese inscriptions, a 12-year Jupiter cycle is used, is mechanically repeated, and the lunar month names which are based on the names of Indian lunar mansions are used.[6] This cycle does not agree with the Indian 12-year Jupiter cycle where year names keep pace with the movements of Jupiter, whose orbital period is not exactly 12 years. The Burmese and Indian cycles coincide in the 7th–8th century A.D., and it may be that the original Burmese cycle diverged from the Indian at that time. There is another Indian system in which the 12-year cycle is mechanically repeated, but this is also different from the Burmese cycle, and coincides with the aforesaid Indian cycle in the 9th century A.D.

The Length of a Year is Indian

The length of a year used in most of Mainland South-East Asian traditional calendars is 365.25875 days. This is evidently a sidereal year just like the year of the Hindu calendar, and not a tropical one.

There are four main schools in Hindu classical astronomy, namely, the Ārdharātrika, Saura, Ārya, and Brāhma. The *Sūrya-siddhānta* is quoted in the

Pañca-siddhāntikā[7] of Varāhamihira (6th century A.D.) belongs to the Ārdharātrika school, and the modern *Sūrya-siddhānta*[8] (ca. 10–11th century A.D.) belongs to the Saura school.

S.B. Dikshit[9] already pointed out that the length of a year used in an astronomical work procured from Siam (Ayutthaya dynasty)[10] is the same as that of the *Sūrya-siddhānta* quoted in the *Pañca-siddhāntikā*. Roger Billard[11] also pointed out that the Indo-Chinese traditional astronomy belongs to the same school as the *Sūrya-siddhānta* quoted in the *Pañca-siddhāntikā*.

The length of a synodic month in the South-East Asian calendar is about 29.5305832 days (more exactly 20760/703 days). This is slightly different from the value used in the *Sūrya-siddhānta* quoted in the *Pañca-siddhāntikā*, which is about 29.5305874 days.

The Burmese Era which starts from the year zero in A.D. 638 is called "khachapañca". Zhang Gongjin[12] supposed that it may be related to the *Pañca-siddhāntikā*, because the word "pañca" (which means five in Sanskrit and Pali) appears in both. This is, however, wrong. The word "kha" means sky or zero, "cha" means six, and "pañca" means five in Pali, and the expression "kha-cha-pañca" means 560, that is the year 560 in Śaka Era which starts from the year zero in A.D. 78.[13] Therefore, the expression "khachapañca" has nothing to do with the *Pañca-siddhāntikā*.

The 19-year Cycle of Intercalation

The 19-year cycle of intercalation was already used in an astronomical work "Souriat" of the Ayutthaya dynasty of Thailand, which was procured by de la Loubere and analyzed by Jean Dominique Cassini in the late 17th century.[14]

In India, the 19-year cycle of intercalation appears in the *Romaka-siddhānta* quoted in the *Pañca-siddhāntikā* (I.9–10 and 15). It may also be noted that the length of a year used there is $365\frac{703}{2850}$ (= 365.2466...) days which is the same as the tropical year used by Hipparkhos. So it may be that the 19-year cycle used there originated in the Western Methonic cycle.

One may suspect that the South-East Asian 19-year cycle was introduced from India, but we should keep in mind that the 19-year cycle was not popular in India, and was not used in the Hindu classical astronomy which includes the Ārdharātrika school etc. There are at least two reasons why the 19-year cycle did not enjoy popularity in India. One is that Hindu classical astronomy uses a period called *yuga*, which consists of 4320000 years, as a basic cycle, but it is not a multiple of 19 years. Therefore, the 19-year cycle cannot be accommodated there. The other reason, which is more important, is that the cycle of intercalation automatically determines the relationship between the length of a year and that of a month. As the length of a month cannot be so inexact in a luni-solar calendar, it should not be so different from 29.53059 days. If the 19-year cycle is used, the length of a year becomes $\frac{235}{19}$ of the length of a month, that is about 365.2468 days. This value can be used as the length of a tropical year, but cannot be used as that of a sidereal year. As the Hindu classical astronomy uses sidereal years, the 19-year cycle cannot be used in this system.

Now, the length of a year used in the South-East Asian calendar is 365.25875 days, and is evidently a sidereal year. Therefore, it is quite strange to find the 19-year cycle being used. In fact, certain adjustments of the year of intercalation were done in the actual South-East Asian calendars.[15]

From the above considerations, I suspect that the basic calendrical system and the 19-year cycle in the South-East Asian calendar are of different origins. It is difficult to suppose that the 19-year cycle was produced in South-East Asia in order to simplify the calendar, because it is much easier to use a cycle of intercalation which is harmonious with the length of year and month used there. It is probable that the 19-year cycle was introduced from certain areas where this cycle was considered to be fundamental and the tropical year was used as the calendrical year. Among the adjoining areas of South-East Asia, only China fits the case. So, we have to consider the possibility that the 19-year cycle was introduced into South-East Asia from China.

Recently, S. K. Chatterjee[16] argued that the South-East Asian 19-year cycle was introduced there by Europeans. This argument, however, cannot be accepted, because no other trace of European influence is found in South-East Asian traditional astronomy.

One piece of information may be added here. A. M. B. Irwin informs us in his *Burmese and Arakanese Calendars* that "Intercalary months are regulated by the Metonic cycle of 19 years, the use of which was propounded in the 10th book of Raja-Mathan, a Hindu astronomer".[17] Htoon-Chan also informs in his *Arakanese Calendar* that "the Arakanese and Burmese had the same method of calculation for the intercalary months and days, known in Burma as the *Makaranta* method, and both adopted the Metonic cycle of 19 years, as propounded in the 10th book of *Raja-mattan*, recognizing 2, 5, 7, 10, 13, 15, ..."[18] If this information is right, it may contradict my hypothesis. However, even if the 19-year cycle is mentioned in an Indian source, it does not mean that it is the only source of the South-East Asian 19-year cycle. This point should be investigated further. The origin of the South-East Asian calendar will be discussed again after examining the method of month reckoning.

Serial Numbers for the Name of Months

In Cambodia, lunar months are usually named after the Indian lunar mansions, just like the case of the Hindu calendar. In Burma, unique month names are used whose origin is not clear.[19]

Among Thai (or Tai) people, lunar months are named by serial number, and there are some variations in this method.[20] According to Central Thai and Lao methods, the first month corresponds to the Mārgaśīrṣa month. According to the Sipsong-panna method, the first month corresponds to the Kārttika month. According to the Northern Thai method, the first month corresponds to the Āśvina month.

In Indian Buddhist and Jaina literature, months are sometimes named by serial number within a season (one-third of a year), but the method of naming all months in a year by serial number does not seem to be Indian.

According to the Sipsong-panna method, the first month corresponds to the 10th month of the ordinary traditional Chinese calendar which follows the "Xia" (夏) reckoning of months. Already in 1938, Dong Yantang (董彦堂)[21] argued that the Tai calendar is based on the Qin (秦) calendar, which started from the 10th month of "Xia" reckoning, but Dong Yantang neglected Indian influence. Zhang Gongjin (張公瑾)[22] pointed out that the Dai (傣) calendar must have accepted the method of numbering the months along with the Chinese cycle of 60 numbers "*ganzhi*" (干支) in the Former-Han (Western-Han) period, and later accepted the Indian calendar along with Buddhism. I think that this is quite possible.

One candidate for the source of the Sipsong-panna method may be the Chu (楚) calendar in the Zhangguo (戰國) period, whose first month is the 10th month of the "Xia" reckoning as mentioned in the Yunmeng Shuihudi (雲夢睡虎地) bamboo slip. It must have been popular in Southern China for a certain period.

It may be mentioned here that the Sipsong-panna method was also used in Cambodia in the 13th century. The *Zhenla-fengtuji* (真臘風土記)[23] of Zhou Daguan (周達觀), who visited Cambodia from 1926 to 1297, tells us that the Cambodian first month, called Kārttika, corresponds to the Chinese 10th month.

We should also note that one Thai inscription of A.D. 1495 mentions three kinds of numbers of months where the Central Thai method is called "Buddhist astrologer's month" while the Northern Thai method is called "Thai month".[24] Another Thai inscription of A.D. 1500 mentions two kinds of numbers of month where the Central Thai method is called "Buddhist month" and the Sipsong-panna method is called "Thai month".[25] These inscriptions suggest that the Sipsong-panna and Northern Thai methods were not considered to be Indian. That the Central Thai method was considered to be "Buddhist" is harmonious with the following words of King Mongkut (Rama IV, 1804–1868) of Thailand:

> "According to the Siamese custom there will be an intercalary month between the eighth and ninth of the Siamese cardinal order, which agrees with the ancient Buddhist political Chronological number ... for the rainy season was considered like the night as being unfavorable to hard work, journeying etc. and the cold or cool season as the dawn of day and which is considered by the common people as the commencement of the day, and hence the custom of numbering their months after the new moon in November. But the Siamese commence their solar year on the 11th or 12th of April as the sun arrives at the point in Aries which marks the sidereal year ..."[26]

The first month of the Central Thai method is exactly the beginning of the cold season in the Indian division of seasons, and the Thai intercalary month is put just before the rainy season. It may be that the Central Thai method is a kind of adjusted method in accordance with the Indian division of seasons, although the origin of the "numbering" itself may not be Indian.

It may be interesting to note that Lithai (King Ruang) of the Sukhothai dynasty of Thailand wrote in his *Traiphum (Three Worlds)*[27] (mid-14th century) that the hot season is from the 1st day of the waning moon of the 4th month to the full moon day of the 8th month; the rainy season is from the 1st day of the waning moon of the 8th month to the full moon day of the 12th month; and the cold season is from the 1st day of the waning moon of the 12th month to the full moon day of the 4th month. This is basically the same as the Central Thai method, but seasons end with a full moon, just the same as in the Indian *pūrṇimānta* method, while months end with a new moon. This fact may suggest that Lithai distinguished the Thai method of month reckoning from the Indian calendar.

In conclusion, I suggest the following tentative hypothesis. The 19-year cycle of intercalation and the Sipsong-panna method of month reckoning were introduced from China, most probably from the Chu calendar, and, after the introduction of the Indian calendar (Ārdharātrika school), the Central Thai and Lao methods of month reckoning were produced in order to adjust to the Indian division of seasons. The 19-year cycle has still been used although it is not harmonious with the Indian calendar.

Historical Development of the South-East Asian Calendar

According to Zhang Gongjin (張公瑾) and Chen Jiujin (陳久金),[28] the *Suding* (蘇定) and the *Suliya* (蘇力牙) were followed before A.D. 1931 for the Dai calendar, but the *Xitan* (西坦) has been followed since A.D. 1931 or so. Their astronomical constants along with those of the four Indian schools are as follows:[29]

	Length of a year	Length of a month	Orbital period of Mars	Orbital period of Jupiter	Orbital period of Saturn
					(days)
Suding, Suliya	365.25875	29.530583	687	4332.33	10766
Xitan	365.258756481	29.530588	686.997	4332.32	10765.77
Ārdharātrika	365.25875	29.5305874	686.9998	4332.321	10766.067
Saura school	365.258756482	29.5305880	686.9975	4332.321	10765.773
Ārya school	365.258680556	29.5305818	686.9997	4332.272	10766.065
Brāhma school	365.2584375	29.5305821	686.9979	4332.240	10765.815

It is clear from this table that the *Suding* and the *Suliya* follow the Ārdharātrika school, and the *Xitan* follows the Saura school. Zhang and Chen indicate that the *Suding* and the *Suliya* do not follow the 19-year cycle of intercalation strictly, but the *Xitan* follows the 19-year cycle, and that the calendrical luni-solar year of the *Xitan* practically keeps pace with the tropical year, although the solar new year's day is calculated by the sidereal year as before.

According to A. M. B. Irwin,[30] the *"Makaranta"* (A.D. 638 and A.D. 1436), which probably follows the "original Surya Siddhanta" (i.e. Ārdharātrika school), was followed in Burma originally, but afterwards the *"Thandeikta"* (A.D. 1738), which chiefly follows the "present Surya Siddhanta" (i.e. Saura school), was used. Irwin writes that the *"Thandeikta"* is said to have been composed in about A.D. 1738 or A.D. 1838. He also writes that the present *Sūrya-siddhānta* is said to have been introduced into Amarapura in A.D. 1786, and translated into Burmese after 50 years.[31] The length of year and month in the "*Makaranta*" and the "*Thandeikta*" is as follows:

	Length of a year	Length of a month
		(days)
Makaranta	365.25875	29.530583
Thandeikta	365.258756477	29.530588

We should note that the Burmese "*Makaranta*" is probably different from the well known Indian Sanskrit astronomical table *Makaranda-sāraṇī* (A.D. 1478), which follows the Saura school.

Irwin tells us that the 19-year cycle of intercalation is followed in the "*Makaranta*", but is not strictly followed in the "*Thandeikta*".

According to this information, the treatment of the 19-year cycle is just the opposite between the Dai and Burmese calendars. The process of calendar reform in these areas should be investigated further.

Conclusion

From the above discussions, it is now clear that South-East Asians accepted both Chinese astronomy and Indian astronomy, and produced their own traditions. However, very little research in the original sources of South-East Asian astronomy has been done. The history of South-East Asian astronomy is definitely one of the most important subjects of future study.

Acknowledgement

I am grateful to Dr. Xu Zelin, Mr. Mineo Nishiyama, and Mr. Takashi Suga who provided me with some references.

Notes

[1] *Suanjing-shishu* (算經十書), Renren-wenku te-325, Taibei, Taiwan-shangwu-yinshuguan, 1974, pp. 1–62.

[2] Pradhan, P. (ed.), *Abhidharmakośabhāṣyam of Vasubandhu*, Patna, K. P. Jayaswal Research Institute, 1975, p. 165.

[3] *The Taishō shinshū daizōkyō* (大正新脩大藏經), vol. 32, Tokyo, The Taisho shinshu daizokyo kanko kai, 1925, pp. 195–97.

[4] Ôhashi, Yukio, "Historical Significance of Mathematical Astronomy in Later-Han China", Kim, Yung Sik and Francesca Bray (eds.), *Current Perspectives in the*

History of Science in East Asia, Seoul, Seoul National University Press, 1999, pp. 259–63.

5 Proceedings of this conference have not been published at present. For Tibetan astronomy, the following article may tentatively be referred to. Ôhashi, Yukio, "Astronomy in Tibet", Selin, Helaine (ed.), *Encyclopaedia of the History of Science, Technology, and Medicine in Non-Western Cultures*, Dordrecht/Boston/London, Kluwer Academic Publishers, 1997, pp. 136–39.

6 For Burmese 12-year cycle, see J. S. F. (= Furnivall): "The Burmese Calendar", *The Journal of the Burma Research Society* (hereafter *JBRS*), vol. 1, part 1, 1911, pp. 96–97; Blagden, C. O., "Epigraphical Notes IV. The Cycle of Burmese Year-names", *JBRS*, vol. 6, 1916, pp. 90–91; de Silva, Thos. P., "12 Year Cycle of Burmese Year Name", *JBRS*, vol. 7, 1917, pp. 263–74; Maung Hla, "The Twelve-year Cycle of Burmese Year-names", *JBRS*, vol. 8, 1918, pp. 270–74; Furnivall, J. S., "The Cycle of Burmese Year Names", *JBRS*, vol. 12, 1922, pp. 80–95; and Luce, Gordon H., *Old Burma-Early Pagán*, vol. 2, New York, J. J. Augustin Publisher, 1970, pp. 330–31. For Indian 12-year cycle, see Cunningham, Alexander, *Book of Indian Eras*, Calcutta, 1883, reprint, Varanasi, Indological Book House, 1970, pp. 26–30; and Sewell, Robert and Sankara Balakrishna Dikshit, *The Indian Calendar*, London, 1896, reprint, Delhi, Motilal Banarsidass, 1995, p. 37.

7 Thibaut, G. and Sudhākara Dvivedī (ed. and tr.), *The Pañcasiddhāntikā*, Benares, 1889, reprint, Varanasi, The Chowkhamba Sanskrit Series Office, 1968; Neugebauer, O. and D. Pingree (ed. and tr.), The *Pañcasiddhāntikā of Varāhamihira*, 2 parts, København, Munksgaard, 1970–71; and Sastry, T. S. Kuppanna (posthumously edited by K. V. Sarma), *Pañcasiddhāntikā of Varāhamihira*, Madras, P. P. S. T. Foundation, 1993.

8 For its English translation, see Gangooly, P. (ed.), *The Sûrya Siddhânta*, translated by Ebenezer Burgess (and W. D. Whitney), Calcutta, 1935, reprint, Delhi, Motilal Banarsidass, 1989.

9 Dikshit, Sankar Balakrishna (tr. by R. V. Vaidya), *Bharatiya Jyotish Sastra*, part 2, New Delhi, India Meteorological Department, 1981, p. 378. (Marathi original was published in 1896.)

10 The astronomical work procured from Siam is presented in de la Loubere (tr. by A. P.), *A New Historical Relation of the Kingdom of Siam*, London, 1693, reprint, Kuala Lumpur, Oxford University Press, 1969, pp. 186–99.

11 Billard, Roger, *L'astronomie indienne*, Paris, École Française d'Extrême-Orient, 1971, p. 74. Also see Eade, J. C., *The Calendrical Systems of Mainland South-East Asia*, Leiden, E. J. Brill, 1995, plates III–V.

12 Zhang Gongjin (張公瑾), "Daili-zhongde jiyuan-jishi-fa" (傣曆中的紀元紀時法), *Zhongguo-tianwen-xueshi wenji* (中國天文學史文集), vol. 3, Beijing, Kexue chubanshe, 1984, pp. 249–73.

13 Taw Sein Ko, "The Early Use of the Buddhist Era in Burma", *JBRS*, vol. 1, part 1, 1911, pp. 31–34; May Oung, "The Burmese Era", *JBRS*, vol. 3, 1913,

pp. 197–203; and Soni, R. L., *A Cultural Study of the Burmese Era*, Mandalay, Institute of Buddhist Culture, 1955, p. 82.

[14] See note 10.

[15] See Eade, *op. cit.* (see note 11), pp. 56 ff.

[16] Chatterjee, S. K., "Traditional Calendar of Myanmar (Burma)", *Indian Journal of History of Science*, vol. 33, 1988, pp. 143–60.

[17] Irwin, A. M. B., *The Burmese and Arakanese Calendars*, Rangoon, Hanthawaddy Printing Works, 1909, p. 3.

[18] Htoon-Chan, *The Arakanese Calendar*, Third edition, Rangoon, The Rangoon Times Press, 1918, p. 1.

[19] For Cambodian months, see Faraut, F. G., *Astronomie cambodienne*, Saigon, F.-H. Schneider, 1910, p. 15. For old Burmese months, see Luce, *op. cit.* (see note 6), pp. 328–29.

[20] Davis, Richard, "The Northern Thai Calendar and Its Uses", *Anthropos*, vol. 71, 1976, pp. 3–32; Eade, *op. cit.* (see note 11), pp. 28–29, and, for Lao calendar, Phetsarath, Prince, "Calendrier laotien", *Bulletin des Amis du Laos*, vol. 4, 1940, pp. 107–140; and Dupertuis, Silvain, "Le Calcul du Calendrier laotien", *Peninsule*, no. 2, 1981, pp. 25–113.

[21] Dong Yantang (董彦堂), "Boyi lifa kaoyuan" (僰夷曆法考源), *Xinan bianjiang* (西南邊疆), no. 3, 1938, pp. 55–71.

[22] Zhang Gongjin (張公瑾), *Daizu wenhua* (傣族文化), Jilin jiaoyu chubanshe, 1986, p. 109.

[23] Xia Nai (夏鼐) (ed.), *Zhenla-fengtuji jiaozhu* (真臘風土記校注), Beijing, Zhonghua shuju, 1981, chapter 13 (pp. 120 and 122).

[24] Eade, J. C., *The Thai Historical Records: A Computer Analysis*, Tokyo, The Centre for East Asian Cultural Studies for Unesco, The Toyo Bunko, 1996, p. 40.

[25] *Ibid.*, p. 51.

[26] "Intercalary Month" Communicated by His Royal Highess T. Y. Chaufa Mongkut, *Siam Repository*, vol. 1, October 1869, pp. 342–43.

[27] Reynolds, Frank E. and Mani B. Reynolds (tr.), *Three Worlds according to King Ruang*, Berkeley, University of California, 1982, pp. 320–21.

[28] Zhang Gongjin (張公瑾) and Chen Jiujin (陳久金), "Daili yanjiu" (傣曆研究), *Zhongguo-tianwen-xueshi wenji*, vol. 2, Beijing, Kexue chubanshe, 1981, pp. 174–284. For Dai calendar, also see Zhang Gongjin, *Daizu wenhua yanjiu* (傣族文化研究), Kunming, Yunnan minzu chubanshe, 1988.

[29] Astronomical constants used in the schools of Hindu astronomy are given in Pingree, David, "History of Mathematical Astronomy in India", *Dictionary of Scientific Biography*, vol. 15, New York, pp. 533–633.

[30] Irwin, *op.cit.* (see note 17), p. 3. The astronomical calculation of the "*Saṁdiṭṭha*", which is probably the same as the "*Thandeikta*", is explained in detail in de Silva, Thos. P., "Burmese Astronomy", *JBRS*, vol. 4, 1914, pp. 23–43, 107–118 and 171–207.

[31] Irwin, *op. cit.*, p. 3. Htoon-Chan, *op. cit.* (see note 18), p. 1, tells that the *Sūrya-siddhānta* was brought from Benares to Mandalay in about 1837 A.D.

A Cartesian in the Kangxi Court
(as Observed in the *Lifa Wenda* 歷法問答)

HASHIMOTO KEIZŌ
Kansai University, Osaka

In 1994 we were able to locate the complete text of the manuscript *Lifa wenda*, preserved at the British Library. Earlier we had also located partial manuscripts preserved at the Vatican Apostolic Library. The *Lifa wenda* was compiled by the French Jesuit, Jean-François Foucquet (Fu Shengze 傅聖澤; 1665–1741), in Beijing between 1712 and 1716. The manuscript is important because it provided the earliest introduction into China of Kepler's Laws. This discovery pushes back the commonly-accepted date of their official introduction into China, 1742, the time of the compilation of the *Lixiang kaocheng houbian* 歷象考成後編. Foucquet, for the first time in China, started to explain Kepler's elliptic theory of the solar (or rather "earth's") movement, and then discussed the development of the theory of elliptic orbits in Europe after Kepler's original proposal of this method. He openly discussed Copernicanism as well, so as to make clear the theory of planetary motions in general. This demonstrates that the theory had been known in the Kangxi 康熙 court at least half a century earlier than has so far been recognized.

It is also significant that the improved astronomical or fundamental numbers are described in the manuscript. They include the table of atmospheric refraction as well as the solar parallax, on top of the obliquity of ecliptic, which was soon adopted in the Imperial edition of the *Lixiang kaocheng* 歷象考成, compiled in 1723. The value of solar parallax would, on the other hand, be improved at the Paris observatory; it is 10", which we also find described in the *Lifa wenda*. At the same time we must emphasize that it was La Hire's *Astronomical Tables* in 1702 that Foucquet relied on for the introduction.

We have already discussed those problems in previous papers, particularly at the *8th International Conference on the History of Science in East Asia in Seoul (1996)*,[1] and here we will not repeat those earlier discussions.

One of the most interesting issues among the contents that the manuscript discusses, however, is the physics of the universe in terms of the vortex theory proposed by René Descartes. We find the topic broached in the treatise on planetary motions. Ole Römer's determination of the velocity of light, as well as the method of the determination of longitude by making use of the eclipses of the first satellite of Jupiter, was also noted. In discussing these topics, the editor follows rather faithfully the content of the *Treatise on Light* (1690) by Christiaan Huygens. We can thus definitely say that the editor was a Cartesian, as was the case with many intellectual figures on the European Continent in the second half of the 17th and early 18th centuries.

Interestingly enough, this topic was also discussed in the *Astronomical Lectures* by Newton's successor as Lucasian Professor of Mathematics at the University of Cambridge, William Whiston. It occurs in Lecture XXI (Prop. VII "To find the Geographical Longitude of Places", and Prop. VII "To determine the Velocity of the Rays of Light"; dated December 14, 1702). I mention this because it allows us to better understand how these astronomical issues were regarded in Europe during the time when French Jesuits were in Beijing.

The Method of Determination of the Geographical Longitude

The first topic we will discuss here is the introduction of the precise determination of geographical longitudes (*Dilitu* 地理圖). We find this topic discussed in the first book of the Treatise on the Five Planets (*Wuwei lizhi* 五緯歷指), a part of which has been found preserved only at the British Library.

Now, the difference of the terrestrial longitudes was determined by timing the eclipses of the first satellite of Jupiter. The problem was to find the difference in mean or local time between a prime meridian and a second place. It was clear that the difference in time is equivalent to the difference in longitude (*Jingdu* 經度). The eclipse of the moon could be made use of for this purpose, but to determine the time of, for example, its precise beginning, was not easy. The occurrence of the eclipses is also not frequent, as the manuscript says.[2] Then we read the introduction of Cassini's method to determine the longitude.

The satellites (*Shiwei* 侍衛) of Jupiter were discovered by Galileo (*Jialile* 嘉理勒) in 1610, and G.D. Cassini, in 1671 determined the diameter of the orbit of the first satellite as 6 times of that of Jupiter and the period as one day 18 hours 29 minutes (about 42 hours and a half).[3] The eclipses occur most frequently and can be used for that purpose conveniently and quickly. During one period there occur four times immersions and emersions altogether. That makes 16 occasions to observe the eclipses in a week.[4]

The text continues the discussion as follows:

> If the satellite is observed with a telescope eighteen feet long, then it comes to be seen more clearly. The more clearly it is sighted, the larger the size of the object becomes, and the more the speed of motion increases. And thus the instant of immersion and the instant of emersion can easily be observed. The rapidity of the immersions and emersions contributes to make it easy to determine the true minutes and seconds of the moment of those eclipses with the pendulum clock (*Zhuizibiao* 墜子表) (invented by C. Huygens).

According to Brown,[5] in Cassini's opinion, the most satisfactory time observations of Jupiter could be made using the immersions and emersions of the first satellite. Six phases of the eclipse should be timed: (1) during the immersion of the satellite when the satellite is at a distance from the limb of Jupiter equal to its own diameter; (2) when the satellite just touches Jupiter; (3) when it becomes entirely hidden by Jupiter's disc; (4) during the emersion of the satellite the instant the satellite begins to reappear; (5) when it becomes detached from Jupiter's disc; (6) when the satellite has moved away from Jupiter a distance equal to its own diameter (Figure 1).

Fig. 1. The planet Jupiter showing the six positions of the first satellite used by the seventeenth-century astronomers to determine the difference in longitude between two places (from Brown, 1949, p. 223).

When Louis-Abel Fontenay, a Jesuit professor of mathematics at the College of Louis le Grand, was preparing to leave for China, hearing of the work done by Cassini and his colleagues, he volunteered to make as many observations as he could without interfering with his missionary duties. Cassini trained him and sent him on his way prepared to contribute data on the longitudes of the Orient and China as well. Eventually, under the order of the Chinese Emperor, the French Jesuits contributed to the preparation of a map of China.[6]

The Determination of the Velocity of the Rays of Light

It is not only the determination of the geographical longitudes, but also physics which is concerned with the subtle matters which Jupiter's satellites raised. Ole Roemer (*Lemo* 樂默; or *Luomoer* 樂默爾) had found the method by which the velocity of light (*Guang-zhi-liuxing* 光之流行) could be determined.

In the explanation of this problem of physics, we can read the strong influence of Cartesianism in the context of the *Lifa wenda*. The problem positively has nothing to do with any astronomical matter (*Lifa* 歷法) at all. It is rather a problem of physical inquiry (*Gewu-qiongli* 格物窮理), which is concerned with the utmost subtlety of matter. The principle of subtle matter was discussed by Huygens, which can be observed in the figures, which were attached on a page of the *Lifa wenda*, Book V-1 (see Figure 2).[7] The discussion has obviously been taken from the *Treatise on Light* by Christiaan Huygens

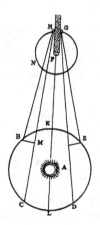

Fig. 2.

(*Xu-ri-ni* 許日尼; 1629–95) of 1690. Having depended on Römer's discovery, he had refuted the assertion of Descartes, who, making use of the lunar eclipse, proved the instantaneousness of the transmission of light (Figure 3).[8]

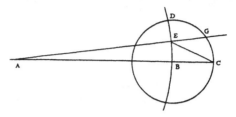

Fig. 3.

But to begin with, Foucquet introduces two ideas concerning the propagation of light. The first is the instantaneous transmission discussed by Descartes, and the other is the successive (*jianci* 漸次) propagation discovered and proved by Ole Römer.[9] According to the representation in the text, the instantaneous transmission (*xunsu* 迅速: lit., "rapidity") is expressed as *busishunqing* 不俟瞬頃 (lit., "to wait no time"), and the latter as *youjiweizhikeji* 有幾微之可計 ("the slightest time span capable of being measured"). Then he starts to emphasize the finiteness of the propagation of light. Eventually he introduces Huygens' theory of light, which differed radically from the corpuscular one accepted by Newton.

Now, Huygens refuted Descartes' assertion and wrote in a paragraph in the *Treatise on Light* as follows:[10]

> To see then whether the spreading of light takes time, let us consider first whether there are any facts of experience which can convince us to the contrary. As to those which can be made here on the Earth, by striking lights at great distances, although they prove that light takes no sensible time to pass over these distances, one may say with good reason that they are too small, and that the only conclusion to be drawn from them is that the passage of light is extremely rapid. Mr. Des Cartes, who was of the opinion that it is instantaneous, founded his views, not without reason, upon a better basis of experience, drawn from the Eclipses of the Moon; which, nevertheless, as I shall show, is not at all convincing. I will set it forth, in a way a little different from his, in order to make the conclusion more comprehensible.

The discussion cited from Huygens' *Treatise on Light* has been found translated and compiled in the *Lifa wenda*, V-2.[11] There we find "Mr. Descartes" transcribed as *Mingshi Gaerde* 名士・噶爾得. The idea can be seen in the drawing of Figure 2. Römer's observation of the eclipse of the satellite by Jupiter proved not only that light takes time for its passage, but also demonstrated how much time it takes.

Let *A* be the Sun, *BCDE* the annual orbit of the Earth, *F* Jupiter, and *GN* the nearest of its satellites, which was more apt for investigation than the other three because of the quickness of its revolution. Let *G* be this satellite emerging from the shadow. Although the diameter of the annual orbit is 24 thousand times that of the Earth in the description of the *Treatise*, it is 34 thousand times that which we read about in the manuscript of the *Lifa wenda*. This value is noted in the text as La Hire's (*Laxier* 臘義爾) determination.[12] This improved value leads to the more precise value of the velocity of light.

Anyway, here we can surely see that it was also Copernicanism which was used as the cosmological premise for this proof. It was unavoidable that Copernicanism be introduced in the text. The period of the revolution of the satellite around Jupiter is shown as $42\frac{1}{2}$ hours even in the Chinese manuscript.[13] In connection with Copernicanism, we must point out it was fully described in another part of the manuscript: the appendices of the first book of the Treatise on Eclipses, which we can find only in the Vatican version.

According to Nissen, the first version of Römer's planetary machine, which, based on the heliocentric model, had been devised in Paris in 1678, must have been brought to China together with a set of ecliptic machines.[14] The machines, as explained in the appendices of the Treatise on Eclipses (*Jiaosi lizhi* 交食歷指) of the Vatican version, were those (or replicas of those) manufactured by Römer himself.

Proceeding to a discussion of the mechanics of planetary motion, Foucquet introduced the medium of ether (*Jingqi* 精氣), relying on Descartes and Huygens' physical arguments.[15] He also mentioned vortices (*Xuanquan-zhi-bo* 旋圈之波, lit., "whirling waves"). In the present context, it is not surprising to find him introducing Cartesian ideas.[16] He also described how those honourable physicists, whatever else they may have produced, had postulated that the subtlest ether filled the universe above and below.[17]

Nocolas Malebranche (1638–1715), the most original Cartesian after Descartes, held that "occasional causes were to be sought in impact, for it is by the motion of visible or invisible bodies that all things happen". Another innovation of Malebranche's was the replacement of Descartes' globules by small vortices, which enabled him to give plausible accounts of the phenomena of heat and light.[18] And his idea of the elasticity of small vortices was derived from the centrifugal force of their rapid internal circulation.

Now, Huygens presented the wave theory of light transmission in terms of the medium of the elastic particles of ether. His presentation in the *Treatise on Light* was, without any essential alteration, introduced in the part of the treatise on the five planets in the *Lifa wenda*. Here, for example, Huygens discusses in the *Treatise* how the rays of light propagate in straight lines, the content of which we can also read in the translated Chinese manuscript (Figure 4).[19]

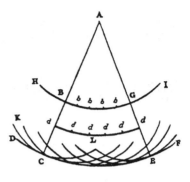

Fig. 4.

As for the elastic particles of ether, in accord with the original text, the Chinese manuscript proves the spread of the movement of waves by the row of equal spheres of hard matter (*jianyingzhiqiu* 堅硬之球). This can be shown in Figure 5, which is taken from Huygens' *Treatise*, and which is obviously missing from the *Lifa wenda*.

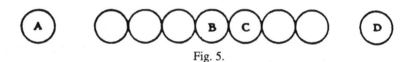

Fig. 5.

Concluding Remarks

The most important contribution of the French Jesuit missionaries to China was the knowledge of geography, which was to result in the mapping of the whole Chinese territory in the Qianlong reign-period as the *Huangyu quanlantu* 皇輿全覽圖. This was, as a matter of fact, a part of the French Academy's plan of making a Planisphere of the Earth, which was originally proposed by Cassini I.

The manuscript shows the magnificent introduction of recent, scientific knowledge by the French missionaries. This fact does not necessarily suggest that the astronomical concerns would have led to a successful result, partly because of the conflict with other "nationalist" Jesuit astronomers. But the introduction of Cartesian physics at least shows essential change over time in trans-cultural scientific contributions during the Qing period.

Notes

[1] K. Hashimoto and C. Jami, "New Evidence on the Transmission of European Astronomy to China: Jean-François Foucquet's *Lifa wenda*", Yung Sik Kim and Francesca Bray (ed.), *Current Perspectives in the History and Science in East Asia*, Seoul National University Press, 1999, pp. 487–96.

[2] *Lifa wenda*, V-1: 22a-b.

[3] *Ibid.*: 25a-b.

[4] *Lifa wenda*, V-1: 26a.

[5] Brown, 1979: p. 222.

[6] *Lifa wenda*, V-1: 32a.

[7] V-1: 33a.

[8] Delambre, 1821: p. 555.

[9] V-2: 37a.

[8] *Treatise on Light*, pp. 4–5.

[11] V-2: 39b-40a.

[12] *Lifa wenda,* V-2: 42a.

[13] V-2: 25b.

[14] Nissen, 1944, p. 32.

[15] V-2: 43a.

[16] Hashimoto & Jami, 1977: p. 178.

[17] *Lifa wenda*, V-2: 43a.

[18] Eric J. Aiton, "The Cartesian vortex theory", Taton & Wilson, 1989, 207–221: p. 218.

[19] Huygens, *Treatise*, p. 19; *Lifa wenda*, V-2: 43b.

References

Andreas Nissen, *Ole Römer*, Copenhagen, 1944.

Christiaan Huygens, *Treatise on Light* (1690), English trans. by S. P. Thompson (1912), New York: Dover Publications, 1962.

William Whiston, *Astronomical Lectures*, London, 1715.

Jean Baptiste Joseph Delambre, *Histoire de l'Astronomie Moderne*, tom. II, Paris, 1821; Facsimile ed. by I. B. Cohen, New York and London, 1969.

L. A. Brown, *The Story of Maps*, original ed., 1949; New York: Dover Publications, 1979.

R. Taton & C. Wilson (ed.), *Planetary astronomy from the Renaissance to the rise of astrophysics. Part A: Tycho Brahe to Newton*, Cambridge University Press, 1989.

K. Hashimoto & C. Jami, "Kepler's Laws in China: A Missing Link?", *Historia Scientiarum*, vol. 6–3, 1997: 171–85.

On the Naksatras in the Chinese Translated Sutras

NIU WEIXING (钮卫星) & JIANG XIAOYUAN (江晓原)
Shanghai Jiao Tong University, Shanghai

Introduction

Intrusions of foreign astronomy occurred three times in Chinese history. They were: Indian astronomy (with Buddhism) from late in the 2nd to early in the 11th centuries; Arabic astronomy (with Islam) from about the 13th to the 15th centuries; and European classical astronomy (with the Jesuits) in the 17th and the 18th centuries. In the period lasting from the end of the Eastern Han to the beginning of the Song Dynasty (about 800 years), occurred the earliest and longest intrusion, that of Indian astronomy. A large quantity of Indian astronomical materials was then introduced into China along with the translation of Buddhist sutras. Being well preserved,[1] the Chinese sutras (Buddhist scriptures in Chinese translation) have now become important first-hand materials in the study of ancient Indian astronomy. Some of these sutras are related to the ancient Indian asterism system of naksatras.

In this paper we will concentrate on the naksatras preserved in the Chinese sutras. All aspects of Indian naksatras, including their names, beginnings, sum, extent, the number of stars in each, their shapes, etc., will be discussed. At the same time, a comparison will be made between Indian naksatras and the Chinese 28 Xius system to provide evidence of similarities. We will also argue that the Indian naksatras by themselves are not definite evidence of astronomy transmission and exchange between India and China.

The Names of the Naksatras

Before Indian naksatras were introduced into China along with Buddhism, the Chinese had a similar system of constellations called the 28 Xius. It was not strange, therefore, that the names of the Indian naksatras were translated into the corresponding 28 Xius. This method of translation can also be found in some of the Chinese sutras (see Table 1).[2]

415

Table 1   Names of Naksatras Translated into Chinese (parallelism)

Four Directions	No.	*DJJ* ch20, ch56; *QYRZJ*	*DJJ* ch41; *XYJ*; *MDJ*
	1	*JIAO* (角)	*MAO* (昴)
	2	*KANG* (亢)	*BI* (毕)
east	3	*DI* (氐)	*ZI* (觜)
seven	4	*FANG* (房)	*SHEN* (参)
naksatras	5	*XIN* (心)	*JING* (井)
	6	*WEI* (尾)	*GUI* (鬼)
	7	*JI* (箕)	*LIU* (柳)
	8	*JING* (井)	*XING* (星)
	9	*GUI* (鬼)	*ZHANG* (张)
south	10	*LIU* (柳)	*YI* (翼)
seven	11	*XING* (星)	*ZHEN* (轸)
naksatras	12	*ZHANG* (张)	*JIAO* (角)
	13	*YI* (翼)	*KANG* (亢)
	14	*ZHEN* (轸)	*DI* (氐)
	15	*KUI* (奎)	*FANG* (房)
	16	*LOU* (娄)	*XIN* (心)
west	17	*WEI* (胃)	*WEI* (尾)
seven	18	*MAO* (昴)	*JI* (箕)
naksatras	19	*BI* (毕)	*DOU* (斗)
	20	*ZI* (觜)	*NIU* (牛)
	21	*SHEN* (参)	*NU* (女)
	22	*DOU* (斗)	*XU* (虚)
	23	*NIU* (牛)	*WEI* (危)
north	24	*NU* (女)	*SHI* (室)
seven	25	*XU* (虚)	*BI* (壁)
naksatras	26	*WEI* (危)	*KUI* (奎)
	27	*SHI* (室)	*LOU* (娄)
	28	*BI* (壁)	*WEI* (胃)

In many of the sutras, however, the names of Indian naksatras were transliterated into Chinese (see Table 2).[3] Compared with the Sanskrit pronunciation of the naksatras,[4] the transliteration is quite precise. There is one exception in which a kind of free translation is used (see Table 3).[5] The two series listed in Table 1 have different beginning naksatras. One of them begins with JIAO Xiu, which is also the first Xiu in the Chinese lunar mansions, and another begins with MAO Xiu. The order of their succession is different.

Table 2   Names of Naksatras in Transliteration

Four Directions	No.	*BXJ* ch4	*DKJ*	HMAI[6]
	1	卯	Qilidijia 讫栗底迦	Krttika
	2	毕	Hulusini 尸嚧呬你	Rohini
east	3	觜	Mielijiasiluo 篾栗伽尸啰	Mrgasirsha
seven naksatras	4	参	Adaluobu 頞达罗补	Ardra
	5	Funaposu 富那婆苏	Naifasufi 捺伐苏	Punarvasu
	6	Fusha 富沙	Busha 布洒	Pushya
	7	Ashilisha 阿失丽沙	Ashilisha 阿失丽洒	Aslesha
	8	Moga 莫伽	Moga 莫伽	Magha
south	9	First-Poqiu 初破求	Former-Lulounu 前发鲁娄拏	Purva-Phalguni
seven naksatras	10	Second-Poqiu 第二破求	Latter-Lulounu 后发鲁娄拏	Uttara-Phalguni
	11	Ashaduo 阿萨多	Hexituo 诃悉多	Hasta
	12	Zhiduoluo 质多罗	Zhiduoluo 质多罗	Citra
	13	Shapodi 萨婆底	Shafudi 沙缚底	Svati
	14	Sushe 苏舍	Bishike 毘释珂	Visakha
	15	Anuluotuo 阿奴逻陀	Anuluotuo 阿奴罗托	Anuradha
	16	Shisetuo 逝瑟吒	Gusetuo 皷瑟侘	Jyeshtha
west	17	Molu 暮罗	Moluo 暮罗	Mula
seven naksatras	18	First-Ashatuo 初阿沙茶	Former-Ashatu 前阿沙茶	Purv-Ashadha
	19	Second-Ashatuo 第二阿沙茶	Latter-Ashatu 后阿沙茶	Uttar-Ashadha
	20	—[7]	Abilishe 阿苾哩社	Abhijit
	21	Shiluopo 失罗婆	Sheluomonu 室罗末拏	Sravana
	22	Tuonisetuo 陀你瑟吒	Dannisetuo 怛你瑟佗	Dhanistha
	23	Satabhisai 舍多毘沙	Shetuobisha 设多婢洒	Satabhisaj
north	24	First-BotuoluoBotuo 第一跋陀罗跋陀	Former-Bodaluobodi 前跋达罗钵地	Purva-Bhadrapada
seven naksatras	25	Second-Botuoluo 第二跋陀罗	Latter-Bodaluobodi 后跋达罗钵地	Uttara-Bhadrapada
	26	Libodi 丽婆底	Jieloulifadi 頡娄离伐底	Revati
	27	Ashipini 阿湿毘腻	Ashuoni 阿说你	Asvini
	28	Poluoni 婆罗尼	Bolaini 跋赖你	Bharani

It is worth mentioning that naksatras preserved in the Chinese sutras were divided into four groups, each consisting of seven naksatras representing one direction: east, south, west and north. This four-part division of the lunar mansion can also be found in the Chinese 28 Xius system, but the order of the four directions is reversed.

To compare pronunciations of the transliterated names, we list the corresponding Sanskrit names of the 28 naksatras in Table 2. It is not difficult to see that the transliteration is quite precise.

A kind of free translation into English is listed in Table 3. Considering the meanings of the Sanskrit naksatras in the Surya-Siddhanta, we found most of the Chinese translations have the exact same meanings as the Sanskrit ones.

Table 3   Free Translation in Comparison with the Surya-Siddhanta

No.	*STJJ*	Comparison with Surya-Siddhanta[8]
1	Appellation (名称)	—
2	Nurture (长育)	ruddy
3	Deer's head (鹿首)	antelope's head
4	Nourishing (生养)	moist
5	Accumulate wealth (增财)	again + good, brilliant
6	Prosperous (炽盛)	nourish, thrive
7	No pilgrimage (不觐)	entwiner, embracer
8	Ground (土地)	mighty
9	Former Virtue (前德)	—
10	Latter Virtue (北德)	—
11	Elephant (象)	hand
12	Colorful picture (彩画)	brilliant
13	Fine beginning (善元)	sword
14	Being good at fighting (善格)	having spreading branches
15	Be pleased with success (悦可)	success
16	Elders (尊长)	the oldest
17	Root (根元)	root
18	Former-Fish (前鱼)	unsubdued
19	Latter-Fish (北鱼)	unsubdued
20	No tolerance (无容)	conquering
21	Acute hearing (耳聪)	hearing, ear
22	Avarice (贪财)	wealthy
23	A hundred poisons (百毒)	having a hundred physicians
24	Former-Footprint (前贤迹)	beautiful foot
25	Latter-Footprint (北贤迹)	beautiful foot
26	Fertility (流灌)	wealthy, abundant
27	Horseman (马师)	the two horsemen
28	Long rest (长息)	bearer away

The Beginning Naksatra

The beginning naksatra varies in different translations of the sutras. In some sutras,[9] the lists of naksatras begin with Citra;[10] its corresponding Chinese Xiu is Jiao, which has been the first one in the Chinese 28 Xius for more than two thousand years. In many other sutras,[11] however, the lists of naksatras begin with Krttika. In a few sutras, the beginning naksatras become Asvini.[12]

The naksatras were essentially a kind of astronomical coordinate system, and the beginning naksatra was the place where the vernal equinox was located. As the initial point of measurement for a celestial body's position, the vernal equinox has a precession on the ecliptic. So the beginning naksatra varies from Krttika to Asvini, which is concordant with the precession of the equinox. This is confirmed by the fact that the sutras taking Asvini as the first naksatra are translated later than those taking Krttika as the first (see Table 4). It should be noted, however, that the period when the sutras were translated into Chinese did not exactly converge with the period when the corresponding sutras emerged in India. But it is reasonable to assume that the sutras, which appeared in India earlier, were translated into Chinese very early as well.

Table 4   A Comparison of Epochs of Translation
(taking Krttika and Asvini as the first naksatras)

The beginning naksatra	Sutras	Translator	Epoch being translated
Krttika	*MDJ*	Zu-lu-yan & Ziqian	230 AD
	STJJ	Dharmaraksa	308 AD
	MSL	Buddhabhadra	416 AD
	BXJ	Prabhamitra	632 AD
	DKJ	Yi-jing	705 AD
	XYJ	Amoghavajra	742–764 AD
Asvini	*NSZ*	Fa-xian	973–1000 AD
	DWYJ	Tian-xi-zai	983–1000 AD

The Sum of the Naksatras

The sum of the naksatras is ordinarily 28. However, a 27 naksatra system from which Abhijit is absent is described in detail in the *XYJ*.[13] This same number of naksatras is also found in the other sutras.[14] In the *XYJ*, each naksatra is divided into 4 portions; each portion is called a "foot", so the

27 naksatras have the total sum of 108 "feet". These 108 "feet" are distributed to the 12 zodiacal signs; each zodiacal sign contains 9 "feet" (see Table 5).

Table 5   Zodiacal Signs and 27 Naksatra System Listed in the *XYJ*

Zodiacal sign	Sum of naksatras' "feet" each zodiacal sign contains		
Leo	Magha 4	Purva-Phalguni 4	Uttara-Phalguni 1
Virgo	Uttara-Phalguni 3	Hasta 4	Citra 2
Libra	Citra 2	Svati 4	Visakha 3
Scorpius	Visakha1	Anuradha 4	Jyeshtha 4
Sagittarius	Mula 4	Purv-Ashadha 4	Uttar-Ashadha 1
Capricornus	Uttar-Ashadha 3	Sravana 4	Dhanistha 2
Aquarius	Dhanistha 2	Satabhisaj 4	Purva-Bhadrapada 3
Pisces	Purva-Bhadrapada 1	Uttara-Bhadrapada 4	Revati 4
Aries	Asvini 4	Bharani 4	Krttika 1
Taurus	Krttika 3	Rohini 4	Mrgasirsha 2
Gemini	Mrgasirsha 2	Ardra 4	Punarvasu 3
Cancer	Punarvasu 1	Pushya 4	Aslesha 4

Yang Jingfeng, who was a commentator on the *XYJ*, tried to explain why only 27 naksatras were listed. He noted:

> There are 28 Xius in the Tang Dynasty. In the Western country (India) the NIU Xiu (abhijit) is omitted. This is because the regent of this naksatra is their God.[15]

It is true that the divinity of Abhijit is Brahma, the Almighty God of India. But we think that the real reason for 27 naksatras appearing in India was the introduction of zodiacal signs and other astronomical knowledge from the west. In Western astronomy the whole ecliptic contains 12 zodiacal signs, 360 degrees, and 21600 minutes. This sum of minutes cannot be divided exactly by 28, but can be divided exactly by 27, so that each naksatra contains 800 minutes, and without any remainder. So the 27-naksatra system is an evidence of Indian traditional astronomy adjusting itself to foreign astronomy.

The Extent (Length) of Each Naksatra

Twenty-eight (or 27) naksatras circle around the heavens, each of them extending for a certain range along the ecliptic. Naksatras were also called lunar mansions in India, so it is not strange that the extent of a naksatra is represented by the length of time when the moon conjoins it. We found that the extent of each naksatra was measured in this way in *DJJ*[16], *MDJ*[17] and *STJJ*[18] (see Table 6). In order to compare the differences in length between Indian naksatras and Chinese 28 Xius, we list the Chinese Xius' lengths below.

Table 6   The Extent of Each Naksatra

No.	Xius	DJJ ch41 (muhurtas)	MDJ (day and night)	STJJ (muhurtas)	Dayanli[19] Chinese Degree (°°) [20]
1	MAO (昴)	30	12 muhurtas	30	11°°
2	BI (毕)	45	1 day and a half	45	17°°
3	ZI (觜)	15	1 day	30	1°°
4	SHEN (参)	45	1 day	15	10°°
5	JING (井)	15	1 day	45	33°°
6	GUI (鬼)	30	1 day	30	3°°
7	LIU (柳)	15	a half day	30	15°°
8	XING (星)	30	1 day	30	7°°
9	ZHANG (张)	30	1 day	30	18°°
10	YI (翼)	15	1 day and a half	45	18°°
11	ZHEN (轸)	30	1 day and 1 night	30	17°°
12	JIAO (角)	15	1 day	30	12°°
13	KANG (亢)	15	1 day	15	9°°
14	DI (氐)	45	1 day and a half	45	15°°
15	FANG (房)	30	1 day and 1 night	30	5°°
16	XIN (心)	15	1 day	15	5°°
17	WEI (尾)	30	1 day and 1 night	30	18°°
18	JI (箕)	30	1 day and 1 night	15	11°°
19	DOU (斗)	45	1 day and a half	45	26°°
20	NIU (牛)	6	1 muhurtas	6	8°°
21	NU (女)	30	1 day and 1 night	30	12°°
22	XU (虚)	30	1 day and 1 night	30	10°°
23	WEI (危)	15	1 day	15	17°°
24	SHI (室)	30	1 day and 1 night	30	16°°
25	BI (壁)	45	1 day and 1 night	45	9°°
26	KUI (奎)	30	1 day and 1 night	30	16°°
27	LOU (娄)	30	1 day and 1 night	30	12°°
28	WEI (胃)	30	1 day and 1 night	30	14°°

The unit of time was usually called muhurta, 30 of which equal one day. This unit of time was used by *DJJ* and *STJJ* as the unit of extent of the naksatras. In the *MDJ*, "day" and "night" were used as the unit of extent. Here, "day" means only daylight. According to Table 6, the extents of naksatras in the *DJJ* and *STJJ* can be separated into three kinds of values. They are 15 muhurtas, 30 muhurtas, and 45 muhurtas. Especially the extent of Abhijit has only 6 muhurtas. For *MDJ*, not all of the values of extents listed in Table 6 are correct; some mistakes must have been made in translation or while making copies of the text, because in another place in the same text we read:

> The Great Brahman, I have told you the 28 Xius; however in these Xius, the next 6 Xius conjoin the moon with 2 days and 1 night. They are BI (毕), JING (井), DI (氐), YI (翼), DOU (斗) and BI (壁); and other 5 Xius conjoin the moon with only 1 day. They are SHEN (参), LIU (柳), JI (箕), XIN (心) and WEI (危). Only NIU (牛) Xiu conjoins the moon with half a day. The rest of the Xius all conjoin the moon with 1 day and 1 night.[21]

It is not difficult to find the difference between the values of extent in the statement above and that listed in Table 6. Obviously there are no extents of 2 days and 1 night in Table 6, although 2 days and 1 night equal 45 muhurtas. In *DJJ* and *STJJ* there are indeed extents of 45 muhurtas. So we think the statement about the Xius' extent in the *MDJ* is essentially correct, only the NIU Xiu's length of half a day is rather large, and should take the same value of 6 muhurtas as in the *DJJ* and *STJJ*. Thus we have the total values of the Xius' extent:

$$6 \times 1.5 + 5 \times 0.5 + 1 \times \frac{6}{30} + 16 \times 1 = 27.7 \text{ (days)}$$

Undoubtedly, this is also the value of days in a sidereal month, just a little bigger than the accurate value 27.32166 days.

According to *STJJ*, there are six naksatras with a length of 45 muhurtas, sixteen of 30 muhurtas, five of 15 muhurtas and one of 6 muhurtas. So we have a total of 831 muhurtas. Divided by 30, which gives the value of a sidereal month as 27.7 days. As we have seen, the lengths of 28 Xius in the *DJJ*, *MDJ* and *STJJ* approximately consist of three kinds of values. However the values of the lengths of the Chinese 28 Xius are so irregular that the widest Xiu JING (井) has 33°°, and the narrowest Xiu ZI (觜) only has 1°°.

Earlier on, we presented a 27 naksatra system. In the 27 naksatra system the length of each naksatras tends to be equalized. We think this equalizing tendency of the lengths, as well as the change in the sum of the naksatras from 28 to 27, was influenced by the intrusion of zodiacal signs from Babylon to India.

The Number of Stars in Each Naksatra

Detailed records about the sum of stars in each naksatra are also given in the Chinese translated sutras.[22] The sums are different from those of Chinese lunar mansions, but they are sustained by separate Sanskrit texts (see Table 7).

Table 7 The Sum of Stars in Each Indian Naksatra and Chinese Xiu

No.	Xius	DJJ ch41	XYJ	MDJ	STJJ	Gargasamhita[23]	Chinese 28 Xius
1	MAO (昴)	6	6	6	6	6	7
2	BI (毕)	5	5	5	5	5	8
3	ZI (觜)	3	3	3	3	3	3
4	SHEN (参)	1	1	1	1	1	10
5	JING (井)	2	2	2	3	2	8
6	GUI (鬼)	3	3	3	3	1	5
7	LIU (柳)	1	6	1	5	6	8
8	XING (星)	5	6	5	5	6	7
9	ZHANG (张)	2	2	2	3	2	6
10	YI (翼)	2	2	2	2	2	22
11	ZHEN (轸)	5	5	5	5	5	4
12	JIAO (角)	1	2	1	1	1	2
13	KANG (亢)	1	1	1	1	1	4
14	DI (氐)	2	4	2	2	2	4
15	FANG (房)	4	4	4	—	4	4
16	XIN (心)	3	3	3	3	3	3
17	WEI (尾)	7	2	7	3	6	9
18	JI (箕)	4	4	4	4	4	4
19	DOU (斗)	4	4	4	4	4	6
20	NIU (牛)	3	3	3	3	3	6
21	NU (女)	4	3	3	3	3	4
22	XU (虚)	4	4	4	4	4	2
23	WEI (危)	1	1	1	1	1	3
24	SHI (室)	2	2	2	2	2	2
25	BI (壁)	2	2	2	2	2	2
26	KUI (奎)	1	32	1[24]	1	4	16
27	LOU (娄)	3	3	2	3	2	3
28	WEI (胃)	3	3	3	5	3	3

According to Table 7, we found that,

(a) There are 16 Xius in each of which the sum of stars presented in the *DJJ*, *XYJ*, *MDJ* and *STJJ* are exactly the same. They are *MAO* (昴), *BI* (毕), *ZI* (觜), *SHEN* (参), *GUI* (鬼), *YI* (翼), *ZHEN* (轸), *KANG* (亢), *XIN* (心), *JI* (箕), *DOU* (斗), *NIU* (牛), *XU* (虚), *WEI* (危), *SHI* (室), *BI* (壁).

(b) The *DJJ* and *MDJ* present almost the same sum of stars in each Xiu, except for *NU* (女) and *LOU* (娄).

(c) Comparing the sum of stars presented in the *Gargasamhita* with that in the four Chinese sutras, there are 15 Xius having the same sum of stars. They are *MAO* (昴), *BI* (毕), *ZI* (觜), *SHEN* (参), *YI* (翼), *ZHEN* (轸), *KANG* (亢), *XIN* (心), *JI* (箕), *DOU* (斗), *NIU* (牛), *XU* (虚), *WEI* (危), *SHI* (室), *BI* (壁). So we can conclude that the sum of stars in the 15 Xius were almost a constant. They did not vary from text to text. Here, *Gargasamhita* is a Sanskrit text dating to the beginning of the Christian era.

(d) *Gargasamhita* has the least difference in sum in comparison with *XYJ*; only *GUI* (鬼), *JIAO* (角), *DI* (氐), *WEI* (尾), *KUI* (奎), and *LOU* (娄) contain different sums.

(e) Chinese Xiu mostly contain more stars than their corresponding Indian naksatras. But *ZI* (觜), *XIN* (心), *JI* (箕), *SHI* (室) and *BI* (壁) contain the same sum of stars compared with the naksatras.

(f) The sum of stars in some of the Indian naksatras varied in different texts. On the other hand, the sum of stars in the Chinese 28 Xiu hardly varied regardless of the text.

The Shape of Each Naksatra

Each naksatra consists of a certain sum of stars, and these stars are considered to form a certain shape against the heavens. Some Chinese sutras present such shapes of naksatras (see Table 8).

According to Table 8, we find the shapes of naksatras presented in different sutras are often identical. Four naksatras have the same shape descriptions in four different sutras. They are *ZI* (觜): "deer's head", *ZHEN* (轸): "hand", *NIU* (牛): "cow's head" and *LOU* (娄): "horse's head". Especially the shape of *NIU* (牛) is rendered as "cow's head", although in Chinese *NIU* (牛) also means "cow".

Table 8   The Shapes of Naksatras

No.	Xius	DJJ ch41	XYJ	MDJ	Sardulakarnavadana
1	MAO (昴)	blade	blade	scattering flower	blade blade
2	BI (毕)	fork	half cart	flying wild goose	cart
3	ZI (觜)	deer's head	deer's head	deer's head	deer's head
4	SHEN (参)	forehead mark	forehead mark	—	forehead mark
5	JING (井)	footprint	house rafter	step	foot (step)
6	GUI (鬼)	fylfot	bottle	vase	saucer
7	LIU (柳)	forehead mark	snake	—	forehead mark
8	XING (星)	river bank	wall	river bend	river bend
9	ZHANG (张)	footprint	pestle	step	foot (step)
10	YI (翼)	footprint	a bonze's sitz	step	foot (step)
11	ZHEN (轸)	hand	hand	hand	hand
12	JIAO (角)	forehead mark	long shade	—	forehead mark
13	KANG (亢)	forehead mark	fire bead	—	forehead mark
14	DI (氐)	footprint	bull's horn	ram's horn	horn
15	FANG (房)	tassel	shade	string of beads	string of pearls
16	XIN (心)	barleycorn	stairs	bird	barleycorn in middle
17	WEI (尾)	scorpion's tail	lion's hair	scorpion	scorpion
18	JI (箕)	bull's horn	cow step	cow step	cow path
19	DOU (斗)	man hold earth	elephant step	elephant step	elephant path
20	NIU (牛)	cow's head	cow's head	cow's head	cow's head
21	NU (女)	barleycorn	—	barleycorn	barleycorn in middle
22	XU (虚)	bird	—	bird	bird
23	WEI (危)	forehead mark	fringe	—	forehead mark
24	SHI (室)	footprint	thill	step	foot (step)
25	BI (壁)	footprint	standing stick	step	foot (step)
26	KUI (奎)	forehead mark	cockboat	—	forehead mark
27	LOU (娄)	horse's head	horse's head	horse's head	horse's head
28	WEI (胃)	triangle	triangle	triangle	vulva

Sardulakarnavadana is a Sanskrit text dated to the beginning of the Christian era.[25] It was translated into Chinese by Zu-lu-yan and Ziqian in 230 AD, and its Chinese title is *Mo-deng-jia Jing* (*MDJ*,《摩登伽经》). From Table 8, we can see that the shapes of naksatras presented separately by *Sardulakarnavadana* and *MDJ* have an identical description, except for *MAO* (昴) and *BI* (毕).

The Junction Star of Each Naksatra

The junction star of each naksatra is another essential parameter of the 28 Xius or naksatras system. The identification of Chinese junction stars has been quite thoroughly completed. As for Indian junction stars, scholars have also obtained a good result. Here two identifications of Indian naksatras are listed in Table 9.

Table 9   A Comparison of Junction Stars in the Indian Naksatras and Chinese 28 Xius

| No. | Junction star of Chinese 28 Xius | | Junction star of Indian Naksatras | | |
	Name	Identification[26]	Name	Identification 1[27]	Identification 2[28]
1	*MAO* (昴)	17 Tau	Krttika	η Tau	η Tau
2	*BI* (毕)	ε Tau	Rohini	α Tau	α Tau
3	*ZI* (觜)	Φ Ori	Mrgasiras	15 Ori	λ Ori
4	*SHEN* (参)	δ Ori	Ardra	λ Ori	α Ori
5	*JING* (井)	μ Gem	Punarvasu	β Gem	β Gem
6	*GUI* (鬼)	θ Cnc	Pusya	ε Cnc	δ Cnc
7	*LIU* (柳)	δ Hya	Aslesa	β Cnc	ε Hya
8	*XING* (星)	α Hya	Magha	α Leo	α Leo
9	*ZHANG* (张)	ν Hya	Purvaphalguni	δ Leo	δ Leo
10	*YI* (翼)	α Crt	Uttaraphalguni	β Leo	β Leo
11	*ZHEN* (轸)	γ Crv	Hasta	η Crv	γ Crv
12	*JIAO* (角)	α Vir	Citra	α Vir	α Vir
13	*KANG* (亢)	κ Vir	Svati	α Boo	α Boo
14	*DI* (氐)	α Lib	Visakha	ι Lib	ι Lib
15	*FANG* (房)	π Sco	Anuradha	δ Sco	δ Sco
16	*XIN* (心)	σ Sco	Jyestha	α Sco	α Sco
17	*WEI* (尾)	μ Sco	Mula	d Oph	λ Sco
18	*JI* (箕)	γ Sgr	Purvasadha	γ Sgr	δ Sgr
19	*DOU* (斗)	Φ Sgr	Uttarasadha	Φ Sgr	σ Sgr

Table 9 *(Continued)*

No.	Junction star of Chinese 28 Xius		Junction star of Indian Naksatras		
	Name	Identification[26]	Name	Identification 1[27]	Identification 2[28]
20	*NIU* (牛)	β Cap	Abhijit	α Lyr	α Lyr
21	*NU* (女)	ε Aqr	Sravana	α Aql	α Aql
22	*XU* (虚)	β Aqr	Dhanistha	α Del	β Del
23	*WEI* (危)	α Aqr	Satabhisaj	λ Aqr	λ Aqr
24	*SHI* (室)	α Peg	Purvabhadrapada	α Peg	α Peg
25	*BI* (壁)	γ Peg	Uttarabhadrapada	α And	α And
26	*KUI* (奎)	ζ And	Revati	ζ Psc	ζ Psc
27	*LOU* (娄)	β Ari	Asvini	β Ari	β Ari
28	*WEI* (胃)	35 Ari	Bharani	33 Ari	35 Ari

According to Table 9, under "identification 1", there are 17 Indian naksatras located in the same constellation together with the corresponding Chinese Xius, they are *JIAO* (角), *DI* (氐), *FANG* (房), *XIN* (心), *JI* (箕), *DOU* (斗), *WEI* (危), *SHI* (室), *LOU* (娄), *WEI* (胃), *MAO* (昴), *BI* (毕) , *ZI* (觜), *SHEN* (参), *JING* (井), *GUI* (鬼) and *ZHEN* (轸), and the five Xius, *JIAO* (角), *JI* (箕), *DOU* (斗), *SHI* (室) and *LOU* (娄) have the same junction stars.

In the same table, under "identification 2", there are 20 Indian naksatras located in the same constellation together with the corresponding Chinese Xius. They are *JIAO* (角), *DI* (氐), *FANG* (房), *XIN* (心), *WEI* (尾), *JI* (箕), *DOU* (斗), *WEI* (危), *SHI* (室), *BI* (壁), *LOU* (娄), *WEI* (胃), *MAO* (昴), *BI* (毕), *ZI* (觜), *SHEN* (参), *JING* (井), *GUI* (鬼) and *LIU* (柳). The five Xius, *JIAO* (角), *SHI* (室), *LOU* (娄), *WEI* (胃) and *ZHEN* (轸) have the same junction stars.

Because the Chinese 28 Xius and Indian naksatras distribute mainly along the ecliptic, and the sum of the ecliptic constellations are finite, it is reasonable to conclude that most of the Indian naksatras and Chinese Xius are located in the same constellation.

Conclusion

The Indian naksatras preserved in the Chinese translated sutras have been discussed in detail in the previous passages. We draw the conclusion that the introduction of the Indian naksatras, along with the translation of Buddhist sutras, extended to the most essential aspects of the Indian naksatras system.

Our research has reached such a level that we can retrieve all the basic knowledge of the Indian naksatras through reading the proper Chinese sutras. However, the sutras were not purely astronomical texts; the knowledge of naksatras presented in the Chinese sutras was rather more qualitative than quantitative. For example, we cannot determine the precise celestial position of each naksatra just on the basis of the translated sutras.

During the period when Indian naksatras were introduced into China, Chinese astronomy had its own well-developed system of 28 Xius, so that the intrusion of the Indian naksatras in this period did have a great impact on the Chinese system. We speculate, however, that the intrusion of Indian naksatras might have greatly changed traditional Chinese astrological fashion. This is an interesting problem that should be considered in another paper or series of papers.

Notes

[1] A large number of Sanskrit sutras were translated into Chinese from late in the 2nd century to early in the 11th century. The translated sutras were compiled and published in in China as well as in Japan and Korea. *Taisho Tripitaka*, which was compiled by Japanese in 1924–34 AD in Tokyo, is the most popular edition of Tripitaka now used by scholars. So the *Taisho Tripitaka* will be our main reference point.

[2] *DJJ*, *TTP* Vol. 13, pp. 138–39, 274–82, 371–73; *XYJ*, *TTP* Vol. 21, pp. 387–91; *MDJ*, *TTP* Vol. 21, p. 404; *QYRZJ*, *TTP* Vol. 21, p. 427.

[3] *BXJ*, *TTP* Vol. 13, pp. 555–56; *DKJ*, *TTP* Vol. 19, pp. 473–74 etc.

[4] Pingree, David, *History of Mathematical Astronomy in India* (short as *HMAI*), *Dictionary of Scientific Biography*, XVI (New York, 1981), p. 535.

[5] *STJJ*, *TTP* Vol. 21, pp. 415–17.

[6] Pingree, David, *HMAI*, pp. 535, 537.

[7] This is noted as "lack of Xu (虛) Xiu" in the original text of *BXJ*. After comparing the pronunciation of naksatras in other sutras, we think the omitted Xiu should be "Niu (牛)".

[8] *Surya Siddhanta*, translated with notes and appendix by Rev. Ebenezer Bugress, Motilal Banarsidass Publishers Private Limited, Delhi, India, 1997, pp. 211–30.

[9] *DJJ*, *TTP* Vol. 13, pp. 138–39, 371; *QYRZJ*, *TTP* Vol. 21, p. 427.

[10] According to Prof. P. C. Senguta, in the oldest Indian sutras, the lists of naksatras began with Citra (*Hindu Astronomy, Heritage of Hindu Culture*, Vol. 3, pp. 341–71).

[11] *DJJ*, *TTP* Vol. 13, p. 274; *BXJ*, *TTP* Vol. 13, p. 555; *DKJ*, *TTP* Vol. 19, p. 473; *XYJ*, *TTP* Vol. 21, p. 388; *MDJ*, *TTP* Vol. 21, p. 404; *STJJ*, *TTP* Vol. 21, p. 415; *MSL*, *TTP* Vol. 22, pp. 500–501.

[12] *DWYJ*, *TTP* Vol. 20, p. 846; *NSZ*, *TTP* Vol. 21, p. 463.

[13] *XYJ, TTP* Vol. 21, pp. 388–90.

[14] *DZDL, TTP* Vol. 25, p. 117; *BXJ, TTP* Vol. 13, pp. 555–56; *NSZ, TTP* Vol. 21, pp. 463–64.

[15] *XYJ, TTP* Vol. 21, p. 394.

[16] *DJJ, TTP* Vol. 13, pp. 274, 275.

[17] *MDJ, TTP* Vol. 21, pp. 404, 405.

[18] *STJJ, TTP* Vol. 21, pp. 415, 416.

[19] *CCCAC* (*The Collection of Chronicles of Chinese Astronomy & Calendar*), edited and published by Zhonghua Shuju (Zhonghua Publishing House), 1975, p. 2225.

[20] 1 Chinese Degree (^{oo}) = 0.9856°. In the traditional Chinese calendar, the total sum of degrees of the ecliptic circle was defined to equal the sum of days in a tropical year. So the sun progress 1^{oo} per day.

[21] *MDJ, TTP* Vol. 21, p. 405.

[22] *DJJ, TTP* Vol. 13, pp. 274, 275; *XYJ, TTP* Vol. 21, pp. 388–90; *MDJ, TTP* Vol. 21, pp. 404, 405; *STJJ, TTP* Vol. 21, pp. 415, 416.

[23] Pingree, D. & Morrissey, P., *On the Identification of the Yogataras of the Indian Naksatras, Journal for the History of Astronomy*, XX (1989), p. 102.

[24] It is noted in the original text of *MDJ* that "KUI Xiu has one big star followed by a group of small stars". *MDJ, TTP* Vol. 21, p. 405.

[25] Pingree, D. & Morrissey, P., *On the Identification of the Yogataras of the Indian Naksatras*, XX (1989), p. 102.

[26] Pannan, *A Chinese History of Star Observation*, Xuelin Press (1989), p. 12.

[27] Pingree, David, *HMAI*, p. 565.

[28] *Surya Siddhanta*, translated with notes and appendix by Rev. Ebenezer Bugress, Motilal Banarsidass Publishers Private Limited, Delhi, India, 1997, pp. 211–30.

Abbreviations in Text

BXJ Bao-xing Tuo-luo-ni Jing, No. 402 sutra of the *Taisho Tripitaka*, translated into Chinese in 632 AD by Prabhamitra.

DJJ Da-fang-deng Da-ji Jing (Sanskrit title is *Mahavaipulyamahasannipatasutra*), No.397 sutra of the *Taisho Tripitaka* (1924–34 AD, Tokyo), translated into Chinese by Narendrayasas in 556–89 AD.

DKJ Da-kong-que-zhu-wang Jing, No. 985 sutra of the *Taisho Tripitaka*, translated into Chinese in 705 AD by Yi-jing.

DWYJ Da-fang-guang Pu-sa Wen-shu-shi-li Geng-ben Yi-gui Jing, No. 1191 sutra of the *Taisho Tripitaka*, translated into Chinese in 983–1000 AD by Tian-xi-zai.

DZDL Da-zhi-du-lun (Sanskrit title is *Mahaprajnaparamitasastra*), No. 1509 sutra of the *Taisho Tripitaka*, translated into Chinese in 401–413 AD by Kumarajiva.

MDJ Mo-deng-jia Jing (Sanskrit title is *Sardulakarnavadana*), No. 1300 sutra of the *Taisho Tripitaka*, translated into Chinese in 230 AD by Zu-lu-yan & Ziqian.

MSL Mo-he Seng-zhi Lu (Sanskrit title is *Mahasanghavinaya*), No. 1425 sutra of the *Taisho Tripitaka*, translated into Chinese in 416 AD by Buddhabhadra.

NSZ Nan-mi-ji-shi Fu-luo-tian Shuo Zi-lun Jing, No. 1312 sutra of the *Taisho Tripitaka*, translated into Chinese in 973–1000 AD by Fa-xian.

QYRZJ Qi-yao Rang-zai-jue, No. 1308 sutra of the *Taisho Tripitaka*, compiled in the early 9th century by Chin Chu-cha.

STJJ She-tou Jian Tai-zi Er-shi-ba-xiu Jing (another edition of *Sardulakarnavadana*), No. 1301 sutra of the *Taisho Tripitaka*, translated into Chinese in 308 AD Dharmaraksa.

TTP Taisho Tripitaka, was compiled by Japanese in 1924–34 AD in Tokyo.

XYJ Wen-shu-shi-li Pu-sa ji Zhu-xian Suo-shuo Jixiong-shiri Shang-e-xiuyao Jing, No.1299 sutra of the *Taisho Tripitaka*, translated into Chinese in 742–64 AD by Amoghavajra.

The Accuracy of *Da Yan Li*'s Calculation of the Solar Eclipse

HU TIEZHU 胡鐵珠

Institute for the History of Sciences
Chinese Academy of Sciences

A solar eclipse was considered a very important phenomenon in ancient China because people believed it was heaven's warning to the Emperor for his mistakes. When a solar eclipse occurred, they had to hold a rite to protect the Emperor. Therefore, the astronomers paid a lot of attention to the calculations of solar eclipses. Four of the seven parts of the ancient Chinese calendars were related to eclipses. This author developed a computer program based on the method of *Da Yan Li* (大衍歷) — one of the most famous calendars in ancient China (8th century) — and calculated the solar eclipses of different time periods. From these calculations, the *Da Yan Li*'s accuracy and scientific value can be gauged.

The Principle and the Computer Program

A few years ago, I finished a chapter about the principle of the *Da Yan Li* in "The History of Chinese Astronomy" (中國天文學史大系), which will be published soon. Here I will simply describe the steps of the solar eclipse calculation in *Da Yan Li*:

(1) Give the mean time of a new moon and calculate the positions of the sun and the moon with respect to their apogee or perigee.

(2) Determine the solar and lunar equation of the center, then calculate the true time of the new moon.

(3) The interpolation methods were used in this step.

(4) Calculate the time difference between the true time of the new moon and the time of the moon passing through the ascendant (or descendent) node, if it is smaller than the eclipse limit, then calculate the magnitude and the time of the eclipse. The lunar parallax was calculated in this step.

431

The computer program was developed according to the principle of the *Da Yan Li*. When given a range of years, this program will calculate the year, month, day name, magnitude, and time of the solar eclipse in that period as well as the sun's right ascensions. Using this program, I compared the solar eclipses recorded in the *Book of History* (書經), the *Book of Songs* (詩經), and the *Spring and Autumn Annals* (春秋), with the time of *Da Yan Li*, and use these with the output of the program. The results are as follows:

The Solar Eclipse in the *Book of History*

The *Book of History* is an ancient book about the history of early China that was written before 200 B.C. A record in this book states that during the *Xia* (夏) Dynasty (the earliest dynasty in China), an astronomical phenomenon occurred on a day of the new moon in September, and the people were all scared. An astronomer named *Xihe* (羲和) was killed because he did not predict this astronomical phenomenon. Most scholars believe that this record refers to a large magnitude solar eclipse. If this is the case, this record would be the earliest recorded solar eclipse in the world. However, because of the uncertainty in the chronology of the early time periods, when this eclipse occurred is still unresolved. What I am interested in is: among all the scholars who noticed this record, *Yixing* (一行, the author of the *Da Yan Li*) is the first one to give the year of the solar eclipse. In *On Da Yan Li* (大衍曆議) he said:

> In September of the fifth year of Emperor *Zhongkang* (仲康), when the year name was *Guisi* (癸巳), the day name was *Gengxu* (庚戌), and the sun's right ascension was two degree in *Fang* (房), a solar eclipse occurred (仲康五年癸巳歲九月庚戌朔日蝕在房二度).

After calculating the solar eclipses from 2200 B.C. to 1900 B.C. by the program, we got two solar eclipses in that time frame which occurred in September when the right ascension of the sun is in *Fang* (*Yixing* believed these two conditions should be the facts). One was on October 13, 2128 B.C. This day and year coincide precisely with what *Yixing* said, but the predicted magnitude is only 0.007. The other was on October 22, 2136 B.C., with a magnitude of 0.65. It seems that *Yixing* chose the first one because he thought it coincided with the number of years that the emperors reigned.

Table 1 The Solar Eclipse of the Spring and Autumn Annals

No.	Date	Day Name	Ancient Record	Da Yan Li's Calculations	
				Time (h)	Magnitude[1]
1*	−719.2.22 (隱公三年二月)	己巳	Solar Eclipse	8.15	13.1 (0.87)
2*	−708.7.17 (桓公三年七月)	壬辰	Total Eclipse	15.85	12.92 (0.86)
3*	−694.10.10 (桓公十七年十月)	庚午	Solar Eclipse	14.2	10.45 (0.70)
4*	−675.4.15 (莊公十八年三月)	壬子	Solar Eclipse	17.43	12.64 (0.84)
5*	−668.5.27 (莊公二十五年六月)	辛未	Solar Eclipse		
6*	−667.11.10 (莊公二十六年十二月)	癸亥	Solar Eclipse	9.21	13.39 (0.89)
7*	−663.8.28 (莊公三十年九月)	庚午	Solar Eclipse	14.74	11.35 (0.75)
8*	−654.8.19 (僖公五年九月)	戊申	Solar Eclipse	14.33	10.07 (0.67)
9*	−647.4.6 (僖公十二年三月)	庚午	Solar Eclipse	17.42	10.01 (0.67)
10	−644 (僖公十五年五月)	壬午	Solar Eclipse		
11*	−625.2.3 (文公元年二月)	癸亥	Solar Eclipse	12.07	10.89 (0.73)
12*	−611.4.28 (文公十五年六月)	辛丑	Solar Eclipse	9.43	7.75 (0.52)
13*	−600.9.20 (宣公八年七月)	甲子	Total Eclipse	14.13	8.02 (0.53)
14*	−598.3.6 (宣公十年四月)	丙辰	Solar Eclipse	7.83	12.89 (0.86)
15	−591.7.14 (宣公十七年六月)	癸卯	Solar Eclipse		
16*	−574.5.9 (成公十六年六月)	丙寅	Solar Eclipse		
17*	−573.10.22 (成公十七年十二月)	丁巳	Solar Eclipse		
18*	−558.1.14 (襄公十四年二月)	乙未	Solar Eclipse	12.18	13.37 (0.89)
19*	−557.5.31 (襄公十五年八月)	丁巳	Solar Eclipse	8.31	3.45 (0.23)
20	−552.8.31 (襄公二十年十月)	丙辰	Solar Eclipse	14.43	11.32 (0.75)

(cont'd overleaf)

Table 1 *(Continued)*

No.	Date	Day Name	Ancient Record	Da Yan Li's Calculations	
				Time (h)	Magnitude[1]
21*	−551.8.20 (襄公二十一年九月)	庚戌	Solar Eclipse	15.24	12.61 (0.84)
22	−551.9.19 (襄公二十一年十月)	庚辰	Solar Eclipse		
23*	−549.1.5 (襄公二十三年二月)	癸酉	Solar Eclipse	7.91	9.99 (0.67)
24*	−548.6.19 (襄公二十四年七月)	甲子	Total Eclipse		
25	−548.7.19 (襄公二十四年八月)	癸巳	Solar Eclipse		
26*	−545.10.13 (襄公二十七年十二月)	乙亥	Solar Eclipse		
27*	−534.3.18 (昭公七年四月)	甲辰	Solar Eclipse		
28*	−526.4.18 (昭公十五年六月)	丁巳	Solar Eclipse	13.6	15.23 (1.01)
29*	−524.8.21 (昭公十七年六月)	癸酉	Solar Eclipse	18.03	7.39 (0.49)
30*	−520.6.10 (昭公二十一年七月)	壬午	Solar Eclipse	12.87	14.3 (0.95)
31*	−519.11.23 (昭公二十二年十二月)	癸酉	Solar Eclipse		
32*	−517.4.9 (昭公二十四年五月)	乙未	Solar Eclipse	10.50	6.50 (0.43)
33*	−510.11.4 (昭公三十一年十二月)	辛亥	Solar Eclipse	8.40	6.57 (0.43)
34*	−504.2.16 (定公五年三月)	辛亥	Solar Eclipse	12.65	13.63 (0.91)
35*	−497.9.22 (定公十二年十一月)	丙寅	Solar Eclipse		
36*	−494.7.22 (定公十五年八月)	庚辰	Solar Eclipse	13.05	14.24 (0.95)
37*	−480.4.19 (哀公十四年五月)	庚申	Solar Eclipse		

The Solar Eclipse in the *Book of Songs*

The *Book of Songs* collected the songs of the 11th to 5th centuries B.C. One of the songs named "October" said:

> In October when the day of the new moon was *Xinmao* (辛卯), a solar eclipse occurred ... (十月之交, 朔日辛卯, 日有食之...).

Yixing said in *On Da Yan Li* that he calculated this eclipse, the result being that it should occur in the sixth year of Emperor You of Zhou (周幽王), a result that was confirmed by the computer program mentioned above. The Gregorian date for this event is September 6, 776 B.C., which coincides with some modern scholar's conclusions. The only difference is that the *Yixing*'s magnitude was 0.58, a value much bigger than the modern calculation of 0.1.

Another date, November 30, 735 B.C., which some scholars think is the date of this solar eclipse, can also be obtained from *Da Yan Li*'s calculation.

The Solar Eclipse of the *Spring and Autumn Annals*

The *Spring and Autumn Annals* was edited about 2500 years ago. In this book, 37 solar eclipses which occurred from 720 B.C. to 481 B.C. were recorded. Table 1 shows the records and the *Da Yan Li*'s calculations for these eclipses.

In Table 1, the "*" means this eclipse could be seen in *Xian* (西安), the capital of China at the time and *Qufu* (曲阜), the city that the solar eclipses were recorded, and their magnitudes are all bigger than 0.2 according to modern calculations.[2] We can see from the table that *Da Yan Li* has 24 of the 37 solar eclipses, and the 23 of them are the same with modern calculations.

The Solar Eclipses in *Da Yan Li*'s Time

Since our calculations of the eclipses as recorded in the *Book of History* and the *Book of Songs* coincided with what *Yixing* said, we knew that the rebuilt calculation method used in the program was reliable. We then used this computer program to calculate the solar eclipses during the time when *Da Yan Li* was used, and compare it with modern calculations.

Table 2 lists 23 solar eclipses. Among them, 19 were predicted by *Da Yan Li*, 3 were not predicted by *Da Yan Li* but were recorded by the ancient astronomer. One was recorded, but according to *Da Yan Li* and modern calculation, it did not occur in *Xian*, the capital of the Tang Dynasty.

Table 2 The Solar Eclipses in 724–761 AD

No.	Date	Day Name	Da Yan Li's Calculations		Modern Calculations[3]		Ancient Records
			Time (h)	Magnitude	Time (h)	Magnitude	
1	开元12年7月 (724.7.25)	戊午	7.69	9.48 (0.63)			No Ecl.
2	开元12年闰12月 (725.1.19)	丙辰	12.12	14.32 (0.95)			Solar Ecl.
3	开元13年12月 (726.1.8)	庚戌	14.78	13.05 (0.87)	17.1	0.11	No Ecl.
4	开元14年12月 (726.12.28)	甲辰	14.43	1.56 (0.10)			
5	开元17年10月 (729.10.27)	戊午	8.17	15.22 (1.01)	6.7	0.9	Almost Total Ecl.
6	开元20年2月 (732.3.1)	甲戌	13.91	11.27 (0.75)			Solar Ecl.
7	开元20年8月 (732.8.25)	辛未	17.54	12.49 (0.83)	18.3	0.62	Solar Ecl.
8	开元21年7月 (733.8.14)	乙丑	9.11	10.21 (0.68)	9.8	0.76	Solar Ecl.
9	开元22年12月 (734.12.30)	戊子			8.4	0.23	Solar Ecl.
10	开元23年闰11月 (735.12.19)	壬午					Solar Ecl.
11	开元25年5月 (737.6.3)	甲戌					
12	开元26年9月 (738.10.18)	丙申	14.45	3.32 (0.22)	14.5	0.49	Solar Ecl.
13	开元28年3月 (740.4.1)	丁亥	7.34	7.99 (0.53)	15.5	0.53	Solar Ecl.
14	天宝1年7月 (742.8.5)	癸卯	14.08	13.91 (0.93)	11.7	0.6	Solar Ecl.
15	天宝5年5月 (746.5.25)	壬子	14.75	8.54 (0.57)			Solar Ecl.
16	天宝6年10月 (747.11.7)	癸卯	16.85	15.6 (1.04)			
17	天宝8年3月 (749.3.23)	乙丑	12.94	8.24 (0.55)	12.9	0.16	
18	天宝11年7月 (752.8.14)	乙巳	11.29	0.2 (0.01)			
19	天宝11年12月 (753.1.9)	癸酉	17.24	10.17 (0.68)			
20	天宝13年6月 (754.6.25)	乙丑	12.06	4.24 (0.28)	11.4	0.82	Almost Total Ecl.
21	天宝15年10月 (756.10.28)	辛巳	15.81	7.76 (0.52)			Total Ecl.
22	乾元2年3月 (759.4.2)	丁卯			17.0	0.92	Total Ecl.
23	上元2年7月 (761.8.5)	癸未	13.19	1.24 (0.08)	9.5	0.98	Total Ecl.

Among the 19 solar eclipses that *Da Yan Li* predicted, nine were visible in *Xian*, which gives a success rate of 47%. If ancient astronomers only predicted the eclipses whose magnitudes were greater than 0.2, then the percentage increases to 56%.

According to modern calculations, there were 12 solar eclipses that could have been seen in *Xian*, and *Da Yan Li* successfully predicted 9 of them. Therefore, *Da Yan Li* could predict 75% of the solar eclipses that were visible in *Xian*.

Compared to modern calculations, the average difference of time between *Da Yan Li*'s predictions and the modern calculations is 0.92 hour. And the average difference of magnitude is 0.33.

Among 16 ancient records, 4 eclipses (No. 2, 6, 7, 12) were recorded but could not be seen in *Xian*. Three of them were predicted by *Da Yan Li*. That means about 75% of the records in the *Twenty-four Histories* were reliable at that time.

Conclusion

Da Yan Li could predict 75% of the solar eclipses that occurred in *Xian*. This rate is pretty high. However, from the perspective of astronomers in the 8th century, only about 50% of the eclipses predicted by *Da Yan Li* could be seen in *Xian*. Therefore, ancient astronomers had to improve their method to increase accuracy, which is why the calendars were still changed frequently after *Da Yan Li*.

Ancient Chinese astronomers mainly used the algebraic system, i.e. approximate empirical formulas made from observations. They didn't have the concepts of the globe and the longitude. So, although the accuracy of the solar eclipse calculation improved after *Da Yan Li*, in theory, the Chinese astronomers still couldn't give precise predictions of the solar eclipse until the appropriate geometrical system was introduced.

Notes

[1] The magnitude of *Da Yan Li* divided by 15 is the value in brackets, it is similar to modern magnitude.

[2] Zhang Peiyu (張培瑜), "Zhongguo shisandai lishi mingcheng kejian rishibiao 中國十三歷史名城可見日食表", *Sanqianwubainian rili tianxiang* 三千五百年日歷天象 (Daxiang chubanshe, 1997).

[3] *Ibid., Xian's local time* is used here (the longitude is 108).

A Date Conversion Table for the Lunisolar and Julian Calendars During the Koryo Dynasty (A.D. 918–1392) in Korea

AHN YOUNG SOOK, YANG HONG JIN,
SIM KYUNG JIN, HAN BO SIK & SONG DOO JONG
Korean Astronomical Observatory

Introduction

Korea does not yet have a systematically organized chronological table of dates, so it would be difficult to identify the exact date relating to any event in antiquity. Although there have been a few conversion studies between lunisolar (太陰太陽歷) and Gregorian calendars for the relatively recent Choson dynasty, one cannot find a conversion table easily for the former dynasties. So far available relevant studies are *Hankuk-sa Yonpyo* (韓國史年表) [1], *Ilgyoeumyangryok* (日交陰陽歷) [2] and *Hankuk Yonryok Daejon* (韓國年歷大典) [3]. *Hankuk-sa Yonpyo* appears to be a well-documented event history from the early Koryo dynasty, but a comparison of the lunisolar and Julian calendars is limited by the number of dates and their distribution. *Ilgyoeumyangryok* contains data from between A.D. 1300 and A.D. 2043, which refers to the *Samjeongjongram* (三正綜覽) [4] of Japan. However, *Samjeongjongram* is now known to have many discordances with *Ilbon Ryokil-Wonjeon* (日本歷日原典) [5] of Japan, which is a more recently published book, and not so reliable. *Hankuk Yonryok Daejon* arranges the record for a wide period from A.D. 1 to A.D. 2000, by cross-checking the Japanese [5] and Chinese almanacs, *Jungkuk Yeonreok Ganbo* (中國年歷簡譜) [6] without checking historical sources.

Most of the old Korean history texts were used to record events along with ganji dates (日辰). The other important factor to keep in mind is that the ancient ideal was that the King should rule the nation according to the heavenly order. Thus the early records have many astronomical observations. The historical record for The Three Kingdoms period (三國時代) does not show the dates in detail, whereas the Koryo dynastys' records have the exact dates for all of the astronomical observations and historical events (including lunisolar calendar and ganji dates). This provides us with the exact dates for historical events, and becomes a base for identifying those events chronologically.

438

Especially for a study of the Koryo dynasty records, one can access the two independent historical records, *Koryo-sa* (高麗史) [7–9] and *Koryo-sa Jeolyo* (高麗史節要) [10], which have about 18,000 and 1,000 ganji dates, respectively. Most data is in good condition. This study utilized *Hankuk Yonryok Daejon* as a starting point and searched the records of *Sega* (世家), *Chonmun-ji* (天文志), *Ohaeng-ji* (五行志) and *Ryok-ji* (歷志) of *Koryo-sa,* as well as the records of *Koryo-sa Jeolyo.*

Thus, we tried to convert the lunisolar calendar of the Koryo dynasty into the Julian calendar, including unpublished data for a chronological arrangement.

Data Acquisition and Analysis

Data Acquisition

Koryo was a nation which ruled the Korean peninsula during the period from A.D. 918 to 1392. Two well-known classical texts for this dynasty are *Koryo-sa* (高麗史) and *Koryo-sa Jeolyo* (高麗史節要), both written during the Chosun dynasty. The former was written during A.D. 1449 (King Sejong, 31st year) and A.D. 1451(King Moonjong, 1st year). It was written by adopting the style of *Sagi* (史記) of Samachun (史馬天), which consists of 4 sections, that is, *Sega* (世家), *Ji* (志), *Pyo* (表), *Yeljeon* (列傳). In this study, we organize all the ganji dates of *Sega* and *Ji. Koryo-sa Jeolyo,* published in A.D. 1452(King Moonjong, 2nd year), is an independent book of *Koryo-sa* (高麗史), which complements the *Koryo-sa* (高麗史).

Koryo-sa (高麗史) and *Koryo-sa Jeolyo* (高麗史節要) were arranged according to ruling period, month, and ganji dates, plus dates for special cases. In cases where there were many issues for a specific month, the first ganji dates only would have month information. In the case of intercalary months (潤月), they only identified the first event without month information, and did not supply month information for the following events. Moreover, if there were no event in the previous month, they identified the specific month and intercalary month.

Table 1 Recorded Date Information in Two History Books

Name of history book	Number of ganji dates	Number of Sag Ilgin	Number of intercalary month	Mis-recorded ganji dates
Koryo-sa	18,761	1,918	128	79
Koryo-sa Jeolyo	966	167	100	23

For many ganji dates, the two historical texts show much overlap. Seventy-five ganji dates from *Koryo-sa Jeolyo* did not appear in *Koryo-sa*. Sag Ilgin (朔日辰) is the ganji of the first day of each lunar month. Properly inserting intercalary months between the ordinary months is the most important task in making the conversion table. Moreover, we identified the erroneous ganji date from the *Koryo-sa* and *Koryo-sa Jeolyo*. In the following section, we will further discuss the accuracy and errors found in the above data.

Data Analysis

In this study, we classified mis-recorded Sag Ilgin into three categories. The 1st case is calculating the wrong Sag Ilgin. Ganji dates repeated 60 as one unit, i.e. (a 60-day cycle) so one month cannot include all of them. Thus, if a text contained the whole series of 60 ganji dates in one month, the record must have been in error. Seventy-nine such errors were identified in *Koryo-sa* and 23 in *Koryo-sa Jeolyo*. These records are given in the appendix titled *The Date Conversion Table of Koryo Dynasty* (高麗時代年歷表) [11].

The second case is when one or two days gap occurred between the Japanese records or their Chinese counterparts. In this case, after checking other Sag Ilgin from the Japanese or Chinese, we carefully chose the correct ones while giving a high priority to *Koryo-sa*. A total of 38 Sag Ilgins are arranged in this way: the ones which have a one day gap accompanying the records of the two nations are 20. Ten of them have the same dates as the Chinese, while seven of them agree with the Japanese. One shows no agreement at all.

The 3rd case is that of the 20 uncertain Sag Ilgins in *Koryo-sa*. Most of these are found after A.D. 1300. Because these records were not clear, we referred to the solar and lunar eclipses, other historical data and our modern calculation of moon phases. The following are some of the troublesome Sag Ilgins.

A.D. 1096: the four Sag Ilgins differ in the case of Japan and China by 2 days. We used the case of *Koryo-sa* (高麗史).

A.D. 1136: In *Koryo-sa* (高麗史), we found Gihae Sag (己亥朔) in April. But an historical record (1123–1144) reads that the King made a journey to *Anwha*-Temple (安和寺) for an anniversary mourning of his father's passing away which was on 10th April of each year. Thus, according to the above record, it would be Musul Sag (戊戌朔) in April. By checking adjacent month calculations, we inferred Musul Sag to be in April.

A.D. 1195: There are records of the Byungsul Sag (丙戌朔) solar eclipse in January in *Koryo-sa Chonmun-ji*. The King was visited by diplomatic representatives of Kum (金) from China. They congratulated the King on his birthday (January 17th) during the period from A.D. 1147 to 1197. Thus it should be Jeonghae Sag (丁亥朔) in January. The other case from this year is the solar eclipse of March recorded for Byungsul Sag (丙戌朔) in *Koryo-sa Sega*. Judging from the solar eclipse event interval, the solar eclipse could not have occurred on January, so it must be another error in the record of *Koryo-sa Chonmun-ji*.

A.D. 1282: Byungsul Sag in August is obviously wrong in *Koryo-sa*, considering the arrangement of nearby ganji dates. Unfortunately there was a discordance between Japan and China. A comparison of nearby Sag Ilgin and our calculations imply Jeonghae Sag.

Data Distribution and Problems

Interesting comparisons can be made between the *Koryo-sa* and *Koryo-sa Jeolyo* data. These two books are known as independent classical texts, so they should supply us with the two independent sources of reliable data. Although numerous records are found in both books, there are missing parts during some specific periods. Figure 1 shows the distribution of the data of *Koryo-sa*. Overall, there are no records for the early Koryo dynasty before A.D. 1010, while no or only a few exist during the period of the early A.D.

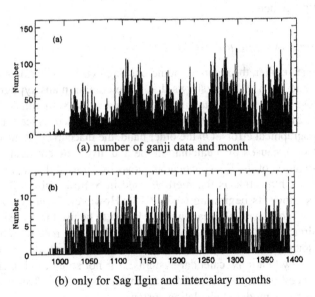

(a) number of ganji data and month

(b) only for Sag Ilgin and intercalary months

Fig. 1. Yearly Distribution of Dates in Kory-sa.

1200s, 1250s and 1330s. The shortage of data before 1010 must be due to the chaotic situation right after the establishment of the new dynasty. The first record of Sag-Ilgin in *Koryo-sa* appears in the year 984, so the data before 984 must be complemented by other records.

The second and third periods of data shortage, the 1200s and 1250s, have obvious historical causes. The 2nd period began with the crisis of A.D. 1135, the Myocheong (妙清) revolt whose confusion continued to A.D. 1170, the Jung Bu Jung (鄭仲夫) revolt and to A.D. 1196, the Choi family's military government. Near its end, A.D. 1231, the Mongolian invasion caused the King to take refuge on Gangwa-Island, which became a temporary capital until the surrender in A.D. 1258. The late part of this period exactly coincided with the shortage of data.

The fourth period of data shortage is the 1330s. The Kings of Koryo visited and stayed in Yuan (元) while a son-in-law ruled Koryo remotely from abroad. Although several kings performed their roles under intervention of Yuan (元), a certain king had to give up his throne due to intervention. We can only infer that the unstable situation may be the reason for the absence of data.

In summary, the lunisolar calendar is relatively well documented for the period since A.D. 1010, although periods of data shortage, before A.D. 1010, and in the early 1200s, 1250s, 1330s, can be indirectly identified by referring to the Chinese and Japanese records, or our eclipsing event calculation or historical judgement.

The Calendar-Making Rule of Koryo

As mentioned earlier, the record of numerous Sag Ilgin's of the Koryo dynasty and solar and lunar eclipses shows that these contain many typographical errors; those of *Sega* and *Chunmun-ji* of *Koryo-sa* are sometimes different from those of other nations by one or two days. The discordances are partly due to typographical errors. On the other hand, the inaccuracy of the eclipses may have been caused by a calculation method involving constants.

Investigation of the Koryo calendar shows that it used the *Seonmyung-li* (宣明曆) of China, that is the method used in Shinra, one of The Three Kingdoms, since its beginning (A.D. 918). However, *Seonmyung-li* was outdated for more than 100 years when it began to be used in Korea and the Chinese already switched to another calendar calculation method. This must be the reason why there was a difference in constants and calculation methods between the two nations' calendar systems. It is not so unusual to find one day differences in Koryo, compared with the Chinese system. This can easily happen even in a modern calculation system.

One must note that there is a large difference between Koryo and Japan, but for a different reason. Japan basically adopted the system of Koryo, although the arrangement of intercalary months was not the same. Especially since A.D. 1300 Koryo maintained good relations with China, so they introduced a new li (歷) system. The calendar dates of Koryo and China agree well, contrary to their frequent discordance with Japan. Perhaps Japan did not introduce the calendar making rule from Koryo promptly.

Around A.D. 1300, the calendar construction system entered a new era, ending the *Seonmyung-li* period. In King Chungreol 7th year (A.D. 1281), an official from Yuan (元) of China delivered *Susi-li* (授時歷), which were not deciphered. The exact calculation method was not clear at that time. Later, in A.D. 1298, when the King stayed in Yuan (元) China, Seongji Choi (崔誠之) tried to learn *Susi-li* but the solar and lunar eclipse and 5 planetary motion prediction methods were still unknown to him. Thus *Seonmyung-li* was still in use in some sense. Finally in A.D. 1336, Gangbo (姜保) who wrote *Susiryeok-Chupbeop Ipsung* (授時歷捷法立成), figured out *Susi-li*. In A.D. 1370, in the period of King Gongmin, Ming (明) China passed out *Datong-li* (大統歷), which was, in fact, not different from *Susi-li* except for the starting epoch. This fact explains well why there was an error in the records of the solar and lunar eclipses in Koryo-sa. One can imagine there would be a large incremental error as time went by. However, the problem of ganji date, and solar and lunar eclipse must be re-investigated within the context of the calendar construction system used at that time.

The Date Conversion Table During the Koryo Dynasty

The conversion table for the lunisolar and Julian calendars is organized based on a lunar month system, which can be easily converted to the Julian calendar. Table 2 shows one page of the conversion table from our investigation. As shown in Table 2, following the style of historical texts, dates are arranged in the order of the King's ruling period and the Sag Ilgin, along with year ganji (歲次) corresponding to a 60 cycle (or 60 ganji). We also identified whether each month was large or small. Large corresponds to 30 days, while small corresponds to 29. The fifth column, month ganji (月建), represents a monthly 60 cycle which is traditionaly used. We also presented the Sag Ilgin of China and Japan. The last column presents the solar and lunar eclipses of the Koryo dynasty and Dangun giweon (Dangi). Here, Dangun was the first historical king of Korea about 4500 years ago. Dangi was the designation of an era used as the beginning year of Korea (i.e. A.D. + 2333).

Table 2　Example of the Date Conversion Table During the Koryo Dynasty

年代 (歲次)陰曆1日日辰	月의 大小	西紀 年月日	율리우스 積日	月建	陰曆1日日辰 中國	日本	備考
顯宗14年							檀紀3356年
(癸亥年) 1月 丙寅朔	小	1023 1 25	209 4732	甲寅	丙寅	丙寅	
2月 乙未朔	小	1023 2 23	4761	乙卯	乙未	乙未	
3月 甲子朔	大	1023 3 24	4790	丙辰	甲子	甲子	
4月 甲午朔	小	1023 4 23	4820	丁巳	甲午	甲午	
5月 癸亥朔	大	1023 5 22	4849	戊午	癸亥	癸亥	
6月 癸巳朔	小	1023 6 21	4879	己未	癸巳	癸巳	
7月 癸亥朔	大	1023 7 21	4909	庚申	癸亥	癸亥	
8月 壬辰朔	大	1023 8 19	4938	辛酉	壬辰	壬辰	
9月 壬戌朔	小	1023 9 18	4968	壬戌	壬戌	壬戌	
閏9月 壬辰朔	大	1023 10 18	4998	閏	(閏)壬辰	(閏)壬辰	
10月 辛酉朔	小	1023 11 16	5027	癸亥	辛酉	辛酉	
11月 辛卯朔	大	1023 12 16	5057	甲子	辛卯	辛卯	甲辰月食
12月 庚申朔	大	1024 1 14	5086	乙丑	庚申	庚申	

Table 2 (*Continued*)

年代 (歲次)陰曆1日日辰	月의 大小	西紀 年月日	율리우스 積日	月建	陰曆1日日辰 中國	日本	備考
顯宗15年 (甲子年) 1月 庚寅朔	小	1024 2 13	209 5116	丙寅	庚寅	庚寅	檀紀3357年
2月 乙未朔	小	1024 3 13	5145	丁卯	乙未	乙未	
3月 戊子朔	大	1024 4 11	5174	戊辰	戊子	戊子	
4月 戊午朔	小	1024 5 11	5204	己巳	戊午	戊午	
5月 丁亥朔	大	1024 6 9	5233	庚午	丁亥	丁亥	
6月 丁巳朔	小	1024 7 9	5263	辛未	丁巳	丁巳	日食,王寅一食
7月 丙戌朔	大	1024 8 7	5292	壬申	丙戌	丙戌	
8月 丙辰朔	大	1024 9 6	5322	癸酉	丙辰	丙戌	
9月 丙戌朔	小	1024 10 6	5352	甲戌	丙戌	乙卯	
10月 乙卯朔	大	1024 11 4	5381	乙亥	乙卯	乙酉	
11月 乙酉朔	小	1024 12 4	5411	丙子	乙酉	乙卯	
12月 甲寅朔	大	1025 1 2	5440	丁丑	甲寅		日食

The newly edited conversion table gives the information only for the first day of each lunar month to avoid enormous volume. Thus, the usual ganji date can be found from the 60 cycle order number table. The wrong ganji dates in *Koryo-sa* and *Koryo-sa Jeolyo* are given separately in the appendix to avoid complexity or difficulty in identifying the date.

All the solar eclipse records of the Koryo dynasty, Julian solar date, and confirmation of eclipses, are also given in the last part.

Discussion

The number of whole months during the Koryo dynasty between A.D. 918 and 1392 are counted as 5875. The lunisolar calendar can be made by inserting 7 additional months every 19 years, so there are more Sag Ilgin and intercalary months to be inserted than in the Julian calendar. Thus the compilation of the conversion of the lunisolar and Julian calendars (陰陽歷 變換) was not so simple.

Among the expected Sag Ilgin during the whole Koryo dynasty, 30.5% of them are shown in historical records, while 73% of the intercalary months appeared in the records. As mentioned earlier, one cannot find any record in *Koryo-sa* and *Koryo-sa Jeolyo* for the period earlier than A.D. 1010. Excluding this period, during the years from A.D. 1010 to 1392, we checked the accuracy of the Sag Ilgin distribution. We found a fairly high accuracy, 40% for the Sag Ilgin of 4737 months and 90% for the intercalary months. Since about 5 of 12 Sag Ilgin of each year can be positioned precisely, this arrangement is enough for a precise 12 Sag Ilgin determination of the lunisolar calendar, considering the fact that the lunisolar month consists of 29 and 30 days. This is certainly not the case when ganji dates for the first days are consecutively given. Nevertheless at least 40% to 80% accuracy is expected from such a number. This number came from the averaging of the Sag Ilgin and intercalary month arrangements. In fact the accuracy increases if one considers other factors. Many first days of each month are recorded as *Sag-il* (朔日) with ganji date in the history books. This habitual characteristic and the other general ganji date provided us an important clue in determining of Sag Ilgin. Other important clues came from the recorded astronomical observational events which became the basic for the precise determination of ganji dates.

The following conclusions can be drawn from the gangi dates of Koryo, closely related to their Chinese counterparts. First, the ganji date distribution for Koryo corresponds fairly well to the Chinese. Especially those after A.D. 1300 closely match with the Chinese. Secondly, the location of intercalary months agree well with the Chinese, but not with the Japanese. Thirdly, we note that the date gangi of Koryo, Japan, and China do not agree with each other. This implies that the calendar construction system of Koryo was independent from the others. Thus, by studying then using the calendar construction systems of antiquity, one can make a more reliable conversion reference.

This newly published conversion table for the lunisolar and Julian calendars of the Koryo dynasty is a significant step since it was based on the records of the actual ganji dates not on the calendar-making calculation method. Thus it can produce Gregorian calendar dates, the system currently in use.

References

[1] 震檀學會. 韓國史年表, 1959, 乙酉文化社, Seoul.

[2] 李殷成, 日交陰陽歷, 1983, 世宗大王紀念事業會, Seoul.

[3] 韓寶植, 韓國年歷大典, 1987, 嶺南大學校出版社部, Daegu.

[4] 日本內務省(編), 三正綜覽, 1965, Tokyo.

[5] 內田正男(編), 日本歷日原典 —第4版, 1994, 雄山閣版社部, Tokyo.

[6] 董作賓(編), 中國年歷簡譜, 1974, 藝文印書館印行.

[7] 韓國學文獻研究所(編), 高麗史, 1990, 亞細亞文化社, Seoul.

[8] 東亞大學校古典研究室(編), 譯註高麗史, 1965, 東亞大學校出版社, Busan.

[9] 社會科學研究員古典研究室(編), 北譯高麗史, 1991, 新書苑編輯部, Seoul.

[10] 韓國學文獻研究所(編), 高麗史節要, 1973, 亞細亞文化社, Seoul.

[11] 沈敬鎮,安英淑,韓甫植,梁洪鎮,宋科鍾,高麗時代年歷表, 1999, 韓國天文研究院.

Mapping the Universe: Two Planispheric Astrolabes in the Early Qing Court

FUNG KAM-WING
University of Hong Kong

Introduction

The astrolabe — the name means "star-finder" — is a portable, usually flat, circular instrument, occasionally spherical, made of brass, on which the positions of the major luminaries in the heavens are depicted relative to the observer's horizon using the principle of stereographic projection. It may well be called the most famous and widespread astronomical instrument of the pre-modern age, both in the East and the West.[1] The Western medieval planispheric astrolabe (Arabic: *asturlâb sathî*; medieval Latin: *astrolabium planisphaerium*) and spherical astrolabe (Arabic: *asturlâb kurî*; medieval Spanish: *astrolabio redonio*) were introduced and used by Islamic astronomical-officials in the observatory of the Astronomical Bureau (*Si tian tai* 司天臺) at the great capital Dadu 大都 during the early Mongolian period in China. According to Song Lian 宋濂's (1310–81) account in *Yuan shi* 元史 (Official History of Yuan Dynasty, 1370), the Marâghah astronomer and instrument maker Cha-ma-lu-ting 扎馬魯丁 (probably refers to Jamâl al-Dîn ibn Muhammad al-Najjârî or Jamâl al-Dîn Muhammad ibn Tâhir ibn Muhammad al Zaydî al Bukhârî) served in the Astronomical Bureau and made a calendar for Kublai Khan (1215–94, r. 1260–94) in the fourth year of the Ziyuan 至元 reign (1267), called the *Wan nian li* 萬年曆 (Ten Thousand Year Calendar). In the same year he made seven different kinds of Islamic astronomical instruments including a portable astrolabe.[2] In addition to this, Wang Shidian 王士點 and Shang Qiong's 商企翁 *Yuan bishujian zhi* 元秘書監志 (Records of the Imperial Secretariat of the Yuan Dynasty, 1342–63) has a detailed bibliographical listings of 242 volumes of Islamic books and related instruments preserved in the northern observatory attached to the Astronomical Bureau in the tenth month of the tenth year of the Ziyuan 至元 reign period (1273). Among those Islamic books and related instruments, there was an Islamic spherical astrolabe and related celestial chart.[3] However, these astronomical instruments were not widely used among the Chinese.

The Translation and Transmission of Texts on the Planispheric Astrolabe in Late Ming and Early Qing

It was very likely that between the years 1583 and 1588, when Jesuit Missionary Matteo Ricci (1552–1610) and Michele Ruggieri (1543–1607) prepared their lexicographical work, *Dizionario Portoghese Cinese* (Portuguese-Chinese Dictionary, Istituto Storico della Compagnia di Gesù, Rome, *MS. Archivum Romanum Societatis Iesu, Jap.-Sin., I, 198*) in Sciaochin (Zhaoqing 肇慶) in northern Guangdong 廣東, Matteo Ricci used translated extracts of Christopher Clavius' (1538–1612) works such as *Treatise on Gnomonics* (*Gnomonices libri octo*, Romae, 1581 [Pé-táng Library at Peking, No. 1301]), *In Sphaeram Ioannis de Sacro Bosco Commentarivs* (Romae, 1570 [Pé-táng Library, No. 1308]) and introduced the basic theory of the Western medieval planispheric astrolabe and other horological instruments to the Chinese people.[4] Father Paul Yang points out that the manuscript of Ricci's Portuguese-Chinese Dictionary contains several summarized explanations and diagram which related to Western scientific instruments. His descriptions are as follows:

> Fol. 17a–23b is an explanation of celestial and terrestrial globes.
> Fol. 170a–171b contains an explanation of the sundial (or astrolabe).
> Fol. 173a, is a sundial diagram.[5]

Ming scholar-official Li Zhizao 李之藻 (1565–1630, baptized Leon in 1610) first called on Matteo Ricci in 1601 during Ricci's second visit to Peking. That was the beginning of an enduring relationship. Between 1605–1607, Li Zhizao translated Christopher Clavius' *Astrolabivm* (Romae, 1593, [Pé-táng Library, No. 1291]) into Chinese with the title *Hungai tongxian tushuo* 渾蓋通憲圖説 under the direction of Matteo Ricci. Several years later, Jesuit Father Sabbathin de Ursis (1575–1620) and Xu Guangqi 徐光啟 (1562–1633) co-authored a treatise on the theory of the planispheric astrolabe entitled *Jianpingyi shuo* 簡平儀説 (1611).[6] Portuguese Jesuit Emmanuel Diaz (1574–1659) even used simplified diagrams of the planispheric astrolabe to demonstrate a quicker way of star-finding and the calculation of time upon queries from Chinese people in his *Tianwen lüe* 天問 (Explanation on Celestial Spheres, 1615).[7]

These translated works exerted great influence upon Chinese scholar-officials. It is of special interest here that a rather obscure Chinese had the opportunity to work with Jesuit missionaries in Peking during this period and compiled an astronomical and horological book entitled *Riyuexingguishi* 日月星晷式 (Sun and Moon Dails and Nocturnals with an appendix on the Planispheric Astrolabe, 1611–22). He might have consulted Christopher

Clavius' *Gnomonices libri octo* and *Astrolabivm* as well as other translated texts.[8] Ming loyalists like Xiong Mingyu 熊明遇 (1579–1649) and Fang Kongzhao 方孔炤 (1591–1655) had cited some portions from these translated texts in their works. In Xiong Mingyu's *Ge-zhi cao* 格致草 (Scientific Sketches, 1648 printed edition), diagrams of the planispheric astrolabe from *Tianwen lüe* were quoted in the section called "Jianping tushuo 平儀圖説" (Diagram and Explanation of planispheric astrolabe).[9] Another example can be found in Fang Kongzhao's voluminous *Zhouyi shilun hebian* 周易時論 合編 (Collectanea of Researches on the Book of Changes, 1660 edition). Fang quoted a number of passages and mathematical tables related to the astrolabe from *Hungai tongxian tushuo* in the chapter called "*Chongzhen lishu yue* 崇禎曆書約" (Summary of Chongzhen Reign Period Treatises on Astronomy and Calendrical Science).[10] Since the Jesuit scientific translations enjoyed a wide circulation in Chinese intellectual circles, Li Quangdi 李光地 (1642–1718), an important early Qing bureaucrat and statesman in the Kang Xi 康熙 reign (1662–1722), had also made inquiries into the construction of planispheric astrolabes in his astronomical work *Lixiang benyao* 曆象本要 (The Original Meaning of Calendrical Science, 1709) which could be traced back to his earlier work entitled *Lixiang yaoyi* 曆象要義 (Essentials of Calendrical Science, 1665).[11]

The Making of Planispheric Astrolabes in the Imperial Palace during the Early Reign of Kang Xi

It is widely known that Emperor Kang Xi (1654–1722, r.1662–1722) had an extremely strong interest in the pursuit of Western exact sciences.[12] He made considerable effort in studying mathematical and calendrical science with Jesuit missionaries, who included Ferdinand Verbiest (1623–88), Thomas Pereira (1645–1708), Antoine Thomas (1644–1709), Jean-François Gerbillon (1654–1707), Joachim Bouvet (1656–1730) and Dominique Parrenin (1665–1741).[13] In the third month of the eighth year of the Kang Xi reign period (1669) when the political crisis between the Jesuits, Yang Guangxian 楊光先 (1597–1669) and the Mohammedans on the divination of hemerology (*xuan ze* 選擇) had just been settled, Father Verbiest was made director of the observatory (*zhili lifa* 治理曆法) in Yang's place and was appointed vice-president of the Directorate of Astronomy (*qintianjian jianfu* 欽天監監副) because of his correct reckoning and divination on heavenly affairs.[14] In the midsummer of the same year, Verbiest made a gilt-silver demonstrational armillary sphere (*yindujin huntianyi* 銀鍍渾天儀) and presented it to the throne.[15] Perhaps because of Verbiest's excellent talent in instrument-making, in the sixth month of this year, Emperor Kang Xi granted him an imperial decree to be in charge of making new astronomical instruments in the Ancient Observatory at Peking.[16]

In his dual capacity as director of the observatory as well as senior administrator in the Directorate of Astronomy, Ferdinand Verbiest devoted himself enthusiastically to a wide range of scientific research, such as physics, medicine, mechanical engineering, geography and cartography.[17] Amongst these disciplines, astronomical and calendrical science figured prominently, and he compiled a number of accurate ephemerides, calendars, and celestical maps. In the eleventh year of Kang Xi's reign (1672), Verbiest finished the *Chido nanbei liang zong xin tu* 赤道南北兩總星圖 (Map of the stars to the north and the south of the equator). In parallel with this celestial cartographical work, another map relating to the planispheric astrolabe was in preparation. The explanatory note of the *Chido nanbei liang zong xin tu* contains the following detailed account:

> Just like constellations in both the southern and northern sides of the [celestial] equator, they are cut across by the equator. Therefore, these constellations are drawn in two different maps. [If] one wants to see the whole map, please refer to *Jianpinggui zongxing tu* 簡平規總星圖. To conduct an observation, place a regula (or an alidade) at the centre of both the southern and northern maps. Besides, place the time circle plate and diurnal arc in the northern map; place the time plate and nocturnal arc in the southern map. Then rotate two maps in which the entire celestial sphere will be displayed. Because the time circle plate and both diurnal and nocturnal arcs are not movable, rotate the maps about the regulas over south and north celestial poles and you will see the stars move from west to east, rising and setting at the horizon. It looks like the celestial sphere revolving. If the map is not movable, you may rotate the time circle plate and the diurnal (or nocturnal) arc. It performs the same function. If you take a level plate, place the star map of north celestial pole on the top of the plate and the star map of south celestial pole underneath the plate. Use a regula (or an alidade) to fix the time circle plates and the diurnal and nocturnal arcs which have already been placed at the north and south celestial poles, so the celestial sphere looks like joining with two celestial hemispheres. It also looks like the celestial sky is in motion over the north and south celestial poles. There are three methods of using this plate and they are also related in the *Jianpinggui zongxing tu*.[18]

In the second month of the 13th year of the Kang Xi's reign (1674), Verbiest completed the building of six large-scale astronomical instruments on top of the Ancient Observatory. He also submitted *Xin zhi lingtai yi xiang zhi* 新製靈台儀象志 (Discourse on the Newly-built Astronomical Instruments in the Observatory) to the throne at the same time.[19] In the middle of this year, he wrote an explanatory treatise on the celestial map of the planispheric astrolabe entitled *Jianpinggui zongxing tujie* 簡平規總星圖解 (Fig. 1) with the help of Liu Wende 劉蘊德 who was also a vice-president of the Directorate of Astronomy. In this treatise, Verbiest describes four methods of using the tablet of horizon of the planispheric astrolabe:

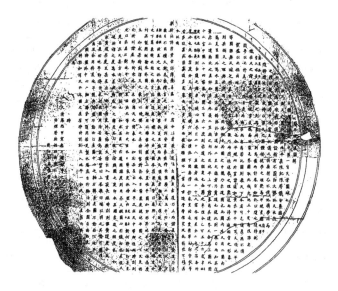

Fig. 1.

All the determinative stars of different lunar mansions in this map are according to exact degrees of the stars observed in the twelfth year of Kang Xi's reign. The method of making circles to observe them is as follows: place a regula in the centre of the map so as to rotate the time circle plate and put the tablet of horizon on it. It has an oval shape. The horizon is in accordance with the altitude of the North Pole. In an attempt to find the time of heavenly bodies in the celestial sky, move the hour scale of the time circle plate until it lies against the current day's fortnightly period (or solar term) and compare it with the heavenly bodies appearing at the top or below as well as four directions of the tablet of horizon. This will be the actual display of the celestial sky of this time of the day. In an attempt to find the time of rising or setting at the horizon and the direction in the zenith of a given star, first set the given star against four directions of the tablet of horizon, and then read off the current day's fortnightly period against the time on the time circle plate. This will be the time for the star which is in the sky with a certain direction. In attempting to find the degrees of right ascension or declination of a given star, use a rule to draw a line from the centre of the map until it cuts against the star, take the degree which is pointed from the fiducial edge of the rule as the degree of declination of the star in the south or north, take the degree of circle of right ascension which is pointed by the sharp tip of the rule as the degree of right ascension of the star. In attempting to find a degree of the ecliptic and a range of degrees of a given star rising or setting at the horizon as well as the times of each fortnightly period, sunrise and sunset, equal and unequal hours of day and night, move the degree of the ecliptic or set the star to cut against the horizon. Then you will get the degree right away.[20]

This map was later presented to Emperor Kang Xi. In the 20th year of his reign (1681), the Emperor made an imperial order to the court craftsmen of the Imperial Workshop (*Zao ban chu* 造辦處) for several pedagogical astrolabes under the direction of Ferdinand Verbiest, two of them known as "Ciqing zhizhi jianpingyi 磁青紙製簡平儀" (ciqing paper-made planispheric astrolabe) and "Yuzhi tongdujin jianpingyi 御製銅鍍金簡平儀" (gilt-copper planispheric astrolabe, midsummer, 1681).[21] For the paper-made planispheric astrolabe, it is interesting to note that paper-made instruments were common in the West, not so much for computation, but for practical instruction. German astronomer Petrus Apianus' (1495–1552) *Astronomicum Caesareum* (Ingolstadii, 1540) contained Apianus' popular paper-made instruments, i.e. equatorial and rotating star chart. "Astronomicum Caesareum" literally means "the emperor's astronomy", even Tycho Brahe (1546–1601) paid a large amount of money to purchase one.[22] Emperor Kang Xi might have had the idea of promoting "the emperor's astronomy and astrolabe". A careful examination of two astrolabes reveals the following details:

(1) Ciqing zhizhi jianpingyi 磁青紙製簡平儀 (ciqing paper-made planispheric astrolabe) Astrolabe diameter: 32.2 cm (Fig. 2).

    The front of the mater is a planisphere of the northern celestial sky while the back is a planisphere of the southern celestial sky. The outer rim of the planisphere is a circle of degree scale with 30 equal divisions, and then a zodiac circle (i.e. the twelve sky houses: Sagittarius, Scorpio, Libra, Virgo, Leo, Cancer, Gemini, Taurus, Aries, Pisces, Aquarius, Capricorn) with equal degree scale. Twenty-four fortnightly periods are also drawn. The stars are very likely according to Verbiest's *Chido nanbei liang zong xin tu*. The rete is of an unusual design that may belong to the type "Gothic Astrolabe".[23] The outer rim of the rete is drawn with Chinese hours circle

Fig. 2.

and mounted with a fixed diurnal arc (with graduations), a vertical limb (with graduations) and a small hours circle with a short movable alidade at the centre pivot. This astrolabe contains a suspension ring and a shackle but no decorated throne which is common for most of the western astrolabes. The rete near the shackle is inscribed with three Chinese characters of "jian ping yi 簡平儀". The back of the astrolabe is similar to its front side. The rete is mounted with a fixed nocturnal arc (with graduations), a vertical limb (with graduations) and a small hours circle with a long movable rule (with names of twenty-four fortnightly periods and graduations) at the centre.

(2) Yuzhi tongdujin jianpingyi 御製銅鍍金簡平儀 (gilt-copper planispheric astrolabe) Astrolabe diameter: 32.1 cm (Fig. 3).

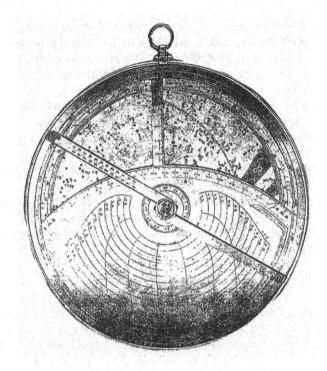

Fig. 3.

The outlook of this astrolabe is similar to the diagram and description that was prepared by Verbiest in his *Chido nanbei liang zong xin tu* and *Jianpinggui zongxing tu*. It consists of two plates: the upper northern diurnal arc plate, and the lower celestial plate. In the upper northern diurnal arc plate, the outer rim is engraved with twelve months (each month 30°) and Chinese hours circle. There is a small hours circle and a long movable rule (with names of twenty-four fortnightly periods but no graduations) in the central part of the plate. A similar device is seen in French globes and planispheres maker Issac Habrecht II's (1589–1633) *Planiglobium coeleste, et terrestre* (Strasbourg, 1628).[24] The centre pivot of this plate is the celestial north pole. The lower part of the plate is engraved with table of lines corresponding with lines of twenty-four fortnightly periods, sunrise and sunset, six intervals of Chinese night-watch. The front side of the lower celestial plate is a planisphere of the northern celestial sky. The outer rim of the planisphere is a Chinese calendar circle, and then a zodiac circle (i.e. the Western twelve sky houses), a 360° circle. Twenty-four fortnightly periods are also engraved. The stars are very likely according to Verbiest's *Chido nanbei liang zong xin tu*. The back side of the lower celestial plate is a planisphere of the southern celestial sky. It has a rete with a fixed nocturnal arc (with graduations) and a vertical limb (with graduations) as well as a small hours circle with a short movable alidade at the centre pivot. In the upper rim near the suspension ring and the shackle, three Chinese characters "jian ping yi 簡平儀" are engraved. In the lower outer rim of the upper diurnal arc plate, there is an epigraph: "*Kang Xi er shi nian sui zai xin you zhong xia yu zhi* 康熙二十年歲在辛酉仲夏御製" (Made at the 20th year of Kang Xi's reign by imperial order).[25]

Concluding Remarks

From the above discussion, it is evident that in the course of reception of Western science, royal patronage played an important role in promoting the making of Western scientific instruments in the imperial court during the early Qing period. Father Verbiest was also around to explain Western scientific knowledge to the emperor. In 1682, Ferdinand Verbiest submitted his Chinese work *Yulan jianpingyi xinshi yongfa* 御覽簡平儀新式用法 (Manual for Imperial Use of New Pattern of Planispheric Astrolabe) to the throne.[26] He also recorded his journey to Eastern Tartary.

(Liaodong 遼東) with Emperor Kang Xi during March and June of the same year as follows:

> (March 23, 1682) He was pleased that I should accompany him, and always be near his person, to take observations, in his presence, of the disposition of the heavens, the height of the pole, the declination of each country, [oblique situation on the globe] and to take with instruments the heights of mountains, and distances of places. He was also desirous of being instructed concerning meteors, and many other matters in physics and mathematics. Toward this end he ordered the necessary instruments to be carried on horses, and recommended me to the Prince his uncle, who is also his Father-in-law, and the second person in the state, being called by a Chinese name signifying colleague in the government.

> (June, 1682) The first day we set out on our return, we were stopped in the evening by a torrent, so large and rapid as to be unfordable. ... When we

got over, the Emperor seated himself on the bank, and me by his side, with the two sons of two Western Regulo's, and the chief *Kolau* (閣老) of Tartary whom he distinguished on all occasions. It being a fine night, and a clear sky, he was pleased that I should name the constellations, that then appeared above the horizon, in the Chinese and European languages, himself naming first those which he knew. Then opening a little (celestial) map, which I had presented him some years before, he sought the hour of night by the star on the meridian, delighting to let everyone see his skill in the sciences.[27]

It is safe to assume that Emperor Kang Xi might have used some portable astronomical instruments such as planispheric astrolabes for his scientific pursuit at the imperial palace or on his journey to Liaodong.

Notes

1   On the principle and history of the astrolabe, see R. T. Gunther (1869–1940), *Early Science in Oxford Vol. II: Astronomy* (Oxford, 1923, reprinted in facsimile, London: Dawsons of Pall Mall, 1967), pp. 182–228 and *The Astrolabes of the World* (2 vols., Oxford, 1932; reprint in one volume, London: The Holland Press, 1976); Willy Hartner (1905–1981), "The Principle and use of the Astrolabe", in Arthur Upham Pope ed., *A Survey of Persian Art* (London and New York: Oxford University Press, 1939), Vol. III, pp. 2530–54, also in *Willy Hartner:Oriens-Occidens* (Hildesheim: Georg Olms Verlagsbuchhandlung, 1968), pp. 287–311; O. Neugebauer, "The Early History of the Astrolabe", *Isis*, Vol. 40 (1949), pp. 240–56; P. H. Van Cittert, *Astrolabes* (Leiden: E. J. Brill, 1954), pp. 1–18; Emmanuel Poulle, *Les Sources Astronomiques (Texts, Tables, Instruments)* [Turnhout-Belgium: Brepols Publishers, 1981], pp. 22–34; Sharon Gibbs and George Saliba, *Planispheric Astrolabes from the National Museum of American History* (Washington: Smithsonian Institution Press, 1984), pp. 12–21; Harold N. Saunders, *All the Astrolabes* (Oxford: Senecio Publishing Company Limited, 1984), pp. 5–20.

2   Song Lian ed., *Yuan shi* (Beijing: Zhonghua shuju, 1976), pp. 998–99. Willy Hartner, "The Astronomical Instruments of Cha-ma-lu-ting, their Identification, and their Relations to the Instruments of the Observatory of Marâgha", *Isis*, Vol. 41 (1950), pp. 184–94, also in *Willy Hartner: Oriens-Occidens*, pp. 215–26; Ma Jian, 馬監, *Huili gangyao* 回曆綱要 (An Outline of Islamic Calendar) [Beijing: Zhonghua shuju, 1955], pp. 18–21, and "Yuan bishujian zhi hui-hui shuji shiyi 《元秘書監志 • 回回書籍》釋義" (The Interpretation of the Islamic Books listed in the *Records of the Imperial Secretariat of the Yuan Dynasty*), in *Huizu shi lunji* 回族史論集 (Collected Essays on the History of Hui People in China) [Yinchuan: Ningxia People's Press, 1983], pp. 193–98; L. A. Mayer, *Islamic Astrolabists and Their Works* (Genève: Albert Kundig, 1956), pp. 1–21; Joseph Needham, *Science and Civilisation in China* Vol. 3: Mathematics and the Sciences of the Heavens and the Earth (Cambridge: Cambridge University Press, 1959), pp. 372–76; Aydin Sayili, *The Observatory in Islam and Its Place in the General History of the*

Observatory (Ankara: Türk Tarîh Kurumu Basimevî, 1960), Chapter VI "The Marâgha Observatory", pp. 187–223; K. Tazaka (田坂興道, 1912–57), *Chugoku ni o ke ru Kaikyo no denrai to so no guzu* 中國における回教傳來その弘通 (Islam in China: Its Introduction and Development) [Tokyo: The Toyo Bunko, 1964], pp. 1540–71; K. Yabuuti (藪內清, *Chugoku no tenmon rekiho* 中國の天文曆法 (Chinese Astronomy and Calendrical Sciences) [Tokyo: Heibonsha, 1969, revised 1990], pp. 202–209, and "The Influence of Islamic Astronomy in China", *Annals of the New York Academy of Sciences*, Vol. 500 (*From Deferent To Equant: A Volume of Studies in the History of Science in the Ancient and Medieval Near East in honor of E. S. Kennedy, 1987*), pp. 547–59; K. Yamada (山田慶兒), *Jyujireki no michi* 授時曆の道 (The Way to the *Shoushi* Calendar) [Tokyo: Misuzu Shobo, 1980], pp. 58–64; K. Miyajima (宮島一彥), "Genshi tenmonshi kisai no isuramu tenmongiki ni tsu ite 『元史』天文志記載のイスラム天文儀器について (New Indentification of Islamic Astronomical Instruments described in the Yuan Dynastic History)", in *Toyo no Kagaku to Gijutsu* 東洋の科學と技術 (Science and Technology in East Asia) [Kyoto: Dohosha, 1982], pp. 407–27; Chen Jiujin 陳久金, *Hui-hui tianwenxue shi yanjiu* 回回天文學史研究 (A Study on the History of Islamic Astronomy in China) [Nanling: Guangxi Science and Technology Press, 1996], pp. 89–105; Ma Mingda 馬明達 and Chen Jing 陳靜, *Zhongguo hui-hui lifa jicong* 中國回回曆法輯叢 (Collected Texts on Islamic Calendar in China) [Lanzhou: Kansu People's Press, 1996], pp. 1–14.

3 Wang Shidian 王士點 and Shang Qiong 商企翁, *Yuan bishujian zhi* (Hangzhou: Zhejiang guji Press, 1992), pp. 129–31.

4 On Christopher Clavius' life and works, see K. W. Fung (馮錦榮), "Christopher Clavius (1538–1612) and Li Zhizao 李之藻 (1565–1630)", in Robert Halleux ed., *Collection of Studies from the International Academy of the History of Science: the Volume on the Spread of the Scientific Revolution outside Western Europe* (Turnhout, Belgium: Brepols Publishers, forthcoming in late 1999), 23 pp. On the discovery and contents of Ricci's *Portuguese-Chinese Dictionary*, see Pasquale M. D'Elia S. I., "Il rpimo dizionario europeo-cinese e la fonetizzazione italiana del cinese", *Atti del XIX Congresso Internationale degli Orientalisti* (Roma: Tipografia del Senato, 1938), pp. 172–78; Paul Yang Fu-mien S. J. (楊福綿), "Linguistique et Historie: The Catholic Missionary Contribution to the Study of Chinese Dialects", *Orbis*, Tome 9, No. 1 (1960), pp. 158–85.

5 Yang Fu-mien, S. J., "The *Portuguese-Chinese Dictionary* of Matteo Ricci: A Historical and Linguistic Introduction", *Proceedings of 2nd International Conference on Sinology, Section on Linguistics & Paleography* (Taipei: Academia Sinica, 1989), pp. 191–242.

6 *Jianpingyi shuo*, in *Tianxue chuhan* 天學初函 (First Collection of Writings on Learning from Heaven), composed by Li Zhizao in 1628; reprinted in Wu Hsiang-hsiang 吴相湘 ed., *Chung-kuo shih-hsüeh ts'ung-shu* 中國史學叢書 (Collectanea of Chinese Historical Studies) 23 (Taipei: Student Book Co. , 1965), Vol. 5, pp. 2719–70; "*Jianpingyi shuo houji* 簡平儀說後記" (Postscript of *Jianpingyi*

shuo), in *Xu Guangqi zhuyiji* 徐光啟著譯集 (Collected and Translated Works of Xu Guangqi) [Shanghai: Shanghai guji Press, 1983], 1a–1b.

7 *Tianwen lüe*, in *Tianxue chuhan*, pp. 2619–2718, in particular pp. 2671–84.

8 Hu Tiezhu 胡鐵珠, "*Riyuexingguishi* tiyao 日月星晷式提要" (An Introductory Note on *Riyuexingguishi*), in Ren Jiyu 任繼愈 ed., *Zhongguo kexue jishu dianji tonghui tianwen juan* 中國科學技術典籍通匯 ——天文卷 (A General Collection of China's Books and Records on Science and Technology — Astronomy) [Zhengzhou: Henan Education Publishing House of China, 1996], Vol. 8, pp. 383–84.

9 *Ge-zhi cao*, 64a–67b. On Xiong Mingyu's biographical studies, see K. W. Fung, "Mingmo Xiong Mingyu *Gezhi cao* neirong tanxi 明末熊明遇《格致草》內容探析" (A Critical Study of Xiong Mingyu's [1579–1649] *Gezhi cao*), *Ziran kexueshi yanjiu* 自然科學史研究 (*Studies in the History of Natural Sciences*, Beijing), Vol. 16, No. 4 (1997), pp. 304–328.

10 *Zhouyi shilun hebian*, Chapter 7, 1a–30a, especially 15b–19a. On Fang Kongzhao's biography, see K. W. Fung, "Mingmo Qingchu shidafu dui *Chongzhen lishu* zhi yanjiu 明末清初士大夫對《崇禎曆書》之研究" (Scholar-officials' Studies on *Chongzhen lishu* during the Ming-Qing Transition), *Bulletin of Ming-Qing Studies* 明清史集刊 (Hong Kong), Vol. 3 (1997), pp. 145–98.

11 *Lixiang benyao*, 46a–48b. On Li Quangdi and Western Learning, see Han Qi 韓琦, "Junzhu he buyi zhi jian: Li Quangdi zai Kangxi shidai di huodong ji qi dui kexue di yingxiang 君主和布衣之間：李光地在康熙時代的活動及其對科學的影響" (Between the Emperor and Mathematician: Li Guangdi's Activity during the Kangxi Reign and Its Influence on Science), *Tsing Hua Journal of Chinese Studies* 清華學報 (Hsinchu), New Series Vol. 26, No. 4 (1996) [Published in Dec., 1997], pp. 421–45.

12 On Emperor Kang Xi's biography, the definitive work in Western languages are Joachim Bouvet (1656–1730), *Histoire de l'empereur de la Chine* (Paris, 1697; La Haye, 1699, 2nd edition; Tientsin, 1940, reprint); Pierre-Martial Cibot (1727–80), "Observations de physique et d'histoire naturelle faites par l' empereur K'ang-hi (康熙幾暇格物篇)", in *Memoire concernant l'histoire, les sciences, les arts, les moeurs, les usages, ec. des Chinois, par les missionnaires de Pe-kin*, Tome IV (1779), pp. 452–84; Jonathan D. Spence, *Emperor of China: Self-portrait of Kang Xi* (New York: Vintage Books, 1974).

13 Eloise Talcott Hibbert, *Jesuit Adventure in China: During the Reign of Kang Xi* (New York: E. P. Dutton and Company, 1941), pp. 70–77, 142–49; Peng Hsiao-fu, " The Kang Xi emperor's absorption in Western mathematics and astronomy ", *Bulletin of Historical Research*, Vol. 3 (1975), pp. 349–422; Pan Jixing 潘吉星, "Kang Xi Di yu xiyang kexue 康熙帝與西洋科學" (Emperor Kang Xi and Western Science), in *Zhongwai kexue zhi jiaoliu* 中外科學之交流 (Scientific Exchange between China and Foreign Countries) [Hong Kong: The Chinese University Press, 1993], pp. 431–53; Han Qi, "The Role of the French Jesuits in China and the *Académie Royale des Sciences* in the Development of the Seventeenth- and Eighteenth-Centuries European Science", in K. Hashimoto,

C. Jami and L. Skar eds., *East Asian Science: Tradition and Beyond* (Osaka: Kansai University Press, 1995), pp. 489–92; Catherine Jami, "From Louis XIV's Court to Kangxi's Court: An Institutional Analysis of the French Jesuit Mission to China", *East Asian Science: Tradition and Beyond*, pp. 493–99, and "From Clavius to Pardies: The geometry transmitted to China by Jesuits (1607–1723)", in Federico Masini ed., *Western Humanistic Culture Presented to China by Jesuit Missionaries (XVII–XVIII centuries)* [Roma: Institutum Historicum S. I., 1996], pp. 175–99; Chen Cheng-yih 程貞一, "Qingdai zhongxi wenhua jiaoliu chuji Kang Xi Di dui tianwenxue diyingxiang 清代中西文化交流初期康熙帝對天文學的影響" (Emperor Kangxi's Influence on Astronomy during the early stage of Cultural Exchanges in the Qing Dynasty), in Li Di 李迪 ed., *Collection of Papers for the Second International Conference on the History of Science & Technology of the Chinese Minority, Aug., 14–18, 1994, Yanji, China* 第二屆中國少數民族科技史國際學術討論會論文集 (Beijing: Shehui kexue wenxian chubanshe, 1996), pp. 1–12. On Ferdinand Verbiest's biography and his contribution to China, see Louis Van Hee, S. J., *Ferdinand Verbiest, E'crivain Chinois (1623–1688)* (Bruges: L. De Plancke, 1913); H. Josson, S. I. et L. Willaert, S. I. ed., *Correspondance de Ferdinand Verbiest* (Bruxelles: Palais des Acadèmies, 1938); Ulrich Libbrecht en Willy vande Walle, *Belgae in China* (Leuven: Ferdinand Verbiest-onderzoesproject, 1985), pp. 38–60; Fu Jen Catholic University, R. O. C. ed. , *Conference Papers on International Conference in honour of Ferdinand Verbiest Commemoration of Tricentenary of his death (1688–1988)* 南懷仁逝世三百週年國際學術討論回論文集 [Taipei: Fu Jen Catholic University, 1987]; Noël Golvers-Ulrich Libbrecht, *Astronoom van de Keizer, Ferdinand Verbiest en zijn Europese Sterrenkunde* (Leuven: Davidsfonds, 1988); Noël Golvers, *The Astronomia Europaea of Ferdinand Verbiest, S. J. (Dillingen, 1687): Text, Translation, Notes and Commentaries* (Steyler Verlag and Leuven, 1993); Roman Augustine, S. V. D. ed., *Ferdinand Verbiest (1623–1688): Jesuit Missionary, Scientist, Engineer and Diplomat* (Steyler Verlag and Leuven, 1994). On Antoine Thomas's biography and his connection with Chinese intellectuals, see Yves de Thomaz de Bossierre, *Un Belge Mandarin a la cour de Chine aus XVIIe et XVIIIe Siècles, Antoine Thomas 1644–1709, Ngan To P'ing-che* 安多平施 (Paris: Les Belles Lettres Cathasia, 1977); K. W. Fung "Min matsu shin sho ni okeru Ko Hyaaka no shogai to chosaku 明末清初における黃百家の生涯と著作" (Huang Baijia [1643–1709]: His Life and Works), *Chugoku Shisoshi Kenkyu* 中國思想史研究 (Kyoto) No. 20 (1997), pp. 61–92, esp. pp. 80–87. On Jean-François Gerbillon's biography, see Isabelle Landry-Deron, *Les leçons de sciences occidentales de l'empereur de Chine Kangxi (1662–1722): Texes des Journal des Pères Bouvet et Gerbillon* (Paris: E. H. E. S. S., 1995).

[14] *Shengzu renhuangdi shilu* 聖祖仁皇帝實綠 (The Veritable Records of the Kang Xi's Reign), in *Qing shilu* 清實綠 (The Veritable Records of the Qing Dynasty) Vol. 4 [Beijing: Zhonghua shuju, 1985], p. 391. For the political dispute on divination of hemerology, see Huang Yi-long 黃一農, "Court

Divination and Christianity in the Kang Xi Era", *Chinese Science*, Vol. 10 (1991), pp. 1–20.

[15] Liu Lu 劉潞 ed., *Qinggong xiyang yiqi* 清宮西洋儀器 (Scientific and Technical Instruments of the Qing Dynasty) [Hong Kong: The Commercial Press, 1998], pp. 8–9; Noël Golvers, "Ferdinand Verbiest, S. J. (1623–88) and the Astronomical Bureau in Beijing", *Review of Culture*, No. 21 (1994), pp. 201–212; Liu Baojian 劉寶建, "Chuanjiaoshi yu Qinggong yiqi zhizao 傳教士與清宮儀器製造" (Missionaries and Instrument-making in the Qing Palace), in Xie Fang (謝方) ed., *Zhongxi chushi* 中西初識 (The First Stage of Encounter Between China and the West) [Zhengzhou: Daxiang chubanshe, 1999], pp. 96–112.

[16] *Shengzu renhuangdi shilu*, p. 406. On the making of astronomical instruments in the Ancient Observatory in Peking, see Chen Zungui 陳遵媯, *Qingchao tianwen yiqi jieshuo* 清朝天文儀器解説 (Explanations on Astronomical Instruments of Qing Dynasty) [Beijing, 1956], pp. 5–35; Yi Shitong (伊世同) and Jerome Heyndrickx, *The Verbiest Celestial Globe in China and Europe* (Leuven: China-Europe Institute, 1989), pp. 6–10.

[17] On Verbiest's cartographical studies, see K. W. Fung, "Li Rui di shengping ji qi *Guanmiao-ju ri ji* 李鋭 (1769–1817) 的生平及其《觀妙居日記》" (Confucian-Mathematician Scholar Li Rui's [1769–1817] Life and his Unpublished Diary *Guanmiao-ju ri ji*), *Wen Shi* 文史 (Beijing), No. 47 (1999), pp. 207–220, in particular p. 208.

[18] Ferdinand Verbiest, *Chido nanbei liang zong xin tujie* 赤道南北兩總星圖解 (preserved in Bibliothèque Nationale, Paris), 1a.

[19] *Shengzu renhuangdi shilu*, p. 600; Nicole Halsberghe, *Xin zhi lingtai yi xiang zhi* 新製靈台儀象志: *Vertoog over de nieuwgebouwde instrumenten op het observatorium, Ferdinand Verbiest, Beijing, 1674* (Unpublished PhD dissertation, Katholieke Universiteit Leuven, 1992) and "Sources and Interpretation of Chapters One to Four in Ferdinand Verbiest's *Xin zhi lingtai yi xiang zhi* ", *Review of Culture*, No. 21 (1994), pp. 213–34.

[20] Ferdinand Verbiest, *Jianpinggui zongxing tujie*, 1b.

[21] Liu Lu ed., *Qinggong xiyang yiqi*, pp. 45–46.

[22] Owen Gingerich, "Apianus' *Astronomicum caesareum* and its Leipzig facsimile", *Journal of the History of Astronomy*, Vol. 2 (1971), pp. 168–77; idem, "Astronomical Paper Instruments with Moving Parts", R. G. W. Anderson, J. A. Bennett and W. F. Ryan ed., *Making Instruments Court: Essays on Historical Scientific Instruments Presented to Gerard L'Estrange Turner* (Aldershot, Hampshire: Variorum, 1993), pp. 63–74.

[23] R. H. van Gent, *The Portable Universe: Two Astrolabes of the Museum Boerhaave* (Leiden: Museum Boerhaave, 1994), pp. 20–28.

[24] Roderick and Marjorie Webster, *Western Astrolabes* (Chicago: Adler Planetarium & Astronomy Museum, 1998), pp. 102–103.

[25] Dong Gao 董誥, *Huang chao li qi tu shi* 皇朝禮器圖式 (Illustrated Description of Sacrificial Vessels, Official Robes and Insignia, Musical Instruments and

Astronomical Apparatus used during the Qing Dynasty, 1759–66) [*Si ku quan shu* 四庫全書 edition], Chapter 3, 16a–17b.

26   Henri Cordier (1849–1925), *Bibliographie des ouvrages publiés en Chine par les Européens au XVIIe et au XVIIIe siècle* (Paris : Ernest Leroux, 1883–1901), p. 61. But this work has not yet been located until today.

27   Du Halde, Jean Baptiste (1674–1743), *A description of the empire of China and Chinese-Tartary, together with the kingdoms of Korea and Tibet* (London : Printed by T. Gardner for Edward Cave, 1738–41), Vol. 2, pp. 268 and 270.

PART VII

Technology/Technique

A Study of the Lacquerish Patina (Qi Gu) on Ancient Bronze Mirrors

ZHOU ZHONGFU, SUN SHUYUN, HAN RUBIN & T. KO

Institute of History of Metallurgy and Materials
University of Science and Technology, Beijing

Introduction

Bronze mirrors are fairly common burial artefacts and many have been unearthed from archaeological sites in China, Japan and Europe. Chinese bronze mirrors are special and interesting in their own right however, and warrant further research. Many scholars are interested in Chinese mirrors not only because of their decoration, style and sophisticated casting technique, but also because of the glossy jade-like and non-corroding surface layer present on some mirrors, known as "qi gu" or "Lacquerish antique patina" by archaeologists and antique collectors.

Up to now the earliest excavated bronze mirrors is from the Qijia Culture in Gansu province. Early bronze mirrors are simple in form and have unstable compositions. All excavated pre-Warring States bronze mirrors have low tin contents and are not "qi gu" mirrors.

The casting technique of bronze mirrors from the Warring States to Song Dynasty (c. 4th century B.C. to 12th A.D.), however, is very sophisticated and the mirrors tend to have stable compositions. Most mirrors from this period are composed of 70% copper, 25% tin and 5% lead. Their casting structure is α phase and ($\alpha + \delta$) eutectoid. Due to their high tin content "qi gu" mirrors when first cast are silvery-white in colour and resemble silver. "Qi gu" mirrors have been found not only from Chinese sites, but also Roman and Japanese sites as well. Due to the formation of a "qi gu" surface layer during burial, the excavated mirrors were found to have resisted corrosion and were well preserved.

"Qi gu" mirrors have been researched by Western scholars such as Collins and Gettens for over sixty years, and the black lacquerish surface of "qi gu" mirrors has been the subject of interest and study by both Western and Chinese scholars for the past two decades. Study of the nature and the formation of "qi gu" attracted many scholars but until recently no convincing results have been obtained.

Investigation of Bronze Mirrors

Since 1993 over 600 mirrors unearthed from various parts of China have been studied, and the natural environments in which the mirrors were found have been investigated. From our investigation, we suggest that the "qi gu" surface layers on bronze mirrors are the result of natural corrosion.

The following are some important observations:

(1) The varying colours of the lacquerish patina layer. "Qi gu" colours range from the commonly seen black and green to gray, yellow, and blue as well as blackish-green, grayish-green and grayish-yellow. Sometimes two or more colours appear on the same mirror.

(2) The cross-sectional surface of some mirrors was covered with the "qi gu" layer, and the colour and the brightness of the "qi gu" surface layer is very close to the surface of the front and back of the mirror.

(3) A special mirror unearthed in Nanyang, Henan province was found with a small pottery circular box adhering to part of its reflecting surface. It was found that the area which had been in contact with the pottery box still retained its original metallic shiny surface, whereas the area which was exposed had developed a black "qi gu" layer.

(4) The majority of "qi gu" mirrors have been unearthed in South China, with just a few found in Central, North and Northwest China. For example, more than 90% high tin/bronze mirrors unearthed in Hunan and Hubei provinces have "qi gu" surfaces, but those found in the Yellow River basin are usually quite badly corroded. The corrosion is less so in mirrors found in the Northwest (Xinjiang and Gansu provinces) where the original metallic colour and shiny surface is usually still preserved. The degree of corrosion appears to be affected by the humidity, type of soil and climate, all of which varies from region to region. This suggests, therefore, that the formation of "qi gu" is the result of natural corrosion and is directly related to not only the environment and climate but also burial conditions.

Examination and Experiments

Optical microscopes, SEM, TEM, XRD, XPS examination and electrochemical and corrosion experiments, have shown that a transparent or translucent layer (about 1–100 µm thick) is present on ancient Chinese mirrors with "qi gu" surfaces, which gives them their shiny or jade-like gloss. Under the surface layer, there is a corroded and partly corroded layer on the metallic core. The transparent or translucent layer is mainly composed of transparent amorphous or nanocrystalline stannic oxide, with a small amount of copper and lead compound and a few trace elements of aluminium, silicon, sulfur, iron, etc. From these observations and examinations it can be suggested that the colour of "qi gu" mirrors is due to the transparency and colour of the surface layer, and the composition colouring of the "corrosion layer" next to the surface layer.

Examination results show that during burial, copper ions and compounds from the mirror surface transfer in solution out into the surrounding soil leaving a tin-rich layer on the surface of the mirror. The copper-rich α phase is more easily corroded than the tin-rich δ phase in the soil conditions where "qi gu" mirrors are unearthed. These phenomena can be illustrated by electrochemical experiments and theories.

Conclusion and Discussion

From archaeological and experimental evidence the following mechanism for the formation of the "qi gu" layer is proposed: When bronze mirrors are buried they corrode and copper is oxidized to cuprous or cupric ions or compounds by electrochemical corrosion. Combined with humic or fulvic acid present in soil or funerary objects, the cupric ions or compounds are then transferred into the surrounding soil. Tin is oxidized to colloidal stannic hydroxide which covers the entire mirror surface through diffusion or by tin transfer. Upon dehydration, this layer turns to noncrystalline stannic oxide. Some of this then recrystallizes to nanocrystalline stannic oxide. Both nanocrystalline stannic oxide and noncrystalline stannic oxide are excellent at preventing corrosion.

Two new interesting phenomena were found during the investigation and research:

Pseudo-photography was observed from the surface of "qi gu" mirrors using optical-microscopy and EMS scattered electric images. This clearly shows that the "qi gu" layer is formed on the original casting structure and not on the treated surface [for example, a mercury-tin polishing compound (Xi Gong Qi)].

It is well known that there are many crackles on the surface of "qi gu" mirrors. Most scholars think that these crackles were formed during burial. But when a "qi gu" mirror was excavated in E-zhou, Hubei province in 1997, it is found that crackles formed even after mirrors had been excavated. After the anti-corroded "qi gu" layer cracked, the corrosion continued and became quite severe on many mirrors. This raises yet another important question — how to best protect and preserve "qi gu" mirrors under some special conditions.

Acknowledgements

This paper reports part of a doctoral dissertation supervised by Professors T. Ko, Hong Yanruo and Han Rubin. They give me a lot of direction and help.

References

1. 夏鼐, "碳—14测定年代和中国史前考古家", 《考古》, 1977年. 第4期, 第217页.
2. 李虎候, "齐家文化铜镜的非破坏鉴定", 《考古》, 1980年. 第4期, 第365–68页.
3. Yao Chuan and Wang Kuang, "A Study of the Black Corrosion-Resistant Surface Layer of Ancient Chinese Bronze Mirrors and its Formation", *Corrosion Australasia*, 10 (1987): 5–11.
4. 张庶元、谭舜, "古铜镜表层组织结构成分的AEM研究", 全国第一次实验室考古学术讨论会论文, 1988年5月4日–7日, 南宁.
5. Masumi Chikashige, "The Composition of Ancient Eastern Bronzes", *Journal of The Chemical Society*, 117, 2 (1920): 917–22.
6. Collins, W. F., "The Mirror-black and Quicksilver Patinas of Certain Chinese Bronzes", *Journal of the Royal Anthropological Institute*, LXIV (1934): 69–79.
7. Barnard N., "Bronze Casting and Bronze Alloys in Ancient China", Monumenta Serica Monograph, XIV, Published jointly by the Australian National University and Monumenta Serica, Canberra, 1961.
8. Massaaki Sawada, "Non-destructive X-ray Fluorescence Analysis of Ancient Bronze Mirrors Excavated in Japan", *Ars Orientalis*, 11 (1979): 195–213.
9. R. J. Gettens, "Some Observation Concerning the Lustrous Surface on Ancient Eastern Bronze Mirrors", *Technical Studies in the Field of Fine Arts*, 3 (1934): 29–37.
10. 徐力、王昌燧、王胜君、吴自勤, "汉镜组织和成分的研究", 《电子显微报》, 1987年4期, 29–34页.
11. W. T. Chase, U. M. Franklin, "Early Chinese Black Mirrors and Pattern-Etched Weapons", *Ars Orientalis*, XI (1979): 215–41.
12. 小松茂、山内淑人, "古镜の化学的研究", 《东方学报》, 第八册.
13. Collins, W. F., "The Corrosion of Early Chinese Bronzes", *Journal of the Institute of Metals*, 45 (1931): 23–55.
14. 孙淑云, N. F. Nennon, "中国古代铜镜显微组织的研究", 《自然科学史研究》, 11卷, 第1期, 1992年, 第54–67页.

15. R. J. Gettens, "The Freer Chinese Bronzes in the Smithsonian Institute, Freer Gallery of Art", *Oriental Studies*, Technical Studies, Washington, II, 7 (1969): 121–39.

16. R. J. Gettens, "Tin-oxide Patina of Ancient High-Tin Bronze", *Bulletin of the Fogg Museum of Art*, 11, 1 (1949): 19–23.

17. 孙淑云、马肇曾、金莲姬、韩汝玢、柯俊，"土壤中腐殖酸对铜镜表面'黑漆古'形成的影响"，《文物》，1992年，第12期，79–89页.

Chinese Influence on the Technological Development of Yayoi Period Japan: Problems of Metal Casting and the Production of Bronze Mirrors

SERGEY LAPTEFF

A. M. Gorky Literary Institute
Moscow

Metal casting came to Japan much later than to neighboring China and Korea. Indeed, metal in China and Korea was being used and molded for more than three millennia while the Stone Age was continuing in Japan [Mei, pp. 234–37; Ri, p. 13]. A huge jump in the technological development of Japan suddenly occurred in the 3rd century B.C. (the Yayoi period), when bronze and almost simultaneously iron started to be produced [Seitetsu, p. 15; Sekino, pp. 81–82]. This process was no doubt influenced from the continent, though the historical problem here is much more complex than just the borrowing of ready-made technologies. To see the results of this process one should look thoroughly at the metal artifacts unearthed in the Japanese archipelago.

It is interesting to note that the most frequently found metal relics in Japan are bronze mirrors. According to calculations made by Prof. Miki Tarô from Komazawa University, approximately 3,500 bronze mirrors have been unearthed in Japan up until 1991, 2,500 of them of Japanese production [Miki, p. 41]. Such a great quantity of metal relics can be compared only with arrowheads and spearheads. Of all these mirrors, only a little more than 300 (figures for 1982) are found in the Yayoi burials, the rest being from tumulus (Kofun) [Iki, pp. 36–37; Okuno, p. 141; Laptev, p. 174]. When one takes into consideration that the time of burial of the mirrors may differ from the time of their production — sometimes up to 300 years — [Kim, p. 100], some part of the Kofun finds can also be attributed to the Yayoi period.

If we look at the structure of mirrors found in the Yayoi burials, it can be seen that the majority were imported from China (76–82% according to different estimations) [Iki, pp. 36–37; Okuno, p. 141; Laptev, p. 174]. Less than one quarter were made by Japanese craftsmen. If we focus our attention to the mirrors found in tumulus (3–6 centuries A.D.), we can see that the situation is the opposite — three quarters of all mirrors are made in Japan (so called bôseikyô, or "imitative mirrors") [Vorob'yov, p. 86].

Table 1   Bronze Mirrors from the Yayoi Burials

Region	Quantity of Unearthed Mirrors	% Towards the Total Quantity of Mirrors, found in Japan
Kyushu	301–319*	93–93, 5
Honshu and Shikoku	21–24*	6, 5–7

Note: *According to different estimations [Iki, pp. 36–37; Okuno, p. 141].

Speaking about mirrors cast in Japan from a technological point of view, it is very important to note their generally high technological level, which makes it difficult to distinguishing their place of production as either Japan or China. That is why even the origin of some big groups of mirrors is disputed [Matsuura, pp. 183–85; Miki, pp. 40–56; Okuno, pp. 136–43; Han, p. 3; Laptev, p. 172; Wang 1982, pp. 630–39; Wang 1981, pp. 346–58; Wang 1984, pp. 468–79].

An illustration of the technical similarity of Japanese and Chinese mirrors is that the main method of determining the mirror's place of production is chemical analysis, and not the examination of the perfection of the mirror's exterior. This is the main reason why bronze mirrors made in Japan deserve the name "imitative mirrors". The chemical composition of Japanese bronze mirrors is also almost the same as of Chinese ones, with only a lower content of tin. Sometimes the lead in Japanese mirrors is thought to be of Chinese origin. On the other hand, if we look at other metal articles found in Japan, for instance bronze arrowheads, we can ascertain that here the situation is quite opposite to that of bronze mirrors. Though 58 different forms of bronze arrowheads have been found in Japan, none of them have an analogue in China or Korea [Vorob'yov, p. 63]. The chemical composition of Chinese and Japanese arrowheads is as follows:

Table 2   Chemical Composition of Japanese and Chinese Arrowheads

Chemical Elements	Japanese Bronze Arrowheads	Chinese Bronze Arrowheads
Copper	96, 48	74, 47
Tin	2, 46	10, 82
Antimony	no	8, 00
Nickel	no	3, 68
Iron	no	3, 03
Arsenic	0, 50	present
Lead	0, 56	no

Note: Vorob'yov, p. 63.

We can see that Japanese arrowheads are cast of almost pure copper, which makes them rather soft and lowers their quality, compared to the Chinese ones. In the making of arrowheads the Japanese were quite independent and did not follow Chinese models. In this case the Japanese approach to technological evolution was gradual, from a very initial step, not as it was with bronze mirrors.

The next question we need to touch on in this paper is how it became possible to make such perfect copies of bronze mirrors in Japan, without any gradual evolution. At first, as is generally known, the Yayoi period was a time of large scale migrations of populations from China and Korea to Japan. According to the estimations of Japanese scholars, the growth of population in Japan during Yayoi and the next, Kofun periods exceeded natural growth by 4.27 times [Iki, p. 13]. Also, from the craniological data researched at Sapporo Medical College [Dodo, pp. 420–21], including dental and some other characteristics, it is clear, that the anthropological type of the Japanese changed greatly during the Yayoi period [Oyamada, pp. 83–99; Hanihara, pp. 399–409; Suzuki, pp. 171–82]. And the latest genetic research leaves us no doubt about the presence of Han ethnic elements in this process (I mean the comparison of genetic configurations from human bone remains, unearthed in Linzi and Takuta-Nishibun) [Yayoijin, p. 3].

If we look on the geography of bronze mirrors during the Yayoi and Kofun periods, we can see the following. According to Table 1 (see above), the majority of the mirrors found in Japan are excavated from Kyushu (93–93.5%). Inside Kyushu island the earliest and largest findings were made in Fukuoka prefecture, in the nortern part of Kyushu, the part closest to the continent and Korea, which was the bridge between Japan and China.

Table 3   Dispersal of Bronze Mirrors in the Yayoi Burials of Kyushu Region

Prefecture	Quantity of Unearthed Mirrors
Fukuoka	236 (242) *8
Saga	35
Nagasaki	32
Ôita	7*
Kumamoto	9
Miyazaki	0
Kagoshima	0

Note: *Figures differ according to different sources [Iki, pp. 36–37; Okuno, p. 141; Laptev, p. 174].

On the contrary, in the Kofun period the situation changes. The main district of mirror findings moves to the Kansai district in Honshu:

Table 4   Quantity of Bronze Mirrors from Tumulus by Region

Region	Quantity of Unearthed Mirrors
Kyushu	46 (45)
Shikoku	25 (23)
Chubu	77 (68)
Kansai	250 (196)
Tôkai	72 (57)
Tôsan	40 (35)
Kantô	94 (84)
Hokuriku	16 (15)
Tôhoku	3

Note: ( ) mirrors remaining unbroken [Laptev, p. 174].

If we examine some inscriptions on the back sides of the mirrors, we can find evidence showing that they were made in Japan probably by Chinese emigrants. Such a number of mirrors with triangular edges and images of spirits and animals, proved by chemical and ornamental analysis have been made in Japan, contain names of Chinese masters (Chen, Wang) [Tokyo, pp. 28–29, 128, 220, 222–23; Umehara, p. 139, Table XXXIV; Wang 1981, p. 354]. An interesting bit of information we find in the inscription on a mirror with triangular edges and images of spirits and animals, and unearthed from Chausuyama tumulus (Osaka prefecture), reads:

> I made the bright mirror really very well. While sailing from the Land under Heaven (China) to the four seas, I brought the used blue copper to the land to the east of the sea.[1]

Indeed, chemical analysis of Japanese bronzes shows us sometimes that the artifacts were made by re-smelting [Vorob'yov, p. 64; Tanabe, p. 43; Yanagida, pp. 21–36].

The analysis of bronze mirrors, unearthed in Japan shows us a kind of progress in mirror making in Japan. Begun from a high technical level, it starts directly from full copying. Its earliest examples are found at the end of Middle Yayoi (1st century A.D.) [Yokoyama, pp. 44–48]. Only from the end of the 2nd century A.D. can we find examples of the creative development of Chinese tradition by emigrant masters in Japan, such as mirrors with triangular edges and images of spirits and animals, mirrors with "T"-shaped patterns, also a bit different from Chinese [Matsuura, pp. 236–40]. Later in the Kofun period we find original Japanese designs and inscriptions (5–6 centuries A.D.), such as famous mirrors with animals and people from Hachiman temple in Suda (Wakayama prefecture) [Zusetsu, p. 13].

Summing up the results, we can underline the following:

1.   Despite significant changes in ornamentation of the back sides of mirrors, their casting technology did not show any visible progress. Due to the direct participation of Chinese masters, the technological level of Chinese originals was soon reached, but no further improvement can be observed.
2.   The peculiarities of casting and molding processes can be shown not only by chemical analysis and technological reconstruction, but also by reading of mirror inscriptions. Inscriptions help us to specify the exact place of production and personify the nationality of the producers and inscribers.
3.   The decisive factor in the dispersion and quick adoption of the bronze mirrors in Japan was large migrational processes which influenced Japanese culture very intensively, making the continental culture an organic part of the national one.
4.   The process of adoption of bronze mirrors in Japan is not a typical one compared with the development of other metal relics, but it shows more clearly the role of continental culture in the Yayoi period's huge technological jump.

Note

¹   For Chinese characters for this inscription see Wang 1981, p. 354.

References

In Japanese

Iki Ichirō. *Nitchū kodai kōryū-wo saguru* (Research of Japanese-Chinese Relations in Antiquity). Fukuoka, 1989.
Kim Sok Hyohg. *Kodai Chō-Nichi kankeishi. Yamato seiken-to Mimana* (History of Korean-Japanese Relations in Antiquity. Political Power of Yamato and Imna), translated by the Korean Scientific Historical Society, Tokyo, 1970.

Matsuura Yūicirō. "Nippon shuddo-no hōkaku "T"-ji kyō" (Mirrors with Rectangular and "T"-shaped Pattern, Unearthed in Japan), *Tokyō kokuritsu hakubutsukan kiyō* (Transactions of Tokyo State Museum), vol. 29 (1993): 177–254.

Miki Tarō. "Gi-no kinenkyō-ni kansuru shiron" (Experimental Discussion about Mirrors, containing Wei Kingdom Dates), *Nippon Rekishi* (Japanese History), no. 157 (June 1991): 40–56.

Motomura Takeaki. "Kofun jidai-no kiso kenkyūkō. Shiryōhen 1, 2" (A Draft of the Fundamental Research of the Kofun Period. Historical Materials, vols. 1, 2), *Tokyō kokuritsu hakubutsukan kiyō*, vol. 16 (1980): 9–197; vol. 26 (1990): 9–282.

Okuno Masao. "Sankakuen shinjiūkyō-wa kokusan data" (Mirrors with Triangular Edges and Images of Spirits and Animals were Made in Japan), *Rekishi Yomihon* (Historical Almanac), vol. 27, no. 5 (April 1982): 136–43.

Oyamada Jōichi. "Sei-Hoku Kyūshū yayoijin-to Kitabu Kyūshū yayoijin-no sikan saizu-ni tsuite" (About the Size of Tooth Crowns of North-West Kyushū Yayoi people and North Kyushū Yayoi People), *Jinruigaku zasshi* (Journal of the Anthropological Society of Nippon), vol. 100, no. 1 (March 1992): 83–99.

Seitetsu Yayoi jidai-kara. "Hiroshima-ken-no iseki. Teisetsu sambyaku nen sakanoboru. Torai bunka-no takasa urazuke" (Iron production beginning in the Yayoi Period. Relics from Hiroshima prefecture. Common Point of View is now Dated 300 years Earlier. Proof of Immigrants' High Level Culture), *Hokkaido Shimbun*, Sapporo, evening edition, no. 18825 (12 January 1995): 15.

Sekino Takeshi. "Chūgoku-no kōkogaku. Nippon kobunka-to-no kanrensei" (Chinese archaeology. The relationship with Japanese Ancient Cultures), *Nihon rekishi*, no. 161 (November 1961): 78–85.

Tanabe Shōzō. "Yamatai koku-no mondai" (Problem of the Location of the State of Yamatai), *Kikan kōkōgaku* (Archaeological Quarterly), no. 54 (February 1996): 40–43.

Tokyō kokuritsu hakubutsukan (Tokyo State museum). *Tokubetsu ten. Nippon-no kōkogaku — sono ayumi-to seika* (Special Exhibition. Japanese Archaeology — its steps and achievements), 4 October–13 November 1988.

Umehara Sueji. *Ō-bei-ni okeru Sina kokyō* (Ancient Chinese Mirrors in Europe and America). Tokyo, 1931.

Yanagida Yasuo. "Namari dōitai hihō-ni yoru seidōki kenkyu-e-no kitai" (Calculations for the Research of Bronze Articles by the Method of Comparing their Lead Content), *Kōkōgaku zasshi* (Archaeological Journal), vol. 75, no. 4 (1990): 21–36.

"Yayoijin-no tairiku toraisetsu 'hokyō'" (The "proof" of Yayoi People's Migration from the Continent Theory), *Chūnichi shimbun*, 19 October 1996, Nagoya Evening Edition; sheet 11, p. 3.

Yokoyama Kunitsugu. "Fukuoka-ken Yoshitake-Takagi iseki" (Relics from Yoshitake-Takagi, Fukuoka prefecture), *Kikan kokogaku*, no. 51 (May 1995): 44–48.

Zusetsu shiryō Nippon-shi (Japanese History through Tables and Other Materials). Nagoya, 1989.

In Chinese

Han Guohe. "Nailiang xin chu sanjiao yuan shen-shou jing de lunzheng" (Dispute over Recently-Found Nara Mirrors with Triangular Edges and Images of Spirits and Animals), *Wenwu bao* (7 August 1998): 3.

Laptev S. "Liang-Han San-guo shidai qingtong jing ji qi dui Riben de yingxiang de tujing he yiyi" (Bronze mirrors of Han and Three Kingdoms Periods and their Influence on Japan — the roots and meanings of this influence), in *Gu tianwen yu Zhonghua chuantong wenhua guoji yantaohui huiyi lunwen ji* (Proceedings of the International symposium on Ancient Astronomy and Chinese Traditional Culture). Nanjing, 1998, pp. 170–77.

Wang Shilun. "Tantan wo guo gudai de tongjing" (About Bronze Mirrors in China), *Kaogu tongxun* (Archaeological Reports), no. 6 (1955): 56–63, illustration 16.

Wang Zhongshu. "Guanyu Riben de sanjiao yuan fo-shou jing. Da Xitian Shoufu xiansheng" (About Japanese Mirrors with Triangular Edges and Images of Buddhas and Animals. Answer to Mr. Nishida Morio), *Kaogu* (Archaeology), no. 6 (1982): 630–39, illustrations 11, 12.

Wang Zhongshu. "Guanyu Riben sanjiao yuan shen-shou jing de wenti" (About the Problem of Japanese Mirrors with Triangular Edges and Images of Spirits and Animals), *Kaogu*, no. 4 (1981): 346–58, illustrations 11, 12.

——————— . "Riben sanjiao yuan shen-shou jing de zunlun" (Generalization of the Problem of Japanese Mirrors with Triangular Edges and Images of Spirits and Animals), *Kaogu*, no. 5 (1984): 468–79, illustrations 7, 8.

In English

Dodo Yukio and Ishida Hajime. "Consistency of Nonmetric Cranial Trait Expression during the Last 2,000 Years in the Habitants of the Central Islands of Japan", *Jinruigaku zassi*, vol. 100, no. 4 (October 1992): 417–23.

Hanihara Tsunehiko. "Dentition of Nansei Islanders and Peopling of the Japanese Archipelago — the Basic Populations in East Asia, IX", *Jinruigaku zassi*, vol. 99, no. 4 (October 1991): 399–409.

Mei Jianjun and Ko Tsun (Ke Jun). *A Comparison of Ancient Metallurgy in India and China, in East Asian Science: Tradition and Beyond.* Osaka, 1995, pp. 233–41.

Ri Sun Jin. *Relics of Tangun's Korea, Democratic People's Republic of Korea*, no. 1 (465) (Pyongyang, 1995): 13.

Suzuki Akira and Takahama Yasuhide. "Tooth Crown Affinities among Five Populations from Akita, Tsushima, Tanegashima, Okinawa in Japan and Middle Taiwan", *Jinruigaku zassi*, vol. 100, no. 2 (April 1992): 171–82.

In Russian

Vorob'yov M. V. *Drevnyaya Yaponiya. Istoriko-arheologicheskiy ocherk* (Ancient Japan. Historical & archaeological outline). Moscow, 1958.

On Liu Xin's Metrological Theory*

GUAN ZENGJIAN

History and Culture College, Zhengzhou University

FU GUIMEI

Medicine Society of Henan Province

Metrology is the technical guarantee of maintaining the normal running of the state machinery. It is also a basis for a country to develop its science and technology. In the process of its development, ancient Chinese society accumulated a wealth of metrological theories. Among them, Liu Xin's theory played an important role. It exerted some influence upon the development of ancient Chinese science.

Liu Xin's metrological theory was chiefly recorded in the book *Records of Temperament and Calendar, History of the Han Dynasty* 《汉书 • 律历志》. During the Yuanshi years (A.D. 1–5) of the West Han Dynasty, Wang Mang 王莽 seized the chief power in the country. In order to make a show of his domination, he called up more than 100 scholars who were familiar with temperament, calendar, weights, and measures to make a large-scale reform of temperament and metrology. Liu Xin was appointed to take charge of the project. On finishing the work, Liu Xin submitted a detailed report to Wang Mang, which described their results. His metrological theory existed chiefly in this report.

Liu Xin is a controversial figure in Chinese history. Though he was a member of the imperial clan of the West Han Dynasty, he helped Wang Mang seize state power and was appointed as Guoshi 国师 (one of the four highest officials of Wang Mang's Xin Dynasty 王莽新朝) after Wang became emperor. When Wang Mang was in danger of falling, Liu Xin wanted to overturn him, but in the end took his own life because Wang Mang discovered his intention. Later people often criticize Liu Xin's personality. But his metrological theory won the approval of many scholars. For example, the famous historian Ban Gu 班固 took his theory as the most detailed and worthy one in this field. When writing the book *History of the Han Dynasty,* he did not neglect it. He recorded it in the chapter "Records of Temperament and Calendar". All discussion about Liu Xin's metrological theory in this paper depend upon that chapter.

Number and Its Function in Metrology

Liu Xin paid great attention to the function of numbers. He wrote:

> What is a number? A number is one, ten, hundred, thousand, ten thousand, etc. Its uses are to count things, to explain the mystery of affairs, and to indicate the inner law of things. 数者, 一、十、百、千、万也, 所以算数事物, 顺性命之理也。

Liu Xin thought numbering was the basis for governing a country because it made the counting of things and affairs possible. He used an ancient book *Yi Shu* 逸书 to support his opinion. The book emphasized that if a monarch wants to control his country well, he must establish arithmetic at first so as to count things of all kinds. Liu Xin's quoting showed that he agreed with *Yi Shu*'s opinion and knew very clearly the importance of metrology to running a country. He thought that metrology was a technical guarantee of maintaining the normal running of the state. Of course his view is right.

Liu Xin had further discussion about the relation between metrology and numbers. He knew that the purpose of metrology was to quantitatively indicate properties of things. Thus numerals are necessary. He said:

> If you want to calculate calendars and temperaments, to design instruments, to deal with cycles and squares, to weigh something in a balance, to measure lengths and volumes, to explore the inner mystery of affairs, you must use numerals. 数者, …夫推历生律制器, 规圆矩方, 权重衡平, 准绳嘉量, 探赜索隐, 钩深致远, 莫不用焉。

His saying certainly is right even when we look at modern science.

But Liu Xin went further into numerical mysticism. He thought if one wanted to indicate quantitative relationships among all kinds of different affairs, it was enough for him to use the sum of 177147 numerals. He said:

> The origin of the numeral is the Huangzhong 黄钟 temperament. The calculation of it begins from 1, then multiplied by 3 again and again till it experiences times related to the Twelve-branches. The result of the calculation is 177147. It can cover all changes of all things. 本起于黄钟之数, 始于一而三之, 三三积之, 历十二辰之数, 十有七万七千一百四十七, 而五数备矣。

If we express his idea in a formula, it is as follow:

zi	chou	yin	mao	chen	si	wu	wei	shen	you	xu	hai
子	丑	寅	卯	辰	巳	午	未	申	酉	戌	亥

$$1 \times 3 \times 3 \times 3 \times 3 \times 3 \times 3 \times 3 \times 3 \times 3 \times 3 \times 3 = 177147$$

Such an idea was consistent with popular philosophic beliefs. At that time, people thought the universe began from Chaos. They called it Yuanqi 元气. Since Yuanqi was the origin of the universe, people called it ONE. They thought elements of the heaven, the earth, and human beings came into existence from Yuanqi. That explains "Yuanqi combines 3 as 1" 元气合三为一. Since all things began with the 3 elements, they had a saying that "3 gives birth to everything". That is the reason numeral 3 is used in the formula. The relation to the Twelve-branches 十二地支 came out of the belief that the Twelve-branches was related to the Twelve Temperaments which can reflect all changes of all things. That is the philosophic background of the formula. Liu Xin was not the first man who raised the matter of the numeral. For example, the numeral 177147 can also be found in *Huainan Zi* 淮南子 that is earlier than Liu Xin's work. What Liu Xin did was to put it into his metrological theory.

Liu Xin's theory set the abstract concept of number in a very high position. His stress was helpful for people in explaining the independence of mathematics and the relations among mathematics and other subjects. It represented great progress in the development of metrological theory in China. But his explanation of the numeral 177147 had no lasting value to the subject of metrology. So it was given up in later dynasties.

The Nature of Temperaments and the Inherent Law of their Formation

Temperaments take on a large role in Liu Xin's metrological theory. Why is this so? Because music had special status in traditional Chinese culture. Rites and music were both emphasized in ancient China. For example, Confucius had said:

> There is nothing better than rite for easing the sovereign and managing people. There is also nothing better than music for changing the old customs and habits. Both of them relate to each other and spread together. 安上治民，莫善于礼；移风易俗月，莫善于乐。二者相与并行。(Ban Gu, *Records of Classical Books, History of the Han Dynasty*, 班固,《汉书·艺文志》)

From the words of Confucius quoted above, we can understand how important music was in ancient Chinese society. On the other hand, to flourish music needs a plentiful knowledge of temperaments as its basis. And in the view of ancient people, temperaments were also the origin of weights and measures. Thus we can understand why temperaments take on a large role in Liu Xin's metrological theory.

Liu Xin's temperament theory mainly dealt with Five Sounds (五声, the five-tone scale), Eight Voices (八音, eight kinds of musical instruments), and the Twelve temperaments 十二律. Five Sounds means Gong 宫, Shang 商, Jiao 角, Zhi 徵, and Yu 羽, which stand for the five-tone scale. The corresponding relations among them are as follow:

Gong	Shang	Jiao	Zhi	Yu
宫	商	角	徵	羽
1	2	3	5	6

Eight Voices means eight kinds of musical instruments. The harmonious combination of them and the tone changes of Five Sounds make wonderful music.

Each of the Five Sounds has its special symbols. Liu Xin said:

> Shang means measure. Objects are not measurable until their formation. Jiao means touch. Things sprout like a horn (in Chinese pronunciation, the word horn is pronounced as Jiao) after they touch the earth. Gong stands for center. It occupies the center and runs around. It sings praise of the beginning of things and brings about living things. It is the guiding tone of the Five Sounds. Zhi is fortune. The larger things are, the more fortune they have. Yu symbolizes room. To preserve objects needs the envelopment of rooms. 商之为言章也，物成孰可章度也。角，触也，物触地而出，戴芒角也。宫，中也，居中央，畅四方，唱始施生，为四声纲也。徵，祉也，物盛大而繁祉也。羽，宇也，物聚藏宇覆之也。

Besides those mentioned above, the Five Sounds have corresponding relations to all kinds of things by means of the Five Elements 五行. These are as follows:

The Special Symbolization of the Five Sounds

5 Sounds 五声	Gong 宫	Shang 商	Jiao 角	Zhi 徵	Yu 羽
5 Elements 五行	earth 土	metal 金	wood 木	fire 火	water 水
5 Chang 五常	faith 信	loyalty 义	benevolence 仁	rite 礼	wisdom 智
5 Shi 五事	thinking 思	talking 言	appearing 貌	watching 视	listening 听
5 Ti 五体	monarch 君	subject 臣	populace 民	affair 事	thing 物

About the formation of the Five Sounds, Liu Xin said:

> The root of the Five Sounds came out of the temperament. The length of the tube of the Huangzhong temperament is 9 Cun 寸, which is used as Gong. From Gong, we can increase it or decrease it so as to determine Shang, Jiao, Zhi and Yu. 五声之本，生于黄钟之律。九寸为宫，或损或益，以定商、角、徵、羽。

This was famous in the history of music as the method of one-third increasing or decreasing 三分损益法. It took the length 9 Cun as a standard and divided it into 3 parts. Then one part was added or deducted. That resulted in new tones. Doing it in the same way successively, the whole of the Five Sounds could be obtained at last.

The one-third increasing or decreasing method came out very early. It was widely used in the practice of music in ancient China. Liu Xin inherited it and put it into his own system.

Liu Xin also discussed the Twelve Temperaments. About the origin of the Twelve Temperaments, he said:

> The Twelve Temperaments were made by Huangdi (黄帝, the Yellow Emperor, a legendary ruler in ancient China). Huangdi asked Ling Lun 泠纶 to go to the west of Daxia 大夏 and the north of Kunlun 昆仑 Mountain to get bamboo that grew in Xiegu 解谷. He selected the bamboo with an even body and cut it into a tube. Then he blew the tube and made it send out the Gong tone of the Huangzhong. He made 12 tubes to imitate the sound of the phoenix. There are 6 sounds of the male phoenix and 6 sounds of the female phoenix. The Huangzhong tube could make all of them. This is the origin of temperaments. 黄帝之所作也。黄帝使泠纶自大夏之西，昆仑之阴，取竹之解谷生，其窍厚均者，断两节间而吹之，以为黄钟之宫。制十二筒以听凤之鸣，其雄鸣为六，雌鸣亦六，比黄钟之宫，而皆可以生之，是为律本。

Those words were not Liu Xin's invention. These were similarly stated in *Lu Shi Chun Qiu* 吕氏春秋. But Liu Xin changed them according to his own theory. For example, Liu Xin abandoned the saying in *Lu Shi Chun Qiu* that the length of the Huangzhong tube is 3 Cun and 9 Fen 分, because he had another idea: that the length of the Huangzhong tube is 9 Cun. At any rate, Liu Xin's act of quotation had its own deep significance. It told people that bamboo tubing was used to make temperaments by sages, which reminds musicians to select bamboo as the material when they are making standard temperaments. Besides, "to imitate the sound of the phoenix" means that temperaments must be in keeping with natural reality. The words quoted by Liu Xin also meant that temperaments had their own inner law of formation.

Those meanings surely were very important for ancient people in developing the theory of temperaments.

The Twelve Temperaments have close relations to natural and social things, especially time and space. The following table is an example.

Relations among the Twelve Temperaments in Time and Space

12 Tem. 十二 律	C 黄钟	#C 大吕	D 太族	#D 夹钟	E 姑洗	F 中吕	#F 蕤宾	G 林钟	#G 夷则	A 南吕	#A 亡射	B 应钟
Lunar month 月份	11 十一	12 十二	1 正月	2 二月	3 三月	4 四月	5 五月	6 六月	7 七月	8 八月	9 九月	10 十月
Direction	N			E			S			W		
方位	zi 子	chou 丑	yin 寅	mao 卯	chen 辰	si 巳	wu 午	wei 未	shen 申	you 酉	xu 戌	hai 亥

The relationship between the Twelve Temperaments and the 12 lunar months was established in *Lu Shi Chun Qiu*. Liu Xin inherited and developed it. His theory influenced people much later. For example, Houqi 侯气 theory was born on the basis of this.

How did Liu Xin decide the lengths of the tubes of the Twelve Temperaments? He did it by means of the method of one-third increasing or decreasing. At first, he stipulated the length of the Huangzhong tube as 9 Cuns, and then he used the method to calculate other temperaments. As a mathematician, he paid attention to the perfection of form in the course of the calculation. In order to keep the form uniform, he even changed the traditional method of one-third increasing or decreasing. The traditional method was decreasing first and then increasing, but increasing again on the *Ruibin* 蕤宾 temperament when one would expect it to decrease in order. Following is the calculation of the traditional method.

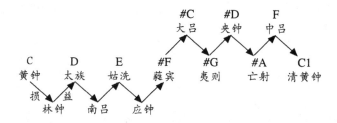

Fig. 1. Traditional method.

And Liu Xin's calculation is shown in Fig. 2.

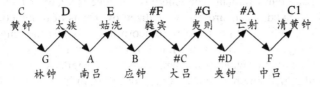

Fig. 2. Liu Xin's method

Later people did not accept Liu Xin's theory because his method damaged the inner completeness of music. Some people who liked Shen Kuo 沈括 even criticized his theory loudly.

Theoretical Basis for the Unity of Weights and Measures

This section describes an important part of Liu Xin's metrological theory. Its main contents involve the basic standard of units of weight and measure.

The guiding thought of Liu Xin in making the basic standard of units of weight and measure was what he called "uniting length, capacity, and weight on the basis of temperament 同律度量衡". According to his metrological theory, temperaments related to all things and determined all their changes. If this is so, it is natural to take temperaments as the basis of the unity of weights and measures. In fact, Liu Xin did not create this concept. For example, Sima Qian 司马迁 had similar ideas well before. But ancient people before him did not describe the relationship among temperaments and the weights and measures clearly. It was Liu Xin who created a model for it. That was the theory of "uniting length, capacity and weight on the basis of temperaments by means of broomcorn millet 乐律累黍说".

Here we shall analyze how Liu Xin applied such a theory to define the standard unit of length. According to the record of the book *Records of Temperament and Calendar, History of the Han Dynasty,* Liu Xin wrote:

> What are the units of length? They are Fen 分, Cun 寸, Chi 尺, Zhang 丈 and Yin 引. Their use is to measure the lengths of objects. Their origin is the length of the Huangzhong temperament. One may take broomcorn millet with moderate size as a standard to examine the length of the Huangzhong tube. The width of a grain of the millet is one ninetieth of the tube. That determines the unit of Fen. Thus the size of millet is 1 Fen. 10 Fen equals 1 Cun. 10 Cun equals 1 Chi. 10 Chi equals 1 Zhang and 10 Zhang equals 1 Yin. That's all for the 5 units of length. 度者, 分、寸、尺、丈、引也, 所以度长短也。本起黄钟之长, 以子谷秬黍中者, 一黍之广度之, 九十分黄钟之长。一为一分, 十分为寸, 十寸为尺, 十尺为丈, 十丈为引, 而五度审矣。

That means the standard units of length come out of the tube of the Huangzhong temperament. The length of the tube is 9 Cun, which provides a standard. Such a standard can be realized by means of broomcorn millet. Since the width of 90 grains of broomcorn millet is just the length of the Huangzhong tube, 1 grain of the millet is obviously 1 Fen. So this kind of broomcorn millet provides a standard unit Fen of length. After the definition of Fen, all other length units could be deduced from it.

Liu Xin's method has its own scientific basis. As we know, the tone produced by a temperament tube actually depends upon its length. The changes of the length of a tube certainly result in tonal changes. The latter can be felt by people. If people find the tune of tube changing, they can take measures to adjust it. Thus the length of the tube selected can be ensured constantly by this method, which qualifies it to be the standard for weights and measures. On the other hand, different people have different understandings of the same tone, and about whether it is the Huangzhong temperament, which leads to the standard being unstable. In order to solve the problem, Liu Xin took broomcorn millet as the medium for defining the standard unit of length. Here the basic standard of length came out of the Huangzhong tube and the function of broomcorn millet was to provide a method for realizing it.

Taking the Huangzhong temperament tube as the standard of weights and measures was an ancient Chinese tradition. But there were different understandings about the length of the tube. Some people thought it was 1 Chi. Others thought it was 9 Cun. Also there were some people who thought it was 8.1 or 3.9 Cun. Since Liu Xin's theory was adopted in the book *Records of Temperament and Calendar, History of the Han Dynasty,* the view that the length of the Huangzhong tube is 9 Cun has been accepted by all official history books. It became a rule that all makers of weights and measures must abide by it in ancient China.

The Huangzhong temperament tube provides not only the standard of length, but also the standard of capacity. Liu Xin said:

> What are the units of capacity? They are Yue 龠, Ge 合, Sheng 升, Dou 斗 and Hu 斛. Their application is to measure capacities. The origin of them is the capacity of the Huangzhong tube. They are defined by length. Taking 1200 grains of broomcorn millet with moderate size to fill the tube, and if the millets are just enough to fill it, we call the capacity of the tube Yue. Double Yue means Ge. 10 Ge equals 1 Sheng. 10 Sheng equals 1 Dou and 10 Dou is 1 Hu. That's all for the 5 units of capacity. 量者，龠、合、升、斗、斛也，所以量多少也。本起于黄钟之龠，用度数审其容，以子谷秬黍中者千有二百实其龠，以井水准其概。合龠为合，十合为升，十升为斗，十斗为斛，而五量嘉矣。

Liu Xin's rule that capacity measures were defined by length was quite scientific. It was the rule that ensured the unity of the standards of length and capacity. In fact, it is unnecessary to use broomcorn millet as a medium to define capacity units under this rule. The use of broomcorn only enhanced the mysteries of the origin of capacity units.

Besides, according to Liu Xin's theory, the Huangzhong temperament can also provide a standard for the unit of weight. He wrote:

> What are the units of weight? They are Zhu 铢, Liang 两, Jin 斤, Jun 钧 and Dan 石. Their application is to measure the weight of objects and to make distribution of things with justice. We can know the weight of objects with them. Their origin is the weight related to the Huangzhong temperament. Since 1200 grains of broomcorn millet are just enough to fill the capacity measure Yue, we define their weight as 12 Zhu. Doubling them creates Liang. That means Liang is equal to 24 Zhu. 16 Liang is 1 Jin. 30 Jin is 1 Jun and 4 Jun is 1 Dan. 权者，铢、两、斤、钧、石也，所以称物平施，知轻重也。本起于黄钟之重，一龠容千二百黍，重十二铢，两之为两，二十四铢为两，十六两为斤，三十斤为钧，四钧为石。

Obviously, Liu Xin's theory has its own basis. He thought the standard of length came out of the Huangzhong temperament and that of capacity was defined by length. If the capacity of a vessel is defined, the weight of some kind of matter the vessel contains is also fixed. So it can be used as a standard of weight. That means the weight standard can also be obtained from the Huangzhong temperament. Thus weights and measures were unified under the Huangzhong temperament in this way.

The Design of Standard Apparatus of Weights and Measures

The meat of Liu Xin's metrological theory is his design of standard apparatus of weights and measures. His thought was that "uniting length, capacity and weight on the basis of temperaments 同律度量衡" offered a way to define the standard unit of weights and measures. But he had to design some standard apparatus for the purpose of examining and determining other apparatus of weights and measures. It is somewhat similar to the 18th century, when Academia Francais determined the length of the one-four hundred millionth of the meridian that passed through Paris as 1 meter, and also had to design a standard ruler for it.

Liu Xin designed two kinds of standard apparatus for length. One is Copper Zhang 铜丈 and the other is Bamboo Yin 竹引. There is a record of his design in the book *Records of Temperament and Calendar, History of the Han Dynasty.* It says:

> The standard ruler Zhang is made of copper. Its thickness is 1 Cun. Its width is 2 Cun and its length is 1 Zhang. We can get length units Fen, Cun, Chi and Zhang from it. Yin 引 is made of bamboo. Its thickness is 1 Fen. Its width is 6 Fen and its length is 10 Zhang. Its normal section is made after a rectangle. The sizes of its thickness and width symbolize Yin 阴 and Yang 阳. 其法用铜, 高一寸, 广二寸, 长一丈, 而分、寸、尺、丈存焉。用竹为身引, 高一分, 广六分, 长十丈, 其方法矩, 高广之数, 阴阳之象也。

We cannot find the standard ruler Yin 引 today because it was made of bamboo and 2000 years have passed since then. But Copper Zhang still exists. It is a material proof of the record mentioned above.

Liu Xin's design of standard apparatus of capacity should also be paid attention to. He combined Yue 龠, Ge 合, Sheng 升, Dou 斗 and Hu 斛 into one and stipulated their whole weight and the size of each part. Thus he realized the unity of length, capacity and weight in one apparatus. His design was described as follows:

> The vessel is made of copper. The inside size of the vessel is slightly larger to contain a square with side of 1 Chi. There is a very small distance from the vertical angles of the square to the inner surface of the vessel. We call it Tiao 庣. Its upper part is Hu and the lower is Dou. Its left ear is vessel Sheng and the right is vessel Ge and Yue. Its shape is like Jue 爵 and its use is to control government salary. ... all of its components are round vessels and the whole weight of them is 2 Jun. ... the sound sent by the vessel fits the Huangzhong temperament regardless of its posture being upward or downward. 其法用铜, 方尺而圆其外, 旁有庣焉。其上为斛, 其下为斗。左耳为升, 右耳为合龠。其状似爵, 以縻爵禄。... 其圆象规, 其重二钧, ... 声中黄钟, 始于黄钟而反覆焉。

The standard vessel designed by Liu Xin still exists in the Palace Museum in Taipei 台北. Its structure is shown in Fig. 3. Because it was issued in the name of Wang Mang's Xin Dynasty, people always call it Xin Mang Good Vessel 新莽嘉量.

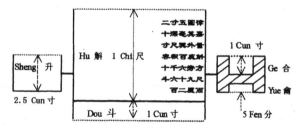

Fig. 3. The structure of the standard vessel designed by Liu Xin.

There are different inscriptions on different parts of Xin Mang Good
Vessel. Each inscription describes the shape, the size, and the volume of the
vessel it belongs to. Here we take the inscription on vessel Hu as an example.
It is as follows:

> It is Good Vessel Hu 嘉量斛 that fits the Huangzhong temperament. It is
> a round vessel. Its inside size is slightly larger to contain a square with
> side of 1 Chi. The length of Tiao Pang 庣旁 is 0.095 Cun and the inside
> area of normal section of the vessel is 162 square Cun. Its depth is 1 Chi
> and its capacity is 1260 cubic Cun, which equals 10 Dou. 律嘉量斛, 方
> 尺而圆其外, 庣旁九厘五毫, 宴百六十二寸, 深尺, 积千六百二十寸, 容
> 十斗。

Here Tiao Pang means the distance from the vertical angles of the square to
the inner surface of the vessel, as shown in Fig. 4. Liu Xin did not tell us the
diameter of the vessel. But we can know it by calculating the data he offered.
The length of the diagonal of the square added to double Tiao Pang is its
diameter. The result is 1.4332 Chi.

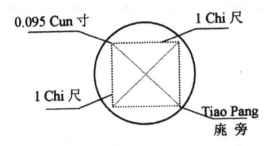

Fig. 4. The sketch of Tiao Pang.

Why did Liu Xin use this method to define the diameter? Because it is a
Chinese tradition. Before him, people did not know how to describe a cycle.
If they wanted to define a cycle then, they had to determine a square at first
and then circumscribe it. That is the so called "circle came out of square and
square came out of rectangle 圆出于方, 方出于矩". Liu Xin succeeded to
this tradition.

It was wise for Liu Xin to define Tiao Pang as 0.095 Cun. The train of
his design was to give the inside area of the vessel first and then to calculate
the diameter. In the course of the calculation, the ratio π must be applied.
Examining the data given by Liu Xin, we find the value of π he used is
3.1547. It was a great achievement because the value of π people used then
was 3. Liu Xin was the first in Chinese history who broke the traditional

concept that the circumference of a circle is 3 times as large as its diameter 周三径一. It is a pity that we know nothing of how he got his result.

Liu Xin also defined the weight of the Xin Mang Good Vessel. Thus we can obtain unit values of length, capacity, and weight from the same apparatus. It is to say that weights and measures were really united in the vessel. Considering those factors, we can certainly say that Liu Xin's design stood at the highest level at that time.

There are still many more things to say about Liu Xin's metrological theory. But here we do not have the room to introduce more of them. What we want to conclude here is that the theory mentioned above was not only fit for the popular philosophic beliefs then, but also very scientific in practice. His theory was the earliest systematic theory of traditional Chinese metrology. It marked the formation of the theories of metrological science in ancient China.

Note

* This paper was supported by the Matsushita International Foundation.

The Preliminary Study of Mechanical Design Methodology in Ancient China*

LIU KEMING 刘克明
Huazhong University of Science and Technology
ZHOU ZHAOYING 周兆英
Tsinghua University
YANG SHUZI 杨叔子
Huazhong University of Science and Technology

The history of engineering design is almost as long as that of man himself. By "engineering" is meant performing the conceiving and planning of engineering and technological systems and turning imagination into realistic technological activities and practices. Machine design is a vital component of mechanical engineering and the foundation of machine building. The earliest monograph that presents a systematic exposition on the ideas and methods of machine design should be *The Rites of the Zhou Dynasty* that appeared in book form following the eastward move of the imperial court of Zhou. The section Kao Gong Ji (《考工记》) in *The Book of Rites of the Zhou Dynasty* is not merely an important record of engineering practice and the departments of technology. As importantly, the many ancient materials it contains make it an invaluable monograph that treats at length ideas and methods of machine design in ancient China. According to Joseph Needham (1900–1995), Kao Gong Ji, Zhou Li (《周礼 • 考工记》) is "the most important document for the study of ancient Chinese technology", a view still held by the scientific community.[1]

The Design Idea of 'Creation by the Intelligent' and 'Exposition by the Ingenious'

For engineering design in ancient China, what was emphasized before everything else was the design idea of "creation by the intelligent". The records in Kao Gong Ji show that China entered the conscious design period in the remote past. In particular, the tendency toward specialization of various production skills as described in this document shows that during the Qin period, the design of products and practice of engineering technology were both performed on definite theories. Moreover, both the design method and design process were represented in the form of state-formulated laws and rules, based on which design and manufacture were carried out using mathematical approaches throughout.

It is stated in the Introduction "总叙" to Kao Gong Ji that "the intelligent create and the skilled expound and their craft was handed down from generation to generation for their descendants to follow". "The intelligent" mentioned here refer to the wise, that is, masterhands or designers with definite fund of technological and professional skills. Here 'creating' means making something new or original. It is thus quite clear that Kao Gong Ji (《考工记》) regards design as a creative activity. It is a process of putting into practice the creative thinking aimed at meeting specific demands within a limited time/space range and under given material conditions. The statement "the skilled expound and their craft was handed down from generation to generation" shows that design and manufacture were already well defined and distinguished from each other in those days.

Meanwhile, it is stressed in the Introduction to Kao Gong Ji (《考工记》) that whatever product which is turned out is the outcome of the coordination among different specialties and a result of collective effort. A typical example is the manufacture of a carriage, which involves nearly all branches of craftsmanship.

The Design Principle of the Dependence of Workmanship on Criteria

Engineering design is an activity with the purpose of meeting social demands by relying on science and technological know-how accumulated in the development of civilization. The specialization of the production departments in society and the standardization of production technology are important scales for measuring the advances of science and technology. The statement in Rites for Youths (少仪) in *The Book of Rites* (《礼记正义》) that "a craftsman should be familiar with the criteria and standards in the process of manufacture indicates that the design and production of all kinds of products during the Qin period strictly abided by definite technological rules and regulations".

What is discussed in detail in Kao Gong Ji (《考工记》) is the criteria for the manufacturing process of all craftsmen. The statement that "a craftsman should be familiar with the criteria and standards in the process of manufacture" includes acquaintance with the correct use of "the compasses and angle square as well as standards for structure-related parameters". The compasses and angle square were extensively used in design and production of articles during the Qin period. It is stated in "Will of Heaven A" of *The Book of Master Mo* (《墨子 • 天志上》) that "holding the compasses and angle square, a vehicle wheel maker can measure the shapes of all square and round things and decide what shape is correct and what incorrect". Sizi (《尸子》) said, "In the old days an artificer by the name of Chui (垂) invented the compasses, the square, the level gage, and the rope so that all the other craftsmen are enabled to follow his example". These records show the roles played by the compasses and angle square during the Qin period. "The criteria for dimension-related items" refer to the choice of material and the standards for the material to be used as well as structural parameters. For instance, in Kao Gong Ji (《考工记》) not only are the grades of carriages given, but also the specific dimensions of all grades of carriages are specified. In Ye Shi (冶氏, a monograph on metallurgy) the composition of alloys for making various kinds of articles is given, or what is also known as "six copper-tin proportioning for making bronze articles". The standards for materials to be used for making bronze articles and the proportioning of raw materials for bronze are scientifically summarized. As it was, the alloys for making most of the bronze articles unearthed were produced in accordance with this regulation and standard.

Regarding the provision "a craftsman should be familiar with the criteria and standards in the process of manufacture", it should be noted that products had to be examined according to regulations so as to ensure the quality of the design. It is recorded in Yu Ren "輿人", carriage makers) in Kao Gong Ji (《考工记》) that "The round shape should conform to a circle drawn with the compasses, the square shape to a square drawn with an angle square, the vertical should be perpendicular to the ground surface from a suspended rope, and the horizontally placed should be paralleled to the water surface. The vertical should look as if grown out of the earth. What intersect and link are like branches and twigs attached into the trunk." All this is a concrete manifestation of craftsmanship conforming to criteria. It is recorded in the Imperial System in *The Book of Rites* (《礼记 • 王制》) that "Those articles that are not up to requirements shall not be sold on the market. Those weapons that are not up to requirements shall not be sold on the market." All this shows that a strict checking system was already an important component of product design.

A Macroscopic View of the Object of Design

The purpose of design is to turn out new products. Such an activity must be made to adapt to the environment. This requires that a designer not just consider the object of design itself but take all relationships of relevance to the product into account. The idea and method of considering an object of design macroscopically were first proposed in Kao Gong Ji (《考工记》), the Introduction of which stated that "a satisfactory product can be turned out only when favorable climate and geographical position, excellent material and consummate skills are all available". Approaching an object of design from the high plane of favorable climate and geographical position, carefully selecting rational manufacturing technology, and taking into consideration the social functions of a product as well as the social estate system — this is one of the characteristics of ancient China's concept of design. As a craftsman, it is necessary to understand the importance of climate and geographical position, that is, "even though the material may be excellent and skills consummate, the product will still not be satisfactory if the climate and geographical position are not favorable". In addition, he should understand the changes the material environment may undergo. With regard to the examination of the season as one of the influencing factors for design, Guo Yu, Qi Yu [《国语 • 齐语》, The States of Qi in Discourses on the (ancient feudal) States] expressly proposed that the "four seasons of a year be taken into account".

To take a macroscopic view of an object of design, it is necessary for a designer to be capable of distinguishing between what is curved and what straight, what is square and what round, as well as of properly using copper, wood, leather, jade and earth to make articles. According to the Introduction "总叙" to Kao Gong Ji (《考工记》), "One who is capable of examining the curved and the straight and the square and round, the obverse and reverse sides of an object as well as the trend of the times is called a craftsman". To repeat, the curved and straight lines, the obverse and reverse sides, and the trend of the times are placed on a par with each other for examination. A material may be curved or straight, the latter case being known without examining while the inspection of what is curved will enable a craftsman to know the grain or texture of a material. The trend may be obverse or reverse, and what is reverse cannot be used to replace what is obverse. After the trend is examined and determined, the correct arrangements for using a material can be made according to its grain or texture. Only in this way will it be possible to attain the design objective of excellent material and consummate skills.[2]

The Design Method of Manufacture Emphasizing the Shape

No engineering design can be carried out without a drawing and the design drawing is an indispensable technical document for guiding production. Performing design by means of drawing is an important mark of the stage of empirical design.

The celebrated saying "an article-maker emphasizes the shape" found in the "Appendix A" to *The Book of Changes* (《周易 • 系辞上》) implies that when manufacturing an article or an implement it is necessary to refer to its shape or drawing. There are numerous records concerning drawings and graphs in Zhou Li (《周礼》, *The Book of Rites of the Zhou Dynasty*), which is the first book of its kind in ancient and medieval China that gives a systematic explanation about the application of graphics. No matter whether it is "an official in charge of mining" as mentioned in Di Guan Si Tu "地官司徒" of *The Rites of the Zhou Dynasty* (《周礼》) who is "responsible for prospecting in the mining area beforehand, making drawings for the miners, and making a tour to see to it that prohibitions are observed" or it is "an official in charge of suburban affairs who, using maps to supervise farming and open up wasteland, has county towns built and boundary lines delimited" or "an official in charge of mausoleums" as mentioned in Chun Guan Zong Bo "春宫宗伯" who "defines the range of tombs and draws the plan of a graveyard" using drawings to "specify standards for size and height of a tomb", it can be seen that as far back as 2,000 or 3,000 years ago, centered on the application of graphs or drawings, the transmission and storage of information, pattern management and filing converged. Performing design and completing the construction of a project by means of drawings was already a mature design method in those days. In addition, the making of different kinds of drawings during the Zhou Dynasty (B.C. 11th century– 256 B.C.) was as a rule based on measurement. A design scheme was regarded as a formal drawing only after it was examined. Drawings for production purposes were duplicated without exception with the copy put on file in the government department concerned. These procedures are all in conformity with the general method of modern engineering design and the principle of management and custody.[3]

Technological Requirements on "Excellent Material" and "Consummate Skills"

The purpose of design is to guarantee the functions of the product and its manufacture as well as to develop a technical system of good performance, low cost, and optimal price.

The technological requirements for "excellent material" and "consummate skills" are specifically proposed in the Introduction to Kao Gong Ji (《考工记》). By "excellent material" we mean material of the best quality while "consummate skills" implies ingenious work. With regard to the design objective of "excellent material" and "consummate skills", there were strict technical standards for manufacture of a specific product. For instance, as stipulated in Kao Gong Ji (《考工记》), there are three steps for bringing an axle up to requirement, that is, "First, it should be made of excellent material; secondly, it should be enduring, and thirdly, it should be flexible." Excellent material and endurability refer to raw material and its quality while flexibility implies that when mounted on a carriage to operate, the machined axle will not get stuck while rotating if its fit clearance with the hub is appropriate. Only when the three preconditions are satisfied will the material be regarded as excellent and the expected design objective attained.

On the other hand, the design objective of skilful craft is also manifested in the product turned out in accordance with specifications, which should have an optimum structural shape with the overall functions guaranteed. A carriage made according to design requirements and "running great distances all day long should not make even elderly riders sitting toward the left side feel tired". Only when this requirement is met shall we be able to say that the objective of skilful craft is attained. There is a statement in Mo Tzu Jie Yong (《墨子 • 节用》 , on practising economy by Mo Tzu) that roughly expresses the same meaning, namely, "A carriage is used to load people or goods and run great distances. The travel should be safe and the running smooth. By being safe is meant none will get injured in an accident and by being smooth is meant quick arrival at the destination. Such are the advantages of a carriage."[4,5]

When we look back at the expositions concerning methods of machine design in ancient and medieval Chinese literature, and investigate and draw on the creativity and confidence of our ancestors in achieving successes in science and technology that are practically peerless anywhere, there is no doubt we will benefit from our predecessors while promoting the advances of the theories of machine design in present-day China and world.

Although ancient machines have already become historical relics, and ancient mechanics itself cannot directly point out the road of development of mechanics today, the creativity manifested in the wonders worked by the Chinese nation under extremely great constraints in the history of science and technology and the wisdom exhibited in their conceptions and methods of machine design have their ever-lasting charm felt to this day.

Note

* Project supported by the National Natural Sciences Foundation of China.

References

[1] Joseph Needham, *Science & Civilization in China*, Cambridge University Press, 1965, Vol. IV, Part II, 11.
[2] Sun Yirang, *Zhou Li Zheng Yi* (《周礼正义》), Zhonghua Publishing House, 1987, Vols. 13–14, 3101–3569.
[3] Lin Yin, *Zhou Li Jin Zhu Jin Yi* (《周礼今注今译》, Annotated Zhou Li), Ancient Books Publishing House, Tianjin, 1988, 174, 227, 228.
[4] Sun Yirang, *Zhou Li Zheng Yi* (《周礼正义》), Zhonghua Publishing House, 1987, Vols. 13–14, 3101–3569.
[5] Wen Renjun, *Zhou Li Kao Gong Ji Yi Zhu* 《考工记译注》, Ancient Books Publishing House, Shanghai, 1993.

Summary of Chinese Ship Navigation in Ancient Southeast Asia

JIN QIUPENG

Institute for the History of Natural Sciences
Chinese Academy of Sciences

China has a long history of marine intercourse with Southeast Asia. This paper argues that the ships which navigated between China and Southeast Asia in classical times were built in Guangdong, Fujian, and Zhejiang provinces, and that they were mostly of the type known as *Fu*. Many marine activities in Chinese history, such as the frequent trade during the Song and Yuan dynasties, Zhen He's voyages in the early period of the Ming Dynasty, and the emigration of Chinese during the Ming and Qing Dynasties, were carried out by this type of ship. Although flat-bottomed *Sha Chuan* (沙船) boats built in Jiangsu also went to Southeast Asia, such cases were not common. Having carefully studied the structure of Chinese *Fu* ships, the author notes that they were sharp-bottomed, and that this made them fairly stable and easy to control while sailing along narrow and reef-laden sea-routes. This type was thus favored for its navigability. The body of the ship was equipped with multi-plank decking (or one layer of thick plank) and the cabin was constructed as waterproof and separate. Such a structure was strong and secure. The dynamic equipment on such ships was sails and sculls. The oblique sail with bright Chinese characters not only could be raised or lowered expediently, but the area of sail could also be easily adjusted according to wind power. The operational device of the ship was the rudder. Rudder and scull were famous inventions by Chinese people in navigation and shipbuilding. Such ships used compasses together with astronomical devices. Living conditions on such ships were also very comfortable, so that merchants both in China and other countries preferred the *Fu* type.

China is not only a continental country, but also an oceanic one. It has a centuries-old history of navigation. In this long history, ships from China have been active on every route in the West Pacific and Indian Oceans, especially in Southeast Asia. This paper deals with the ships navigating only in this later area. From the viewpoint of the history of sailing in this region, ships built in Northern India before the Song Dynasty were the first to flourish along the sea routes of the South China Sea, Southeast Asia, the Indian Ocean, and the Arabian Sea. Since the Tang and Song Dynasties, however, and especially since the Bei Song Dynasty, Chinese shipbuilding and sailing developed rapidly. Chinese vessels then took the place of foreign ones in Southeast Asia step by step.

In the history of shipbuilding in ancient China, various ship-types have been created to meet complicated geographical conditions on various rivers and seas. Despite the variety of ship-types in Chinese history, in the main, Chinese vessels can be classified as having either sharp or square sterns, and sharp or flat bottoms. *Fu Chuan* (福船) is a famous representation of the sharp stern and flat bottom type, mostly sailing on routes away the Southern Ocean. *Sha Chuan* is a famous representation of the square stern and flat bottom type, which mostly sailed on routes north from the Yangtze River. During the Ming and Qing Dynasties, *Sha Chuan* ships sailed to Southeast Asia. But as a whole, the most common type that sailed in Southeast Asia were not *Sha Chuan* but *Fu Chuan* made in Fujian. *Fu Chuan* was named after Fujian. This type is as high as a building, with the broad pointed bottom supporting a high sharp stern and square poop and baffles on both sides. The cabin is constructed as waterproof and in isolation. With a deep sea gauge because of the sharp bottom and ease at cutting a feather because of the sharp stern, the ship was fairly stable and easily changed directions while sailing on narrow and reef-laden sea-routes.

Here we will deal with the problem of the type of Zheng He's treasure ships (宝船) in the Ming Dynasty (first half of 15th century). In the past, these were regarded as *Sha Chuan* types, mainly because they were made in the treasure shipyard in Nanjing. The boats made by shipyards along the Yangtze River were generally *Sha Chuan*. But in recent years, more and more researchers in the history of Chinese shipbuilding agree that the treasure ships were of the *Fu* type. *Fu Chuan* became the most popular type on the sea routes of Southeast Asia and was appreciated by travelling Chinese and foreign merchants due to its excellent structure and performance. It is said in vol. 176 of *San Guo Bei Meng Hui bian* 《三朝北盟汇编》 that ships made in Fujian are the best ones. It is also recorded in the vocabulary entry for *Fu Chuan* in vol. 116 of *Wu Bei Zhi* 《武备志》 that the shipbuilding technology in Fujian was mature. The *Fu Chuan* type is described in another vocabulary entry as 'the ship for passenger transportation' (vol. 34 of *Xuan He Feng Shi Gao Li Tu Jing* 《宣和奉使高丽图经》 written by Xu Jing during the later part of the Bei Song Dynasty). There is yet another vocabulary entry for '*Fu Chuan*' in vol. 116 of *Wu Bei Zhi* 《武备志》.

A *Fu Chuan* boat built in the Nan Song Dynasty has actually been excavated in the Quanzhou gulf, and another built in the Yuan Dynasty has been excavated in Xin'an 全罗南道 in South Korea. Additionally, it was described in Marco Polo's *Travel Notes* 《马可波罗游记》, which narrated the construction of a fleet to escort Princess Kuo Kuo Zhen (阔阔真) to Iran. It is also described in the notes written by Arabian travelor *Yin Ben Ba Tu Ta* (伊本拔图塔) in the Yuan Dynasty. We can acquaint ourselves with the basic structure, features and the evolutionary status of *Fu Chuan* through these materials.

It is the keel that determines the shape and the structure of the hull of *Fu Chuan* ships. The keel is in the middle of the bottom of the hull, constructed by connecting three sections: stern keel, main keel and poop keel. The hull spreads in an arc towards both sides forming the keel. The hull boards are linked by tenons, clinched by bamboo nails, iron nails and curium nails, with seams sealed by twines and tung oil ashes and resulting in scalariform. The hull is commonly constructed using multiple boards. The bottom of a ship built in Houzhu (后渚) in Quanzhou is double boarded and the sides are triple boarded. It is said in Marco Polo's *Travel Notes* that, (the body) is constructed by double boards, that is, there is a board covering the soleplate on each part. The bottom cabin was constructed as waterproof and isolated, that is, it was divided into several independent parts. The bottom cabin was divided into 13 parts by 12 boards in the Houzhu ship of Quanzhou, and 8 parts by 7 boards in the Xin'an ship. Waterproof isolated cabins are an important invention of Chinese ship-construction. The craft originated in about the Han Dynasty and was gradually applied in *Sha* ships, *Fu* ships and *Guang* ships thereafter. The principle is to improve the security of sailing by fully using its buoyancy. The ship would float using buoyancy produced by water expelled by the ship. By using the system of waterproofing and isolation, only a few compartments would fill with water if the hull was penetrated. So the water pushing inside could be isolated, and not threaten the whole. Thus the ship would not sink if the weight of the water inside did not surpass the buoyancy. If the damage was not so serious and the water was not so much, it could be repaired after moving the cargo out of the flooded parts, without affecting continual navigation. Otherwise, it could sail to nearby ports or land and be repaired after abandoning the cargo to lighten the weight. The system of waterproofing and isolation thus improved stability so as to increase the security of the crew and cargo (Picture 1: Yangzhou Ship Model of Tang Dynasty, Picture 2: Quanzhou Sea-going Ship of South Song Dynasty). But the advantage of waterproofing and isolation were not just that. Since the clapboards were made of thick and hard boards nailed close together on the hull, they increase the transverse intensity of the ship body. The utilization of clapboards instead of ribs simplified the ship-manufacturing technique. In addition, the division of the interior increased the loading and unloading efficiency and allowed the easy separation and management of the goods of different merchants.

Picture 1. Yangzhou Ship Model of Tang Dynasty

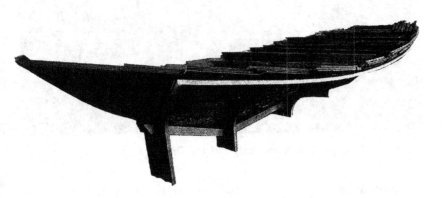

Picture 2. Quanzhou Sea-going Ship of South Song Dynasty

As for the superstructure, it no longer existed in the excavated Quanzhou and Xin'an ships, but can be ascertained from other historical materials. It was said in Marco Polo's *Travel Notes* that, in the ships made in Quanzhou, "there are about 60 rooms on the deck or in the cabin, more or less according to the boat's size. Travellers can live comfortably in it". It is also said in Yin Ben Ba Du Ta's Travel Notes 《伊本拔图特游记》 that, "there are four floors in Chinese boats with bedrooms and living rooms above them accommodating travelling merchants".

The helm is the device that controls direction when sailing. It was a critical invention in shipbuilding technique in ancient China, and was used only by the Chinese for a fairly long time. It is not yet determined when the helm came into being. In 1976, a timbal was unearthed in the No. 1 tomb of the Xihan Dynasty in Luo Bo Gulf in Gui county, Guang Xi province. It included a decoration of a dragon boat with a well-developed helm. The helm was in common use by the Donghan Dynasty of the 1st and 2nd centuries B. C. It availed the lever principle, which allowed the adjustment and control of sailing directions and routes by turning the helm. There was a hole behind the helm, which allowed it to hang on landing (Picture 3: Guangxi Gui Xian Design of the Dragon Boat of Xihan Dynasty).

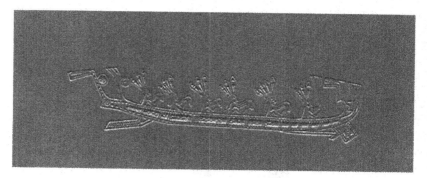

Picture 3. Guangxi Gui Xian Design of the Dragon Boat of Xihan Dynasty

With a developing recognition of its function, the shape and configuration of the helm was continually improved. One momentous improvement was the invention of the vertical helm, that is, the rudder-post was extended downward into water vertically instead of slanting from the stern. The joint of the helm and rudder-post was moved to the side from the center (Picture 4: Vertical Helm in Drawing of Zhen Qian in Tang Dynasty). The creation of this advanced device of a stern helm is closely

Picture 4. Vertical Helm in Drawing of Zhen Qian in Tang Dynasty

related to the special structure of the stern of Chinese ships. Chinese ships were always turned up at both the fore and tail. The stern assumed a crescent form, so that the end of it became the proper place for installing the vertical helm. A poop was built on the stern as the specific site for operating the helm, which could be hung up and stored in the poop when berthing. The vertical helm was used widely in sea boats. It was praised in *Ling Wai Dai Da* 《岭外代答》 written by Zhou Qufei of the Song Dynasty, who noted that the helm seems really the most valuable treasure to sailing ships. Because of the varied depth of water levels along sea routes, the helm evolved into an elevator that could be easily raised and dropped. The height of the helm could be adjusted with the depth of the sea water. Helms in large boats were as great as the long tiller of 11.7 meters in the treasure ships, which carried Zhen He to the Occident in the early Ming Dynasty (Picture 5: The Helm of Zheng He's Treasure Ship). It was too great to be raised and dropped only by helmsmen, so a winch was installed in the poop to deal with it. There were several helms in most large boats. The helms in the ship which went to Korea in the year of Xuanhe were divided into two

Picture 5. The Helm of Zheng He's Treasure Ship

classes, big and small ones, which were adapted to the depth of the water, together with accessorial helms. There were four helms in sea boats sent to Ryukyu in the Ming Dynasty, three of which were spare ones saved for accidents. When meeting with storms, the crew could put the largest helm into deep water, where it would not be affected by the flows and vortices produced in the stern. It could not only increase the efficiency of the helm, but also decrease the transversal float of the ship, thus stabilizing it.

The dynamic equipment in ships were sail and scull. Chinese sails were decorated with bright characters. Bamboo strips were added at certain intervals as transversal strengthening materials in the horizontal direction of the sail. Both ends of the strips were fixed by canvass ropes hung from the crossyard, which formed a sail yard easily raised and dropped. The awnings were knotted with each strip to the edges of the sail yard, which smoothed the awnings so as to get the best effect of wind power. These kinds of sails had two advantages: because they had transversal strengthening materials and the distance between the strips was not great, fabric of high quality was not required. Leafs of bamboo and other plants could be used to fabricate the awnings. These cheap and practical sails were widely used in ancient China. Both faces of the sail were fabricated by thin bamboo strips with leafs of bamboo, palms or reeds between them. This kind of awnings could also prevent sails from being torn or broken by the wind. Even when there were holes in the awnings, they were highly efficient in channelling wind power. Every horizontal bamboo strip was fastened by ropes and was linked together at a certain distance. Having these bamboo strips as its backbone, the whole awning could be folded in two ways. One was into scalariform according to the intervals between bamboo strips. A pullet was located at the top of the mast holding the awning. Ropes went through the pullet and reached the hands of sailors. The pullet, canvass ropes, and the awning consisted of an organic unity facilitating the pulling-up and dropping-down of the sail. Areas of the awning could be controlled by pulling the rope through the pullet in order to manipulate wind power from different directions in the course of the voyage. When dropping down the sail, one action would be enough — free the canvass rope and the awning would drop down and fold itself by gravity (Picture 6: Sailing Ship in the Picture of 'He Gong Qi Ju Tu Shuo' 《河工器具图说》).

Picture 6. Sailing Ship in the Picture of 'He Gong Qi Ju Tu Shuo' 《河工器具图说》

The most common sails were approximately trapezia. The top border of the canvass was skew and the bottom border was an arc. When the sail encountered the wind, an upward force would be generated. If the wind power was exceedingly large on the sea, the upward force acting on the awning might turn the boat upside down. In order to enhance security in sailing, the wind's acting center must be lowered and the force diminished. So the sea boats were always equipped with canvass shaped in the way mentioned above: narrowing at the top border and widening at the bottom border. At the same time an appendix was attached beneath the canvass to lower the center of wind power more efficiently. The scull was a propelling apparatus which was invented by the Chinese. When drawing the scull back and forward, a propelling force was generated from differences in pressures. In this way, the remittent force was changed into the continuous force by drawing the scull. Such a labor-saving but more efficient propelling tool as the scull was usually used when the ship was about to set off from or arrive at a harbor. And when there was no evident wind power, sailors would resort to the scull to run their ship. In volume 34 of *Xuan He Feng Shi Gao Li Tu Jing* 《宣和奉使高丽图经》, Xu Jing (徐兢) wrote that at that time the ordinary ship at sea was equipped with 10 sculls to facilitate voyaging (Picture 7: Scull of the Song Dynasty).

Picture 7. Scull of the Song Dynasty

A Traditional Architectural Heating System in China: The Manchurian *kang* and *huodi*

GUO QINGHUA
University of Melbourne

Introduction

A *kang* or *huokang* is a living and sleeping platform, literally a 'heated bed'. It is constructed of brick or adobe and consists of three parts: a fireplace, *kang* proper, and chimney. Beneath the flat surface of the *kang* are flues where hot air from the fireplace is conducted through to the room. The *kang* usually occupies one half to one third of a room area. If the entire floor of a room is constructed and heated in this manner, it is called *huodi* (lit. 'heated floor'). This architectural phenomenon appeared in northern China, typically in Manchuria, (which was defined by the Manchus as including Northeast China), the northern Korean peninsula, and parts of eastern Russia from the 12th century onwards.[1]

Originally the term *kang* meant 'to dry'; its use can be traced back to A.D. 121, when it first appeared in a Chinese dictionary.[2] The word *huodi* did not come into general use until the Qing dynasty (1644–1911) when the Manchus occupied the throne of China.[3] In fact, heated floors had been in existence one thousand years before the Qing, and were recorded at the turn of the 5th century.[4] The *huokang* and the *huodi* are closely linked technically, but each developed in conjunction with a particular way of life: sitting on the floor or furniture. Why and how did the Manchurians utilize two different heating systems? How did these two different but related innovations develop? What were their origins? Many questions remain to be answered, as this system has never been the subjects of study before. This paper intends to look into these questions, based on observations and investigations of the *kang* and the *huodi* in northern China, with a focus on Manchuria. The paper will consider the heating systems within two contexts: their placement within the Manchurian imperial palace; and evidence of their vernacular use based on architectural remains and archaeological excavations across northern China.

Form, Structure and Technique

Shenyang, situated north of the 42°N, was the capital of the Manchu state (1625–43).[5] The climate there has four distinct seasons, and its winters are especially long and cold. In Shenyang palace, both the *kang* and the *huodi* were the primary means for heating. According to form and type, the *kang* can be classified in two categories: I-shaped *kang* and U-shaped *kang*.

The I-shaped *kang* is the most commonly used type in northern China (the present provinces of Gansu, Shannxi, Shanxi, Henan, Hebai, Liaoning, Jilin, Heilongjiang and Inner Mongolia). The U-shaped *kang* was developed in relation to specific Manchurian activities and could be found only in Manchurian buildings. Both can be seen in the emperor's Fresh-peace Hall (Qingning Gong) in the Shenyang palace. The building is five bays by eleven purlins in plan (22.5 m × 16.5 m) with a veranda along the elevations. The entrance off the central axis divides the building into two parts: a one-bay bedroom in the east and a four-bay living room in the west. The east bay is further divided longitudinally into two bedrooms each with an I-shaped *kang* (the southern room for winter and the northern for summer), whereas the living room has a *kang* placed along the south, west and north walls forming a U in plan.

I-shaped kang

This feature includes a fireplace located outside on the building platform. The *kang* is fueled with charcoal so that it has no chimneys. To permit ventilation small passages were cut into the sides of the platform, often ornamental, the hot air flowed out of them which was aptly called 'dragon's breath'.

U-shaped kang

The U-shaped *kang* in the Fresh-peace Hall has two indoor fireplaces and a chimney. For Manchurians the west end of a room was sacred, and the large west space of this building was used for religious rites. The Manchus drew spiritual support from Shamanism, a religion involving the worship of natural phenomena and the spirits of ancestors.[6] Shamanist practice included dance, which was regularly performed in this building especially at the end of the year. The U-shaped *kang* defined a central space for the dance and provided people with a large, warm place to sit. Two cooking ranges attached to the *kang* at both ends were used to cook meat as part of the ritual. Between the south and the north *kang*, the west *kang* (designed to lead away smoke) is much narrower than the others. Above this an ancestral shrine is displayed. The chimney is about two meters away from the rear of the building. This is a traditional design, derived from sedge-thatched houses. To avoid fire sparks, the chimney had to be located away from the building.

Structure of the Heated Bed: kang

The dimensions of the *kang* correspond to the proportions of a human body: 1.7 m high on average, 1.8–2.0 m in width, and usually full-room long. The *kang* proper is a pounded earth platform upon which there are adobe flues with an inward slope of 2 : 100 to facilitate air flow to a chimney. A flue consists of several channels 200–280 mm wide and 150–250 mm high. Designed to allow the flow of heated air along the whole length and to retain more heat inside, the flue has two basic arrangements in plan: linear flue and grid flue. Each has many varieties.

Two ash pits, made between the fireplace and the *kang* proper and between the *kang* and the chimney respectively, are designed to collect ash and to prevent the flues from blocking up. The pits are covered with slabs for ash removal. A regular removal of ashes is required, for which the frequency depends on the type of fuel used. Fuels vary from firewood, straw to coal depending on the materials available in the region. The first pit is built with an adjacent slope channel to direct the hot air from the fireplace over this pit to the flue, and to prevent back flow. The second pit has a different function: to hold the hot air before it flows out from the chimney. The bottom level of the chimney is lower than that of the flue so that cold air from outside is trapped within and prevents a counter flow to the *kang*. Moreover, a plate is set close to the ground level of the chimney to adjust the circulation of air. The *kang* is paved with bricks and coated with plaster. A mixture of clay and straw is commonly used as the first coat applied directly to the *kang*, clay with sand as the second coat, then lime with hemp fiber and finally oiled papers, or bed-covers made of mat or felt. The edge of the *kang* is finished with a long timber trim to stop objects from sliding off, as well as to protect the edge from wearing off.

In common dwellings, a kitchen range is connected to the *kang*. Thus heating is combined with cooking resulting in very good energy efficiency. The cooking ranges are characteristic of localities and vary according to fuels used (for instance, underfloor coal-fired fireplaces). The *kang* permits the conservation of energy: its surface temperature is about 40°C and can largely be maintained overnight. The *kang* is used as a bed at night, and bedding is laid out for sleeping but put away in the morning. The *kang* provides a large warm platform upon which people manage many household activities during the day.

Structure of Heated Floor: huodi

The *huodi* is a full-floor *kang* consisting of a fireplace with or without a chimney depending on the fuels used. Our knowledge about the inner structure of the Manchurian *huodi* is largely obtained from the Forbidden City in Beijing,[7] since no substructures of the Shenyang palace have come to light. On the surface the Manchurian *huodi* seems to be similar to the traditional Korean floor heating system, *ondol*, but there is a marked difference between the two.

The Manchurian *huodi* has a double-flue system one on top of the other. The lower layer of the flues is linear and the upper is a grid in plan. Between the two are holes to bypass the fire; the holes are smaller, closer to the fireplace and vice-versa. The second layer is made by regularly arranging circular bricks cut from square bricks and covered with floor bricks. The hot air travels through the lower flues, then the upper ones, to heat the floor. The *huodi* was heated from an underfloor fireplace, which was approached by a subterranean entrance at the top of the building platform, or its rear or side walls. Because of its heavy charcoal consumption, it required adequate working spaces for refueling.

The Korean *ondol* is made by digging the ground about 30 cm deep to make channels which are then covered with stone slabs. Upon the channels are floor stones with a graveled bedding of 5 cm in depth in between. The floor is then plastered with sandy mud and finally finished with oiled paper. The Korean people in Yanbian, Jilin province, China, still build their floors in this way.

The following information found in the Shenyang palace deserves attention: (1) Cooking was separated from heating. Outdoor fireplaces below ground level were employed for better circulation and hygiene compared to the indoor ones, also for segregating servants. (2) Charcoal was used for heating to reduce air pollution in the palace so that no chimney was needed. Evidence for the consumption of charcoal has been found in almost all fireplaces and in storehouses located along the north palace wall together with kitchens and granaries. (3) The buildings with *kang* or *huodi* are all residential. In the official buildings stoves and pots were used. The houses with *kang* of either the first or the second type were all constructed between 1625 and 1632, whilst those with *huodi* were built in the 1740s and the 1780s.[8] (4) The double-flue *huodi* with one flue over the other was a later technique.

Formation and Development of the Heating System

The origin of the *huodi* may be traced back to the ancient method *zhidi,* and the *kang* to *huoqiang.* Fortunately, archaeological excavations have taken place elsewhere in the region, and provide us with valuable information about their formation and development.

Archaeological Evidence

Recent archaeological excavations have revealed that at the Neolithic building remains called Xinle, in Shenyang, some floor surfaces were treated by burning so that they turned into baked clay.[9] That is, by repeatedly broiling the ground (*zhidi*) the floor became a baked floor which made an effective moisture barrier.[10] This finding confirmed another major archaeological discovery made earlier at the Neolithic settlement site of Banpo, Xi'an, where the sleeping area was defined by rising the floor some 10 cm above ground level. In the 1980s, the Beijing Silicate Institute conducted extensive tests and came to the conclusion that floors were entirely prepared by pounding clay, and some surfaces were treated by burning for heating purposes.[11] The floor was probably heated before sleeping as a Tang poet, Meng Jiao, wrote in his poem, *Handi Baixing Yin*: 'No fuel to heat the floor to sleep, standing and crying with cold at midnight instead'.

The earliest evidence of heated platforms was unearthed in four building remains of the first century at Tuanjie county, Heilongjiang province. It has been identified as an L-shaped low wall, about one meter in width with a single flue on the west and the north sides of each rectangular dwelling. The wall was made of adobe and cobblestones and covered with stone slabs. From the fireplace, the flue gradually went upwards toward the chimney.[12]

Unlike the first-century remains, heated walls of the fourth-century had double flues, suggesting that the structure represents more than just a wall, but a masonry bench. The example was found in a palace building dated to A.D. 392 in Ji'an, Jilin province.[13] The palace has an L-shaped adobe bench with a pair of parallel flues.[14] It was structurally more complex than that, having a single flue, and functionally closer to the *kang*. While the surface was made of stone slabs, the flues were formed with earthenware terracotta. In the middle of the building is an open fire pit. Heated walls can still be seen in Northeast China, where they are made either high as a dividing wall (*huoqiang*) within a building or low as a sleeper wall (*dilong*) under a wooden bed.

Dated *kang* remains of considerable interest have been discovered further north at Longquanfu Palace of Bohai, a Tang vassal (699–926, present Ninghai, Heilongjiang province).[15] The building is 30 m east-west and 17.3 m south-north in plan. It has three rooms with a veranda all around.

Each room except for the middle one has an L-shaped *kang* with two parallel flues connected to a huge stone chimney at the back about five meters apart from the building.

A grouping of *kang* was found at a Song pottery manufactory remains (Song Dynasty, 960–1126) in Yaozhou near Xi'an. The *kang* were in a dormitory of pottery-makers and in workshops which dried earthenware prior to firing. In one building a *kang* constructed of heat-resistant masonry is 2.7 m by 1.68–1.82 m in plan and 0.38 m in height with a cooking range.[16]

Literature Evidence

Chinese historical books record heating facilities. The *keng*, open fire pits of the Gaogouli period (or Koguryo in Korean), are mentioned in the *Old Historical Book of the Tang Dynasty (Jiu Tangshu, 945).*[17] It reads: 'the Gaogouli people lived in mountains. They used straw to roof their houses, whereas only temples, shrines and palaces were covered with tiles. The common people made long fire pits (*changkeng*) to warm themselves in winter.' This is repeated in the *New Historical Book of the Tang Dynasty (Xin Tangshu, 1061).*[18] As the Chinese character for *keng* has the word representing earth for its radical, fire was made in a long fire pit in the ground. The pit was situated in the middle of the building and was surrounded by stones, which were used to keep the fire in place and to retain the heat. The *ondol* may have been invented to reserve the heat over-night by simply covering the long fire pit with slabs of stone on which people slept. In this connection, the *keng* may be the *ondol*'s origin.

A heated floor was described in the book *Commentary on the Waterways Classic* (Shui Jing Zhu, Chapter *Baoqiu Shui*, 5th century). It was utilized in Guanji Si Temple near present Fengrun, Hebei province: 'In the temple, there is a grand hall, high and wide to allow a thousand monks (to chant scriptures). The platform of the hall was made of stones arranged as a network of channels covered with floor bricks laid with mortar. Fires were made at the platform sides outdoors, while the flames and heat were retained inside, which warms the entire room.'

The Manchurian *kang* was recorded in Chinese historical books too, such as the *Collected Records of the Northern Alliance during Three Reigns*[19] and the *History of the Great Jin:*[20] 'Built on hillside terraces, houses were made of wood and roofed with thatch or shingle, or bark surrounded by wooden fences. The door faced east. The bedroom had an earth bed on three sides. Fire was set underneath, and people ate and lived on the *kang*.'

Hypothesis

The *kang* is the kind of innovation that comes from a combination of ideas, and it must owe its techniques to pottery. It might have been first evoked by the example of pottery stoves, then by replicating kilns. The models of stoves together with kitchen utensils as part of tomb furnishings (*mingqi*) for burial purposes are widely recovered from ancient tombs in China (the dead were served as the living; nobility and the rich had elaborate tombs). They are scaled models of actual pieces in various styles of the 2nd century B.C.

From the Neolithic period, Chinese started to produce pottery and also invented porcelain and complex glazing techniques. By making ceramics and bronze, ideas and skills concerning the application of heat were developed. All these may well have marked the beginning of the *kang*. Yet let us not jump to the conclusion that the *kang* was formed in China at that early date. The formation of architectural heating systems was a complex process, which has been related to the daily mode of living. Archaeological evidences and literary documents suggest that the heated floor appeared in China in about A.D. 500. The Chinese ate, sat, and slept on the floor from ancient times. Furniture represented by a low table, big wooden bed, and armrest were excavated in Henan, Hunan and Guangdong provinces, dating back to the Warring Sates (453–221 B.C.). In the Northern-and-Southern Dynasties (317–589 A.D.), furnishing as a concept (and related pieces of high furniture) entered with Chinese through trade, commerce, and migration. This led to changes in the manner of sitting — from floor to chair. A new life style came to the fore. Evolved in northern China, the *kang* inherited the bed feature and developed into a household object of special technique. That is to say the *kang* as 'high furniture' was created by raising a platform above ground level, with low tables, screens and cabinets still in their assigned positions. The *kang* was definitely a revolution in architectural design in so far as furniture was introduced for the first time. However, in the Tang dynasty (618–907) two sedentary modes became equally popular, which are evidenced by various paintings, murals and reliefs. The ancient sedentary lifestyle — seated with legs crossed or knees bent — is still kept by Koreans, Mongolians and Japanese. Probably the Korean heated floor, the *ondol*, has ensured the continuation of this custom in Korea. According to Chang Kyong-cho,[21] the *ondol* in Korea was firmly established about 350 years ago in the middle of the Li dynasty (1392–1909), though many early remains of the Gaoli period (or Koryo in Korean, 918–1392) have been found in northern Korea, such as at Dong-po-gun, Pyung-an-buk-do, Shinuiju-si and Man-wal-dae.[22] The heated floor can also be seen in a type of Mongolian tent used as lamaseries in Mongolia. However, the open fire pit located between floor mats (*tatami*) has always been utilized in Japan, where severe weather is also a factor.

For the Manchurians, fundamental changes in the daily mode of living occurred in the 12th century. The big heated bed utilized by the Manchus suggests Chinese influence. The traditional Manchurian architecture was constructed of rough-hewn logs on maintains, and its walls were mud plastered to resist the wind. The house always faced south to make good use of sunlight and radiated heat. Upon moving away from the mountains, the log houses gradually gave way to tile-and-platform timber structures, at first among state buildings, temples and halls of the gentry, and later among urban dwellings. The heated platforms cater to the needs not only of high society but also of ordinary folk. Tiles are resistant to water penetration and platforms keep people away from excess damp. More importantly, they are both fireproof. For Manchurians, the new life style emerged by compromising with the old: elevating themselves on a raised *kang*.

High furniture was further adopted, the *kang* heating system were maintained but modified. While enlarging the size of the *kang*, the heated floor emerged in Qing architecture. The heated floor did not replace the *kang*, but co-existed with it and even added variety to Manchurian heating. In other words, when the *kang* took a new form the older form continued to co-exist with it, though the heated floor had many advantages over the heated bed: it did not require people to take off boots inside of houses, and it was suitable for employing wooden beds. These are the special characteristics of Manchurian architecture that distinguish it from Korean. For Koreans, it is a custom to remove one's shoes on entering houses, and the Japanese custom is similar. Houses covered half with *ondol* floors and half with wooden floors are the best solution to different requirements between winter and summer in the Southern Korean peninsula.

Since ancient times, the *kang* has been widely utilized over a vast geographical area in Northern China. It gave people protection and made it possible for them to live in a cold climate. As a critical architectural component, the *kang* has been found in cave dwellings and pit houses. Without it the oldest forms of habitation might not have been habitable by uncountable thousands of people in the thick loess reaches of the Yellow River, such as Hebei, Henan, Shanxi, Shannxi, Gansu provinces, and Inner Mongolia. All *kang* look surprisingly alike. The Manchurian *kang* shares many of the same characteristics as those in China. Thus the proposal that the *kang* originated in northern China seems reasonable.

As time went on many variations were developed, such as the heated partition wall (*huoqiang*), the sleeper wall (*dilong*) and stave bed (*huochuang*). The technique was developed extensively during the Qing. The 'water *kang*' used in public baths represents another variations of a type. The *kang* has been in use through to the present day. It obviously plays a major role in architectural history. The formation and development of the *kang* are still open to debate, but it is clear that the *kang* is a Chinese innovation worthy of study.

Notes

[1] Before 1635, Manchuria was called Nuzhen. Zhang Bibo, *History of the Minorities in North China* (Zhongguo Gudai Beifang Minzu Wenhua Shi). Harbin: Heilongjiang Publishing House, 1993.

[2] Xu Shen, *Analytical Dictionary of Characters* (Shuowen Jiezi), Chapter 10. A.D. 121.

[3] *Houdi* was mentioned in the classics, *Dream of the Red Chamber* (Honglou Meng), by Cao Xueqin, 1763.

[4] Li Daoyuan, *Commentary on the Waterways Classic* (Shui Jing Zhu), late 5th or early 6th century. Chapter *Baoqiu Shui*.

[5] Qinghua Guo, 'Shenyang: The Manchurian Ideal Capital City and Imperial Palace (1625–1643)', to be published in *Urban History*, 2000.

[6] Fu Yuguang and Meng Huiying, *Manchurian Shamanism* (Manzu Samanjiao Yanjiu). Beijing: Beijing University Press, 1991.

[7] Bai Lijuan, 'Technique of Floorheating System', *Traditional Chinese Architecture and Gardens* (Gujian Yuanlin Jishu) 4 (1996), 14–15.

[8] When the Qing ruler moved the government to Beijing in 1644, Shenyang retained a special status as an auxiliary capital, and its palace received eleven visits from the Manchurian emperors.

[9] Shenyang Office for Preservation of Antiquities, (1) 'Excavation of Neolithic Village at Xinle, Shengang', *Kaogu Xuebao* 2 (1985), 209–22. (2) *Kaogu Xuebao* 4 (1978), 449–66.

[10] Institute of Archaeology of Academia Sinica, *The Neolithic Village at Banpo, Xi'an* (Beijing, 1963), 16–17, 27.

[11] Miao Jisheng *et al.* 'Preliminary Investigation of Cementing Material Used in Ancient China', *Journal of the Chinese Silicate Society* (Guisuanyan Xuebao) 2 (1981), 234–40.

[12] (1) Heilongjiang Relics Investigation Team and Archaeological School of Jilin University, 'Report of Excavation of Cultural Relics at Tuanjie, Dongning', in the *Proceedings of the 1st Conference of the Archaeologists Society of Jilin*, 1979. (2) Zhang Taixiang, 'Architectural Remains at Tuanjie, Tongning', *Guangming Daily* (Guangming Ribao), 23 July 1978.

[13] Jilin Museum, 'Gaogoli Architectural Remain Cleared up at Ji'an, Jiling', *Archaeology* (Kaoku) 1, 1960.

[14] 'Building Remains at Dong Taizi, Ji'an', *Cultural Relics* (Wenwu) 1, 1960.

[15] Wei Cunji, 'Bohai Architecture', *Heilongjiang Cultural Relics* (Heilongjiang Wenwu Congkan) 4 (1984), 36–43.

[16] Institute of Archaeology, Shaanxi Province and Yaozhou Kiln Museum, *The Yaozhou Kiln Site of the Song Period* (Beijing, 1998), 29–31.

[17] Liu Xu, *Old Historical Book of the Tang Dynasty* (Jiu Tangshu, Passage Gaoli), 945. Chapter 199.

[18] Ou'yang Xiu, *New Historical Book of the Tang Dynasty* (Xin Tangshu, Passage Gaoli), 1061. Chapters 145 & 220.

[19] Xu Mengshin, *Collected Records of Northern Alliance during Three Reigns* (Sanchao Beimeng Huibian), 1196. Chapter 3.

[20] Yuwen Maozhao, *The History of the Great Jin* (Da Jin Guo Zhi). Song Dynasty. Chapter 39.

[21] (1) Chang Kyong-cho, *Traditional Korean Architecture* (Hanguk ui chontong konchuk). Seoul: Munye Chulpansa, 1993. (2) Veritable Records of the Jungjong (Jungjongsilrok), Chosun era.

[22] Li Dianfu, *Archaeological Research of Northeast China* (Dongbei Kaogu Yanjiu). Zhenzhou: Classical Book Publisher, 1994.

The Influence of Chinese Techniques on the French Silk Industry as Shown by French Patents from 1791 to 1860[1]

MAU CHUAN-HUI

Ecole des Hautes Etudes en Sciences Sociales, Paris

The end of the 19th century was the turning point for the Chinese in their traditional thinking about their silk industry. From that time on, the Chinese were no longer proud of the quality of silk produced in China, and began to consider modifying Chinese silk-making techniques after the model of their Occidental counterparts.[2] Today, this analysis is shared by almost all modern historians.[3] The contribution of Chinese silk manufacturing techniques to those of the West have been largely overshadowed in historical accounts by Western criticism. However, the West continued to praise ancient Chinese techniques previous to the Song dynasty. And, we should not overlook the fact that Chinese know-how contributed to the development and success of Western silk manufacture.

This paper aims to understand the adaptation of Chinese silk techniques by the French silk industry through an examination of the French records of patents from 1791 to 1860. I will focus mainly on three issues: what were the needs and context of the French society that motivated inventors to appropriate Chinese silk techniques; which techniques did the French inventors choose; in what ways and to what extent did these adaptations differ from Chinese techniques.

The patent system in France is an ideal source of information not only because a wide range of inventions in France during this period were recorded, but also because in these patent requests, inventors would sometimes reveal the source of their inspiration and the reasons why they invented them. Although these patents may not always have technical value, I believe that by dint of their sheer volume and frequency, certain trends can be inferred. The period I cover begins in 1791, when the system for patents was created, and ends in 1860, just before the implementation of free trade between England and France.[4]

This article is divided into three sections. Section I presents the historical background of the patent system in France. Section II briefly discusses my research methodology. In Section III, I will discuss the results of my analysis.

I. Description of the Patent System in France[5]

The origins of the French patent system can actually be traced to England.[6] By the end of the 18th century, a patent system had already been established in England.[7] Its success encouraged English inventions and began to threaten French industry. In fact, as French inventors were attracted to England to register their discoveries and inventions,[8] a trade deficit with England began to develop. Therefore, in order to protect "highly-talented men with neither credit nor fortune",[9] the French Parliament set up its own patent system in 1791.[10] In 1801, a *Société d'Encouragement pour l'Industrie Nationale* was founded[11] — another institution copied from the English Society of Arts — which frequently awarded prizes to different inventions according to the need for development in particular industries.

II. Methodology

To determine whether a given French silk technique had its inspiration from Chinese silk techniques, one must study in detail the individual requests for a patent in French silk techniques. Because there are tens of thousands of patent requests during this period, what is at issue is the identification of specific requests. To accomplish this, I examined three sources: the catalogue of patents, request files, and the *Bulletin de la Societe d'Encouragement*[12] (BSE).

For each request, the Catalogue of Patents provides a short description of the patent. When retrieving the description, I used "silk", "silk cloth" and "silkworm" as key words at the initial stage.[13] Within this group, I then looked for any direct references to China. This yielded the first group of files I examined, which we shall call the "Directly-Influenced Group". I then determined a second group from the descriptions of those patents whose technique could reasonably be considered to have Chinese origin. If a particular technique had not already existed or had not been conceived in France, but was found in China and introduced to France, then it would be included in the "Indirectly-Influenced Group".

A few patent descriptions directly mention in their title that they aim to imitate Chinese processes or products. For these patents, we can consult the files to analyse in which manner they adopted Chinese techniques or products.

Nevertheless, there are many patents for which their inventors did not mention the source. In these cases, we must search through the files one at a time to compare French inventions with parallel Chinese processes to determine if the inventors had to any extent been influenced by the Chinese. These analyses are very delicate as a French silk worker is equally capable as his Chinese counterpart of profiting empirically. In such cases, we must credit his invention to independent discovery.

In this regard, the periodical *Bulletin de la Société d'Encouragement*, as well as correspondences and certain published works translated from Chinese or on Chinese techniques are useful for clarifying the historical circumstances. At the moment we do not possess many biographies of the inventors, which might have provided us with information to prove their sources of inspiration. Except for these difficulties, patent descriptions often give us their opinion on Chinese techniques and the reason why these techniques are interesting.

III. Results of Analysis

The following analysis will be done on two levels: quantitatively and qualitatively. According to the nature of the inventions, I have divided them into four categories: sericulture (preparation of silkworms, feeding and silkworm raising); silk making (silk reeling and threading); silk products (construction of weaving looms and new cloths, and finishing); silk boiling and dying.[14]

Quantitative Results

Among 65,242 patents of all kinds requested, there are 836 concerning the silk industry, 1.28% of the total. According to the statistical curve, we can divide this period into three phases: (a) Germinal stage: 1791 to 1815; (b) Development stage: 1816 to 1840; and (c) Transformative stage: 1841 to 1860.

(a) Germinal stage:

> At the beginning of the patent system, there were not many requests. During this period, patents concerning the silk industry took on an important role[15] and reached the highest percentage: 25% in 1797. Nevertheless, the frequency with which these requests were made was not stable. Those requests concerning the silk making and silk products sectors[16] covered almost all requests and represented a fair importance. There is only one patent influenced by Chinese techniques which I classified under the "Indirectly-Influenced Group".

(b) Development stage:

> From 1816 onwards, the number of patent requests increased steadily; a phenomenon common to silk making, silk products and finishing. In

1827, for the first time a sericulture patent request appeared. One patent in the "Directly-Influenced Group", and three belonging to the "Indirectly-Influenced Group" were requested.

(c) Transformative stage:

After a lapse from 1848 to 1852, silk making patents and finishing patents witnessed a remarkable growth in 1853. Nevertheless, silk products patents experienced a light downturn. I identified two patents in the "Directly-Influenced Group" and 10 in the "Indirectly-Influenced Group".

Qualitative Results

(a) The Directly-Influenced Group:
There are three patents we can classify as the "Directly-Influenced Group".

Process for producing "Crêpe de Chine" requested by Dugas Frère & Co., 1816[17] Crêpe de Chine refers to thin silk with a taffetas weaving pattern. It corresponded to the fashion for light silk cloth, and at the same time answered to the threat coming from English products, since the end of the 18th century, most noticeably English gauze. This event re-ignited interest in the French silk industry in the use of Chinese raw silk.[18]

The dye "Chinese Black" (Noir de Chine) requested by Conte, 1845[19] This is a suggestion to replace certain natural ingredients in dying processes with industrial products, for example, sodium and mordant of iron. This patent was one of the results of the *Mission Lagrenée* to China from 1843 to 1846.[20] During the Mission's investigation in China, a lot of useful information concerning Chinese silk techniques had been sent to France. After the Mission was finished, several exhibitions were organized and works published. Several prizes, especially those relating to dying processes, had been offered by the French Government because the field had been the weakest one in all of the French silk industry.[21]

A spinning wheel for a silk filature by Chorel-Mieton, 1856[22] In his patent request, Chorel-Mieton debated the advantages and weaknesses of the Chinese spinning wheel: the Chinese spinning wheel produced a thread of superior quality to that of French spinning wheels, but its speed of weaving was by comparison slower.

In response, Chorel-Mieton's invention combined the best aspects of both machines.

(b) The Indirectly-Influenced Group:

As a criterion, we will select just those which were the first to have been influenced by Chinese techniques: This choice will reduce the quantity of inventions, and eliminate those which aimed only to modify French inventions that had already adapted Chinese techniques.

Silk making   This field covers killing moths in the cocoons, silk reeling and silk twisting. From the description of patents, we can only deduce Chinese influence by the similarity between Chinese techniques and these patents. We can number five patents probably influenced by Chinese techniques: one concerning chrysalis killing,[23] and four, silk reeling.[24]

Silk making was the earliest field to be influenced by Chinese silk techniques. In the Germinal stage, the one patent that we find aimed to modify Vaucanson's silk reeling machine. Its inventor, Tabarin, added a smoke column to keep smoke from dirtying the silk being reeled.[25] The idea of constructing a smoke column which would conduct smoke outside, thus conserving the heat and preventing smoke dirtying the silk being reeled had been mentioned in *Nongzheng Quanshu*,[26] which had been imported into France in the 18th century.

In the Development stage, one patent concerning the method for killing moths in cocoons using steam was requested. The inventor combined the Chinese way with European techniques to introduce steam through a pipe.[27] Another patent was the invention of Balay Jr. and Vignal. The inventor applied a Chinese concept in order to dry the silk as soon as possible after it has been reeled.[28] To achieve this result, the inventor installed a heat-conducting pipe under the reeled silk.

In the Transformative stage, I will just say that two requests were made. Like the one under the Directly-Influenced Group requested by Chorel-Mieton, they were the result of comparison, analysis and adaptation of their initial source of inspiration — Chinese techniques.

Finishing   This field consists of silk boiling, dying and fabric-printing. Two patents were requested in 1822[29] and 1844[30]; they appear to have been influenced by Chinese techniques concerning both silk cloth boiling and dying. The first patent was for boiling and dying silk goods; the second version aimed to boil and dye the silk product at the same time.

Due to the fashion factor and for technical reasons, these processes did not exist in France before the 19th century. Previously, except for gauze production, the French silk manufacturers used to throw, boil and dye the raw silk before weaving because figured silk was more in demand and the know-how to weave raw silk and to dye silk fabric was lacking.

As already mentioned, at the end of the 18th century, light silk cloth was in vogue and the French silk manufacturers needed to find a less costly process.

Sericulture   This field was the last one to figure in French patent records, and only after sericulture came to be considered an industry in France. The first patent, requested in 1827, was also one that had adapted Chinese techniques. From 1840 to 1860, six patents were requested: two concerning installation of silkworm farms ; one, silkworm feed (processed mulberry leaves); and three, silkworm breeding.

During the 17th and 18th centuries, some Chinese works relating to silk techniques had been imported into France by Jesuit missionaries, and several investigations were made and reported to the French *Academie des Sciences*, resulting in some published works and analyses.[31] But those concerning sericulture had been neglected and were not to be taken up again until the end of the 18th century.[32] Many members of learned societies tried to apply Chinese techniques in their silkworm breeding experiences or to analyze substances used in Chinese sericulture (for example, lime used in disinfection of silkworm farms) and then published the results in their associations.[33] At the same time the French government was importing Chinese white silkworm eggs to introduce them into French breeds, but the better quality and lower prices of Chinese silk made this experiment far from cost-effective, so the project was abandoned.[34] It was not resumed until 1806 when the Continental Blockade against England started.

In 1835, after the publication of the experiments by d'Arcet and Camille Beauvais, Beauvais called attention on Chinese techniques and requested the French government to translate the three Chinese works imported into France in the previous century.[35] In his publication, d'Arcet praised the Chinese silkworm breeding theory and techniques, but criticized their imperfect utensils. He also invented a modern silkworm farm according to Chinese theory. From then on, numerous patents influenced by Chinese silk techniques swiftly appeared. This phenomenon coincided with the French sericulture policy and the *Mission Lagrenée*, which brought back a substantial amount of information on Chinese silk work.

Conclusion

The silk industry occupied an important place at the creation of the patent system, and then gradually lost its dominance among other French industries. Within the textile industry, the cotton and wool industries got the upper hand because their quality and productivity saw improvements which in certain regards, made those fabrics comparable to or even better than silk.

At the beginning of the creation of the French patent system, Chinese influence was perceptible through the adoption of tools, and was limited to silk making.

During the Development stage, riding on scientific and technical changes, sericulture began to be listed as an industry sector in itself. During this period, Chinese silk techniques were first analyzed and experimented with, and then adapted mechanically and sometimes chemically. Up to this point, inspiration of French manufacturers by Chinese techniques still had a somewhat concrete character.

Before the Transformative stage, information about Chinese silk techniques had been imported into France, but the real study in scope and in depth had been concentrated in the period from the end of the 18th century to the first quarter of the 19th century. The second wave of study of Chinese silk techniques dates from the beginning of the French Mission Lagrenée in China from 1843 to 1846.

Some traditional Chinese techniques might have possessed ingenious ideas but employed imperfect apparatus, so that the quality and productivity had to depend on the craftmen's own ingenuity and skilfulness. Doubtless, Chinese techniques and know-how had not only influenced the French silk industry, but also paper, porcelain, ink and cotton manufacture. During the Transformative stage itself, as French mechanization matured and scientific and technical means became available, Chinese influence turned more toward concept acquisition. Isidore Hedde, delegate of the French silk industry for the Mission Lagrenée, said that "even though the Chinese system has a primitive simplicity, it solves a problem of optical effect which concerns both dioptrics and mechanics",[36] and this helps explain why Chinese silk techniques could still provide inspiration to the French silk industry all the way up to the middle of the 19th century, when French mechanization was approaching maturity.

At the same time, French manufacturers also made efforts to collect and analyze techniques in other areas of the world where silk work existed.

Notes

[1] This paper is based on "Les échanges technologiques, stylistiques et commerciaux entre les industries françaises et chinoises de la soie de la fin du XVIIIᵉ siècle au début du XXᵉ siècle", a doctoral thesis in progress at Ecole des Hautes Etudes en Sciences Sociales, Paris, France.

[2] *Cf.* Ma Jianzhong (馬建忠), "Fumin Shuo" (富民説, Theory for enriching the people), in *Shikezhai jiyan jixing* (適可齋記言記行); Xue Fucheng (薛福成),

"Chouyang chuyi" (籌洋芻議, A modest proposal on foreign affairs), etc. The arguments criticizing the inferior quality of Chinese silk extended to the beginning of the 20th century. *Cf.* Tsing Tung Chun (曾同春, Zeng Tongchun), *De la production et du commerce de la soie en Chine*, Lyon, Bosc frères et Riou impr., 1927, pp. 1–2.

[3] A majority of modern historians working on the silk industry in pre-modern China explained the reasons for mechanization in Chinese silk industry in this way. Nevertheless, a few historians reasoned differently. For example, Lillian M. Li had quoted from C. F. Remer, *The Foreign Trade of* China, to the effect that "Chinese silk is either excellent or rather poor in quality, ... Japanese raw silk is of a more uniform quality, but the best Chinese silk is said to be superior to the Japanese product ". *Cf. China's Silk Trade: Traditional Industry in the Modern World, 1842–1937*, Cambridge, Harvard University Press, 1981, pp. 84–85.

[4] The foundation of the patents system aimed to help French national industries in their competition with English merchandise. The contradiction presented by the free-trade spirit lead to the modification in the patent system.

[5] For more information, see Yves Plasseraud & Francois Savignon, *L'Etat et l'invention*, Paris, La documentation française, 1986; Caron Francois (organizer), *Les Brevets, Leur utilisation en histoire des techniques et de l'economie*, Table ronde CNRS, 6–7 December 1984.

[6] The Patent system is not the first type of method to encourage and favor specific crafts and manufactures. In fact, a system of industrial privileges, trading licenses, and tolerances were used since the Middle Age. These practices will not be discussed here because they had the adverse effect of hindering invention and development. To learn more about this system, see "Abrégé du systeme de la législation anglaise relatif aux privileges exclusifs, ou patentes accordees aux inventeurs et importeurs", in *Description des machines et procédés spécifiés dans les brevets d'invention, de perfectionnement et d'importation, dont la durée est expirée*, Paris, Impr. de Mme Huzard, 1811, pp. vij–xij and Yves Plasseraud & François Savignon, *L'Etat et l'invention: histoire des brevets*, Paris, Documentation française, 1986, 264p.

[7] The patent system was created in 1623 in England. Its influence on English industries is discussed in the book entitled *Inventing the Industrial Revolution. The English Patent System (1660–1800)*, by Christine McLeod, Cambridge, Cambridge University Press, 1988.

[8] Law of 7 January 1791 and Law of 25 May 1791.

[9] *Cf. ibid.*

[10] Without preliminary examination in the patent request processes, this system could not really offer any protection. So there existed two phenomena: some inventors having created some important inventions did not want to make requests; or, a lot of requests were done for some "inventions" which were not valuable. But the prizes offered by the Société d'Encouragement interested the inventors.

[11] This establishment was recognized in 1824 by Imperial orders as an establishment for public use.

12 This bulletin gave an account of sessions of the Conseil d'Administration, descriptions of the foreign techniques useful for French industries, and some correspondence on the new models, machines, etc. 1801–1946.

13 I also considered some particular glossaries used for silk production.

14 The patent requests contain files relating to standardization of silk making, which did not exist in Chinese silk industry. So I will not refer to this section in this paper.

15 From 1791 to 1795, there were no silk patents requested. The same thing for 1798, 1799; 1803 to 1805; 1812, 1814 and 1815.

16 There were four patents relevant to the finishing process and one aimed to measure the humidity contained in silk.

17 Institut National de la Propriété Industrielle (INPI), Dugas Frère & Co., 29.07.1816, *Procédé de fabrication d'une étoffe de soie nommée crêpe de la Chine.*

18 Regarding the menace of English gauze that appeared in France since the last quarter of the 18th century, some concluded that English gauze benefitted from a better quality owing to the superior quality of Chinese white raw silk: "Cette supériosité dans les gases anglaises viens de ce que cette nation tire directement de la Chine des soyes plus blanches que les notres, et qu'ils ont l'art de mieux files". The French Government was solicited to protect French products and to encourage introduction of Chinese white silkworms into French sericulture. *Cf.* Archives de la Chambre de Commerce de Paris, X-4.00(2), *Soie et Gazes "manufactures de Paris"*, Lettre pour Monsieur le Contrôleur général, le 17 7bre 1784.

19 Inpi, Conte, 06.12.1845, *Procédé de teinture en noir sur soie, dit «noir de Chine».*

20 A lot of information on and samples of Chinese silk work were sent into France by Isidore Hedde, who was the delegate of the French silk industry. Several exhibitions had been organized and works published after their return from China. See Hedde Isidore, *Description méthodique des produits divers recueillis dans un voyage en Chine; Exposition des produits de l'industrie sérigène en Chine ...* , Saint-Etienne, impr. de Théolier aîné, 1848. IV-400p.

21 The mordant dying processes were introduced into France via England in the middle of the 18th century.

22 Inpi, 26349 (12.02.1856), Chorel-Mieton, *Appropriation du rouet molette ou chinois au moulin français pour le moulinage des soies.*

23 Inpi, Chateau-neuf et Grand-boulogne, 16.04.1828, *Four à vapeur isolée, dans lequel on éteint la chrysalide des vers à soie.*

24 Inpi, Tabarin, le 17 fructidor an IV (03.09.1796), *Tour propre au tirage de la soie.* This patent consisted in a modification of the Vaucanson silk reeling machine.

25 *Description ...* , t. 4, p.19

26 *Nongzheng Quanshu* juan, 31.

27 There was another patent concerning the same objective: Inpi, Giraud, 04.06.1840, *Procédé dit "hygrodome" ou étouffoir dessiccatif, destiné à donner la mort et à sécher la chrysalide du ver.*

28 Inpi, Balay fils et Vignal, 15.11.1837, *Appareil destiné à sècher les brins de soie à la sortie des cocons pendant le dévidage.*

[29] Inpi, Giraud, 11.10.1822, *Fabrication d'étoffe et rubans avec la soie grège, et mécanisme propre à les décruer après leur confection, et à leur appliquer en même temps toute espèce de couleurs.*

[30] Inpi, Morel, 16.11.1844, *Teinture et cuite, par un seul bain, des soies en noir et noir-bleu dit "noir Morel".*

[31] We can quote the translation of Père Dentrecolles (1664–1741) published in Du Halde J. B. (Père), *Description géographique, historique, chronologique, politique, et physique de l'Empire de la Chine et de la Tartarie chinoise enrichie des cartes générales et particulières du Tibet, et de la Corée; et ornée d'un grand nombre de Figures et de Vignettes gravées en Taille-douce, Avec un avertissement préliminaire, où l'on rend compte des principales améliorations qui ont été faites dans cette Nouvelle Edition*, La Haye, chez Henri Scheurleer, 1736 (1ère éd. 1735), 4 vol.

[32] *Cf. Mémoires concernant l'Histoire, les sciences, les arts, les moeurs, les usages, etc. des Chinois par les missionnaires de Pe-kin*, Tome 11, 1786, p. 73, "le feu P. d'Incarville (1706–1757) en ayant vu des champs … , en parla dans ses lettres. MM. de Machault & de Trudaine lui demanderent des graines du *Siao-lan*, & des détails sur la manière de le cultuiver, d'en tirer d'indigo, & d'employer cet indigo dans la teinture. … M. de Jussieu écrivit ici il y a près de trente ans, que le *siao-lan* avoit fort bien réussi au jardin du Roi … On nous a demandé de nouveau, en 1777, quel est le procédé chinois pour préparer l'indigo".

[33] In this regard, Blancard declared in *Société Royale d'Agriculture* that he had experimented a successful silkworm raising. *Cf. Mémoire de la Société Royale d'Agriculture*, 1789. In the same paper, he mentioned the publication of the work of Macquer, entitled *Dictionnaire de Chymie* and the experiment of Marigues, in *Histoire du Dauphiné*, in 1781.

[34] *Cf.* Bardel, 'Mémoire sur l'amélioration des soies blanches, et description d'un nouveau procédé pour l'étouffage des cocons', in *BSE*, N° CXXVIII, février 1815, p. 35.

[35] *Résumé des principaux traités chinois sur la Culture des Mûriers et l'Education des vers à soie*, 桑蠶輯要, published in Paris in 1837. Three translated Chinese works were imported into France by Jesuits before the middle of the 18th century.

[36] "Le système chinois, d'une simplicité primitive, résout un problème d'effet d'optique qui intéresse à la fois la doptrique et la mécanique", *cf.* Hedde Isidore, *op. cit.*, p. 145.

Takamine Jokichi and the Transmission of Ancient Chinese Enzyme Technology to the West

H. T. HUANG 黃興宗
Alexandria, VA, USA

When we talk of technology transfer in the last hundred years, we tend to think of the traffic as flowing entirely from West to East. Actually, even in the 20th century significant bits of Chinese or East Asian technology were also being transmitted to America and Europe. Of these the most influential and yet least appreciated is the use of microbial enzymes in food processing and related industries.

For the people who live in the industrialized countries, processed foods are now a familiar feature of their daily diet. But few realize that many common processed foods are manufactured with the aid of microbial enzymes. In fact, these enzymes are employed on such a large scale that their production is now an industry in its own right.[1] The types of enzymes produced and their principal applications are listed in Table 1. The function of the enzymes is usually the hydrolysis of complex food materials into smaller units and eventually into their constituent monomers. For example, the amylases hydrolyze starch into glucose, the pectinases hydrolyze pectins into galacturonic acid, and the proteinases hydrolyze proteins into amino acids. From our perspective, it is noteworthy that most of the enzymes listed are derived from organisms from three genera of moulds, namely, the *Aspergillus, Rhizopus* and *Mucor*. The products include consumable foods, such as fast cooking oatmeal, beer, whiskey, apple juice, lactose free milk, and processed cheese, as well as corn syrup, glucose, maltodextrins, fructose, and soy hydrolysates which are ingredients used in the manufacture of packaged foods such as cookies, breads, rolls, and other baked goods.

I first gained a working knowledge of the enzyme industry in the U.S. in the early 1950s when I worked as a research chemist at the Enzyme Research Department of Rohm and Haas Co., in Philadelphia, PA. The

Table 1   The Use of Microbial Enzymes in Food Processing

Microbial Enzyme	Substrate	Process	Products
Amylases*	Starch	Hydrolysis 1,4 & 1,6 Glucose	Corn (rice) syrup Maltodextrins
G-F Isomerase	**Glucose**	**Isomerisation**	**Fructose**
Amylases*	Grains (Starch)	Mashing	Beer Whiskey
Amylases*	Oats	Hydrolysis Oatmeal	Quick cooking
Pectinase*	Fruits	Clarification Wine	Clarified fruit juice
Lactase*	**Milk (Lactose)**	**Hydrolysis**	**Lactose free milk**
Esterase#	**Milk**	**Hydrolysis**	**Italian cheese**
Proteinase# (rennet)	**Milk**	**Coagulation of protein**	**Cheese**
Proteinase*	Wheat flour Soy flour	Baking Hydrolysis	Enhanced dough Soy hydrolysate

Notes: * Aspergillus (or Rhizopus), # Mucor

enzymes printed in bold letters in Table 1 were then unknown. If we delete these we would have the state of the art in the 1950s represented in Table 2. Besides Rohm & Haas, enzymes were also produced by the Wallerstein Co. of New York, and Takamine Laboratory, Clifton, New Jersey. All three companies were founded close to the turn of the 20th century. Rohm & Haas pioneered the application of pancreatic enzymes for the bating of hides; Wallerstein introduced the use of papain for the chill-proofing of bottled beer, and Takamine was best known as the producer of the digestive aid Takadiastase which was sold widely by Parke Davis. But by the 1950s all three companies were heavily engaged in the development of new applications of microbial enzymes.

Table 2 The Use of Microbial Enzymes in Food Processing, 1951

Microbial Enzyme	Substrate	Process	Products
Amylases*	Starch	Hydrolysis 1,4	Corn (rice) syrup Maltodextrins Glucose
Amylases*	Grains (Starch)	Mashing	Beer Whiskey
Amylases*	Oats	Hydrolysis	Quick cooking Oatmeal
Pectinase*	Fruits	Clarification	Clarified fruit juice

* Aspergillus (oe Rhizopus)

Rohm & Haas Co., Philadelphia, PA		Pancreatin	Baiting of Leather
Wallerstein Co., Brooklyn, NY		Papain	Chillproffing of Beer
Takamine Laboratories, Clifton, NJ		Takadiastase	Digestive Aid

In addition to pancreatin, Rohm & Haas also marketed sheep rennet for use in the making of cheese. By an ironic twist of fate, twenty years later, in the early 1970s I became associated with Wallerstein Co., which had by then been acquired by Baxter International, and played an active role in the commercial development of a protease from *Mucor miehei* as a microbial substitute for animal rennet.[2] It is estimated that at least a third of all the cheeses made in the world today are based on the use of microbial rennets.

My professional colleagues knew of the pioneering role that the Takamine Laboratory had played in promoting the use of microbial enzymes in food processing in the U.S., but we all thought that this innovation was based on recent discoveries in the biochemistry of enzymes in America and Europe. Few of us suspected that the innovation had any connection with *koji*, a microbial culture used in Japan for centuries in the making of alcoholic drinks and soy condiments.

My interest in this relationship was kindled in the 1980s when I began a study of the history of food technology and nutrition in China. I learned that in Japanese Kanji *koji* is written as 麹 (i.e. *qu* in Chinese). Thus, it seems reasonable to me that the modern enzyme industry may actually have been descended from the *qu* of ancient China. *Qu* is a complex culture of moulds, yeast and bacteria, grown on a cooked grain medium. It holds a unique and pervasive presence in the traditional food technology of China. The constituent organisms of *qu* are involved not only in the fermentation of alcoholic drinks and vinegar, but also in the making of fermented soybean, soy paste, soy sauce, fish paste and fish sauce as well as in the preservation of vegetables, fruits and meat. Modern microbiological studies show that the principal organisms in *qu* are grain moulds of the genera Aspergillus, Rhizopus and Mucor.

The origin of *qu* is obscure. It was a known entity in the early Zhou (1000 B.C.), but it could have been in existence much earlier, perhaps even before the legendary Xia dynasty (2000 B.C.). According to the *Jiu Gao* 酒誥 (Wine Edit), c. 300 A.D., of Jiang Tong 江統 *qu* was first obtained when steamed rice was inadvertently left in the open and became mouldy.[3] This view is supported by two pieces of evidence. First, rice was already cultivated extensively and pottery steamers were known around 6000 B.C. in the Hemudu culture near Hangzhou. Second, this is precisely the way a rudimentary *qu* was prepared and used in making wine in the 1950s by the aborigines in Taiwan. Thus, the earliest *qu* probably had rice as the growth substrate. Rice is still the principal substrate in South China today. But during the Zhou, as barley and wheat (which were considered inferior grains) grew in importance as cultivated crops in North China, they were adopted as the preferred substrates for making *qu*.[4]

The art of using *qu* or *koji* to make fermented drinks, i.e. *sake*, was brought to Japan from South China, probably towards the end of the Han dynasty.[5] Their employment in the making of fermented soy condiments, such as *Jiang*, arrived later during the Tang, accompanied by the famous agricultural treatise, the *Qimin Yaoshu* 齊民要術 (Important Arts of the People's Welfare) of 544 A.D. But the Japanese soon developed their own versions of soy condiments such as *miso* which is quite unlike the fermented *jiang* 醬 (soy paste) of China. By the 19th century the mouldy grain had been an integral part of the food system for so long that most Japanese presumably assumed that it was an indigenous development. After the opening of Japan to the West, the Japanese studied *koji* with the methods of modern microbiology and identified its principal mould as *Aspercillus oryzae*.

The technology of *qu* was also transmitted, presumably by Chinese immigrants in more recent centuries, to Indonesia. It is called *raggi* by the natives or *peh-khak* (白麴) by the Chinese settlers. In the 1890s Dutch scientists determined that the fungi in *raggi* were species of *Mucor* and *Rhizopus*. They tried to exploit the amylolytic activity of these fungi for converting grains to alcohol commercially in Seclan, France and Antwerp, Belgium, but their attempts were unsuccessful. These early experiments might have remained forgotten in the annals of East-West technological transmission had it not been for the pioneering efforts of a Japanese scientist, Takamine Jokichi, who at about the same time was performing similar experiments across the Atlantic in the U.S.

Takamine Jokichi was born in Japan in 1854, the year that Commodore Matthew Perry opened Japan to the West.[6] He graduated from the University of Tokyo in 1879, and was sent by the Japanese Government for postgraduate study in Glasgow, Scotland. Upon his return in 1883, he worked at the Department of Agriculture and Commerce. In 1884 he came as a representative of Japan to the Cotton Exposition in New Orleans, Louisiana. There he fell in love with Caroline Field Hitch, the daughter of a Civil War colonel. He proposed. The Hitch family gave their approval on the condition that he prove he could provide her with economic security. He returned to Japan and rose quickly in the bureaucracy of the Department to become the Acting Chief of the Patent Bureau. In 1887 he came back to New Orleans and married Caroline in a French Quarter wedding. During their honeymoon they visited fertilizer plants in North Carolina and he studied patent law in Washington, D.C. Upon their return he established the first superphosphate plant in Japan. Two sons were born, Jokichi Jr. in 1888 and Eben in 1890.

In his travels in the U.S. Takamine learned how the brewers and distillers in the U.S. used malt to hydrolyze grains into sugar so they could be fermented into alcohol. He realized that the Japanese *koji* was much more active than malt for the hydrolysis of grains. He studied the production of 'diastase' (amylolytic enzymes) by the *koji* mould *Aspergillus oryzae* when he returned to Japan. The results convinced him that replacing malt with *koji* enzymes would be a great improvement in the manufacture of whiskey. In 1890, with the help of Caroline's parents, Takamine made the fateful decision of moving his family to the United States and started the Takamine Ferment Company in Peoria, Illinois to produce diastase for a local distillery. The initial results were highly promising.

However, Takamine's apparent success presented an unwelcome threat to the malt producers in Peoria. They incited local xenophobia. One night in 1994 the distillery which supported his experiments was burnt down. He moved his family to Chicago, where he continued to promote his diastase preparation. Fortunately, by this time he had obtained a patent on the use of his diastase as a cure for dyspepsia.[7] Park Davis agreed to market the product under the brand name, Takadiastase. In 1897 he moved his family and business to New York, where he became interested in isolating the active principle from the adrenal gland of sheep. With the help of a young Japanese assistant, he isolated crystalline adrenaline (i.e. epinephrine) and patented its pharmaceutical applications in 1900.[8] The commercial success of Takadiastase and epinephrine soon made Takamine a wealthy man. He owned a townhouse on Riverside Drive, New York, and a country house in Sullivan County, New York. He founded Sankyo Pharmaceutical Company of Tokyo and Takamine Laboratory in Clifton, N.J. which became a pioneering center of research and development in industrial enzymes in the U.S. He was an influential figure in U.S.–Japan trade and cultural exchange. In 1911, when he learned that Mrs. William Howard Taft was interested in beautifying the Tidal Basin in Washington, D.C. he helped to arrange a gift of 3000 cherry trees from the Mayor of Tokyo to the people of the U.S.

Takamine died in 1922. He was buried in Woodlawn Cemetery, New York. His success soon prompted Rohm and Haas and Wallerstein Co. to become interested in the production and marketing of microbial enzymes. By the 1950s, microbial enzymes had become a firmly established segment of U.S. industry. It was soon to take root in Europe and Japan. The commercial success of grain mould enzymes was to inspire the discovery and commercialization of new enzymes from organisms that are far removed from those found in *koji* or *qu*. These innovations have aided, for better or for worse, the great expansion of the consumption of processed foods in the industrialized countries in the latter part of the 20th century.

In retrospect, it is easy to see why Takamine succeeded in the U.S. while his contemporaries failed in Europe. The principal reason is, I think, that his approach of utilizing the amylase activity in *koji* was more creative and flexible than that employed by his European contemporaries, who used the whole culture to duplicate the Chinese process for converting grains to alcohol. By contrast, Takamine's strategy was to isolate the enzyme activity and then use it in existing Western processes which require the hydrolysis of starch to glucose. He understood that the activity of the culture was extra cellular and found that it could be extracted with water and then precipitated by the addition of alcohol. What he did was highly innovative for the 1890s when the question of whether enzyme activity could exist outside the living cell was still a matter of controversy.[9] He also made major improvements in the *koji* process, which greatly reduced the cost of production. First, he lowered the cost of the substrate by using wheat bran instead of rice as the growth medium.[10] Secondly, he hastened the rate of growth of the mould by incubating the inoculated medium on a rotating drum.[11] He studied the U.S. market to ascertain the potential applications of his product in the food and related industries. When he failed to convince the distillers to replace malt with mould enzymes, he developed a new use for his product, that of a digestive aid. The commercial success of the latter not only provided him with economic security but also the means with which to support the development of other applications for microbial enzymes in U.S. industry. Finally, he was incredibly confident and persistent in the pursuit of his goals. He successfully wooed and married an American girl in an age when such an interracial marriage was frowned upon on both sides of the Pacific. To move his family from the security of his Japanese environment to an unknown venture in a foreign land was an act of remarkable faith and courage. We marvel at his dedication and determination in carrying on even in the face of fierce xenophobic opposition. Through his pioneering efforts he became the founder of the modern enzyme industry.

Because the products of this industry are used by manufacturers rather than consumers, they are practically invisible to the general public. And yet, in the developed countries today, enzymes derived from or inspired by *qu* or *koji* may be touching the life of the average citizen in unsuspecting ways. Indeed, each time he consumes a cup of lactose-free milk, drinks a glass of apple juice, gulps down a tankard of beer, gobbles a slice of bread, eats a bowl of fast cooked oatmeal, inserts a slice of processed cheese in a sandwich, sprinkles grated Romano cheese on spaghetti, or pours corn syrup on a pancake or waffle, he may be the beneficiary of an ancient Chinese technology that was transmitted to the West by Takamine Jokichi barely a hundred years ago. It is fitting that at this conference we should celebrate his remarkable life and pay tribute to his achievement in transplanting an East Asian technology to the West and transforming it into a modern industry that affects the daily lives of millions of people around the world.

Notes

[1] For a survey of industrial enzymes see William M. Fogarty ed., *Microbial Enzymes and Biotechnology*, Applied Science Publishers, Essex, England, 1983.

[2] Peppler, H. J., J. G. Dooley and H. T. Huang, *J. Dairy Science*, 59 (1976), 859–62.

[3] *Jiu Gao* 酒誥 (Wine Edit), c. 300 A.D., of Jiang Tong 江統 reads as follows: 酒之所興，肇自上皇。或云儀狄，一曰杜康。有飯之盡，委之空桑。郁積成味，久蓄氣芳。本出于此，不由奇方。

[4] For further details on the origin of *qu* see H. T. Huang, *Science and Civilization in China, Vol. 6, Part 6, Fermentations and Food Science*, Cambridge University Press (Forthcoming).

[5] Shinoda Osamu, "On the Origin of Japanese Sake Brewery", in *The Japanese and Cultures of the Southern Areas,* Committee for the Commemoration of Takeo Kanaseki's Seventieth Birthday, Tokyo (1967), pp. 551–74.

[6] Bennet, J. W. "Preface", in *Takamine, Documents from the Dawn of Industrial Biotechnology*. Miles, Inc., Elkhart, Indiana (1988), pp. vii–xvi. I thank Bill Shurtleff for drawing my attention to this book and Celeste Aaron for generously providing me with a xerox copy.

[7] Takamine Jokichi, *Preparing and making taka-koji, U.S. Patent 0,525,820* (1894). See also Takamine Jokichi, "Diastatic Fungi and their utilization", *Amer. Journ. of Pharmacy* (1898), 70, pp. 137–41.

[8] _____ , in "Adrenalin, the Active Principle of the Suprarenal Glands and its Mode of Preparation", *Amer. Journ. of Pharmacy* (1901), 73, pp. 523–31; *German Patent 131,496* (1901).

[9] In fact, the question was not settled until 1897 when Büchner showed that a cell-free extract of yeast was able to carry out the fermentation of glucose to alcohol.

[10] Takamine Jokichi, *Process of making diastatic enzyme, U.S. Patent 0,525,823* (1894).

[11] _____ , *Process for producing a diastatic product. U.S. Patent 1,054,324* (1913).

The Modernization of the Petroleum Industry in Japan

ISHIDA FUMIHIKO 石田文彦 & ISHII TARO 石井太郎
Joetsu 上越 *University of Education, Japan*

Introduction

It is said that the modern oil industry began when Drake first succeeded in drilling oil wells with the cable tool system in the U.S.A. in 1859. In Japan, with the increase in demand for lamp oil, the oil industry began after the *Meiji* (明治) Restoration of 1868. In those days, 90% of lamp oil, namely kerosene, depended upon imports. In 1879, the Japanese government asked Lyman, an American, to survey the oil resources of Japan, and his conclusion was that they were not very promising.[1] So the government did not specify the oil industry as a target of her industrial policy. As a result, the oil industry was promoted as a local industry of Niigata (新潟) Prefecture, where 80 to 90% of crude oil in Japan was produced, on a private basis. In 1914, the First World War broke out. This war triggered the spread of the internal combustion engine, and consequently demand for oil as fuel increased and oil grew into a nation-wide industry. In this paper, we deal with the development of the technologies used for drilling oil wells and refining crude oil, from 1868 to 1924.

Oil Well Drilling

In the former period of *Meiji*, oil wells were dug by hand (Fig. 1).[2,3] Miners used picks, and they were endangered by a lack of oxygen, gushes of gas and water, and the collapse of well walls. Fresh air was brought inside the well using a blowing apparatus called the *tatara* (踏鞴). Wooden boards were used to prevent the collapse of walls, and the outside of the boards was covered with clay so as to prevent water from gushing in. The depth of the wells were around 150 meters. Although major companies stopped employing this method around 1910,[4] it was continued by smaller businesses because of the low cost.

In 1893, the traditional *kazusa-bori* (上総堀) method that had been used to drill water wells was introduced in the oil industry (Fig. 2).[5] In this method,

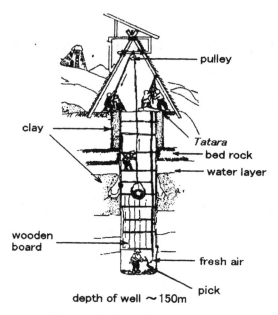

Fig. 1. Hand-dug well method

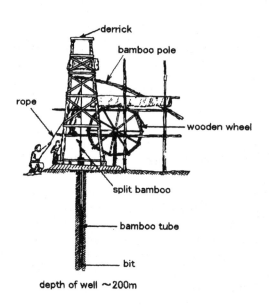

depth of well ～200m

Fig. 2. *Kazusa-bori* (上総堀) method

drilling is carried out by crushing the rock bed with an iron bit. A bamboo pole is used as the working beam, its elasticity being utilized in drilling the well. The bit is lowered down the well or raised up to the surface by the rotation of a wooden wheel. This method allows the workmen to work on the ground and is inexpensive. It was therefore used to drill wells of about 200 meters in depth.

In 1891, the *Nihon* Oil Company introduced the cable tool system from the U.S.A. (Fig. 3).[6] In this system, drilling is carried out by the iron bit, weighing about 2 tons, hanged from a cable attached to the top end of the working beam. The beam is operated by a steam engine and the rock bed of the well is crushed with the bit. The crushed rocks were taken out of the well with a bailer or sand pump. The well was shielded with casing pipes so as to prevent the well wall from collapsing and water from gushing in. Crude oil was pumped up through the oil-collecting pipe. The use of the steam engine as motive power converted well drilling from manual labour to mechanical work. This conversion permitted the drilling of wells as deep as 500 meters. Since the cable tool system requires the bit to be pulled out every time the powdered rock is removed, it takes one year in order to drill a well of 1,000 meters in depth, with a high risk of failure. Accordingly an improvement in the drilling speed was demanded.

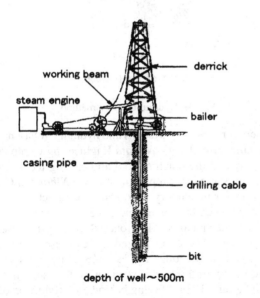

depth of well～500m

Fig. 3. Cable tool method

The *Nihon* (日本) Oil Company purchased a hydraulic rotary machine from the U.S.A. in 1912 and employed two American engineers (Fig. 4). They drilled a well of 1,000 meters in depth in 2 months.[7] The hydraulic rotary system conducts drilling while rotating the steel bit that is attached to the lower end of the drilling pipe. Muddy water is circulated inside and outside the drilling pipe to remove the powdered rock. Therefore, the work efficiency was improved. As a result, a drilling speed of 5 times that of the cable tool system was obtained, allowing drilling wells of more than 1,000 meters in depth. This rotary system has been adopted until now.

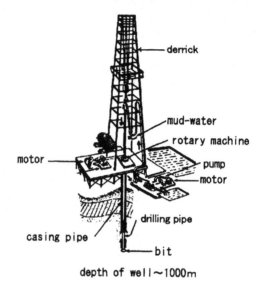

depth of well ~1000m

Fig. 4. Hydraulic rotary method

The machinery for cable tool systems and hydraulic rotary machines were imitated and manufactured by the *Niigata* Engineering Company.[8] In those days, Japan already had the machine tool and casting technology to make these. This company started as a repair plant of the *Nihon* Oil Company, and developed into a major machinery maker that manufactured oil machines, machine tools, internal combustion engines, and so on. The cable tool system uses casing and oil-collecting pipes. The rotary drilling system uses in addition to these drilling pipes. These are forged or seamless steel pipes of 2 to 15 inch diameter.[9] They account for ⅕ to ½ of the drilling expense.[10] However, the domestic production of steel pipes, the most important parts, was realized only around 1934,[11] falling behind the production of other parts due to the delayed formation of the Japanese steel industry. In Japan, iron had been produced with native *Tatara* methods. It was in 1904 when pig iron was first produced from the blast furnace at the government-managed *Yawata* (八幡) Steel Works,[12] and it was in 1913 when seamless pipes began to be manufactured at the *Nihon* Steel Pipe Company.[13]

The yield of oil rapidly increased from around 1897 along with the increase in the number of wells drilled by machine (Fig. 5).[14] In 1907, the number of wells drilled by machine was 1,251, the number of wells dug manually was 208, and the number of wells drilled with *Kazusa-bori* was 163 (figures for *Niigata* Prefecture). The machine-drilled wells accounted for 77% of the total. The *Nihon* Oil Company and the *Houden* (宝田) Oil Company led in the introduction of machines, and their yields of crude oil accounted for 84% of the total production in Japan in 1907 (Fig. 6).[15] As a consequence, the two major companies dominated the production of crude oil, which was then monopolized by one company as a result of amalgamation in 1921.

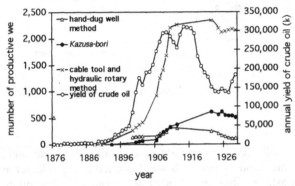

Fig. 5. Yield of crude oil and number of oil wells in *Niigata* (新潟) Prefecture

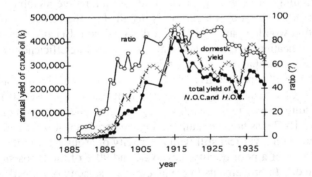

Fig. 6. Ratio of the total yield of *Nihon* (日本) Oil Company and *Houden* (宝田) Oil Company to domestic yield

In the meantime, drilling technology was developing, repeating a pattern. It was composed of the introduction of technology, the invitation of engineers from the U.S.A., and the manufacture of equipment by imitation. The domestic yield of crude oil increased gradually to three hundred thousand barrels by the end of the *Meiji* Era and then remained nearly constant (Fig. 7).[16] At present, Japan depends on imports for more than 99% of its crude oil.

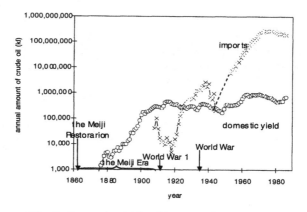

Fig. 7. Domestic yield and imports of crude oil in Japan

Oil Refining

Crude oil was distilled to separate it into gasoline, kerosene, and gas oil by using differential boiling points. It was liquefied in a condenser and further cleaned by washing with sulphuric acid and caustic soda. The oil that remains in the still is used as heavy oil, or it is distilled again to obtain lubricant, etc. In the beginning, oil distilling was carried out in a native alembic used for spirits (Fig. 8).[17] A cast iron-cauldron is installed on the furnace and a rice-cooker with a wooden pail is placed on this cauldron. Because the still had an integrated heating section and cooling section, cooling efficiency was quite bad, and refining capacity was about 50 litres.

A vertical still separated from the cooling section, namely the condenser, was introduced around 1873 (Fig. 9),[18,19] and it was widely used as it could be manufactured easily in the neighbouring foundry. The oil refining industry until around 1897 was home manufacturing, and crude oil was refined mainly for kerosene with a vertical still with a capacity of around 1 kiloliter.[20] This refined oil was of a poor quality, however, and 90% of the kerosene in use was imported.[21] Improvements in the quality and capacity of refined oil were required with the rapid increase in the production of crude oil from around 1897, and a shift to factory production began.

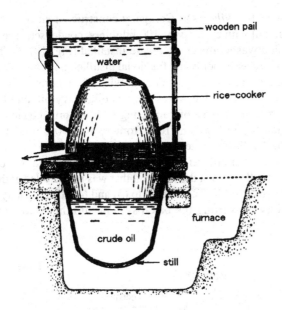

Fig. 8. Native alembic

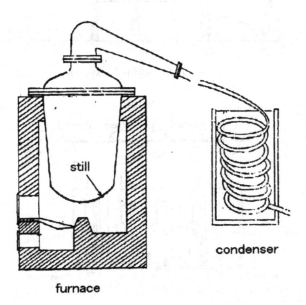

Fig. 9. Vertical still

In 1899, the *Nihon* Oil Company constructed an oil refinery where a horizontal still with capacity of 100 kiloliters, condenser, receiver, air-blown agitator, and washing tank, etc. were installed (Fig. 10).[22,23] The receiver is divided into several rooms and fractional distillation is carried out by changing the pooling room for each specific gravity of evaporated oil. The crude oil was refined into gasoline, kerosene, gas oil, heavy oil, and lubricant. High quality kerosene, capable of competing with imports, began to be produced. Self-sufficiency in kerosene rose from 10% to 50% in 1915.[24] In 1915, a continuous distillation system was introduced from Russia (Fig. 11).[25] A plural number of stills at different temperatures were installed. Crude oil was first produced at still (1), then the crude oil that was not distilled at this temperature was transferred to still (2), and then to (3) and (4). The capacity of oil refining increased by converting the batch system into the continuous distillation system.

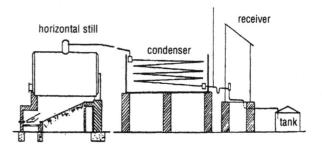

Fig. 10. Batch distillation system

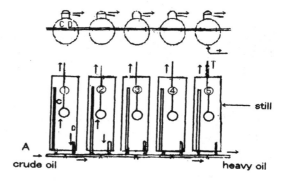

Fig. 11. Continuous distillation system

In the First World War, the internal combustion engine came into wide use and the demand for gasoline, light oil and heavy oil increased rapidly (Fig. 12).[26] On the other hand, the demand for kerosene for lamps showed a continuous decrease on account of the spread of electric lamps. As a result, the demand for fuel oil began to exceed that of kerosene in 1915.

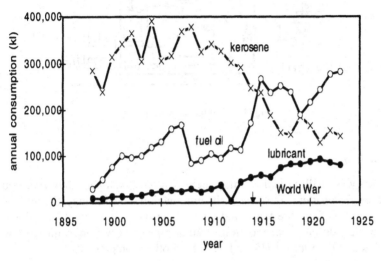

Fig. 12. Consumption of kerosene, fuel oil, and lubricant in Japan

The domestic yield of crude oil reached a peak in 1915 and remained nearly constant, while on the other hand, imports increased rapidly as shown in Fig. 7. As a result, there was a plan to convert the economic structure of the Japanese oil industry to one centreing on the refining of imported crude oil along the coast of the Pacific Ocean, where site conditions were good. The *Nihon* Oil Company also had difficulties in catching up to the age of fuel oil using conventional refining facilities designed mainly for producing kerosene. Therefore, the company constructed a refinery for imported crude oil at *Tsurumi* on the coastline of the Pacific Ocean in 1924. *Tsurumi* refinery introduced the technology of the Dubbs cracking process from the U.S.A. (Fig. 13).[27] This process was so named because it involved the actual "cracking" of hydrocarbon molecules. The refining system was converted from distillation to thermal decomposition. Crude oil is decomposed into gasoline in a reaction chamber at high temperature and pressure, and the mass production of gasoline became possible.

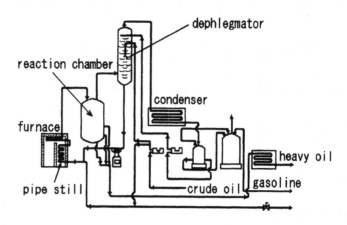

Fig. 13. Dubbs cracking process

Conclusion

As oil well drilling and refining technologies developed, they repeated a pattern. The pattern consisted of the introduction of technology, the invitation of engineers from the West, and the production of equipment by imitation. However, domestic production of steel pipe, a very important part, was realized only around 1934, falling behind other components.

Notes

[1] *Koubu syou*, ed. (工部省), *Nihon yuden tisitu sokuryou syo* (日本油田地質測量書), *Koubu syou* (工部省), 1877.

[2] Kazutaka Ito (伊藤一隆), *Nihon sekyu si* (日本石油史), *Nihon* (日本) Oil Co. Ltd., 1914, pp. 113–35.

[3] *Nihon* (日本) Oil Co. Ltd., ed., *Nihon sekiyu hyaku nen si* (日本石油百年史), *Nihon* (日本) Oil Co. Ltd., 1988, p. 136.

[4] *Houden* (宝田) Oil Co.Ltd., ed., *Houden 25 nen si* (宝田二十五年史), *Houden* (宝田) Oil Co. Ltd., 1920, p. 117.

[5] Ito, *op. cit.*, pp. 208–221.

[6] Ito, *op. cit.*, pp. 173–76.

[7] Ito, *op. cit.*, pp. 548–55.

[8] Yosihiko Yamasita (山下良彦), *Niigata tetukou jyo 40 nen si* (新潟鉄工所40年史), *Niigata* (新潟) Engineering Co. Ltd., 1934, pp. 101–102.

[9] Satio Nakamura 中村幸夫 ("*Sekiyu saikutu you kikai kigu no tokusyu zairyou oyobi sono kakou hou* 石油採掘用機械機具の特殊材料及其加工法)", *Nihon kougyou kaisi* (日本鉱業会誌), No. 530, 1929, p. 458.

[10] *Nihon, op. cit.*, p. 135.

[11] Seiiti Kojima (小島精一), *Nihon koukan kabushiki kaisha 30 nen si* (日本鋼管株式会社三十年史), Nihon (日本) Steel Co. Ltd., 1942, p. 465.

[12] *Nihon Kougakukai* (日本工学会) and *Keimeikai* (啓明会), ed., *Meiji kougyou si Tetu kou hen* (明治工業史鉄鋼篇), *Gakujyutu bunken fukyuukai* (学術文献普及会), 1969, p. 152.

[13] Kojima, *op. cit.*, p. 460.

[14] Figures for the domestic yield of crude oil are based on the following material. *Tuusyou sangyou daijin kanbou tyousa toukei bu, ed.* (通商産業大臣官房調査統計部), *Honpou kougyou no suusei 50 nen si* (本邦工業の趨勢50年史), *Tuusyou sangyou tyousa kai* (通商産業調査会), 1963, p. 226, *Tuusyou sangyou daijin kanbou tyousa toukei bu, ed.* (通商産業大臣官房調査統計部), *Honpou kougyou no suusei 50 nen si* (本邦工業の趨勢50年史), *Zoku hen* (続篇), *Tuusyou sangyou tyousa kai* (通商産業調査会), 1964, pp. 158–72, 190–91. The number of oil wells is based on the following material. *Toukyou koutou syougyou gatukou, ed.* (東京高等商業学校), *Etigo sekiyu gyou tyousa houkoku* (越後石油業調査報告), *Toukyou koutou syougyou gatukou* (東京高等商業学校), 1903, p. 22, *Toyoji Monma* (門間豊次), *Hokuetu sekiyu gyou hatutatu si* (北越石油業発達史), *Kouhou sya*, 1909, *Furoku* (附録), pp. 10–11, *Nagaoka syoukou kaigijyo, ed.* (長岡商業会議所), *Toukei nenpou* (統計年報), *Nagaoka syoukou kaigijyo* (長岡商業会議所), 1927, p. 166.

[15] *Nihon, op. cit.*, pp. 100, 154, 206, 958–59.

[16] The domestic yield of crude oil is based on the following material: *Nihon, op. cit.*, pp. 958–59, 980. The imports of crude oil are based on the following material: *Nihon, op. cit.*, pp. 960–61, Kyuhei Kobayasi (小林久平), *Sekiyu oyobi sono kougyou* (石油及其工業), *Jyou kan* (上巻), *Maruzen* (丸善) Co. Ltd., 1938, p. 26.

[17] Ito, *op. cit.*, pp. 52–53.

[18] *Nihon, op. cit.*, p. 57.

[19] Kyuhei Kobayasi (小林久平), *Sekiyu oyobi sono kougyou* (石油及其工業), *Jyou kan* (上巻), *Maruzen* (丸善) Co. Ltd., 1938, p. 459.

[20] Kyuhei Kobayasi (小林久平), "*Etigo niokeru seiyu hou* (越後に於ける製油法)", *Kougyou kagaku zatusi* (工業化学雑誌), 1899, pp. 703–736.

[21] *Nihon, op. cit.*, p. 980.

[22] Ito, *op. cit.*, pp. 284–88.

[23] Kobayasi, *op. cit.*, pp. 463, 466.

[24] _____ , p. 33.

[25] _____ , pp. 475–78.

[26] *Nihon, op. cit.*, pp. 980–81.

[27] _____ , pp. 240–41.

PART VIII

Scientific Literature

Ancient Vietnamese Manuscripts and Printed Books Related to Science, Medicine and Technology (Inventory, Classification and Preliminary Assessment)

NGUYEN DIEN XUAN
The Institute of Sino-Nom Studies
Hanoi

This paper presents the preliminary results of the cataloguing and classification of ancient and medieval Vietnamese books and manuscripts on technology and sciences written in classical Chinese and Nom.

As you may know, Vietnam is an agricultural country where the development of sciences and technology was relatively slow. The knowledge related to technology and the sciences in many cases was transmitted orally. Also, books and manuscripts on technology and science were regarded as less important in comparison with books on Confucian and Buddhist philosophy and religion. In the past, people who had both scientific expertise and literary education and were thus able to write books were very few. That is why the information found in books and manuscripts by the old Vietnamese is very precious for the studies of the history of science, technology and medicine in Vietnam, and, more broadly, in East Asia.

The author of the present paper, together with Ms. Chu Tuyet Lan investigated the books and manuscripts on science and technology found in the libraries in Vietnam and abroad. These documents mainly include: traditional medicine and pharmaceutics, cartography, mathematics, astronomy and calendar, astrology, geomancy, agriculture, handicrafts, technology, and botany. The purpose of this research is to make available to the international scholarly community data on the extant sources on Vietnamese science. The documents that we have catalogued (printed and hand-written books and maps) include microfilms available in libraries both domestic and abroad. As for overseas documents, we draw considerably on the bibliographies which were recently published in the *Sino-Nom Review*. As for the Vietnamese libraries which we did not visit, we provide our data on the basis of the bibliographies and publications in specialized scholarly journals, as well as papers presented at recent conferences. However, the resulting bibliography on ancient books and manuscripts on science, technology and medicine that we have compiled is certainly not complete. And it should be mentioned that there are several domestic libraries that we have not visited, and certainly there are many Vietnamese books preserved in private collections in Vietnam as well as in libraries abroad that we are not aware of.

In what follows we are going to introduce the extant Vietnamese books and manuscripts related to science, technology and medicine, classified according to the fields.

Medicine and Pharmaceutics

Documents on traditional medicine and pharmaceutics have the largest number and probably the greatest value among the documents on science and technology. According to Prof. Phan Ngoc, the reason for the high esteem for medicine in traditional Vietnam is due to the fact that the work of healers embodied the humanistic ideals of Confucian scholars. These documents are a treasury of knowledge on traditional medicine and pharmaceutics summarized on the basis of rich practical experience of the Vietnamese medical doctors in the framework of interaction with traditional Chinese medicine and pharmaceutics. The authors of books on medicine and pharmaceutics include professional and amateur writers and their extant works consist of hereditary and popular therapies. Among these the *Nam duoc than hieu* 南藥神效 (*Nan yao shen xiao*) (on the divine efficiency of the southern drugs), the *Hong Nghia giac tu y thu* 洪義覺斯醫書 (*Hongyi juesi yishu*) by Tue Tinh (慧靜) and the *Hai Thuong Y tong tam linh* 海上醫宗心領 (*Haishang yizong xinling*) by Hai Thuong Lan Ong Le Huu Trac (海上懶翁黎有晫) are the most outstanding books on traditional medicine and pharmaceutics. The difficulty we encountered is that medicine and pharmaceutics are mainly practical work, and, therefore, modern medical libraries are not interested in the old books on traditional medicine. For example, the Central Traditional Medicine hospital (on Nguyen Binh Khiem street, Hanoi) has only a few classical Chinese books. However, the number of books on traditional medicine and pharmaceutics in the Institute of Sino-Nom Studies is slightly less than 400. The Institute of Sino-Nom Studies obviously has the largest collection of the books on medicine and pharmaceutics in Vietnam but, unfortunately, it lacks specialists in this field. The present Traditional Medicine Association has 230,000 members in the whole country. Because each of the members is attached to a particular sphere, and treatment is their way of making a living, they most probably possess a considerable number of old books. This part of the treasury of ancient books on medicine and pharmaceutics is beyond our reach.

With several hundred book titles and hundreds of thousands of manuscript pages, immersion in works on traditional medicines in Vietnam is a worthy investment of time, resources, and expertise. The inventory, classification and preliminary assessment of the extant books are the first tasks in promoting research on classical Vietnamese medicine. The future possible output of this research might include practical applications of traditional medical knowledge.

Another task is to broaden cooperation on a history of medicine and pharmaceutics with all countries in the region and even worldwide. At ASEM 2 organized in London, the proposal of Vietnam to cooperate on the history of medicine and pharmaceutics between Asian and European countries was enthusiastically approved. This event is a hopeful sign for the future progress of cataloguing, assessment, research and application of traditional medical and pharmaceutical heritage in East Asia. The study of traditional medicine in the whole region of East- and Southeast Asia will better reveal the specific characteristics of Vietnamese traditional medicine and pharmaceutics. More information on traditional Vietnamese medicine is provided in the papers of Chu Tuyet Lan and Michelle Thompson.

Maps

Despite their small quantity, the maps we have found are of considerable value. Prof. Tran Nghia's article published in the *Sino-Nom Review*, No. 2, 1990, provides the most complete inventory and classification of the ancient maps found in the library of the Sino-Nom Institute. According to Tran Nghia's article, the Institute of Sino-Nom Studies has 27 atlases and 14 books having maps inside. There are also 8 atlas volumes in France. Among the above-mentioned documents, there are 14 maps composed during the Le dynasty (1428–1527 and 1533–1788), 1 compiled during the Tay Son dynasty (1788–1802), and 30 maps dating to the Nguyen dynasty (1802–1945). There are also 4 maps of uncertain publication date. In his article, Prof. Tran Nghia provides information concerning the authors, materials, colours of the maps, as well as the types of data and patrons. As far as the types of data are concerned, there are regional maps, national maps and local maps. As far as the patrons are concerned, there are maps of geographical situations, maps of arteries and maps of trading roads.

According to our survey, there are 74 ancient maps in libraries and archives both domestic and foreign. Vietnamese ancient maps help us know about old Vietnam's cartographic tradition and the progress of cartographic science. Moreover, the treasury of Vietnamese ancient maps bears witness to our ancestors' territorial consciousness. It is the symbol of deep tradition and Vietnam's distinctive identity.

Astronomy and Calendar

Our ancient books and manuscripts on astronomy are rare. There are 27 documents concerning astronomy. The astronomy described in these documents combines Chinese source material and the records of folk experience. These documents present the knowledge and experiences of Vietnamese people over ten thousand generations and are written in easy-to-read folk verses.

The chapter named *Thien van khao* 天文考 (*Thian wen kac*) in *Su hoc bi khao* 史學被考 (*Shi xue bei kac*) written by Dang Xuan Bang (1828–?) is one of the most representative works of ancient astronomy influenced by the Chinese tradition. The documents on astronomy are a treasury of accumulated experiences to serve our people's life. These books were closely connected with cultivation and breeding, two common agricultural activities in traditional Vietnam.

Along with astronomy, calendars also constitute an important part of our treasury of ancient books on science and technology. Documents on calendar relevant to research on history, customs, and traditional culture have drawn the considerable attention of researchers.

According to Prof. Le Thanh Lan of the Institute for Information Technology, the most ancient calendar in Vietnam is found in the treatise *Bach trung kinh* 百中經 (*Bai zhong jing*), housed in the library of the Institute of Sino-Nom Studies (call No. A2873). This is the Le-Trinh calendar computed for the years from 1624 to 1785 in *Bach trung kinh*. The first part of this book, containing the calendar for the years from 1624 to 1738, is of great value because it was published in the Le dynasty between 1739 and 1746. The other book entitled *Kham dinh van nien thu* 欽定萬年書 (*Qin ding wan nian shu*) (call No. R2200) housed in the National Library in Hanoi, (call No. R2200) is also of great value because it contains the calendar of the Le-Trinh dynasty from 1544 to 1630, the calendar of the Nguyen Lords in the South from 1631 to 1801, and the calendar of the Nguyen dynasty from 1802 to 1903. This book was reprinted in one of the two years 1849 or 1850. In addition, there is the calendar of the Le-Trinh dynasty from 1740 to 1788 and a calendar of the Tay Son dynasty from 1789 to 1801 published in the treatise *Lich dai nien ky bach trung kinh* 歷大年紀百中經 (*Li dai nian ji bai zhong jing*) (call No. A1273) found in the library of the Institute of Sino-Nom Studies.

Research on documents on the calendar not only helps historians to revise historical data but also to compare documents on Vietnamese calendars with those on Chinese calendars, thus making it possible to find out particular features of the Vietnamese calendar, as well as to describe its development over time. Nowadays, as the Vietnamese State Calendar Committee is doing research to revise and improve the present lunar calendar, it is indispensable to investigate documents on traditional calendars.

Mathematics

Documents on mathematics listed in our inventory consist of 20 book titles. The Institute of Sino-Nom Studies has 21 copies.

Mathematics developed in Vietnam very early. According to *Viet su thong giam cuong muc* 越史通鑑綱目 (*Yue shi tong jian gang mu*), Vietnamese people had mathematical examinations to choose mandarins from the Ly dynasty (1010–1225). The first examination took place in 1077. Ee also has evidence that mathematic examinations took place in 1261, 1404 and 1762. The treatises *Toan phap dai thanh* 算法大成 (*Shuan fa da cheng*) by Luong The Vinh and *Lap thanh toan phap* 立成算法 (*Li cheng shuan fa*) by Vu Huu are the most famous documents on Vietnamese mathematics.

Luong The Vinh, also known as Thuy Hien (睡軒), was born in 1441 in Vu Ban district, Nam Dinh province. He was the first doctoral candidate when only 22 years old. Because of his talent in mathematics, he was said to be the best in measurement (called Trang Luong in Vietnamese). He spent a lot of time in measuring and calculating the field surface in order to draw general principles which were passed down to later generations.

Vu Huu was born exactly in the same year as Luong The Vinh (1441). He received a doctor's degree (the second in the palace exam) in 1463. He was once the Minister of the Five Ministries. He and Luong The Vinh were known as the best mathematicians in the country, their works especially concerned with the measurement of field surface. The earliest extant document on mathematics is the *Toan phap dai thanh* 算法大成 (*Shuan fa da cheng*) by Luong The Vinh. Two copies of the treatise are found in the library of the Institute of Sino-Nom Studies. More information on Luong's treatise is provided in A. Volkov's paper presented at this conference. Another document is the *Dai thanh toan hoc chi minh* 大成算學指明 (*Da cheng shuan xue zhi ming*) written by Pham Gia Ky (1791–?) and revised by Pham Gia Chuyen (19th century). This treatise is also stored in the library of the Institute of Sino-Nom Studies.

Documents on Vietnamese mathematics were rich in content. For example, the *Toan phap dai thanh* 算法大成 (*Shuan fa da cheng*) by Luong The Vinh dealt with the extraction of the square root in great detail. The *Y trai toan phap nhat dac luc* 意齋算法一得錄 (*Yi zhai shuan fa yi de lu*) by Nguyen Huu Than (19th century) devotes the whole of tome 8 to a discussion on the calculation of the volumes of solid figures and the cube root. The majority of other printed books deal with geometry and metrology. These books obviously reflected the agricultural production and the needs of measurement and calculation of our ancestors.

Geomancy

Geomancy is a popular cultural practice in East Asian countries. Nowadays, there are 84 treatises on geomancy found in Vietnam. The Institute of Sino-Nom Studies has 70 treatises. A considerable number of these books were credited to the authorship of a certain Cao Bien 高駢 (China) and Ta Ao 左幼 (Vietnam).

Documents on geomancy had instructions for building graves, houses, temples, palaces and villas. ... So far, the documents on geomancy in Vietnam have not been studied. In addition to this, Vietnam has no specialists working on the history of geomancy. Only some researchers in architecture and art history show concern for geomancy, mainly related to problems in their fields. However, documents on geomancy are not only limited to the specialized fields mentioned above. Knowledge of geomancy is also expressed from early times on stelea inscriptions in many temples, pagodas, and shrines.

Documents on geomancy instruct people to construct buildings in favourable positions according to natural conditions, including the configurations of the land, water flows and so on. This suggests that the purpose of geomancy was to a certain extent close to that of present environmental science. Therefore, a comparison of environment studies with the received tradition of geomancy could probably be useful in providing solutions for ecological problems.

Agriculture

Vietnam is an agricultural country. Vietnamese people's cultivation standard is very high within the region and the continent of Asia. Knowledge of agriculture is very rich and results from the experiences and intelligence of ten thousand generations. And yet books on agriculture are not numerous. According to our evaluation, there are 74 extant treatises including documents on land registers, dikes, taxes, local products, testament, public fields, agricultural astronomy, and so on. At present, the Institute of Sino-Nom Studies has 73 documents concerning agriculture. Besides, there are 513 land registers with thousands of pages on private and public fields, especially on the changes of fields in 20 provinces in the North and Centre of Vietnam in the 18th and 19th centuries. Studying these documents, we realize that almost no work dealt with the technique of earth working, cultivating, choosing seeds/breeds, fertilizing, killing insects, irrigation, or planting and breeding. Obviously, agriculture knowledge was transmitted orally from generation to generation. It was a main traditional transmission method of the old Vietnamese. Speaking about agriculture, we cannot help mentioning the chapter *Pham Vat* 品物 (*fin wu*) in the book named *Van Dai Loai Ngu* 雲臺類語 (*Yun tai lei yu*) by Le Quy Don (1726–84). This chapter dealt with sticky rice, ordinary rice, fruits, vegetables and some experiences of agriculture. In 1994, the Institute of Sino-Nom Studies published the book entitled "Traditional Agriculture in Sino-Nom: Documents", Education Publishing House, Hanoi, 1994. Since then, it has been considered the most complete Sino-Nom inventory of the heritage on agriculture written in Han and Nom.

Traditional Handicrafts

Handicrafts quickly developed and there are many handicraft villages not only in Hanoi and neighbouring provinces but all over the country. Each has its own founder with a particular history and festival. And yet, there is almost no document of technical instructions. Nowadays, according to our statistics, there are 13 books on handicrafts. It should be mentioned that some valuable handicrafts require a thorough study. They are: the technique of food processing, weaving, ceramic making, mining, metallurgy and paper making.

Conclusion

The total number of the ancient books and manuscripts on science, technology and traditional medicine and pharmaceutics in both domestic and overseas libraries and archives equals 681 titles with nearly 1,000 call-numbers (that is, there exist several copies of some treatises). The majority are treatises on traditional medicine and pharmaceutics. The other disciplines are represented by a much smaller number of books. This treasury incorporates the knowledge of science, technology, and medicine of the Vietnamese people accumulated from generation to generation.

Unfortunately, so far, only a small part has been explored. At this conference, we hope that the international scholarly community will give us further assistance and co-operation in the exploration, research, and application of knowledge of science, technology, and traditional medicine and pharmaceutics in Vietnam.

References

Le Thanh Lan. *Some More New Discoveries on the Vietnamese Chronology*. Hanoi: Vietnamese International Conference, 15–17 July 1998.

Nguyen Thi Oanh. "The Sino-Nom Bibliography in Indo-Chinese Archives in Japan". *Han-Nom Review*, No. 4 (21) (1994).

Phan Ngoc. *Vietnamese Culture Identity*. Hanoi: Information & Culture Publishing House, 1998.

Traditional Agriculture on Sino-Nom Documents. Hanoi: Education Publishing House, 1994.

The Study of Science and Technology in Vietnamese History. Hanoi: Social Sciences Publishing House, 1979.

Tran Nghia. "Ancient Atlas of Vietnam". *Han-Nom Review*, No. 2 (9) (1990).

Tran Nghia. "Han-Nom Books in the Libraries of the United Kingdom". *Han-Nom Review*, No. 3 (24) (1995).

Tran Nghia & Francois Gros. *Di san Han-Nom Thu Muc De Yeu — Catalogue des Livres en Han-Nom*. Hanoi: Social Sciences Publishing House, 1993 (3 vols.)

Tran Nghia & Nguyen Thi Oanh. "Sino-Vietnamese Books at Four Major Archives in Japan". *Han-Nom Review*, No. 1 (38) (1999).

External and Internal in Ge Hong's Alchemy

EVGENY TORCHINOV

St. Petersburg State University

The problem of the shift from external alchemy (*wai dan* 外丹) to internal alchemy (*nei dan* 內丹) is important for understanding the history of *Dao*ism as well as for elucidating some crucial questions in the history of science in China. Briefly speaking, the practices of inner alchemy (such as visualization, breathing control, different types of contemplation, etc.) are much older than the techniques of laboratory alchemy. Indeed, they compose the very core of the mainstream of *Dao*ist practical methods and techniques. Nevertheless, those techniques and methods obtained their systematic unity only by borrowing the technical language, terminology, and theoretical background of external alchemy. The Six Dynasties (*Liu chao* 六朝) period is of extreme importance here. First of all, it was a time of maturity for external alchemy, when it flourished among *Dao*ists of all branches and trends. Secondly, this is when there appeared the first signs of the formation of the inner alchemical tradition in the midst of the laboratory alchemy of *wai dan*.

The aim of this paper is to present some evidences of the process of transition from external to internal alchemy using the materials of Ge Hong's (葛洪) *Baopuzi nei pian* (包朴子內篇). This classical and well-known work is usually treated as being dedicated purely to external alchemy. This is only partly true. Even in such a practical and experience-oriented work as *Baopuzi nei pian* (henceforth BPZNP), sprouts of inner alchemical attitudes and approaches found their way to expression.

The most interesting part of BPZNP for our purpose is certainly Chapter 18, *Di zhen* (地真), or "Earthly Truth". The contents of this chapter may be summarized as follows:

1. Metaphysics of the *Dao* (道). *Dao* (the Way as the first cosmological and/or ontological principle) is described here not only as *Xuan yi* (玄一), the hidden and unrevealed substance (analogous to *Deus Absconditus* of the theistic apophatic mysticism) but also as the self revealing principle of *Zhen yi* (真一), immanent in the very nature of empirically existing things. If the hidden Mysterious *Dao* has no

form, or image, the manifested *Dao* of the True One has an image of its own. It can be supposed (though Ge Hong does not write so explicitly) that signs of the presence of the True One can be found in everything, and are "signatures" of the *Dao* (probably the specific presence of the True One in some substances spiritualizes them, *ling* 靈, or *shen* 神); this spirituality in its turn makes such substances suitable for the preparation of different elixirs. Briefly speaking, it is but a kind of especially subtle *pnuema*, *qi* 氣.

2. Paraphysiology. Nevertheless, Ge Hong speaks in detail about the manifestation of the True One within the human body, where the mystical signatures of *Dao* are cinnabar fields (*dan tian* 丹田). Probably, Ge Hong is the first writer to speak about three cinnabar fields (earlier texts mentioned only one *dan tian*, the centre in the lowest part of abdomen, beneath the navel). Ge Hong describes cinnabar fields in metaphoric language. Here Ge Hong uses the term *shou yi* (守一 literally: "preservation of the One") which was the earliest designation of different *Dao*ist meditative and contemplative practices directly connected with inner alchemy (the practices of *shou yi* are rather well known from such comparatively early texts as the Classic of the Great Equanimity *Taipingjing* 太平經).

3. Ge Hong enumerates the following aims of *shou yi* practices: protection from demoniac attacks and influences; protection from armed enemies; protection from illness and infections. It can be supposed, therefore, that the function of these practices was purely protective. But some passages from Chapter 18 of BPZNP relate the contemplative techniques of the Preservation of the One to the leading theme of Ge Hong's discourse (i.e., obtaining longevity and immortality). For example:

> The only method of the prolongation of life and attainment of the state of immortality is but the way of Gold and Cinnabar; the only method to preserve one's body and to cut off evil influences is [contemplation] of the True One. Therefore the ancients treated such affairs extremely seriously.

This passage describes the *shou yi* practices as complimentary to the "Great Work" of the way of external alchemy.

Some fragments of the second part of this chapter are even more interesting, not only for their contents but for their composition and structure. The semantic beginning of this part of the chapter is Ge Hong's statement regarding the metaphysical relations between the manifested *Dao* of the True One and the hidden *Dao* of the Mysterious, or the Mysterious One (*xuan yi*).

Ge Hong proclaims the equal importance of purely meditative practices connected with the realization of the Mysterious One (described in the opening chapter of BPZNP) and inner magic of the True One. Nevertheless, he states that the True One practices are simpler than the Mysterious One practices. Moreover, the preservation of the True One (*shou zhen yi* 守真一) is the simplest way to preserve or keep the Mysterious One as well, because of their ontological unity (the manifested *Dao* is an "eye" through which the hidden *Dao* "contemplates" the Universe).

The practices of the preservation of the One are the methods of obtaining different supernatural powers (such as multiplication of the bodily form or contemplation of the *hun-po* 魂魄 souls within one's body).

The passage which follows seems to be irrelevant to the themes of the preceding section. Here Ge Hong in a rather eloquent manner speaks about the art of alchemy (making the Great Medicine *da yao* 大藥 or Golden Cinnabar *jin dan* 金丹) as hard work demanding great efforts and laborious behaviour. But in reality it is just an introduction to a new evaluation of the practices of preserving the One: the alchemical work leading to immortality is hard; it takes plenty of time to fulfil. Therefore the adept must do his best to keep his body in good health, protected against sickness as well as against demoniac attacks and malevolent influences of evil spirits and ghosts. Here Ge Hong mentions the *shou yi* practices together with the contemplation of the inner spirits of the body (*si shen* 思神) which must also protect the body against all destructive forces.

The next theme of Chapter 18 is the parallel between the human body and the state. In the first part of the chapter Ge Hong had given a highly symbolical description of the human body with its subtle energetic centres (here the body obtained an image of the sacred mount of Kun lun 昆侖 with its palaces and chambers of immortals; astral imaginary of constellations was also important for this passage). At the concluding part of the chapter Ge Hong simply in a rather traditional way makes analogy between parts of the body and functions of the state. His conclusion: to master one's own body is the same as to master the state; pneumata (*qi*) of the body is the same as common people (*min* 民) in the state. The *Dao*ist practitioner must nourish the pneumata like a lord of the state who must take care of his subjects. Here Ge Hong states that the presence of the True One in the body as a result of the cultivation of pneuma gives piece and stability to "three and seven", that is souls of *hun* and *po*. It will lead to the prolongation of life (*nian ming yan* 年命延) and the elimination of all evil (*bai hai que* 百害郤). The *shou yi* practices are extremely helpful (even to a greater degree than the amulets and charms described in Chapter 17 of BPZNP) for exorcisms in the wilderness of remote mountains and forests where the *Dao*ists prefer to cultivate their alchemical skill.

Therefore, it can be said that Ge Hong evaluates the inner practices of *shou yi* as having only subsidiary character. They are important in providing the practitioner of external alchemy (the principal method) with safety and ease. Nevertheless, they are necessary for the alchemical adept, and only fools are able to ignore them: "If only three gates of four are locked, the robbers can enter the building. And what can be done if all four gates are opened!" It is substantial that Ge Hong looks for a kind of harmony between external and internal methods of *Dao*ist cultivation. The leading role of the external methods still exists but the function of inner cultivation becomes very important too.

Here it seems reasonable to examine the elements of inner cultivation within the frame of external laboratory alchemy as such. It is impossible to divide technical, magical, and ritualistic aspects in the alchemical approaches of Ge Hong. He denies the idea of the automatic or mechanical effect of the elixirs, combining the technical and chemical procedures with fasting, prayers and purification (Chapter 4 *jin dan pian* 金丹篇). Everywhere in BPZNP Ge Hong stresses the importance of such practices as gymnastics (*dao yin* 道引), control over pneumata (*xing qi* 行氣), and sexual techniques (*fang zhong zhi shu* 房中之術) all of which were closely related to the formation of the system of inner alchemy. Certainly, Ge Hong was sure that all those methods could not lead the adept to his final goal, that is, immortality, but nevertheless believed that all of them were extremely valuable, helpful, and even necessary as subsidiary and additional means to prolong the adept's life or to protect him from evil and harmful influences.

In another words, Ge Hong was a master of external alchemy, which was thought by him to be the highest way to immortality but (1) this external alchemy included in itself some elements of the inner doing (purifications, sacral bathing, fasting, prayer, meditation, etc.) and (2) he believed in the great efficacy of the inner practices as subsidiary means of a macrobiotic and protective character.

Other than in Chapter 18, the term *shou yi* occurred only two times in Chapters 3 and 5 of BPZNP. The first case (Chapter 3) is a verse from an unknown classic of immortals (*xian jing*): "Those who eat medicines and keep/preserve the One (*shou yi*) can obtain the longevity of Heaven; those who practice 'returning of semen' (*huan jing* 還經) and 'embryonic breath' (*tai xi* 胎息) can prolong their life making it unlimited (*wu ji* 無極)."

The second case (Chapter 5) is the following: "The cause of death is a deficiency: old age, harm derived from illness or inner venoms or the influences of the bad pneuma or cold and wind. Because of this there exist means and methods of gymnastics, control over pneumata, returning of semen to nourish the brain, diet regulations as well as principles of rest and action, eating of the medicines, contemplation of spirits (*si shen*) and preservation of the One ... "

It is obvious that here *shou yi* is mentioned in a list of other inner practices of a subsidiary kind and palliative importance, however useful and effective they may be. It can therefore be said that BPZNP has room for the inner practices but that all of them are allowed to play only secondary roles.

Nevertheless, it is important to note that Ge Hong's treatise is one of the earliest signs of the beginning of a shift from purely external to combined and even purely internal alchemy. In more radical terms, it is possible to suppose that the element of the inner practices was included in the laboratory alchemy from its very beginning, but that the religious and cultural situation of the Six Dynasties period produced some important conditions for actualization of these hidden internal elements, their development and gradual formation into the system known to us as "inner alchemy" (*nei dan*). And Ge Hong's classic stands at the beginning of this process which became of crucial importance in the subsequent history of the *Daoist* religion.

On Some Trends in Contemporary Critiques of Shen Gua 沈括 and His Works

COLETTE DIÉNY

Centre National de la Recherche Scientifique, Paris

Since the first punctuated edition of Shen Gua's *Mengqi bitan* (MQBT) 夢溪筆談 by Hu Daojin 胡道靜 appeared in 1956, numerous articles have been published by authors applying their varying competences to the many different aspects of the works of Shen Gua. Obviously they are of unequal interest: some are hardly more than enumerations of the many subjects dealt with by Shen Gua in fields such as astronomy, biology or natural phenomena. I shall pass over these in silence. Others are the result of deep reflection and relate to certain trends in contemporary inquiry.[1] In studying these articles, I benefited from a summary of all the jottings of the MQBT compiled by my research group on Shen Gua, under the direction of Prof. P.E. Will of the Collège de France. We are in the first stage before conducting a complete translation into French of the MQBT. Members of the group are specialists in the history of astronomy, calendar making, biology, economics, music, and the fine arts.

The Chinese author whose writings on Shen Gua are the most numerous is, of course, Hu Daojing. Prof. Nathan Sivin, the well-known author of the article on Shen Gua in the *Dictionary of Scientific Biography*,[2] speaks very aptly of his "unflagging labors in China"! Hu Daojing has contributed interesting data concerning the history of the transmission of Shen Gua's works in his article *Mengqi bitan buzheng*.[3] He recalls the disaster of the proscription of books during the Song dynasty. A quotation of a passage from the *Songshi*[4] clearly shows the determination of Cai Jing 蔡京, when he was Head of the Literary Council, to banish all heretical doctrines by forbidding the reading or possession of the works of certain men of letters — his political opponents — including Shen Gua. His unrelenting efforts as well as the damages of war, according to Hu Daojing, could explain why of the 38 works attributed to Shen Gua, 29 appear to be lost, while 3 are partially and 6 entirely conserved.

In his enumeration of the editions of the MQBT that have come down to us, Hu Daojing mentions a fairly recently discovered edition (1965), dating from the Dade era of the Yuan: the Dadeben 大德本. For a detailed account of this discovery and a precise description of this edition, one must read the article by Li Zhizhong 李致忠 entitled *Yuan Dade Mengqi bitan*.[5] It is thought to have been found by chance in Hong Kong during the sale of rare editions (*shanben*). Thanks to the intervention of the "venerated president" Zhou Enlai, it escaped purchase by a foreigner and returned home to the mother country. It is kept today in the Beijing National Library.

Taking as a source the Song encyclopaedia *Shishi leiyuan* 事實類苑, Li Zhizhong maintains that the original edition of the MQBT had 30 chapters. An awkward rearrangement may explain today's illogical division into 26 chapters. According to him, the oldest known edition, dated 1166, seems to have disappeared, contrary to what is affirmed in the *Sung Bibliography* published in 1978 by Yves Hervouet. This remark should incite prudence in those who seek to comprehend some of Shen Gua's ideas through the classification of his jottings. However, Prof. Fu Daiwie, in a brilliant recent article in which he studies the classification of the MQBT and compares it to the taxonomy in the *Youyang zazu* 酉陽雜俎 of the Tang, vigorously contests the existence of an original edition in 30 chapters.[6]

To end this rapid survey of articles concerning different editions of the MQBT, I would mention the article of Hu Daojing "Shen Gua wenwu er san shi" ("Two or Three Remarks on Shen Gua's Works") of 1980,[7] in which he says that the first edition in movable type of the MQBT would be a Japanese edition, based upon that of Mao Jin 毛晉, the famous bibliographer and printer of the 18th century. Only two copies of it still exist; one is kept at the Library of the Imperial Palace, the other at the Library of the Kansai University in Osaka. Hu Daojing received this information from Prof. Sakade Yoshinobu 出祥伸.

Among the three works of Shen Gua that have been partially preserved, the *Mengqi bitan wanghuai lu*, a book on agriculture, has been remarkably restored by Hu Daojing, who collected fragments from seven different sources. The article that he has dedicated to this work deserves to be presented here, at least briefly.[8] This "Collection of things one might forget" — such is the meaning of the title according to Hu Daojing — belongs to the genre *sanju xitong* 山居系統, books in which scholars retired in the mountains would record their experiences, particularly about agricultural matters. The first part of the book deals with bamboo cultivation, and on this subject Shen Gua must have relied on the *Qimin yaoshu* 齊民要術, but added some useful specifications. The method which he describes has become famous and has influenced the art of gardening since the end of the Ming dynasty. Another part of the book concerns the cultivation of medicinal plants, in particular a certain digitalis (*dihuang* 地黃) and a polygonum (*huangjing* 黃精), an herb considered by the Taoists to be a food of the immortals. Shen Gua proposes a new theory about their harvesting which can also be found in the MQBT (jotting 485). In 1985, Hu Daojing referred again to the importance Shen Gua attaches to bamboo and medicinal plants in an article about his other contributions to the science of agriculture, particularly technical developments in water management and irrigation.[9]

The famous description given by Shen Gua of the first method of movable-type printing has inspired many articles. Indeed, no other document confirms or relates in detail the invention of Bi Sheng 畢昇 described in the MQBT, and many questions remain. The first one is: who was Bi Sheng? In 1990, the discovery of the tomb of a certain Bi Sheng and his descendants at Yingshan 英山 (Hubei), had raised hopes that this problem might be solved. Sun Qikang 孫啟康 in 1993 and Wu Xiaosong 吳曉松 in 1994[10] maintained that the family of Bi Sheng was originally from Yingshan. Wu devised an ingenious deciphering of the date on the tombstone, but failed to convince Zhang Xiumin 張秀民, the well-known author of many books on the history of printing in China, who recently, with his nephew Han Qi 韓琦, devoted a whole book to the history of movable type in China.[11] A controversy about Bi Sheng arose in 1993 and 1994 between Zhang and Sun in the journal *Zhongguo yinshua*.[12] Zhang believes that the Bi Sheng of the tombstone is not the person mentioned by Shen Gua. The printer Bi Sheng would have been a native of Hangzhou, which would explain how Shen Gua could have learned of his invention.

A further question concerning the identity of Bi Sheng has arisen. The MQBT mentions another person of distinction named Bi Sheng 畢升 and presented as an old blacksmith (*lao duan* 老鍛). Stanislas Julien and many others, among them Wang Guowei 王國維, Hu Shi 胡適, Needham (1976) and Feng Hanyong 馮漢鏞 (1983) thought that this blacksmith, who was also an alchemist, was none other than the printer.[13] But Zhang Xiumin and Wu Xiaosong have shown that they are two distinct persons, whose given name was written differently but pronounced in the same way.

As to the essential question, that of the invention itself, Shen Gua speaks of movable-type characters of clay. Many authors, including Luo Zhenyu 羅振玉 and Hu Shi, doubted that Bi Sheng could have made and used characters of clay. Feng Hanyong, who supposed the printer and the blacksmith-alchemist to be the same person, thought he had used characters made of a special clay, *liu yi ni* 六一尼, a mixture of different substances employed by the Taoist alchemists. The most interesting contributions concerning this problem have sought to show that these movable characters of clay were indeed used in later periods. For Wu Xiaosong, such is the case for Chinese, Korean, and Japanese works of the 17th and 18th centuries. He also mentions two recent discoveries of fragments of sûtras, some of which were found in 1965 in a temple of the Northern Song at Wenzhou 溫州 (Zhejiang), while others, printed in the 12th and 13th centuries, were found in 1987–89 in a temple at Wuwei 武威 (Gansu). The irregularities of the inking, the shape of the characters and their alignment show beyond doubt that they were printed with clay type. As for Zhang Xiumin, he mentions what should be the oldest work (1193) printed with clay type: the *Yutang zaji* 玉堂雜記 of Zhou Bida 周必大. It is a Taiwanese scholar, Huang Kuanzhong 黃寬重, who discovered in the collection of the complete works of Zhou Bida a statement by the author that he had his book printed according to "Shen Gua's method" with movable clay type.

Along with printing, cartography is another subject dealt with by Shen Gua that has attracted the attention of many scholars. Much has been written about the relief map he had made after his inspection in the Hebei Xilu and presented to the emperor Shenzong (jotting 472). Still more has been written about the *Shoulingtu* 守令圖 atlas which he compiled in 1076 on imperial order. On the other hand, Cao Wanru 曹婉如, in an article in 1980,[14] has endeavoured to demonstrate that Shen Gua is the author of the famous map engraved on stone in 1136, about forty years after his death, and kept today in the Museum of Steles in Xi'an: the *Yujitu* 禹跡圖 (map of the Tracks of Yu the Great). No document indicates its origin. Qing scholars thought that it came from the Tang *Hainei huayitu* 海內華夷圖 of Jia Dan 賈耽, dated 801. In fact, some details of the *Yujitu* could be dated between 1081 and 1094, the period of the *Shoulingtu*, compiled between 1076 and 1087. The Xi'an stele must have been made from a map drawn by Shen Gua between 1081 and 1082, while he was staying in the Shaanxi. There exists another stone-engraving of this map, kept in Zhenjiang, which was erected in 1142, perhaps from a copy of the original map in Shen Gua's possession when he came to live in the Jiangsu in 1088.

The same article deals with jotting 575 on cartographic techniques, especially the representation of distances when adopting a bird's-eye view. Shen Gua's text seems to enumerate seven processes. Cao Wanru, while cleverly correcting the text (she substitutes a character *zhi* 之 for the character *qi* 七) and clarifying the syntax, reduces the number to six, the same as those of Pei Xiu 裴秀 in the third century (circa 267). She also proposes an ingenious interpretation of the most difficult of these technical terms: *bang yan* 傍驗, altogether different from that of Li Qun 李群 and Needham.[15] Most of these remarks had already been made by Hu Daojing in 1964.[16] Concerning another of the six technical terms, *hutong* 互同, the same Hu Daojing, in an article in 1979,[17] puts forward a fascinating hypothesis: *hutong* should be understood in the sense of *denggaoxian* 等高線 (the line that joins points of equal elevation), our modern contour line. Unfortunately, the maps of the *Shoulingtu* have disappeared, and we have no supporting documents.

A special interest in archaeology developed in the Song era. In studying the vestiges of the past, one becomes conscious of the passage of time, to which Shen Gua was sensitive. In an article which appeared in 1971 and which was revised and reprinted in 1979,[18] Xia Nai 夏鼐 sets forth Shen Gua's personal ideas on archaeology. Shen Gua severely criticizes some more or less fanciful works such as the famous *Sanlitu* 三禮圖 in which the author reinvents certain artifacts mentioned in the three classic texts on rites. Shen Gua, says Xia Nai, regards such imaginary reconstructions as unreliable and draws his own conclusions from archaeological evidence. He seems sceptical, for instance, about the decoration of ritual vases and bells described in the jottings 319 and 320 of Chapter 19. On the other hand, Xia Nai notes, rather than describing artifacts superficially as archaeologists of this time would do, Shen Gua tried to understand their fabrication and function. Concerning bells, Lothar von Falkenhausen[19] notes in his book on "suspended music" of the ancient Chinese, that Shen Gua was the first thinker to reflect on the difference between round and almond-shaped ones. He states that when a bell is round, its sound is long and undulating; when it is narrow, its sound is short and less resonant (jotting 536).

Shen Gua also wrote on ancient weapons, such as a particular crossbow with divisions to improve aiming accuracy (jotting 331) or swords made of different kinds of steel. Xia Nai praises Shen Gua for his interdisciplinary approach when dealing with archaeology. That he arrived alone at this very modern conception, Xia explains, was because Shen Gua was versed in many fields of science, such as optics, geometry and metallurgy.

Not only was the study of ancient weapons important to him, but as a follower of Wang Anshi, Xia Nai adds, he took an active part in promoting the production and improvement of armament. Yet, curiously, he wrote nothing on fire-arms.

But a text by Hu Daojing does shed light on Shen Gua's military thinking.[20] It was through his mother, née Xu 許, that he became familiar with the military arts. Indeed, his uncle Xu Dong 許洞 was the author of a work entitled *Huqianjing* 虎鈐經 (the classic of the tiger's key), inspired by the ideas of Sun Wu 孫武 and Li Quan 李筌, which was a forerunner of the *Wujing zongyao* 武經總要. Hu also quotes a passage in a report that Shen Gua addressed to the emperor during his campaign against the Xi Xia in 1081. Shen Gua exposes the idea that victory over a mighty opponent is certain if one has a deep sense of the *yi* 義, that is, of the reality and evolution of a forthcoming situation. In order to reach this understanding, Shen Gua says that he relies on his *xin* 心, and that it was by his trust in the *yi* that he succeeded, during his embassy of 1075, in convincing the enemy to accept his proposals.

Let us now go beyond the frame of military thinking. Many authors have wondered how to characterize Shen Gua's work and its underlying thought. Among them, I have limited myself to a choice of five specialists: Hu Daojing, Wen Renjun 聞人軍 who contributed two articles in *Shen Gua yanjiu*;[21] Terachi Jun 寺地遵;[22] Sakade Yoshinobu;[23] one of the translators of the complete edition of the MQBT into Japanese; and Nathan Sivin.

Shen Gua's encyclopaedic curiosity and his qualities as a scientist compel at once the admiration of his readers. He was a great observer. After studying, with precision and perseverance, the movement of the North Star around the celestial North Pole, he observed that the magnetic needle does not point exactly north (Terachi, pp. 101–102). He insists on the same precision when he engages in a study of capacity measurement under the Han and the Song. His aim is to refute the prevailing opinion that, under the Han, people had an extraordinary resistance to drunkenness (Terachi, *ibid.*). He displays his ingenuity in correcting clepsydras (Terachi, p. 104 and Chen Meidong 陳美東[24]) or devising instruments to measure the flow of Bian canal 汴渠 (Sakade, p. 7). In the preface to his *Su Shen liangfang* 蘇沈良方, he states that he believes only what he has himself verified. He has no unconditional belief in the classics, and he takes pains to go and check on the spot a geographic datum of the "Yugong" 禹貢 in the *Shujing* in order to justify his own explanation (Sakade, p. 5).

He also criticizes the unthinking observance of old customs. In this respect, Terachi Jun mentions his remarks on the gathering of medicinal plants. It was customary to collect them only in the second and third months. Shen Gua considered it necessary to fix different dates, according to the species, the parts of the plants to be taken, the location and the weather. Wen Renjun gives yet another example of Shen Gua's critical mind in his attitude towards those who blindly apply the theory *wuyun liuqi* 五運六氣: this theory of the "five revolutions and six breaths", which could be applied to both meteorology and therapeutics, was used without regard to the complexity of different situations. But Shen Gua believed in the validity of this theory, derived from the ancient *Suwen* 素問, and he recounts proudly how he himself had been able, during a drought, to foresee the coming rain despite the incredulity of those around him (jotting 134).

What were the determining influences that played upon such a rich personality? Hu Daojing puts forward the difficult material situation of the family; the Shen had no course open to them but to take up a bureaucratic career. This is also underlined by Terachi Jun. In his numerous appointments in the provinces and in the capital, Shen Gua had to tackle all kinds of concrete problems. According to Sivin, the strongest influence in his life was the bureaucracy. On the other hand, Sakade Yoshinobu points out that he lived in a society in which science was developing. The time seemed ripe, therefore, for his research and, except for the hostility of Cai Jing, his work was well received. Zhu Xi and other thinkers of the Southern Song have borrowed from it, as did, later on under the Ming, Fang Yizhi 方以智 when he was looking for similarities between Chinese tradition and Western science. Finally, one should stress his self-confidence and courage when, for instance, he accepts a difficult mission to the Khitan territories or proposes improvements to the calendar, the clepsydras, and the maps.

But was he only a "supertechnician"? For Terachi Jun, he was just an "able official" (*nengli* 能吏). Sakade Yoshinobu reproaches him for his lack of system and his failure to construct a synthetic philosophy of nature. However, Hu Daojing and Wen Renjun insist rather on the fact that, with his very Chinese conviction that the *dao* is unceasing change and motion, he pays great attention to the passage of time, as attested by his observations in the fields of geology (erosion, marine fossils) and medicine. He advises doctors to consider the evolution of a disease in both the long and the short term (underlined by Hu Daojing with a quotation from jotting 123). Another traditional conception is the idea of "resonance" (*xianggan* 相感). Sakade Yoshinobu illustrates this by Shen Gua's explanation of the crystallization of salt (jotting 422): it is a resonance phenomenon between two energies or *qi* (of the wind and the salt). Yet, Shen Gua can easily deviate from this theory when it is at variance with his observations. Thus, he concludes that tides are influenced by the moon, rather than by sunrises and sunsets (jotting 544).

An interesting example also exists of the way in which Shen Gua endeavours to conciliate innovation and tradition. To justify an innovation of cardinal importance, the introduction of a solar calendar, he evokes the necessity of having a calendar in keeping with the traditional practices of imperial justice. In order that capital punishments continue to take place in winter, as is the rule, it is necessary to prevent a shift in seasons, allowing for the differences between the solar and lunar months.

I should also like to refer to a problem dealt with several times by critics: the attitude of Shen Gua towards phenomena which seem inexplicable, such as certain effects of lightning (metals melted but no damage to thatch, wood or lacquer [jotting 347]) and the power of some so-called longevity drugs like cinnabar, which can also cause death. In such a case, he is sometimes satisfied with ancient beliefs. For example, he compares the magnetic needle, always pointing south, to the cypress always facing westwards, because the cypress, like the West, is traditionally associated with the colour white (jotting 437).

In his search for explanations, Sakade Yoshinobu notes that Shen Gua, like his contemporary Cheng Yichuan 程伊川, believes each phenomenon must have its special principle (*li* 理). But perhaps Shen Gua did not imagine the existence of a universal metaphysical principle, since he never deals with this question, contrary to the philosophers of the Southern Song.

Finally, I should recall his attachment to the philosophy of Mencius, about whose work he wrote a commentary which has been preserved. Hu Daojing explains this partiality by Shen Gua's political ideas: he was a follower of Wang Anshi, himself the author of a commentary on the *Mengzi*. Indeed, Mencius declared that "the most honourable thing is the people" (*min wei gui* 民為貴). But the affinity of Shen Gua with Mencius was certainly deeper. Terachi Jun notes that for Shen Gua the *xin* 心 (the heart, the mind) is in essence celestial or divine. Therefore, its authority transcends the authority of the senses. Terachi illustrates this idea with several quotations from Chapters 19 to 32 of the *Changxing ji* 長興集. Mencius had established a bond between the *xin* and Heaven. The philosophers of the Southern Song would investigate this idea more thoroughly and find in it a way to understand the *li*.

Wen Renjun, on the basis of another passage of the *Changxing ji* (ch. 22), goes so far as to suggest a possible Buddhist or Taoist influence on Shen Gua's thought (p. 132). In any case, Shen Gua seems to have a certain sense of the supernatural; he was interested in all manner of strange phenomena, although his attitude towards divination and prescience was ambiguous. Thus, it would be presumptuous to judge Shen Gua by the standards of contemporary science. We should try rather, as Sivin suggests, to understand him. I would like to quote Sivin's passage in closing: "I would have to say that I failed to find the internal unity of Chinese science that I was looking for in the mind of Shen Kua. By way of compensation, I did learn the importance of an issue that I hadn't paid enough attention to before, that is, the relations of the sciences to other kinds of knowledge."[25]

Notes

[1] At a lecture which he gave for our group in 1997, Wang Qianjin 汪前進 of the Academia Sinica gave us copies of a number of articles recently published in China concerning Shen Gua's works. Many of them would have been hard to find outside China. I extend to him my warmest thanks.

[2] A revised version of this article of 1975 was published in 1995 in Sivin's collection of essays *Science in Ancient China*, ed. Variorum, Brookfield, Vermont. In his bibliography, Sivin has included a number of recent monographs that will be useful for further research on Shen Gua.

[3] Hu Daojing, *Mengqi bitan* buzheng 夢溪筆談補證. *Zhonghua wenshi luncong* 1979/3, pp. 111–35.

[4] *Songshi* 宋史, j. 356, pp. 11213–17 (ed. Zhonghua shuju).

[5] Li Zhizhong, "Yuan Dadeben *Mengqi bitan*", *Shehui kexue zhanxian* 1978/4, pp. 351–52.

[6] Daiwie Fu, "A Contextual and Taxonomic Study of the 'Divine Marvels' and 'Strange Occurrences' in the *Mengxi bitan*", *Chinese Science* 11, 1993–94, pp. 3–35.

[7] Hu Daojing, "Shen Gua wenwu er san shi" 沈括文物二三事. *Zhonghua wenshi luncong* 1980/2, pp.151–67.

[8] Hu Daojing, Shen Gua de nongxue zhuzuo *Mengqi wanghuai lu*, 沈括的農學著作夢溪忘懷錄. *Wenshi* 1963/10, pp. 221–25.

[9] Hu Daojing, Shen Gua zai nongye kexue shang de chengjiu he gongxian 沈括在農業科學上的成就和貢獻. *Nongshu nongyeshi wenji* 1985/6, pp. 160–74.

[10] Wu Xiaosong, Bi Sheng mudi faxian ji xiangguan wenti chubu tantao 畢昇墓地發現及相關問題初步探討. *Zhongguo keji shiliao* 15, 1994/1, pp. 3–18. About Sun Qikang, see Zhang Xiumin & Han Qi (below, note 11), p. 7.

[11] Zhang Xiumin & Han Qi, *Zhongguo huozi yinshua shi* 中國活字印刷史, Beijing: Zhongguo shuji chubanshe, 1998.

[12] *Zhongguo yinshua* 中國印刷, see Zhang Xiumin & Han Qi, *op. cit.*, p. 7.

[13] Joseph Needham, *Science and Civilisation in China*, vol. V: 3, pp. 187–88. About Feng Hanyong, see Zhang Xiumin & Han Qi, *op. cit.*, p. 4.

[14] Cao Wanru, "Lun Shen Gua zai dituxue fangmian de gongxian" 論沈括在地圖學方面的貢獻. *Kejishi wenji* 3, 1980/8, pp. 81–84.

[15] Li Qun, *Mengqi bitan xuandu* 孟溪筆談選讀. Beijing: Kexue chubanshe, 1975, pp. 58–81. J. Needham, *Science and Civilisation in China*, vol. 3, p. 576.

[16] Hu Daojing, "Gudai ditu cehui jishu shang de 'qi fa' wenti" 古代地圖測繪技術上的七法問題. *Zhonghua wenshi luncong* 5, 1964, p. 236.

[17] Hu Daojing, "*Mengqi bitan* buzheng" 夢溪筆談補證. *Zhonghua wenshi luncong* 1979/3, p. 120.

[18] Xia Nai, "Shen Gua he kaoguxue" 沈括和考古學. *Kaoguxue he kejishi* 1979, pp. 15–18.

[19] Lothar von Falkenhausen, *Suspended Music. Chime-Bells in the Culture of Bronze Age China*. Berkeley: University of California Press, 1993, p. 76.

[20] Hu Daojing, "Shen Gua junshi sixiang tansuo — lun Shen Gua yu qi jiufu Xu Dong de shicheng guanxi" 沈括軍事思想探索 — 論沈括與其舅父許洞的師承關係. *Shen Gua yanjiu* 沈括研究, Zhejiang renmin chubanshe, 1985, pp. 27–35.

[21] Wen Renjun, "Shen Gua kexue sixiang tansuo" 沈括科學思想探索. *Shen Gua yanjiu*, 1985, pp. 124–42.

[22] Terachi Jun, "Shin Katsu no shizen kenkyû to sono haikei" 沈括の自然研究とその背景. *Hiroshima daigaku bungakubu kiyô* 廣島大學文學部紀要, 27/1, 1967, pp. 99–121.

[23] Sakade Yoshinobu, "Shin Katsu no shizenkan ni tsuite" 沈括の自然觀について. *Tôhôgaku* 東方學, 39, 1970, pp. 74–87.

[24] Chen Meidong, "Woguo gudai louhu de lilun yu jishu — Shen Gua de *Fulouyi* ji qita" 我國古代漏壺的理論與技術 — 沈括的浮漏議及其它. *Ziran kexueshi yanjiu* 1982/1, pp. 21–33.

[25] *Science in Ancient China, op. cit.*, VII, p. 50.

A History of *Da-Yan Qiu-Yi Shu* in the West

WANG XIAOQIN

Institute for the History of Natural Sciences
Chinese Academy of Sciences

Introduction

It was not easy for Qin Jiushao's *Da-Yan Qiu-Yi Shu* to be understood in the West. In 1852, Alexander Wylie published in *North-China Herald* a famous paper titled "*Jottings on the Science of the Chinese: Arithmetic*", in which he explained the solution of Sun Zi's famous remainder problem and the first problem of *Shu-Shu Jiu-Zhang* by means of *Da-Yan Shu*. He did this without showing explicitly how to find the *Cheng Lü*, i.e., the solution of the congruence $ax \equiv 1 \pmod{b}$. In translating this paper, the German scholar K. L. Biernatzki misunderstood the *Da-Yan Shu*, confusing the *Cheng Lü* with the *Ji*, which is the least positive residue of $a \pmod{b}$. So did the French Mathematician Olry Terquem in translating Biernatzki's paper. L. Matthiessen, a German mathematician, judiciously corrected Biernatzki's mistake and pointed out the identity of *Da-Yan Shu* with C. F. Gauss' rule in *Disquisitiones Arithmeticae*. L. Matthiessen also offered an explanation of the case in which the moduli are not relatively prime in pairs. However, he did not know the *Qiu-Yi Shu*. It was Y. Mikami who first explained this using modern notations in his *Development of Mathematics in China and Japan*. Over a century passed before the Belgian scholar U. Libbrecht exhaustively studied the *Da-Yan Qiu-Yi Shu* in his *Chinese Mathematics in the Thirteenth Century*.

The Chinese Remainder Theorem in elementary number theory is as follows: if k_i satisfies

$$K_i \frac{M}{m_i} \equiv 1 \pmod{m_i} \quad (i = 1, 2, \ldots, n),$$

where $M = m_1 m_2 \ldots m_n, m_1, m_2, \ldots, m_n$ are relatively prime in pairs, then the solution for the system of the congruence of the first degree

$$N \equiv r_i \pmod{m_i} \quad (i = 1, 2, \ldots, n) \tag{1}$$

is

$$N \equiv \sum_{i=1}^{n} k_i M_i r_i \pmod{M}.$$

This theorem can be found in any general treatise on the history of Chinese mathematics. So far, however, no special study has been made of how it was understood by Western scholars. Li Yan, one of the precursors in the field of the history of Chinese mathematics, briefly referred to the work of L. Matthiessen, M. Cantor, Y. Mikami and L. Van Hée. [1] Some later historical writings simply mentioned that A. Wylie introduced the solution of Sun Zi's remainder problem (i.e. *"Wu Bu Zhi Shu"*) and *Da-Yan Shu* to the West in 1852, and L. Matthiessen pointed out the identity of Qin Jiushao's solution with the rule given by C. F. Gauss in his *Disquisitiones Arithmeticae* in 1874. Since then it has been called the Chinese Remainder Theorem in Western books on the history of mathematics. [2–6] This paper discusses the history of the theorem as it was understood in the West.

Little was known about Chinese mathematics before the first half of the 19th century although J. E. Montucla (1725–99) dealt with the subject in one chapter of the classic *Histoire des Mathematiques*. [7] As the writings of 18th-century missionaries like A. Gaubil (1689–1759) were the only relevant literature available to the author, his work was mainly devoted to Chinese astronomy. As to pure mathematics, Montucla mentioned nothing more than simple mensuration, the Pythagoras Theorem, and spherical trigonometry. During the next half century, Montucla's works and the relevant writings of the Jesuits were the principal sources for Western scholars on the state of Chinese mathematics. [8]

In the 1830s, the French sinologist E. Biot (1803–1850) and Italian mathematician G. Libri (1830–69) made a study of Cheng Dawei's *Suan Fa Tong Zong*, which had been transmitted to France in the 18th century. Libri believed that it was the only Chinese mathematical work known in Europe, which had not been contributed by the missionaries. [9] In March 1839, Biot published the general list of *Suan Fa Tong Zong* in *Journal Asiatique*, in which the author introduced Sun Zi's remainder problem without giving its solution:

"On demande un nombre tel qu'en divisant par 3, il reste 2; par 5, il reste 3; par 7, il reste 2." [10]

In 1852, Alexander Wylie (1815–87), a British missionary who came to China in 1847, published in *North-China Herald* a famous paper titled

Jottings on the Science of the Chinese: Arithmetic in which he introduced Sun Zi's problem and the rule for working it out:

> "Given an unknown number, which when divided by 3, leaves a remainder of 2; when divided by 5, it leaves 3; and when divided by 7, it leaves 2; what is the number? Ans. 23.

> Dividing by 3 with a remainder of 2, set down 140; dividing by 5 with a remainder of 3, set down 63; dividing by 7 with a remainder of 2, set down 30; adding these sums together gives 233, from which subtract 210, and the remainder is the number required." [11]

Wylie explained the rule by means of Qin Jiushao's *Da-Yan Shu*:

Ding Mu (Fixed Parents):	3 5 7
Yan Mu (Extension Parents):	$3 \times 5 \times 7$
Yan Shu (Extension Numbers):	35 21 15
Ji (Remainders):	2 1 1
Cheng Lü (Multiplying Terms):	2 1 1
Yong Shu (Use Numbers):	70 21 15
Yu Shu (Original Remainders):	2 3 2
Required Number:	$N = 70 \times 2 + 21 \times 3 + 15 \times 2 - 3 \times 5 \times 7 = 23$

However, Wylie did not explicitly explain the *Qiu-Yi Shu* which is used to find the *Cheng Lü*. In the above solution, there are two remainders equal to 1, which are directly taken to be *Cheng Lü*, as for the third 2, Wylie wrote:

> "This Remainder being more than unity is then submitted to a subsidiary process termed 求一 *Keu Yih* "Finding unity", which is the alternate division of the extension parent[1] and Remainder by each other, till the remainder is reduced to 1; the result in the present instance is 2 which is the Multiplying term". [11]

Wylie did not make clear here how the *Cheng Lü* are found. Neither did he when he translated the first problem of *Shu Shu Jiu Zhang*, i.e. "*Shi Gua Fa Wei*".

There are some mistakes and defects in *Jottings*. For example, Wylie took Yi Xing, the famous monk and astronomer in the Tang dynasty, for the first to apply the *Da-Yan Shu* to astronomy. It is unsatisfactory that he chose the first problem of *Shu Shu Jiu Zhang*, which is different in nature from other indeterminate problems involving the system of the congruence of the first degree, to introduce as an application of *Da-Yan*

Shu. It was the direct cause of the misunderstandings of later Western scholars.

Misunderstanding of K. L. Biernatzki and O. Terquem

In 1853, Henry Shearman (?–1856), the editor of *North-China Herald*, reprinted *Jottings* in *Shanghai Almanac and Miscellany* for the year 1853, a copy of which was obtained by the German scholar Biernatzki by chance. Having read Wylie's paper, Biernatzki was deeply interested in it and translated it into German. Biernatzki's translation of the solution of Sun Zi's problem is faithful to Wylie's original text at first:

> "Man multiplicire die drei Divisoren 3, 5, und 7, wodurch man die Zahl 105 erhält, welche **Yen mu** oder 'Stamm-Erweiterung' heifst. Diese dividire man durch die 'bestimmte Stammzahl' oder **Ting mu**, hier die Zahl 7, so ist der Quotient 15 die 'Erweiterungszahl' oder **Yen su**. Diese Erweiterungszahl 15, dividirt durch 7, läfst 1 als Überschufs, welches der 'Multiplicator' oder **Tsching suh** ist; mit dem Multiplicator 1 aber vermehrt, giebt sie als Product die 'Hülfszahl', oder **Yeng su**, 15. Dadurch ist erklärt, dafs es oben heifst: Für 1, durch 7 gewonnen, schreibe 15".[2] [12]

Biernatzki goes on:

> "Auf dieselde Weise werden die andern Hulfszdahlen gesucht, namlich:
>
> > 105/5 = 21, d.i. die Erweiterungszahl;
> > 21/5 läfst den Rest 1, der Multiplicator;
> > 21 ×1 = 21, d.i. die Hülfszahl.
>
> Daraus erklart sich das obige: Für 1, durch 5 gewonnen, schreibe 21. Endlich
>
> > 105/3 = 35, d.i. die Erweiterungszahl;
> > 35/3, läfst den Rest oder **Ki** 2, d.i. der Multiplicator;
> > 35 ×2 = 70, d.i. die Hülfszahl,
>
> oder wie es vorhin hiefs: für 1, durch 3 gewonnen, schreibe 70". [12]

Here we see that Biernatzki's translation deviates from Wylie's original text, omitting the name *Qiu-Yi Shu*, the rule for finding the *Cheng Lü*, and confusing the *Ji* with the *Cheng Lü*. The reason why Biernatzki misunderstood the *Da-Yan Shu*, is that Wylie did not completely explained the *Qiu-Yi Shu*, and that each of the three *Ji*'s 1, 1 and 2 in the solution of Sun Zi's problem is exactly equal to its corresponding *Cheng Lü*. So is each of the four *Ji*'s 1, 1, 1, and 3 in the *"Shi Gua Fa Wei"* problem.

There is another mistake in Biernatzki's translation. In Wylie's original text, after introducing the solution of Sun Zi's problem, Wylie referred to Yi-Xing (Yih-Hing), and next to Yi Jing (Yih King). Biernatzki carelessly equated Yih-Hing to Yih King, and took Qin Jiushao's works for the annotation of Yi Xing's *Da Yan Li*, so that the *Da-Yan Shu* was attributed to Yi Xing. This mistake was faithfully repeated by M. Cantor, O. Terquem, H. Hankel, L. Matthiessen, Y. Mikami, L. E. Dickson, D. E. Smith, G. Sarton, Yushkevitch, etc. This error also caused some Indian scholars to believe that Yi Xing learned the method of solving the indeterminate equation $Ax - By = C$ from the Indians. [13–16] Despite the mistakes, Biernatzki's paper deeply interested scholars, both in and outside Germany, who had known nothing about Chinese mathematics before.

In 1858, the German historian of mathematics Moriz Cantor (1829–1920) published *Zur Geschichte der Zahlzeichen*. According to Biernatzki, Cantor asserted that the *Da-Yan Shu* is erroneous and that the Chinese were less advanced than other contemporary cultural groups in the investigation of the indeterminate analysis. [17]

Shortly after Biernatzki's translation was published, the French mathematician Olry Terquem (1782–1860) translated it into French and his translation, *Arithmétique et Algébrbre des Chinoise*, was published in *Nouvelles Annales de Mathematiques* in 1862, two years after his death. Due to Biernatzki's misunderstanding, Terquem erroneously generalized the solution of Sun Zi's problem:

A number divided by p_1, the remainder is s_1; divided by p_2, the remainder is s_2; divided by p_3, the remainder is s_3. To find this number. If

$$p_1 p_2 = p_3 + r_3;$$
$$p_2 p_3 = p_2 + r_2;$$
$$p_1 p_3 = p_1 + r_1,$$

then the required number is

$$N = p_1 p_2 r_3 s_3 + p_1 p_3 r_2 s_2 + p_2 p_3 r_1 s_1.$$

Therefore, only when r_i satisfies

$$r_i^2 \equiv 1 \pmod{p_i}, \quad i = 1, 2, \ldots, n,$$

this solution can satisfy the given conditions. In Sun Zi's problem, the three remainders $r_1 = 2$, $r_2 = r_3 = 1$ exactly satisfy the above condition. However, when it is not satisfied, the solution is evidently erroneous.

Terquem omitted the *"Shi Gua Fa Wei"* problem because he could not understand it. After introducing other indeterminate problems of *Shu Shu Jiu Zhang*, Terquem wrote: "The Indians had a similar rule named *Kuttaka*. But it is impossible for the Chinese to learn it from the Indians." [18]

After Terquem, perpetual secretary of the French Academy of Sciences, the famous mathematician Joseph Bertrand (1822–1900) felt it necessary to bring Wylie's researches before Western mathematicians after he read Biernatzki's translation [19] and retranslated Biernatzki's paper. The translation appeared in *Journal des Savants* in 1869.

The Crucial Study of L. Matthiessen

Jottings was made known to Western scholars through translations of Biernatzki, Terquem and Bertrand. It was the most important literature on Chinese mathematics in the West in the second half of the 19th century and the first decade of the 20th.

The translation of Biernatzki was an important reference of the German historian Hermann Hankel (1839–73), who touched Chinese mathematics in his posthumous works *Zur Geschichte der Mathematik in Alterthun und Mittelalter*. Hankel asserted that the Chinese *Da-Yan Shu* was identical with the Indian *Kuttaka* and that a very intimate interconnection exists between Indian and Chinese mathematics. [17]

In 1874, the German mathematician Ludwig Matthiessen (1830–1906) corrected Biernatzki's mistakes and proved the identity of the Chinese *Da-Yan Shu* with Gauss' rule in a letter written to Cantor. [20] Gauss' general method of solving (1) was embodied in the first chapter of his *Disquisitiones Arithmeticae* (1801). If $\alpha, \beta, \gamma, \delta, \ldots$ satisfy

$$\alpha \equiv 1 \ (\mathrm{mod}\ m_1) \equiv 0 \ (\mathrm{mod}\ m/m_1),$$
$$\beta \equiv 1 \ (\mathrm{mod}\ m_2) \equiv 0 \ (\mathrm{mod}\ m/m_2),$$
$$\gamma \equiv 1 \ (\mathrm{mod}\ m_3) \equiv 0 \ (\mathrm{mod}\ m/m_3),$$
$$\delta \equiv 1 \ (\mathrm{mod}\ m_4) \equiv 0 \ (\mathrm{mod}\ m/m_4),$$
$$\cdots\cdots\cdots\cdots\cdots\cdots\cdots\cdots\cdots$$

where $M = m_1 m_2 m_3 \ldots$, then the solution of (1) is

$$N \equiv \alpha r_1 + \beta r_2 + \gamma r_3 + \delta r_4 + \ldots \quad (\mathrm{mod}\ M). \ [21]$$

In a paper written in the following year, Matthiessen compared the *Da-Yan Shu* with the *Kuttaka* and concluded that the two were different. In a paper published in 1881, Matthiessen again dealt with *Da-Yan Shu*, pointing out that Biernatzki mistook the *Ji* for the *Cheng Lü* and denoted its solution in modern algebraic notations as in Table 1:

Table 1   Matthiessen's Introduction of the Solution to the Sun Zi's Problem

Bestimmte Primzahlen	Reste	Divisionen	Multiplicatoren	Hülfszahlen
$m_1 = 3$	$r_1 = 2$	$5 \cdot 7 \cdot k_1 = 1 \ (\mathrm{mod}\ 3)$	$k_1 = 2$	$5 \cdot 7 \cdot k_1 = 70$
$m_2 = 5$	$r_2 = 3$	$3 \cdot 7 \cdot k_2 = 1 \ (\mathrm{mod}\ 5)$	$k_2 = 1$	$3 \cdot 7 \cdot k_2 = 21$
$m_2 = 7$	$r_3 = 2$	$3 \cdot 5 \cdot k_3 = 1 \ (\mathrm{mod}\ 7)$	$k_3 = 1$	$3 \cdot 5 \cdot k_3 = 15$

$$N = 2 \cdot 2 \cdot 5 \cdot 7 + 3 \cdot 1 \cdot 3 \cdot 7 + 2 \cdot 1 \cdot 3 \cdot 5 - 3 \cdot 5 \cdot 7 \cdot n = 233 - 105 \cdot n.$$

Matthiessen gave Gauss' method and showed its identity with the Chinese rule, pointing out that α, β, γ, δ, ... in the solution are the *Young Shu*'s in the *Da-Yan Shu*.

As Matthiessen could rely on no literature other than Biernatzki's paper, he mistook Qin Jiushao for the commentator of *Sun Zi Suan Jing* and *Da Yan Li*.

The four moduli $m_1 = 1$, $m_2 = 2$, $m_3 = 3$, $m_4 = 4$ in the "*Shi Gua Fa Wei*" problem are not relatively prime in pairs, which attracted the attention of Matthiessen. Qin Jiushao obtained:

Ding mu:	1	1	3	4
Yan Shu:	12	12	4	3
Cheng Lü:	1	1	1	3
Fan Yong:	12	12	4	9

Having found the *Fan Yong Shu*'s, Qin Jiushao proceeded to find the *Ding Yong Shu*'s:

"The second *Yuan Shu* 2 having been reduced to 1 at the beginning, the *Yan Mu* 12 is consequently added to the second *Fan Yong*, and the *Fan Yong Shu* then became the *Ding Yong Shu*". [22]

So we have

Ding yong:	12	24	4	9

The above data were given in both Wylie's *Jottings* and Biernatzki's translation. Based on these data, Matthiessen concluded that "Yi Xing"'s general method of solving the system of the congruence of the first degree

$$N \equiv r_i \pmod{m_i}, \quad i = 1, 2, \dots, n, \tag{2}$$

where the moduli $m_1, m_2, \dots, m_n$ are not relatively prime in pairs, is: finding the L.C.M. of the moduli $m_1, m_2, \dots, m_n$, $M = 1^p \cdot 1^q \cdot \dots \cdot 2^r \cdot 3^s \cdot$ $\cdot 5^t \cdot \dots = \mu_1 \cdot \mu_2 \cdot \dots \cdot \mu_n$, where $m_i \equiv 0 \pmod{\mu_i}$, $i = 1, 2, \dots, n$; from the congruence

$$\alpha_i \equiv 0 \pmod{M/\mu_i} \equiv 1 \pmod{\mu_i}, \quad i = 1, 2, \dots, n$$

finding the *fan Yong Shu*'s α_i and the *Ding Yong Shu*'s $\alpha_i (1 + m_i - \mu_i)$, then the solution of (2) is

$$\sum_{i=1}^{n} \alpha_i r_i \pmod{M}$$

or generally

$$\sum_{i=1}^{n} \alpha_i (1 + m_i - \mu_i) r_i \pmod{M}.$$

Matthiessen also gave the condition on the solvability of (2):

$$r_p \equiv r_q \pmod{\delta(m_p - m_q)}$$

where δ denotes G.C.D. Not being satisfied with Biernatzki's translation, (which embodied the *Tui Ji Tu Gong* problem) in Vol. 1 of *Shu Shu Jiu Zhang*, without giving the data, Matthiessen added certain numbers to it. The following is the assumed problem:

> To find the number of completed units of work and the same number x of units to be performed by each of four sets of 2, 3, 6 and 12 workmen, such that after certain whole days' work, there remain 1, 2, 5, 5 units not completed by the respective sets.

This problem can be solved by means of the following system of congruence:

$$x \equiv 1 \pmod{2} \equiv 2 \pmod{3} \equiv 5 \pmod{6} \equiv 5 \pmod{12}$$

Matthiessen gave the solution in detail as in Table 2. Hence the require number is:

$$x = 1 \cdot 24 + 2 \cdot 36 + 5 \cdot 16 + 581 - \mu_1 \cdot \mu_2 \cdot \mu_3 \cdot \mu_4 \cdot n = 581 - 12n.$$

The smallest positive solution is $N = 17$. Because $17 = 8 \cdot 2 + 1 = 5 \cdot 3 + 2 = 2 \cdot 6 + 5 = 1 \cdot 12 + 5$, the finished units of work is $8 \cdot 2 + 5 \cdot 3 + 2 \cdot 6 + 1 \cdot 12 = 55$ units.

Table 2   Matthiessen's Solution of the Problem of Work

Origin Number	Fixed Number	Remainder	Division	Multiplicator	Expansion Use Number	Fixed Use Number
$m_1 = 2$	$\mu_1 = 1$	$r_1 = 1$	$1 \cdot 3 \cdot 4 \cdot k_1 \equiv 1 \pmod 1$	$k_1 = 1$	$\alpha_1 = 12$	$\alpha_1(1 + m_1 - \mu_1) = 24$
$m_2 = 3$	$\mu_2 = 1$	$r_2 = 2$	$1 \cdot 3 \cdot 4 \cdot k_2 \equiv 1 \pmod 1$	$k_2 = 1$	$\alpha_2 = 12$	$\alpha_2(1 + m_2 - \mu_2) = 36$
$m_3 = 6$	$\mu_3 = 3$	$r_3 = 5$	$1 \cdot 1 \cdot 4 \cdot k_3 \equiv 1 \pmod 3$	$k_3 = 1$	$\alpha_3 = 4$	$\alpha_3(1 + m_3 - \mu_3) = 16$
$m_4 = 12$	$\mu_4 = 4$	$r_4 = 5$	$1 \cdot 1 \cdot 3 \cdot k_4 \equiv 1 \pmod 4$	$k_4 = 3$	$\alpha_4 = 9$	$\alpha_4(1 + m_4 - \mu_4) = 81$

Though Matthiessen's explanation of the finding of *Ding Yong Shu* from the *Fan Yong Shu* is correct in mathematical principle, and accords with the data of *Shi Gua Fa Wei*, it was not Qin Jiushao's original meaning. In the *Shu Gua Fa Wei* problem, the reason why Qin Jiushao used the *Ding Yong Shu* instead of *Fan Yong Shu* is that the sum of the *Fan Yong Shu* 37 is "senseless, and the divining straws are too small in number to be concealed" [22]. None of the other indeterminate problems in *Shu Shu Jiu Zhang* is solved with Matthiessen's method. Like Cantor, Matthiessen knew nothing about Qin Jiushao's *Qiu Yi Shu* and wished to find Wylie's original paper, but in vain. According to a letter from Wylie in 1874, *Jottings* could not be found any longer in Shanghai. [20]

In 1880, Cantor published his *Vorlesungen über Geschichte der Mathematik*. According to Matthiessen's researches, the author gave the complete solution of Sun Zi's problem, corrected his own erroneous viewpoint, and praised the inventor of this method as of the most fortunate ingenuity. [23] However, owing to Biernatzki's omission of *Qiu Yi Shu*, he thought that the *Cheng Lü* is probably determined through *trial and error*.

The Work of 20th-century Western Scholars

It was Y. Mikami who first explained Qin Jiushao's *Qiu-Yi Shu* and applied it to Sun Zi's problem. In *The Development of Mathematics in China and Japan*, Mikami gave the method of finding the *Cheng Lü* corresponding to the *Ding Mu* 3 and *Ji* 2 and expressed it in Figs. 1–3. Mikami completely translated Qin's *Qiu-Yi Shu* and explained it in modern algebraic notations. Let the *Ji* be a, *Ding Mu* b, $a < b$. Suppose the successive quotients of the Euclidean algorithm are $q_1, q_2, ..., q_{2k}$ and the last remainder is $r_{2k} = 1$. Let

$$A_1 = 1q_1, A_2 = A_1q_2 + 1, A_3 = A_2q_3 + A_1, A_4 = A_3q_4 + A_2, ..., A_{2k} = A_{2k-1}q_{2k} = A_{2k-2},$$

then the last one is the required *Cheng Lü*. [24]

1 celestial element (*Tian Yuan*)	2 residue (*Ji*)	1 celestial element	2 residue	2	2 remainder of the residue
	3 base-number (*Ding Mu*)	1 reduced number	1 remainder of base	1 reduced	3 remainder of the base

| Fig. 1 | Fig. 2 | Fig. 3 |

In 1913, the Belgian missionary L. Van Hée (1872–1951) published *Les Cent Volailles ou l'Analyse Indéterminée en Chine* in *T'oung Pao*, [25] in which the author explained the solution of Sun Zi's problem in modern notations. According to a paper of D. E. Smith, published in *Bibliotheca Mathematica*, titled "The Ganita-Sara-Sangraha of Mahaviracayra", [26] Van Hée asserted that the *Da-Yan Shu* is similar to *Kuttaka* (perhaps he was swayed by Wylie), and from this he rashly concluded that the *Da-Yan Shu* was the product of transmission from India. In the paper in question, Smith gave Mahavira's three examples of indeterminate equations:

(1) $(6x \pm 10)/9$ = integral;
(2) $ax + by + cz + dw = p$;
(3) $x + y + z + w = n$.

But he did not give their solutions.

That Van Hée dispatched this subject in such a summary way, without deep inquiry into the *Da-Yan Shu* and rigorous comparison between the Chinese and Indian methods, shows how shallow his understanding was. In the following year, Van Hée published *Bibliotheca Mathematica Sinensis Pe-Fou* in *T'oung Pao*, [27] in which he explained the methods of finding the *Yan Shu, Ji* and the *Cheng Lü*. The explanation, however, seems to be a plagiarism of Mikami's works.

In 1920, the American mathematician L. E. Dickson published *History of the Theory of Number*. In the 2nd volume of this book, the author dealt with the history of the Chinese remainder problem. [28] Using the relevant writings of Matthiessen and Mikami, he introduced the Chinese contributions.

However, he erroneously attributed the solution of (2) to Yi-Xing and that of (1) to Qin Jiushao. Let the congruence be $ax \equiv 1 \pmod{b}$, where $a < b$. Dickson's introduction of *Da-Yan Shu* is not accurate: suppose the successive quotients of the Euclidean algorithm are $q_1, q_2, q_3, \ldots$, the remainders are $r_1, r_2, r_3, \ldots, r_n = 1$. Let

$$A_1 = q_1, A_2 = A_1 q_2 + 1, A_3 = A_2 q_3 + A_1, A_4 = A_3 q_4 + A_2, \ldots,$$

then the required *Cheng Lü* is $x = A_n$. Clearly, Dickson deviated here from Mikami's original text, neglecting the fact that n should be even in the *Da-Yan Shu*. In fact, when n is odd, A_n is the solution of the congruence $ax \equiv -1 \pmod{b}$.[3]

In 1920, the Italian historian of mathematics G. Loria (1862–1954) questioned the invention of *Da-Yan Shu* by Sun Zi, "because Sun Tsu gives not the slightest indication that he regards this problem as more interesting or valuable than its trivial companions". [29] In his *Storia delle Matematiche dall'Alba della Civilta al Secolo XIX*, Loria introduced the solution of Sun Zi's problem, and admitted its identity with Gauss' rule, but like Cantor, he believed that the *Cheng Lü* is probably determined by *trial and error*, though he had read Mikami's famous works. During the next decades, the *Da-Yan Qiu-Yi Shu* received little attention from Western scholars. In *A History of Japanese Mathematics* published in 1914 by D. E. Smith (1860–1944) and Y. Mikami and the second volume of *History of Mathematics* published in 1925 by Smith, only Sun Zi's problem was mentioned. [30,31] Several years later, Smith wondered whether there were indeterminate problems before the 8th century, when Yi-Xing wrote on the *Da-Yan Shu*. [32]

G. Sarton (1884–1956), the famous American historian of science, accepted Wylie's viewpoint that the *Da-Yan Shu* is similar to *Kuttaka*. [33]

Though Mathiessen discussed the solution of (2), he did not explicitly explain how the *Yuan Shu* m_i, which are not relatively prime in pairs, are reduced to the *Ding Shu* μ_i, which are relatively prime in pairs. The British scholar K. Mahler determined μ_i by means of factorizing m_i into primes: let the factorization of m_i be

$$m_i = p_1^{a_{i1}} p_2^{a_{i2}} \cdots p_t^{a_{it}} \quad (i = 1, 2, \ldots, n),$$

denote

$$a_\tau = \max_{i=1,2,\ldots,n} a_{i\tau} \quad (\tau = 1, 2, \ldots, t)$$

and

$$a_{i,\tau} = a_\tau \quad (\tau = 1, 2, \ldots, t),$$

take

$$\alpha_{i\tau} = \begin{cases} a_\tau \ (i = i_\tau) \\ 0 \ (i \neq i_\tau) \end{cases} \quad (\tau = 1, 2, ..., n)$$

then

$$\mu_i = p_1^{\alpha_{i1}} p_2^{\alpha_{i2}} \cdots p_l^{\alpha_{il}} \quad (i = 1, 2, ..., n)$$

are the *Ding Shu*'s that satisfy

(1) $\mu_i \mid m_i \quad (i = 1, 2, ..., n)$;
(2) $(\mu_i, \mu_j) = 1, i \neq j$;
(3) L.C.M. $(\mu_1, \mu_2, ..., \mu_n) =$ L.C.M. $(m_1, m_2, ..., m_n)$.

Mahler proved the sufficient and necessary condition under which (2) is solvable.

Though Mahler tried to reproduce what he believes is the mathematical content of this old Chinese method, [34] it is clearly not what the ancient Chinese mathematicians actually did to determine the *Ding Shu* by means of factorization.

J. Needham's introduction of *Da-Yan Shu* in the third volume of his *Science and Civilization in China* is unsatisfactory. Like Wylie, whom Needham highly praised, Needham did not mention the *Qiu-Yi Shu* but merely said that the finding of the multiplying terms is the principal process or the most important part in the operation. [35] About the relation between the *Da-Yan Shu* and Kuttaka, Needham's conclusion is the same as that of Wylie and Sarton. Needham, as well as his cooperator Wang Ling, also attempted to probe into the relation of the *Da-Yan Shu* to Yi-Xing's *Da-Yan Li*, but they did not arrive at a definite conclusion because they did not know how Yi-Xing solved the problem involving the following system of congruence:

$$1110343x \equiv 44820 \ (\text{mod } 60 \times 3040) \equiv 49107 \ (\text{mod } 89773).$$

They asserted that the method of solving the *Zhi Li Yan Ji* problem, the third one in Vol. 2 of *Shu Shu Jiu Zhang*, is most approximate to that of Yi-Xing. [36]

In 1963, the famous American historian of mathematics D. J. Struik published in *Mathematics Teacher*, [37] an article in which he introduced Sun Zi's problem and gave its solution with modern notations. He thought that *Da Yan* class was the most interesting feature of *Shu Shu Jiu Zhang*. He also gave two problems: (1) $N \equiv 1 \ (\text{mod } 2) \equiv 3 \ (\text{mod } 3) \equiv 1 \ (\text{mod } 4)$;

(2) $N \equiv 32 \pmod{83} \equiv 70 \pmod{110} \equiv 30 \pmod{135}$, and showed the solution of the system of congruence in which the moduli are not relatively prime in pairs.

In the 1970s the Belgian scholar U. Libbrecht published *Chinese Mathematics in the Thirteenth Century*, in which the author made an exhaustive study of the *Da-Yan Qiu-Yi Shu*. This book deeply interested Chinese historians of mathematics.[4]

One hundred and twenty years passed from Wylie's first introduction of the *Da-Yan Qiu-Yi Shu* till U. Libbrecht's exhaustive study of it. Wylie's precursory work played a very important role in the history of *Da-Yan Qiu-Yi Shu* in the West and was the starting point of it being understood. Biernatzki's translation made Wylie's work known in the West. But Biernatzki's confusion of the *Ji* with the *multiplying terms* gave rise to misunderstanding of the *Da-Yan Shu* by Cantor and Terquem. Matthiessen's remarkable researches played the most important role in this history. He correctly understood the *Da-Yan Shu*, proved its identity with Gauss' rule, and consequently laid the foundation of the term "Chinese Remainder Theorem". Though Matthiessen also offered an explanation of the case in which the moduli are not relatively prime in pairs, he did not know the *Qiu-Yi Shu*, which is used to find the *Cheng Lü* because of the unavailability of further relevant literature. Therefore he could not completely know about Qin Jiushao's method of solving the system of congruence of the first degree. Mikami introduced the method of finding the *Cheng Lü*, and translated Qin's *Da-Yan Qiu-Yi Shu*. His English work was conducive to the understanding of the *Da-Yan Qiu-Yi Shu*. The later Western historians of mathematics such as Van Hée, Loria, etc. debased or were suspicious of the *Da-Yan Shu*, exerting negative influence. Mahler offered a proof of the case in which the moduli are not relatively prime in pairs, but his proof does not accord with Qin Jiushao's original meaning. Needham's introduction of the *Da-Yan Qiu-Yi Shu* was not novel. Only Libbrecht's work put to an end to the history of interpreting the *Da-Yan Qiu-Yi Shu* in the West.

Notes

[1] In fact it should be the Fixed Parent (*Ding Mu*).

[2] Wylie's original text is: "Multiplying together the three divisors 3, 5, and 7, gives 105 for the 衍母 *Yen-moo* 'Extension parent'. Divide this by the 定母 *Ting-moo* 'Fixed parent' 7, the quotient 15 is the 衍数 *Yen-soo* 'Extension number'. Divide this again by 7, and there is an overplus of 1, which is the 乘率 *Ching-suh* 'Multiplying term'; by which, multiply the Extension number 15, and the product

15 is the 用数 *Yung-soo* 'Use number', or as it is given above, — for 1 obtained by 7, set down 15".

³ The congruence $ax \equiv \pm 1 \pmod{b}$ is equivalent to the indeterminate equation $by \equiv ax \mp 1$. By the Euclidean algorithm, we have: $b = q_1 a + r_1 \ (0 < r_1 < a)$, $a = q_2 r_1 + r_2 \ (0 < r_2 < r_1)$, $r_1 = q_3 r_2 + r_3 \ (0 < r_3 < r_2)$, $r_2 = q_4 r_3 + r_4 \ (0 < r_4 < r_3)$, ..., $r_{n-2} = q_n r_{n-1} + r_n \ (r_n = 1)$. To use Euler's notation, $b = [r_{n-1}, q_n, q_{n-1}, ..., q_3, q_2, q_1]$, $a = [r_{n-1}, q_n, q_{n-1}, ..., q_3, q_2]$. When the number of the terms $q_1, q_2, q_3, ..., q_{n-1}, q_n$, r_{n-1} is even, $y = [q_n, q_{n-1}, ..., q_3, q_2]$, $x = [q_n, q_{n-1}, ..., q_3, q_2, q_1]$ are the solutions of $by = ax + 1$; when it is odd, y and x are the solutions of $by = ax - 1$. Cf. [21].

⁴ About 11 of the 30 papers embodied in [38] cited this work, and one is a review of it.

References

1. Li Yan. "The Past and Future of the *Da-Yan Qiu-Yi Shu*". In *Zhong Suan Shi Lun Cong* (I), 1933. 60–121.

2. Qian Baocong. *A History of Chinese Mathematics*. Beijing: Science Press, 1964. 77–78.

3. Zhong Wai Shu Xue Jian Shi Bian Xie Xiao Zu. *A Concise History of Chinese Mathematics*. Ji Nan: Shan Dong Education Press. 1986. 186–88.

4. Shen Kangshen. *An Introduction to Ancient Chinese Mathematics*. Shanghai: Shanghai Education Press, 1986. 286–87.

5. Liu Dun. *Da Zai Yan Shu*. Shen Yang: Liao Ning Education Press, 1995. 271.

6. Li Wenlin & Yuan Xiangdong. "Chinese Remainder Theorem". In *Ancient Chinese Achievements in Science and Technology*. Beijing: Chinese Youth Press, 1996. 106–115.

7. J. E. Montucla. *Histoire des Mathématiques* (Vol. I), Paris, 1799. 450.

8. G. Loria. *Storia delle Mathematiche dall'Alba della Civilta al Secolo XIX* (Vol. I). Seconda Edizione, Ulrico Hoepli, Milano, 1950. 146, 152.

9. J. C. Matzloff. *Histoire des Mathématiques Chinoises.Masson*, 1988.2.

10. E. Biot. "Table Générale d'un Ouvrage Chinoise Intituté *Suan-Fa Tong-Tsong*". *Journal Asiatique*, 7 (1839): 193–217.

11. A. Wylie. "Jottings on the Science of the Chinese: Arithmetic". *North-China Herald*, Aug. 21; Sept. 11, 18, 25; Oct. 16, 23; Nov. 6, 13, 20, 1852.

12. K. L. Biernatzki. "Die Arithemetik der Chinesen". *Journal für die Reine und Angewandte Mathematik*, 52 (1856): 59–94.

13. S. Ganguli. "India's Contribution to the Theory of Indeterminate Equations of the First Degree". *Journal of the Indian Mathematical Society*, 19 (1931): 110–20, 129–42, 153–68.

14. S. N. Sen. Study of Indeterminate Analysis in Ancient India. *Proceedings of the 10th International Congress of the History of Science*, Ithaca, 1962. Paris, 1964. 493–97.

15. A. K. Bag. "The Method of Integral Solution of Indeterminate Equations of the Type: by = ax ± c in Ancient and Medieval India". *Indian Journal of History of Science*, 12 (1977): 1–16.

16. A. K. Bag. *Mathematics in Ancient and Medieval India*. Delhi, 1979. 215–16.

17. U. Libbrecht. *Chinese Mathematics in the Thirteenth Century: The Shu-Shu Chiu-Chang of Ch'in Chiu-shao*. Cambridge, 1973. 314–15, 359–60.

18. O. Terquem. "Arithmetique et Algebre des Chinois". *Nouvelles Annales de Mathematiques*, 2e Serie, 1 (1862), 35–44; 2 (1863): 529–40.

19. H. Cordier. "The Life and Labours of Alexander Wylie, Agent of the British and Foreign Bible Society in China. A Memoir". *Journal of the Royal Asiatic Society*, 19 (1887): 351–68. Repr. in *Chinese Researches*, Shanghai, 1897. 7–18.

20. L. Matthiessen. "Ueber das sogenannte Restproblem in den chinesischen Werken Swan-king von Sun-tsze und Tayen lei schu von Yih-hing". *Journal für die reine und angewandte Mathematik*, 91 (1881): 254–61.

21. C. F. Gauss. *Disquisitiones Arithmeticae*, in *Werke* (I), Georg Olms Verlag, Hildesheim, 1981. 20–21, 26.

22. Qin Jiushao. *Shu Shu Jiu Zhang*. In Guo Shu Chun ed., *Zhong Guo Ke Xue Ji Shu Dian Ji Tong Hui*, Mathematics (Vol. I). Zheng Zhou: He Nan Education Press, 1993. 446–47.

23. M. Cantor. *Vorlesungen uber Geschichte der Mathematik* (I), Leipzig, 1922, 687.

24. Y. Mikami. *The Development of Mathematics in China and Japan*. Leipzig: B. G. Teuber, 1913. 66–68.

25. L. Van Hée. "Les Cent Volailles ou l'Analyze Indeterminee en Chine". *T'oung Pao*, 14 (1913): 203–210, 435–50.

26. D. E. Smith. "The Ganita-Sara-Sangraha of Mahaviracarya". *Bibliotheca Mathematica*, 3. Folge, 9 (1908–1909): 106–110.

27. L. Van Hée. "Bibliotheca Mathematica Sinensis Pe-Fou". *T'oung Pao*, 15 (1914): 111–64.

28. L. E. Dickson. *History of the Theory of Numbers* (Vol. II). Chelsea Publishing Company, New York, 1952. 58–60.

29. G. Loria & R. B. McClenon. "The Debt of Mathematics to the Chinese People". *Scientific Monthly*, 12 (1921): 517–21.

30. D. E. Smith & Mikami. *A History of Japanese Mathematics*. Chicago: The Court Publishing Company, 1914. 10.

31. D. E. Smith. *History of Mathematics* (vol. II). Ginn and Company, Boston, 1925. 380.

32. D. E. Smith. "Unsettled Questions Concerning the Mathematics of China". *Scientific Monthly*, 33 (1931): 244–50.

33. G. Sarton. *Introduction to the History of Science* (Vol. II, Part II). Robert E. Krieger Publishing Company, Huntington, New York, 1975. 626.

34. K. Mahler. "On the Chinese Remainder Theorem". *Mathematische Nachrichten*, 18 (1958): 120–22.

35. J. Needham. *Science and Civilization in China* (vol. 3), Cambridge, 1959. 42: 120–21.
36. Wang Ling. The Date of the Sun Tzu Suan Ching and the Chinese Remainder Problem. *Proceedings of the 10th International Congress of the History of Science*, Ithaca, 1962. Paris, 1964. 189–492.
37. D. J. Struik. "On the Ancient Chinese Mathematics". *The Mathematics Teacher*, 1963, 56(6): 424–32.
38. Wu Wenjun, *Qin Jiushao and Shu Shu Jiu Zhang*. Beijing Normal University Press, 1987.

Glossary

Cheng Lü	乘率
Dan Yan Li	大衍历
Dan-Yan Qiu-Yi Shu	大衍求一术
Dan-Yan Shu	大衍术
Ding Shu	定数
Ding Yong Shu	定用数
Fan Yong Shu	泛用数
Ji	奇
Qin Jiushao	秦九韶
Qiu-Yi Shu	求一术
Shi Gua Fa Wei	蓍卦发微
Shu Shu Jiu Zhang	数书九章
Suan Fa Tong Zong	算法统宗
Tui Ji Tu Gong	推计土功
Wu Bu Zhi Shu	物不知数
Yan Mu	衍母
Yan Shu	衍术
Yong Shu	用数
Yuan Shu	元数
Zhi Li Yan Ji	治历演纪